Lecture Notes in Networks and Systems

Volume 428

The series "Lecture Notes in Networks and Systems" publishes the latest developments in Networks and Systems—quickly, informally and with high quality. Original research reported in proceedings and post-proceedings represents the core of LNNS.

Volumes published in LNNS embrace all aspects and subfields of, as well as new challenges in, Networks and Systems.

The series contains proceedings and edited volumes in systems and networks, spanning the areas of Cyber-Physical Systems, Autonomous Systems, Sensor Networks, Control Systems, Energy Systems, Automotive Systems, Biological Systems, Vehicular Networking and Connected Vehicles, Aerospace Systems, Automation, Manufacturing, Smart Grids, Nonlinear Systems, Power Systems, Robotics, Social Systems, Economic Systems and other. Of particular value to both the contributors and the readership are the short publication timeframe and the world-wide distribution and exposure which enable both a wide and rapid dissemination of research output.

The series covers the theory, applications, and perspectives on the state of the art and future developments relevant to systems and networks, decision making, control, complex processes and related areas, as embedded in the fields of interdisciplinary and applied sciences, engineering, computer science, physics, economics, social, and life sciences, as well as the paradigms and methodologies behind them.

Indexed by SCOPUS, INSPEC, WTI Frankfurt eG, zbMATH, SCImago.

All books published in the series are submitted for consideration in Web of Science.

For proposals from Asia please contact Aninda Bose (aninda.bose@springer.com).

Bibudhendu Pati · Chhabi Rani Panigrahi ·
Prasant Mohapatra · Kuan-Ching Li
Editors

Proceedings of the 6th International Conference on Advance Computing and Intelligent Engineering

ICACIE 2021

Springer

Editors
Bibudhendu Pati
Department of Computer Science
Rama Devi Women's University
Bhubaneswar, India

Chhabi Rani Panigrahi
Department of Computer Science
Rama Devi Women's University
Bhubaneswar, India

Prasant Mohapatra
University of California
Davis, CA, USA

Kuan-Ching Li
Department of Computer Science
and Information Engineering
Providence University
Taichung, Taiwan

ISSN 2367-3370 ISSN 2367-3389 (electronic)
Lecture Notes in Networks and Systems
ISBN 978-981-19-2224-4 ISBN 978-981-19-2225-1 (eBook)
https://doi.org/10.1007/978-981-19-2225-1

This Springer imprint is published by the registered company Springer Nature Singapore Pte Ltd.
The registered company address is: 152 Beach Road, #21-01/04 Gateway East, Singapore 189721, Singapore

Preface

This volume contains the papers presented at 6th International Conference on Advanced Computing and Intelligent Engineering (ICACIE) 2021: The 6th ICACIE (www.icacie.com) is held during December 23–24, 2021, at Bhubaneswar Institute of Technology in collaboration with Rama Devi Women's University, Bhubaneswar, India. There were a total of 247 submissions, and each qualified submission was reviewed by a minimum of two Technical Program Committee (TPC) members using the criteria of relevance, technical quality, originality, and presentation. The TPC committee accepted 60 full papers for oral presentation at the conference, and the overall acceptance rate is 24%.

ICACIE 2021 was an initiative taken by the organizers which focuses on research and applications on topics of advanced computing and intelligent engineering. The focus was also to present state-of-the-art scientific results, to disseminate modern technologies, and to promote collaborative research in the field of advanced computing and intelligent engineering. Researchers presented their work in the conference through virtual as well as online mode due to COVID-19 pandemic and had an excellent opportunity to interact with eminent professors, scientists, and scholars in their area of research. All participants were benefitted from discussions that facilitated the emergence of innovative ideas and approaches. Many distinguished professors, well-known scholars, young researchers, and industry leaders were participated in making ICACIE 2021 an immense success. We had many invited talks by professors, research scholars, and industry personnel in emerging topics of advanced computing, sustainable computing, and machine learning.

We express our sincere gratitude to the Patron Prof. Amit Kumar Mishra, Chairman, Bhubaneswar Institute of Technology, for allowing us to organize ICACIE 2021 and his unending timely support toward organization of this conference. We would like to extend our sincere thanks to Prof. Binod Kumar Pattanayak and Mr. Seeven Amic, Special Session General Chairs of ICACIE 2021, for managing the special session and offering their valuable guidance during review of papers as well as in other aspects of the conference. We thank all the Technical Program Committee members and all reviewers/sub-reviewers for their timely and thorough participation during the review process. We appreciate the time and efforts put in by the members of the local organizing team at Bhubaneswar Institute of Technology, Bhubaneswar, India, and administrative staff, who dedicated their efforts to make ICACIE 2021 successful. We would like to extend our thanks to Er. Subhashis Das Mohapatra and Mr. Sanjeev Cowlessur for designing and maintaining ICACIE 2021 Web site and extending their support for managing the sessions in virtual as well as online mode.

Bhubaneswar, India — Bibudhendu Pati
Bhubaneswar, India — Chhabi Rani Panigrahi
Davis, USA — Prasant Mohapatra
Taichung, Taiwan — Kuan-Ching Li

Contents

Editors and Contributors

About the Editors

Dr. Bibudhendu Pati is Associate Professor in the Department of Computer Science at Rama Devi Women's University, Bhubaneswar, India. He received his Ph.D. degree from IIT Kharagpur. Dr. Pati has 23 years of experience in teaching, research, and industry. His areas of research include wireless sensor networks, cloud computing, big data, Internet of Things, and networks virtualization. He is Life Member of the Indian Society of Technical Education (ISTE), and Senior Member of IEEE, Member of ACM, and Life Member of Computer Society of India (CSI). He has published several papers in leading international journals, conference proceedings, and books. He is associated with organizing different reputed international conferences.

Dr. Chhabi Rani Panigrahi is Assistant Professor in the Department of Computer Science at the Rama Devi Women's University, Bhubaneswar, India. Dr. Panigrahi has received her doctoral degree from Indian Institute of Technology Kharagpur, India. Her research interests include software testing and mobile cloud computing. She has more than 20 years of teaching and research experience, and has published several scholarly articles in reputed journals and conferences of international repute. Dr. Panigrahi is Author of a text book and a few edited books. She is Life Member of the Indian Society of Technical Education (ISTE) and Member of IEEE and the Computer Society of India (CSI).

Prof. Prasant Mohapatra is serving as Vice Chancellor for Research at University of California, Davis. He is also Professor in the Department of Computer Science and served as Dean and Vice-Provost of Graduate Studies during 2016–2018. He was Department Chair of Computer Science during 2007–2013. In the past, Dr. Mohapatra has also held Visiting Scientist positions at Intel Corporation, Panasonic Technologies, Institute of Infocomm Research (I2R), Singapore, and National ICT Australia (NICTA). Dr. Mohapatra received his doctoral degree from Penn State University in 1993, and received an Outstanding Engineering Alumni Award in 2008.

He is also the recipient of Distinguished Alumnus Award from the National Institute of Technology, Rourkela, India. Dr. Mohapatra received an Outstanding Research Faculty Award from the College of Engineering at the University of California, Davis. He received the HP Labs Innovation awards in 2011, 2012, and 2013. He is Fellow of the IEEE and Fellow of AAAS. Dr. Mohapatra's research interests are in the areas of wireless networks, mobile communications, cybersecurity, and Internet protocols. He has published more than 350 papers in reputed conferences and journals on these topics. Dr. Mohapatra's research has been funded through grants from the National Science Foundation, US Department of Defense, US Army Research Labs, Intel Corporation, Siemens, Panasonic Technologies, Hewlett Packard, Raytheon, and EMC Corporation.

Prof. Kuan-Ching Li is Professor in the Department of Computer Science and Information Engineering at the Providence University, Taiwan. He has been Vice-Dean for the Office of International and Cross-Strait Affairs (OIA) at the same university since 2014. Professor Li is the recipient of awards from Nvidia, the Ministry of Education (MOE)/Taiwan and the Ministry of Science and Technology (MOST)/Taiwan, and also guest professorships from various universities in China. He received his Ph.D. from the University of Sao Paulo, Brazil in 2001. His areas of research include networked and graphics processing unit (GPU) computing, parallel software design, and performance evaluation and benchmarking. He has edited 2 books: Cloud Computing and Digital Media and Big Data published by CRC Press. He is Fellow of the IET, Senior Member of the IEEE, and Member of TACC.

Contributors

Abd El-Hafeez Tarek Department of Computer Science, Faculty of Science, Minia University, EL-Minia, Egypt;
Computer Science Unit, Deraya University, EL-Minia, Egypt

Abhinav Kumaar K. Department of Mechatronics Engineering, Sri Krishna College of Engineering and Technology, Coimbatore, India

Ali Lone Nawaz Department of Computer Science, Islamic University of Science and Technology, Kashmir, India

Anand Revika CSE Department, Indira Gandhi Delhi Technical University for Women, Delhi, India

Anisha M. Department of Biomedical Engineering, Rajalakshmi Engineering College, Thandalam, India

Aravind Priyamvadan V. Madras Institute of Technology Campus, Anna University, Chennai, Tamil Nadu, India

Arjun B. R. Department of Communication Engineering, School of Electronics Engineering, Vellore Institue of Technology, Tamil Nadu, Vellore, India

Arun Vignesh N. Department of Electronics and Communication Engineering, Gokaraju Rangaraju Institute of Engineering and Technology, Hyderabad, India

Arunachalam S. Department of Mechatronics Engineering, Sri Krishna College of Engineering and Technology, Coimbatore, India

Arunnehru J. Department of Computer Science and Engineering, SRM Institute of Science and Technology, Vadapalani, Chennai, Tamil Nadu, India

Baby Sangeeth Sajan Department of Computer Science and Engineering, Amrita Vishwa Vidyapeetham, Amritapuri, India

Bagate Rupali Department of Information Technology, Army Institute of Technology, Pune, Maharashtra, India

Bala Vineel Kollipara Department of Computer Science and Engineering, Amrita School of Engineering, Amrita Vishwa Vidyapeetham, Bengaluru, India

Banu L. Jarina Department of Mathematics, Ayya Nadar Janaki Ammal College, Sivakasi, Tamil Nadu, India

Barick Suvojit Department of Computer Science and Engineering, Siksha 'O' Anusandhan (Deemed to Be University), Bhubaneswar, Odisha, India

Bashir Rumaan Department of Computer Science, Islamic University of Science and Technology, Kashmir, India

Bavan Kalyan R. Department of Mechatronics Engineering, Sri Krishna College of Engineering and Technology, Coimbatore, India

Behera Anukampa ITER, S'O'A (Deemed to Be) University, Bhubaneswar, India

Behera Sitesh Bangalore, India

Benderskaya Elena N. Peter the Great St. Petersburg Polytechnic University, St. Petersburg, Russia

Bhambhu Lekha Chitkara University Institute of Engineering and Technology, Rajpura, Punjab, India

Bhattacharya Sharmistha Tripura University, Agartala, India

Bheekharry Normada Université des Mascareignes, Beau Plan, Pamplemousses, Mauritius

Bhushan Bharat Sharda University, Greater Noida, India

Bidwe Ranjeet SCTR's Pune Institute of Computer Technology, Pune, Maharashtra, India

Bisoy Sukant Kishoro C.V. Raman Global University, Bhubaneswar, Odisha, India

Chakraborty Ivy Infosys Limited, Bangalore, India

Chaudhari Vrushali JSPM's Rajarshi Shahu College of Engineering, Pune, Maharashtra, India

Choubey Dilip Kumar IIIT, Bhagalpur, Bihar, India

Das Avirup Advanced Computing and Microelectronics Unit, Indian Statistical Institute, Kolkata, India

Das Satyapragyan Department of Computer Science and Engineering, Siksha 'O' Anusandhan (Deemed to Be University), Bhubaneswar, Odisha, India

Dave Bhavi Department of Computer Engineering, Dwarkadas J. Sanghvi College of Engineering, Mumbai, Maharashtra, India

Deshmukh Savali School of Computer Science and Engineering, Vellore Institute of Technology—AP University, Amravati, India

Deshpande Ashwini M. Department of Electronics and Telecommunication Engineering, MKSSS's Cummins College of Engineering for Women, Pune, India

Devi S. Prasanna Department of Computer Science and Engineering, SRM Institute of Science and Technology, Chennai, Tamilnadu, India

Dhal P. K. Department of Electrical and Electronics Engineering, Vel Tech Rangarajan Dr, Sagunthala R&D Institute of Science and Technology, Chennai, India

Dhek Gaurav Singh Department of Information Technology, Army Institute of Technology, Pune, Maharashtra, India

Dhilip Kumar G. Madras Institute of Technology Campus, Anna University, Chennai, Tamil Nadu, India

Dinesha H. A. NCET Bangalore, Affiliation to VTU Belagavi, Beedaganahalli, India

Doomun Rubeena Open University of Mauritius, Réduit, Mauritius

Ezhilan M. Department of Communication Engineering, School of Electronics Engineering, Vellore Institue of Technology, Tamil Nadu, Vellore, India

Gandhi Khushi ECE Department, Indira Gandhi Delhi Technical University for Women, Delhi, India

Geetha M. Department of Computer Science and Engineering, Manipal Institute of Technology, Manipal Academy of Higher Education, Manipal, Karnataka, India

Ghalyan Riya CSE Department, Indira Gandhi Delhi Technical University for Women, Delhi, India

Ghosh Bravish Department of Computer Science and Engineering, ITER Siksha 'O' Anusandhan (Deemed to Be University), Bhubaneswar, Odisha, India

Ghosh Debashis Department of Electronics and Communication Engineering, Indian Institute of Technology Roorkee, Roorkee, Haridwar, Uttarakhand, India

Giri Kaiser J. Department of Computer Science, Islamic University of Science and Technology, Kashmir, India

Giri V. K. M. M. M. University of Technology Gorakhpur, Gorakhpur, Uttar Pradesh, India

Gokul M. Department of Biomedical Engineering, Kalasalingam Academy of Research and Education, Krishnankoil, Srivilliputhur, India

Gokuldhev M. Vel Tech Rangarajan Dr. Sagunthala R&D Institute of Science and Technology, Chennai, India

Gomaa Mamdouh M. Department of Computer Science, Faculty of Science, Minia University, EL-Minia, Egypt

Guha Thakurta Parag Kumar Department of Computer Science and Engineering, NIT Durgapur, Durgapur, West Bengal, India

Gupta Mehul Department of Computer Science and Engineering, Kongu Enginnering College, Perundurai, India

Hamed Ibraheem Eliwa Entesar Department of Computer Science, Faculty of Science, Minia University, EL-Minia, Egypt;
College of Science, King Faisal University, Al-Hofuf, Saudi Arabia

Hooda Sakshi GGSIPU, New Delhi, India

Hrithik K. Department of Mechatronics Engineering, Sri Krishna College of Engineering and Technology, Coimbatore, India

Inamdar Gayatri Department of Electronics and Telecommunication Engineering, MKSSS's Cummins College of Engineering for Women, Pune, India

Jena Pradeep Kumar Vardhaman College of Engineering, Hyderabad, India

Jenita Hermina J. Department of Computer Science and Engineering, Er. Perumal Manimekalai College of Engineering, Hosur, India

Joshi Aparna Department of Information Technology, Army Institute of Technology, Pune, Maharashtra, India

Jyothi N. M. Department of Computer Science and Engineering, Koneru Lakshmaiah Education Foundation, Vaddeswaram, Guntur (DT), Andhrapradesh, India

Jyothi Upadhya K. Department of Computer Science and Engineering, Manipal Institute of Technology, Manipal Academy of Higher Education, Manipal, Karnataka, India

Kabir Kazi Amrin Department of Computer Science and Engineering, NIT Durgapur, Durgapur, West Bengal, India

Kalaivani S. Department of Communication Engineering, School of Electronics Engineering, Vellore Institue of Technology, Tamil Nadu, Vellore, India

Kankaria Riddhi Department of Electronics and Telecommunication Engineering, MKSSS's Cummins College of Engineering for Women, Pune, India

Katage Siddhi Department of Electronics and Telecommunication Engineering, MKSSS's Cummins College of Engineering for Women, Pune, India

Khansama Rasmi Ranjan C.V. Raman Global University, Bhubaneswar, Odisha, India

Khatri Kanika Manav Rachna International Institute of Research and Studies, Faridabad, India

Khoyratty Bushra Department of Software Engineering, Pamplemousses, Mauritius

Komarasay Dinesh Department of Computer Science and Engineering, Kongu Enginnering College, Perundurai, India

Krishna Chitirala Abhinay Department of Computer Science and Engineering, Amrita School of Engineering, Amrita Vishwa Vidyapeetham, Bengaluru, India

Krishna Das K. S. Department of Computer Science and Engineering, Amrita School of Engineering, Amrita Vishwa Vidyapeetham, Amritapuri, India

Kumar Betchoo Nirmal Department of Management, Faculty of Business and Management, Université Des Mascareignes, Beau Bassin-Rose Hill, Republic of Mauritius

Kumar Jena Pradeep GITA, Bhubaneswar, Odisha, India

Kumar Jha Pramod Centre for Advanced Systems, DRDO Hyderabad, Hyderabad, India

Kumar Mohapatra Pradyumna Vedang Institute of Technology, Bhubaneswar, Odisha, India

Kumar Rout Saroja Vardhaman College of Engineering, Hyderabad, India; GIFT, Bhubaneswar, Odisha, India

Kumar Verma Rajesh Infosys Limited, Bangalore, India

Kumar Santosh S'O'A University Bhubaneswar, Bhubaneswar, Orissa, India

Kumar Saurabh C.V. Raman Global University, Bhubaneswar, Odisha, India

Madhankumar S. Department of Mechatronics Engineering, Sri Krishna College of Engineering and Technology, Coimbatore, India

Madhusudhanan S. Department of Computer Science and Engineering, Koneru Lakshmaiah Education Foundation, Vaddeswaram, Guntur (DT), Andhrapradesh, India

Majumder Bibek Tripura University, Agartala, India

Mallika Garg Department of Electronics and Communication Engineering, Indian Institute of Technology, Roorkee, Uttarakhand, India

Mane Deepak JSPM's Rajarshi Shahu College of Engineering, Pune, Maharashtra, India

Manivannan S. Department of Electrical and Electronics Engineering, Dr MGR Educational and Research Institute University, Chennai, Tamil Nadu, India

Mann Suman Maharaja Surajmal Institute of Technology, New Delhi, India

Mohamed El Koshiry Amr Faculty of Specific Education, Minia University, EL-Minia, Egypt;
College of Education, King Faisal University, Al-Hofuf, Saudi Arabia

Mohapatra Pradyumna Kumar Vedang Institute of Technology, Bhubaneswar, Odisha, India

Monish Singam S. Department of Communication Engineering, School of Electronics Engineering, Vellore Institue of Technology, Tamil Nadu, Vellore, India

Monisha R. Department of Biomedical Engineering, Kalasalingam Academy of Research and Education, Krishnankoil, Srivilliputhur, India;
Department of Biomedical Engineering, Rajalakshmi Engineering College, Thandalam, India

Murugesan Manojj Department of Computer Science and Engineering, Kongu Enginnering College, Perundurai, India

Nair Aishwarya Infosys Limited, Bangalore, India

Nair Arun K. Department of Computer Science and Engineering, Amrita School of Engineering, Amrita Vishwa Vidyapeetham, Amritapuri, India

Narayan Panda Ravi Vardhaman College of Engineering, Hyderabad, India

Narwal Bhawna IT Department, Indira Gandhi Delhi Technical University for Women, Delhi, India

Nayakavadi Rutuja JSPM's Rajarshi Shahu College of Engineering, Pune, Maharashtra, India

Nayan Navneet Department of Electronics and Communication Engineering, Indian Institute of Technology Roorkee, Roorkee, Haridwar, India

Nikesh Shah Samit Department of Computer Engineering, NMIMS Mukesh Patel School of Technology Management and Engineering, Mumbai, Maharashtra, India

Nithya A. N. Department of Biomedical Engineering, Rajalakshmi Engineering College, Thandalam, India

Panchal Harsh Instrumentation and Control Engineering Department Institute of Technology, Nirma University, Ahmedabad, Gujarat, India

Panchal Nisarg Institute of Technology, Nirma University, Ahmedabad, Gujarat, India

Panda Bijayalaxmi Department of Computer Science, Rama Devi Women's University, Bhubaneswar, India

Panda Ravi Narayan GIFT, Bhubaneswar, Odisha, India

Pandey Anand Department of Information Technology, Army Institute of Technology, Pune, Maharashtra, India

Pandhe Sanjana JSPM's Rajarshi Shahu College of Engineering, Pune, Maharashtra, India

Panigrahi Chhabi Rani Department of Computer Science, Rama Devi Women's University, Bhubaneswar, Odisha, India

Parhi Manoranjan Department of Computer Science and Engineering, ITER Siksha 'O' Anusandhan (Deemed to Be University), Bhubaneswar, Odisha, India

Parikh Harsh Institute of Technology, Nirma University, Ahmedabad, Gujarat, India

Pati Abhilash Department of Computer Science and Engineering, Siksha 'O' Anusandhan (Deemed to Be University), Bhubaneswar, Odisha, India

Pati Bibudhendu Department of Computer Science, Rama Devi Women's University, Bhubaneswar, Odisha, India

Patil Mayur JSPM's Rajarshi Shahu College of Engineering, Pune, Maharashtra, India

Peeroo Swaleha Université des Mascareignes, Pamplemousses, Mauritius

Pirbhai-Jetha Neelam Centre for Digital Humanities, Université des Mascareignes, Beau Bassin-Rose Hill, Mauritius

Pradhan Pyari Mohan Department of Electronics and Communication Engineering, Indian Institute of Technology Roorkee, Roorkee, Haridwar, Uttarakhand, India

Rahul Shah Deep Department of Computer Engineering, NMIMS Mukesh Patel School of Technology Management and Engineering, Mumbai, Maharashtra, India

Rajesh Menon Pranav Department of Communication Engineering, School of Electronics Engineering, Vellore Institue of Technology, Tamil Nadu, Vellore, India

Rajesh M. Department of Computer Science and Engineering, Amrita School of Engineering, Amrita Vishwa Vidyapeetham, Bengaluru, India

Rajesh S. Department of Mechanical Engineering, R. M. K. Engineering College, Tiruvallur, India

Ramsamy Shyama Centre for Digital Humanities, Université des Mascareignes, Beau Bassin-Rose Hill, Mauritius

Ramyachitra D. Department of Computer Science, Bharathiar University, Coimbatore, India

Rao Thokala Mohan Department of Electronics and Communication Engineering, Siddhartha Institute of Engineering and Technology, Hyderabad, India

Rao B Dinesh Manipal School of Information Sciences, Manipal Academy of Higher Education, Manipal, Karnataka, India

Rathore Naincy Department of Information Technology, Army Institute of Technology, Pune, Maharashtra, India

Ravishankar M. VVIET Mysuru, Affiliation to VTU Belagavi, Mysore, India

Revathi M. Department of Computer Science, Bharathiar University, Coimbatore, India

Sai Akhilesh Chunduri Department of Computer Science and Engineering, Amrita School of Engineering, Amrita Vishwa Vidyapeetham, Bengaluru, India

Samantaroy Rojalin Vedang Institute of Technology, Bhubaneswar, Odisha, India; Vardhaman College of Engineering, Hyderabad, India

Sathya S. Department of Electronics & Communication Engineering, Gojan School of Business and Technology, Chennai, Tamilnadu, India

Sekhar Mohanty Sumant Department of CSE, Siksha 'O' Anusandhan (DU), Bhubaneswar, Odisha, India

Selvaraj G. Department of Mathematics, Ramco Institute of Technology, Viruthunagar, Tamil Nadu, India

Shah Alam Md. Department of Accounting and Information Systems, University of Rajshahi, Rajshahi, Bangladesh

Shah Kunal SCTR's Pune Institute of Computer Technology, Pune, Maharashtra, India

Shah Saloni KLE Technological University, Hubli, Karnataka, India

Shah Seema Department of Computer Engineering, NMIMS Mukesh Patel School of Technology Management and Engineering, Mumbai, Maharashtra, India

Sharma Ankit Institute of Technology, Nirma University, Ahmedabad, Gujarat, India

Sharma Ankit K. Instrumentation and Control Engineering Department Institute of Technology, Nirma University, Ahmedabad, Gujarat, India

Shekharappa Gouda Tanush VVIET Mysuru, Affiliation to VTU Belagavi, Mysore, India

Siji Rani S. Department of Computer Science and Engineering, Amrita School of Engineering, Amrita Vishwa Vidyapeetham, Amritapuri, India

Sikdar Pragna Labani Department of Computer Science and Engineering, NIT Durgapur, Durgapur, West Bengal, India

Singh Aman Department of Information Technology, Army Institute of Technology, Pune, Maharashtra, India

Singhal Shefali Manav Rachna International Institute of Research and Studies, Faridabad, India

Singhal Vidushi Manav Rachna International Institute of Research and Studies, Faridabad, India

Singh Bipin Institute of Engineering & Technology Lucknow, Lucknow, Uttar Pradesh, India

Singh Debabrata S'O'A University Bhubaneswar, Bhubaneswar, Orissa, India

Sinha Ritesh Kumar C.V. Raman Global University, Bhubaneswar, Odisha, India

Solapure Rishikesh SCTR's Pune Institute of Computer Technology, Pune, Maharashtra, India

Soni Kanal Information Technology Department, Gandhinagar Institute of Technology, Ahmedabad, Gujarat, India

Sri Chandan Krosuri Department of Computer Science and Engineering, Amrita School of Engineering, Amrita Vishwa Vidyapeetham, Bengaluru, India

Srivastava Kriti Department of Computer Engineering, Dwarkadas J. Sanghvi College of Engineering, Mumbai, Maharashtra, India

Surekha S. Department of Biomedical Engineering, Kalasalingam Academy of Research and Education, Krishnankoil, Srivilliputhur, India

Surya Siddarthan S. Madras Institute of Technology Campus, Anna University, Chennai, Tamil Nadu, India

Talapatra Soumyasree Advanced Computing and Microelectronics Unit, Indian Statistical Institute, Kolkata, India

Tripathi Deepshikha Department of Information Technology, Army Institute of Technology, Pune, Maharashtra, India

Tripathy Sushreeta Department of CA, Siksha 'O' Anusandhan (DU), Bhubaneswar, Odisha, India

Trivedi Abhilash Department of Information Technology, Army Institute of Technology, Pune, Maharashtra, India

Tukaram Dethe Department of Computer Science and Engineering, Amrita School of Engineering, Amrita Vishwa Vidyapeetham, Bengaluru, India

Ullas S. Department of Computer Science and Engineering, Amrita School of Engineering, Amrita Vishwa Vidyapeetham, Bengaluru, India

Uma Maheswari B. Department of Computer Science and Engineering, Amrita School of Engineering, Amrita Vishwa Vidyapeetham, Bengaluru, India

Vaijayanthi S. Department of Computer Science and Engineering, SRM Institute of Science and Technology, Chennai, Tamil Nadu, India

Valliyammai C. Madras Institute of Technology Campus, Anna University, Chennai, Tamil Nadu, India

van Greunen Darelle Nelson Mandela University, Gqeberha, South Africa

Vasudevan Ananthu Department of Computer Science and Engineering, Amrita School of Engineering, Amrita Vishwa Vidyapeetham, Amritapuri, India

Venkasatyasai Phani Ramaaditya Garimella Department of Computer Science and Engineering, Amrita School of Engineering, Amrita Vishwa Vidyapeetham, Bengaluru, India

Verma Arunima Institute of Engineering & Technology Lucknow, Lucknow, Uttar Pradesh, India

Vishnukumar A. Department of Information Technology, Vel Tech High Tech Dr.Rangarajan Dr.Sakunthala Engineering College, Avadi,Chennai, Tamilnadu, India

Vora Harsh Kamlesh Department of Computer Science and Engineering, Manipal Institute of Technology, Manipal Academy of Higher Education, Manipal, Karnataka, India

Yadav Narendra Kumar Department of Information Technology, Army Institute of Technology, Pune, Maharashtra, India

Youssef Aktham Department of Computer Science and Engineering, Siksha 'O' Anusandhan (Deemed to Be University), Bhubaneswar, Odisha, India

Machine Learning Applications in Healthcare

Respiratory Disease Diagnosis with Cough Sound Analysis

S. Monish Singam, Pranav Rajesh Menon, M. Ezhilan, B. R. Arjun, and S. Kalaivani

Abstract Mel frequency cepstral coefficients are one of the most prominent sets of primary features of an audio signal which are used for speech detection and cough analysis. This paper presents a new method that can overcome some of the common problems faced by using MFCCs for cough detection. In the proposed method, the most prominent part of the cough sample (HCP) is extracted and used to obtain the MFCC vectors of that particular window. These HCP MFCC vectors work as a standard comparison index for all cough samples to detect any respiratory disorders. The evaluation of the proposed method is done using 40 samples of COVID-19 patients of which 20 are positive and 20 are negative. The accuracy of the proposed method is compared with that of the standard MFCC method for the same set of samples. The proposed HCP MFCC method produces results that are 7.84% more accurate than the standard method. By bringing a standard set of comparing features that can work for almost all use cases, this method can be used as a quick identifying tool for various respiratory diseases.

Keywords Highest cough point (HCP) · Machine learning (ML) · COVID-19 · Mel frequency cepstral coefficients (MFCCs)

S. Monish Singam (✉) · P. Rajesh Menon · M. Ezhilan · B. R. Arjun · S. Kalaivani
Department of Communication Engineering, School of Electronics Engineering, Vellore Institue of Technology, Tamil Nadu Vellore, India
e-mail: monishsingam.s2019@vitstudent.ac.in

P. Rajesh Menon
e-mail: pranavrajesh.menon2019@vitstudent.ac.in

M. Ezhilan
e-mail: ezhilanm.2019@vitstudent.ac.in

B. R. Arjun
e-mail: arjunb.r2019@vitstudent.ac.in

S. Kalaivani
e-mail: kalaivani.s@vit.ac.in

B. Pati et al. (eds.), *Proceedings of the 6th International Conference on Advance Computing and Intelligent Engineering*, Lecture Notes in Networks and Systems 428,
https://doi.org/10.1007/978-981-19-2225-1_1

1 Introduction

Going to the hospital has always been a hesitant decision, especially now during the COVID-19 pandemic. 1–1.2 billion people suffer from respiratory problems globally [1]. The process of obtaining a COVID test during this period can be worrying as it involves the risk of meeting other COVID-positive patients waiting to get tested. Not only is the process painful and time taking, but it is also an additional load on the already burdened healthcare sector. A cost-effective and easy way for the early diagnosis will help people from places where lab diagnosis is impossible. The proposed paper investigates a novel method for analyzing and diagnosing a patient dealing with respiratory conditions using their cough sample.

The first symptom of respiratory diseases such as the common cold, lung infections, asthma, chronic obstructive pulmonary disease (COPD), pulmonary fibrosis, and lung cancer can be a persistent cough [2]. Cough sound contained information about the pathophysiological mechanisms of coughing, which indicates the structure of the tissue [3]. We can extract this information using MFCC vectors and differentiate them between disease and no disease using machine learning algorithms. The proposed paper defines a new standard, set/segment of features of the cough signal, which can extract the information while also tackling problems faced while applying machine learning algorithms.

The organization of the rest of the paper is as follows. The background is mentioned in Sect. 2 and has sub-sections such as mel spectrogram, MFCCs, and feature vectors. HCP or highest common point is presented in Sect. 3. This section defines the HCP concept in detail. Section 4 methodology is written, explaining cough sample extraction using MFCCs, model building, and training and making a decision tree for the model followed by results and discussion.

1.1 Motivation

As the number of infected individuals has increased, there is a lot of pressure in offering medical and healthcare services. Due to this burden, there is a backlog and deprivation of other medical procedures as well [4]. This research could help ease this burden on the medical sector and simplify the diagnosis process for respiratory diseases.

1.2 Contributions

Previous methods to determine diseases using cough samples have not developed a standard set/segment of feature vectors to extract from the cough sample. Doing this saves the time required to preprocess the extracted features from the cough signals.

It also defines a standard set of features of cough samples that are needed to detect diseases. Doing this could reduce wastage of data and give consistent results in real-world use cases.

2 Background

For cough recognition, cough features are extracted and inputted into a model classifier [5]. The spectrogram is obtained to extract features from a sound signal, and feature vectors from the spectrogram can be obtained by quantizing functions. However, in the case of human beings, the perception of sound is different. Therefore, to improve the performance of a machine learning classifier, the scale used to measure cough signals is altered into the logarithmic scale. This will make the extracted feature vectors more accurate inputs for the machine learning algorithm.

Feature vectors extracted to describe and differentiate the cough signal are mel frequency cepstral coefficient vectors or MFCCs.

2.1 Mel Spectrogram

Sound signal can be represented visually as a spectrum of frequencies called a spectrogram. The x-axis of a spectrogram represents time, and the y-axis represents the frequency. In a spectrogram graph, each point is represented by a different color. These colors show how present a certain frequency is at a certain point in time. However, the regular frequency spectrum is linear in nature, but the way human beings perceive frequency is quite different. Human beings have better resolution of sounds/notes in the lower frequencies when compared to the higher frequencies. This means that it is easier to perceive the difference between sounds/notes of lower frequencies in comparison to sounds/notes of higher frequencies. This is because humans perceive frequency logarithmically. Therefore, a normal spectrogram will not be able to represent sounds the way humans perceive them. Through the trial-and-error method over the years, researchers came up with the mel scale. The mel scale is used as an alternative to the hertz scale which is used to understand the difference between sounds. The mel scale unlike the hertz scale perceives sounds in a logarithmic scale that is very similar to human perception. Log frame improves performance while doing audio classification [6].

$$m = 2595 \times \log(1 + f \div 500) \tag{1}$$

$$f = 700\left(10^{\frac{m}{2595}} - 1\right) \tag{2}$$

Equation (1) shows the conversion of a signal from Hertz scale (f) to mel scale. Equation (2) shows the conversion of a signal from mel scale to hertz scale.

Using the mel scale frequencies, the mel spectrogram can be determined. The mel spectrogram is a matrix containing the different mel bands and frames. Since the mel scale frequencies are remarkably similar to the way humans perceive sound, the mel spectrogram can show a visually more relevant graph which can help determine different sounds and noises. In the mel spectrogram, the x-axis represents time, the y-axis represents the different mel bands, and the different colors in the graph tell us about how present a certain mel band is at a particular instance of time. The mel spectrogram is a more accurate way to classify audio and is extensively used in audio ai research.

2.2 *MFCCs*

Cepstrum is another version of spectrum which is used to identify speech and sound signals. Cepstrum can differentiate between the signal produced by the vocal track and vocal cords. This provides a better way to compute and analyze speech signals.

If $F(x(t))$ represents discrete Fourier transform of $x(t)$ which is a spectrum of the signal $x(t)$, then the cepstrum $C(t)$ is given by:

$$C(t) = F^{-1}[\log(F(x(t))] \tag{3}$$

where F^{-1} is the inverse Fourier transform.

To obtain mel frequency cepstral coefficients (MFCCs), a set of functions are applied to the waveform (Fig. 1).

Mel spectrogram is obtained by mel scaling. Mel spectrogram is then converted into cepstral coefficients using a discrete cosine transform. This helps in identifying the sounds using a set of constants. The discrete cosine transform will give a set of coefficients known as MFCCs. These MFCCs are primary features of sound that are used to uniquely identity sounds.

Multiple sets of MFCCs are obtained for a waveform depending on the number of windows used. For example, a one-second waveform and a 20-ms window give 50 different sets of MFCCs. Each window contains a specific amount of MFCCs. For speech processing, 12–13 MFCCs from the front are more than sufficient as per the analysis.

Fig. 1 Process of obtaining MFCCs. This figure explains all the steps needed to obtain the MFCCs of a given waveform

2.3 *Feature Vectors*

Feature vectors are a set of dimensions used to classify different subjects similar to how x, y, and z-axes represent the relative position of objects in the Cartesian plane. Similarly, for the classification of sounds, MFCCs are used as feature vectors. 12 MFCCs are enough to classify a sound sample. Hence, 12 MFCCs + 1 energy feature vector is obtained per window to analyze and classify various cough samples.

3 HCP

Speech signals originate in the vocal tract. Different syllables or sounds are produced depending on the shape of the vocal tract. Initially, speech signals are in the form of glottal pulses which are noisy signals generated by the vocal cord. The vocal tract acts as a filter on the glottal pulse and produces the speech signal.

While coughing, the vocal cords will open widely in order to allow additional air to flow into the lungs. The epiglottis closes off the windpipe, simultaneously abdominal, and rib muscles contract. This will increase the pressure behind the epiglottis. Air is now forcefully expelled, producing a rushing sound as it moves past the vocal cord. This rushing air removes any irritant present in the throat, making it possible to breathe comfortably again.

The glottis behaves differently under different pathological conditions, making it possible to distinguish between coughs due to various diseases [7]. MFCC vectors detect these variations. By extracting MFCCs and using a machine learning algorithm it is easier to differentiate these minute differences which are present in coughs of people with respiratory disease.

The explosive phase, intermediate phase, and voiced phase are the three different phases of a cough signal [8, 9]. The maximum amount of air is expelled in the explosive phase of cough through the vocal cord contains a large amount of important information about the glottal pulses. Therefore, these glottal pulses carry information about the irritant present.

A standard set/segment of feature vectors can help the machine learning algorithm accurately differentiate between cough samples. This will allow the algorithm to compare coughs based on real differences which need to be detected instead of determining based on unnecessary or irrelevant factors. The explosive phase of the cough produced sounds similar in most of the cases. At this stage of the cough, the shape of the vocal tract will be similar in humans. Therefore, we extract the loudest window of cough from this portion of the signal. Here, the variation in the vocal cords can be detected and compared with multiple other samples with higher accuracy. This point is called the highest cough point (HCP).

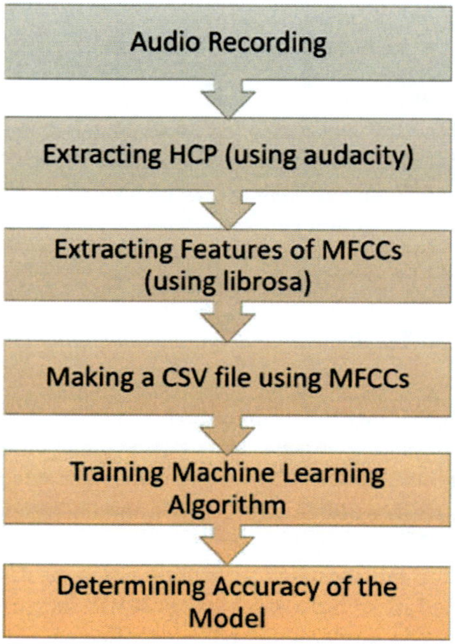

Fig. 2 Methodology block diagram. This diagram shows all the steps involved to determine the accuracy of the proposed model. First step is to record the audio for obtaining samples. In the next step, HCP window can be extracted from cough samples. Following extraction of HCP window MFCCs feature vectors can be extracted. Once feature vectors are extracted, they can be added to .csv file to train the machine learning algorithm. Once model is train, we can test its accuracy

4 Methodology

4.1 Block Diagram

See Fig. 2.

4.2 Extraction of HCP

Audacity (software used to work with audio signals) is used to extract the HCP. Once the cough sample is uploaded, it is observed that multiple coughs are present. From most prominent cough, a window of 20 ms with the highest amplitude is extracted. This 20 ms audio sample is the HCP (Fig. 3).

Frequency of 22,050 is set as default for all samples. The samples are exported in .wav format.

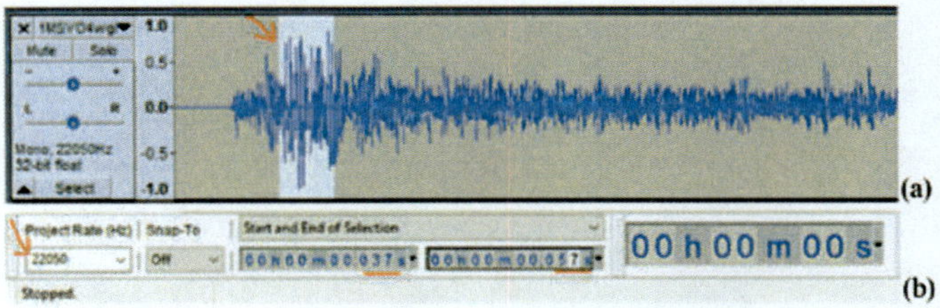

Fig. 3 **a** Selecting HCP window in audacity. **b** Setting frequency at 22,050 Hz. These images show how the HCP window is extracted from the given cough sample

4.3 Extracting the Features of Cough Using Mel Frequency Cepstral Coefficient

MFCCs are primary features that are extracted using standard signal processing techniques in both the time and frequency domain.

The librosa package for audio signal processing in python is used to extract the MFCC vectors from the HCP. The function librosa.feature.mfcc extracts 13 MFCCs including one energy vector.

A .csv file is created using the extracted MFCCs features. This .csv file can be used to train and test the machine learning model.

4.4 Model Building and Training

Decision tree classifier is used for training the machine learning model. Its training time is faster compared to neural network algorithms. It can handle high dimensional data with good accuracy. Machine learning uses csv file format which is split into training set and data set by algorithm. Once model is trained, its accuracy on train and test data is found using the accuracy_score command.

4.5 Decision Tree for Proposed Model

To detect diseases, the primary goal is to detect any obstruction in the vocal cord while coughing. While coughing a large amount of air is exhaled which carries information about the irritation that is present in the throat. Hence, if only the MFCCs of the highest window are extracted, a standard set of feature vectors containing more information that is easily comparable can be obtained. This set of features are called the highest cough point MFCCs which are extracted from the highest point of the most prominent cough (Fig. 4).

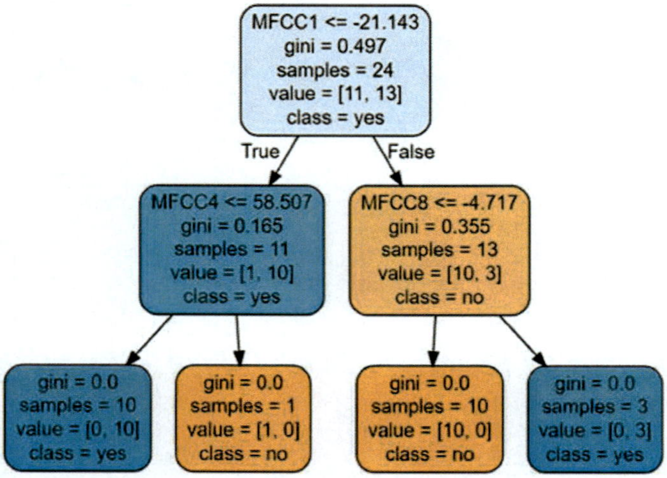

Fig. 4 Decision tree for proposed model. This figure shows a visual representation of the decision tree used by proposed model to detect COVID-19

In the proposed paper, a comparison between MFCCs obtained using HCP and MFCCs of the entire sample (industry standard) is done. The key advantages that the HCP MFCC vectors have are that they are easily comparable, they carry the part of cough which contains more relevant information about the disease, and background noise is negligible.

5 Results and Discussions

To compare both the methods, the average of ten consecutive accuracy results produced by the machine learning algorithms are taken. These results are taken from 40 cough samples obtained (20 COVID-19 positive and 20 COVID-19 negative). A CSV file is made using the HCP of the set of coughs samples; this is compared with another CSV file which is obtained using the standard method. The table below shows the comparison of accuracies obtained from the samples by both methods (Table 1).

The table shows different combinations of HCP MFCCs being compared with standard MFCCs. The parameter column contains four different cases which are used to compare both the methods. These cases are different combinations of data, which are used as inputs for machine learning. The other columns represent the average accuracy results obtained from the method with HCP and the method without HCP. There are two sets of accuracies in order to avoid distortion in the result. It is evident from the table that the HCP MFCCs produce more accurate results compared to standard MFCCs.

Table 1 Table containing values of accuracy

Parameter	Average accuracy with HCP set 1 (%)	Average accuracy without HCP set 1 (%)	Average accuracy with HCP set 2 (%)	Average accuracy without HCP set 2 (%)
13 HCP MFCCs versus 13 MFCCs	63.29	63.12	59.98	49.375
13 HCP MFCCs versus 20 MFCCs	69.995	49.99	63.29	58.75
13 HCP MFCCs versus current industry standard	69.97	60.6	61.63	65.625
All variables with HCP MFCCs versus all variables without HCP MFCCs	62.6	53.7	70.625	57.5

The values in this table show the accuracy of the models used. The parameter column represents the different parameters in which the accuracies of the model with HCP and the model without HCP are compared

From the results obtained above, it is seen that the extraction by HCP method on an average gives 7.84% better accuracy of detecting COVID-19 from a cough sample. The HCP method is able to avoid background noise while only carrying relevant information.

6 Conclusion

Further research can be conducted on this method to compare the cough sounds of various other respiratory diseases. Research can be conducted to find the optimal window length for the defined method. Research could be conducted on the defined method with larger datasets to understand the saturation point of the machine learning algorithm.

Through this paper, a new method to detect respiratory diseases has been identified. This can be used by a simple application on smartphones to detect respiratory diseases in any remote place in the world. The HCP method is widely applicable as it can be used to compare a wide range of coughs. This is because altering the length of the sample to extract a set amount of MFCCs, which can be used in a machine learning algorithm, is moot.

References

1. Rudraraju, G., Palreddy, S. D., Mamidgi, B., Sripada, N. R., Padma Sai, Y., Kumar, V. N., & Haranath, S. (2020). Cough sound analysis and objective correlation with spirometry and clinical diagnosis. *Informatics in Medicine Unlocked, 19*. 100319. https://doi.org/10.1016/j.imu.2020.100319; Clerk Maxwell, J. (1892). A treatise on electricity and magnetism (3rd ed., Vol. 2, pp. 68–73). Oxford: Clarendon.
2. Grace, M. S., Dubuis, E., Birrell, M. A., & Belvisi, M. G. (2013). Pre-clinical studies in cough research: Role of transient receptor potential (TRP) channels. *Pulmonary Pharmacology & Therapeutics,26*(5), 498–507. https://doi.org/10.1016/j.pupt.2013.02.007
3. Korpáš, J., Sadloňová, J., Vrabec, M. (1996). Analysis of the Cough sound: An overview. *Pulmonary Pharmacology, 9*(5–6), 261–268. ISSN 0952-0600. https://doi.org/10.1006/pulp.1996.0034
4. Parida, S., Mohanty, A., Nayak, S. C., Pati, B., & Panigrahi, C. R. (2021). Study and impact analysis of COVID-19 pandemic clinical data on infection spreading (p. 225). In *Data science for COVID-19: Volume 2: Societal and medical perspectives.*
5. Michalak, N., Sng, O., Wang, I., & Ackerman, J. (2020). Sounds of sickness: Can people identify infectious disease using sounds of coughs and sneezes? *Proceedings of the Royal Society B: Biological Sciences.,287*, 20200944. https://doi.org/10.1098/rspb.2020.0944
6. Panigrahi, C., Pati, B., Rath, M., & Buyya, R. (2021). *Computational modeling and data analysis in COVID-19 research.* CRC Press.
7. Higenbottam, T. (2002). Chronic cough and the cough reflex in common lung diseases. *Pulmonary Pharmacology & Therapeutics, 15*(3), 241–247. ISSN 1094-5539.
8. Ramasamy, V., et al. (2021). Application of deep learning strategies to assess COVID-19 patients. In *Computational intelligence techniques for combating COVID-19. EAI/Springer innovations in communication and computing*. Springer. https://doi.org/10.1007/978-3-030-68936-0_2
9. Morice, A. H., Fontana, G. A., Belvisi, M. G., et al. (2007). ERS guidelines on the assessment of cough. *European Respiratory Journal,29*(6), 1256–1276.

A Deep Learning-Based Model for Arrhythmia Detection Using Feature Selection and Machine Learning Methods

Santosh Kumar, Bharat Bhushan, Lekha Bhambhu, Debabrata Singh, and Dilip Kumar Choubey

Abstract Arrhythmia is one of the diseases that affects many people around the world. Deep learning provides an efficient tool to detect arrhythmia disease. A convolutional neural network (CNN) is an emerging technique used often for feature extraction in the medical domain. In this paper, AlexNet, VGG-16, VGG-19 models are used as the feature extraction method, and the selected feature is supplied as input to four well-known classifiers such as decision tree, kNN, LDA, and SVM for arrhythmia detection. Furthermore, an experiment is conducted with the combination of proposed CNN model where mRMR is used as feature selection method. Finally, the result of experiment is compared with different machine learning algorithms where LDA shows the efficiency in term of classification accuracy. The classification accuracy of the proposed model is recorded as 99.46%. The performance of the proposed model is higher in terms of classification accuracy compared to previous work on arrhythmia detection.

Keywords Convolutional neural network · Feature selection · Arrhythmia · Machine learning

---

S. Kumar (✉) · D. Singh
S'O'A University Bhubaneswar, Bhubaneswar, Orissa, India
e-mail: santoshkumar@soa.ac.in

D. Singh
e-mail: debabratasingh@soa.ac.in

B. Bhushan
Sharda University, Greater Noida, India

L. Bhambhu
Chitkara University Institute of Engineering and Technology, Rajpura, Punjab, India

D. K. Choubey
IIIT, Bhagalpur, Bihar, India
e-mail: dkchoubey.cse@iiitbh.ac.in

B. Pati et al. (eds.), *Proceedings of the 6th International Conference on Advance Computing and Intelligent Engineering*, Lecture Notes in Networks and Systems 428,
https://doi.org/10.1007/978-981-19-2225-1_2

1 Introduction

Arrhythmia is one of the leading causes of the death around the world. The term cardiac arrhythmia implies as irregular heart rates that causes of higher death rate and required early treatment to save the life. Therefore, early arrhythmia detection is so important aspect of prevention and medical treatment of patents to save their life. Deep learning (DL) is one of the emerging techniques in biomedical application. Convolutional neural network (CNN) a subdivision of deep learning achieves higher performance compared to several machine learning algorithms.

Isin [1] presented a DL model for the arrhythmia detection. In this model, deep CNN is used as feature extractor, and extracted features are used as input to backpropagation algorithm for the classification. The proposed model obtained the accuracy of 92%. Acharya [2] developed a CNN-based model for detecting different type of arrhythmia heartbeats. The proposed method was used to train by augmented data with 94% of accuracy on original dataset and 93.47% on noise-free ECGs. Warrick [3] proposed a method combination of CNN and long short-term memory (LSTM) unit used for cardiac arrhythmias detection. The proposed structure used no explicit feature selection and 83% accuracy on held-out test data obtained in terms of F-measure and 80% on the hidden dataset. Mousavi [4] presented the deep CNN and sequence-to-sequence model for arrhythmia detection. The proposed method obtained the best results compared to existing methods. Dindin [5] represented the deep learning-based multichannel architecture that based on the topological data analysis. The purposed method shows the comparative performance compared with various naïve methods for arrhythmia classification. Sannino [6] presented a DL model for arrhythmia classification. The proposed model shows the efficiency in terms of specificity, sensitivity, and accuracy over state-of-the-art methods. Sainath [7] represents a model term as CLDNN based on combination of CNNs, LSTM, and DNN for arrhythmia detection. The experimental result shows the substantial improvement of 6% in WER over strongest model LSTM. Li et al. [8] have proposed CNN- and LSTM-based learning model to incorporate with spatial temporal intonation. The proposed method achieves 87.40% in terms of accuracy for human activity analysis. Ozal et al. [9] proposed the convolutional auto-encoder (CAE)-based nonlinear compression structure to reduce the signal size of arrhythmia beats. The proposed model used the LSTM with CAE network to reduce the arrhythmia beats. Finally, 99% of accuracy validated the model. Mathews [10] represented the model that is combination of deep belief networks (DBN) and restricted Boltzmann machine (RBM) for ECG classification. The presented model achieved the high average recognition that validated its efficiency. Zal et al. [11] have proposed 1D-CNN model for arrhythmia detection. Compared to previous results, the proposed model succeeded to obtain the high accuracy of 91.33%.

The main motivation behind using deep learning model is its efficiency and performances in image processing and ECGs wave-based datasets. Recently, deep learning-based CNN models have been applied successfully in biomedical engineering. The past studies show the better outcome of CNN models compared the several state-of-the-art models on any diseases. Cardiac arrhythmia is one of the leading causes reported higher mortality rate worldwide. There is significant presence of DL models in solving complex classification problem such as arrhythmia detection. The past studies also reveal the better results of hybrid CNN model compared to CNN model. In this paper, pre-trained CNN hybrid model with different combination of AlexNet, VGG-16, and VGG-19 is used to detect arrhythmia.

In this paper, CNN-based models such as AlexNet, VGG-16, and VGG-19 are constructed for arrhythmia detection. In this work, a pre-trained CNN model with default parameter is constructed for feature extraction on arrhythmia dataset. The extracted feature feeds to machine learning (ML) algorithm such as decision tree (DT), K-nearest neighbor (kNN), linear discriminant analysis (LDA), and support vector machine (SVM) for classification task. Furthermore, mRMR is used for feature extraction and extracted feature supplied once to ML algorithm. In addition, feature extracted by the mRMR is added to CNN-based models.

This paper is organized briefly as follows: In paper organization, Sect. 2 explains the CNN model, ML algorithms, and the proposed methods. Section 3 discusses the experiments and results. Lastly, we have concluded this study with valid remarks in Sect. 4.

2 Methods

In first phase, CNN models such as AlexNet, VGG-16, and VGG-19 are considered with similar layered architecture with 279 features for arrhythmia detection. In second phase, four different combinations of CNN model are constructed. The dimensions of deep features are later reduced by mRMR feature selection methods, and its outcome features are supplied to ML algorithms such as DT, kNN, SVM, and LDA for the classification.

2.1 Dataset

Machine learning datasets are widely used to analyze the performance of the newly proposed algorithm. In this paper, the main focus is to detect arrhythmia on MIT arrhythmia dataset. Arrhythmia: This dataset contains 452 patterns, and 206 linear valued and 73 nominal attributes are taken from 279 attributes. The primary goal is to identify the occurrence of cardiac arrhythmia from 16 different classes. Class 01 represents the normal ECG, classes 02–15 apply to various arrhythmia classes, and class 16 is denoted as unclassified ones.

2.2 Convolutional Neural Networks

CNN is a class of deep neural network architecture, mostly applied in the field of image classification. CNN consists of multiple convolution units, pooling unit, and finally a fully connected network. Basic idea of this deep architecture is to massively compute, combine, and extract an abstract feature map establishing nonlinear relationships between the input data and ground truth. CNN automatically learns features from raw input features. In comparison to other deep learning models, CNN requires less preprocessing steps and shows impressive ability in analyzing spatial information. Thus, CNNs are most commonly applied for manipulating image data. Recently, CNN has also been applied for various biological applications such as gene expression prediction, protein classification, and clinical datasets classification. In this work, AlexNet, VGG-16, and VGG-19 models are considered due to their higher performance in image-based classification problem. AlexNet is fully connected architecture along with convolution and pooling layers. It consists of 227×227 pixels of image with 5×5 filter size. VGG-16 is similar to AlexNet including 16 layers of convolution and fully connected layers with 224×224 pixel size of image and 3×3 size of filter image. VGG-19 consists of 16 and 3 layers of convolution and full connected layers, respectively. It uses 3×3 filter size of images. In these three models, FC8 activation layer is used for feature extraction.

2.3 Machine Learning Method

The field machine learning addresses the various problems and their exact solution by help of computer program. The performance of machine learning program depends upon the algorithm selection and their efficiency in specific domain. These algorithms are classified mostly in supervised and unsupervised learning techniques used for prediction- and classification-related problems. Furthermore, most of the ML algorithms are implemented in Python, and evaluation is done by scikit-learning tools. In this paper, several machine learning algorithms such as decision tree, kNN, SVM, and LDA are applied to evaluate the performance of deep learning model.

DT is tree-based supervised learning model for solving classification and regression problem. The decisions are carried out after breaking the datasets in smaller subsets as decision nodes and leaf nodes, while leaf node is considered for classification. The advantage of decision tree is its handling capability for both numerical and categorical data. In this work, C4.5 algorithm is used which was developed by Ross Quinlan [12, 13] in 1986 and 1993.

kNN [14] classifier is a lazy learning algorithm mostly considered for classification problem. It does not learn from training datasets instead storing the dataset to perform some action for classification. The datasets are observed based on the distance between two coordinates X and Y. The commonly used distance metric is Euclidean distance. Other distance measures can be used depending on the available data. Once the K-nearest neighbors are obtained, the test object is assigned a class y based on its nearest neighbors' classes. The majority vote is the most popular rule used to search for the predominant class in the neighborhood. kNN is preferred due to its simplicity and efficiency with noisy and large training data.

SVM [15] is widely used supervised learning algorithm for the classification problem. SVM constructs the hyperplane in infinite dimension and categories the class that is linearly not separable. Classes are differentiated with their position on hyperplane. The dimensions depend on the number of features in datasets. As an example, a dataset having three features, then constructed hyperplane will be two-dimensional planes.

Linear discriminant analysis (LDA) [16] is one of the popular dimension reduction techniques used in machine learning. The best results can be possible on two and multi-class problem. The features are linearly separated from training dataset and categories in their assigned class.

2.4 *Feature Selection*

Minimum redundancy maximum relevance (mRMR) [17] selects the features that are maximally dissimilar in case of minimum redundancy, and maximum relevance is to maximize total relevance. The discrete and continuous features are defined in Eqs. 1 and 2.

$$\min W_i, W_i = \frac{1}{|s|^2} \sum_{i,j \in s} I(i, j) \tag{1}$$

$$\min W_c, W_c = \frac{1}{|s|^2} \sum_{i,j \in s} |C(i, j)| \tag{2}$$

where $I(i,j)$ and $C(i,j)$ are mutual information and the correlation between f_i and f_j, respectively. In case of maximize relevance, discrete and continuous features are defined in Eqs. 3 and 4.

$$\max V_i, V_i = \frac{1}{|s|^2} \sum_{i \in s} I(h, i) \tag{3}$$

$$\max V_c, V_c = \frac{1}{|s|^2} \sum_{i} F(i, h) \tag{4}$$

where h is the target class and $F(I, h)$ is the F-statistic.

2.5 Proposed CNN Method

In the proposed approach, CNN model is deployed over arrhythmia dataset. The process starts with CNN model with the original feature set where FC8 activation layer was used. Furthermore, mRMR feature selection method was applied to extract the relevant features. Finally, CNN model was evaluated by machine learning algorithm. In the proposed model, a separate CNN model is used with default parameter. No pre-training on CNN model is conducted, and only FC8 model is used. In the second stage, mRMR algorithm is applied to extract the relevant features from 279 features. In the third stage, combination of two models with 100 features and combination of three model with 150 features were constructed. Then, the model was evaluated with machine learning algorithm such as DT, kNN, LDA, and SVM, and finally comparison was made on it. Figure 1 shows the proposed architecture for the arrhythmia detection.

In this work, we have evaluated the performance of CNN model using standard metrics such as F-score, specificity, sensitivity, precision, and accuracy.

$$\text{Accuracy} = \frac{\text{TP} + \text{TN}}{\text{TP} + \text{FP} + \text{FN} + \text{TN}} \tag{5}$$

$$\text{Precision} = \frac{\text{TP}}{\text{TP} + \text{FP}} \tag{6}$$

$$\text{Recall} = \frac{\text{TP}}{\text{TP} + \text{FN}} \tag{7}$$

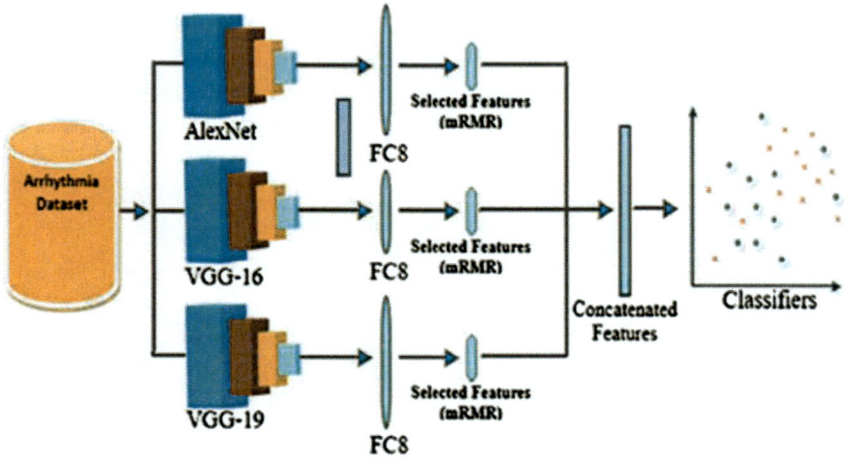

Fig. 1 Illustration of proposed CNN model

$$\text{Specificity} = \frac{\text{TN}}{\text{TN} + \text{FP}} \tag{8}$$

$$F1 = 2 * \frac{\text{Precision} * \text{Recall}}{\text{Precision} + \text{Recall}} \tag{9}$$

3 Results

The experiment was conducted on Python with Intel i7 processor @ 2.5 GHZ with 8 GB of RAM. The proposed model is implemented with GPU support. In first stage, the arrhythmia dataset is evaluated with various CNN model (AlexNet, VGG-16, and VGG-19). The experiment was carried out with original 279 features with default parameter of CNN model. In this stage, dataset was partitioned in 70–30% rations as training and test dataset, respectively. Then, pre-trained CNN model with FC8 layers is constructed to extract the features. Moreover, various ML algorithms are applied to evaluate the classification accuracy of arrhythmia where SVM obtained the high accuracy. Table 1 shows the performance of CNN model in terms of accuracy, sensitivity, specificity, precision, and F-score. Figure 2 illustrates the outcomes of CNN model for training and validation accuracy and loss.

In second stage, mRMR is used to extract the 50 relevant features for original dataset. We have considered 30% data as test dataset where hold-out validation is performed. Here, four models are constructed such as Model-1 (AlexNet and VGG-16), Model-2 (AlexNet and VGG-19), Model-3 (VGG-16 and VGG-19), and Model-4 (AlexNet, VGG-16 and VGG-19) where three models are combination of two DL algorithm and one model having combination of three DL algorithms. The

Table 1 Outcome of CNN models on arrhythmia data using four different classifiers

Model	Classifier	Features	Accuracy	Sensitivity	Specificity	Precision	F-score
AlexNet	DT	279	85.34	84.56	87.38	87.55	85.63
	kNN		91.12	86.38	97.44	97.85	91.29
	LDA		95.18	94.34	96.38	96.25	95.44
	SVM		93.54	92.25	95.63	95.38	93.69
VGG-16	DT	279	89.14	90.48	88.11	87.82	89.04
	kNN		92.53	89.68	95.64	95.38	92.37
	LDA		93.27	93.30	94.40	94.53	93.67
	SVM		95.26	96.02	95.72	95.10	95.69
VGG-19	DT	279	90.52	89.43	91.02	91.14	90.63
	kNN		94.61	91.43	97.27	98.13	94.07
	LDA		93.48	93.71	93.82	94.52	93.65
	SVM		95.37	95.71	96.45	96.75	95.76

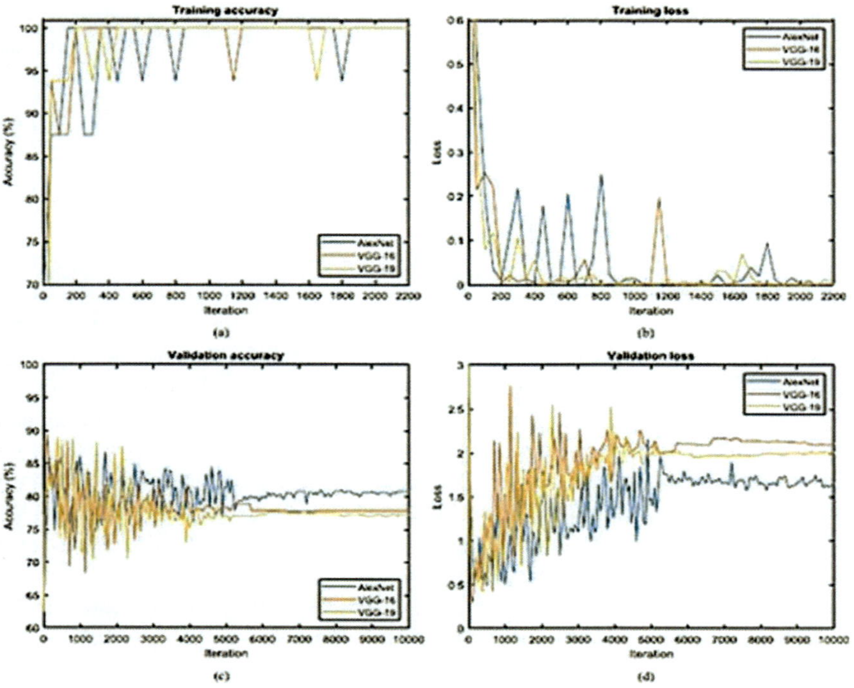

Fig. 2 Results of CNN model for training and validation accuracy and loss

number of features considered in these four models is 100, 100, 100, 150 features. The CNN model having 150 features is managed to achieve high accuracy of 99.46% over LDA algorithm. Table 2 shows the obtained data in percentage as 99.46, 99.61, 99.19, 99.14, and 99.32, respectively, in terms of accuracy, sensitivity, specificity, precision, and F-score.

4 Conclusions

This study tries to classify the arrhythmia dataset in two classes as normal and arrhythmic on basis CNN deep learning models. The proposed method works in two different stages. In first stage, various CNN models (AlexNet, VGG-16, and VGG-19) are used as feature extractor. The extracted features are finally fed as input to machine learning algorithm such as DT, kNN, LDA, and SVM for classification task. The best results came when we combined three deep learning algorithm and using mRMR algorithm for feature extraction. The number of extracted features (i.e., 150) supplied as input to ML algorithm. LDA shows the best performance in terms of

Table 2 Classification result obtained after applying mRMR method on features of CNN

Model	Features	Classifier	Accuracy	Sensitivity	Specificity	Precision	F-score
Model-1	100	DT	95.10	94.45	96.24	96.31	95.07
		kNN	96.31	97.25	96.09	96.04	96.37
		LDA	99.17	99.27	99.34	99.22	99.17
		SVM	99.10	99.32	99.25	99.15	99.39
Model-2	100	DT	97.42	96.80	97.48	97.55	97.43
		kNN	96.33	97.25	95.34	95.03	96.40
		LDA	99.25	99.30	99.09	99.29	99.25
		SVM	99.15	99.16	99.26	99.56	99.31
Model-3	100	DT	96.28	96.15	96.06	96.36	96.35
		kNN	96.15	97.29	96.12	96.61	96.76
		LDA	99.20	99.15	98.38	98.18	99.09
		SVM	98.08	99.62	99.24	98.29	98.20
Model-4	150	DT	98.22	97.28	98.36	98.33	98.10
		kNN	97.18	99.36	96.44	96.13	97.54
		LDA	99.46	99.61	99.19	99.14	99.32
		SVM	99.08	99.37	99.16	99.12	99.27

accuracy of 99.46%. The proposed model gives the highest accuracy compared to previous work on similar arrhythmia dataset. In the future work, we can check the performance of proposed model on different available arrhythmia datasets.

References

1. Isin, A., & Ozdalili, S. (2017). Cardiac arrhythmia detection using deep learning. *Procedia computer science, 120*, 268–275.
2. Acharya, U. R., et al. (2017). A deep convolutional neural network model to classify heartbeats. *Computers in Biology and Medicine, 89*, 389–396.
3. Warrick, P. A., & Homsi, M. N. (2018). Ensembling convolutional and long short-term memory networks for electrocardiogram arrhythmia detection. *Physiological Measurement, 39*(11), 114002.
4. Mousavi, S., & Afghah, F. (2019). Inter-and intra-patient ecg heartbeat classification for arrhythmia detection: A sequence to sequence deep learning approach. In *IEEE international conference on acoustics, speech and signal processing (ICASSP)*.
5. Dindin, M., Umeda, Y., & Chazal, F. (2020). Topological data analysis for arrhythmia detection through modular neural networks. In *Canadian conference on AI*.
6. Sannino, G., & De Pietro, G. (2018). A deep learning approach for ECG-based heartbeat classification for arrhythmia detection. *Future Generation Computer Systems, 86*, 446–455.
7. Sainath, T. N., et al. (2015). Convolutional, long short-term memory, fully connected deep neural networks. In *IEEE international conference on acoustics, speech and signal processing (ICASSP)*.
8. Li, C., et al. (2017). Skeleton-based action recognition using LSTM and CNN. In *IEEE international conference on multimedia & expo workshops (ICMEW)*.

9. Yildirim, O., et al. (2019). A new approach for arrhythmia classification using deep coded features and LSTM networks. *Computer Methods and Programs in Biomedicine, 176*, 121–133.
10. Mathews, S. M., Kambhamettu, C., & Barner, K. E. (2018). A novel application of deep learning for single-lead ECG classification. *Computers in Biology and Medicine, 99*, 53–62.
11. Yıldırım, Ö., et al. (2018). Arrhythmia detection using deep convolutional neural network with long duration ECG signals. *Computers in Biology and Medicine, 102*, 411–420.
12. Quinlan, J. R. (1986). Induction of decision trees. *Machine Learning, 1*(1), 81–106.
13. Quinlan, J. R. (1993) C4. 5: Programming for machine learning. *Morgan Kauffmann, 38*, 48.
14. Cover, T., & Hart, P. (1967). Nearest neighbor pattern classification. *IEEE transactions on information theory, 13*(1), 21–27.
15. Mustaqeem, A., Anwar, S. M., & Majid, M. (2018). Multiclass classification of cardiac arrhythmia using improved feature selection and SVM invariants. *Computational and Mathematical Methods in Medicine*.
16. Hastie, T., Tibshirani, R., & Friedman, J. (2009). The elements of statistical learnin. Cited on 33.
17. Peng, H., Long, F., & Ding, C. (2005). Feature selection based on mutual information criteria of max-dependency, max-relevance, and min-redundancy. *IEEE Transactions on Pattern Analysis and Machine Intelligence, 27*(8), 1226–1238.

Epilepsy Prediction Using Time Series Classifiers

C. Valliyammai, S. Surya Siddarthan, V. Aravind Priyamvadan, and G. Dhilip Kumar

Abstract Epilepsy is a very common neurological disease that affects millions of people worldwide. Though there is no definite cure for epilepsy, predicting the onset of epilepsy in the preictal period will ensure proper health care which is administered during the ictal period. The proposed epilepsy prediction model uses the time series deep learning classifiers called the echo state network (ESN) and the InceptionTime network. The surface and intracranial EEG data from the University of Bonn's Time Series EEG dataset are used for training the proposed models. The experimental results prove that the proposed deep learning model achieves an accuracy of 94.6% using ESN and 100% using InceptionTime network.

Keywords Deep learning · Time series classifier · Echo state network · InceptionTime classifier

1 Introduction

The word epilepsy is derived from the Greek word *epilambanein* which means "to be seized." Epilepsy affects nearly 50 million people worldwide between 4 and 10 per 1000 people have active epilepsy at a given time. Globally, estimates suggest five million people are affected each year, and in higher income nations, the proportions are estimated to be 49 per 100,000 people, and in lower income nations, the numbers can soar as high as 139 per 100,000 people. Epilepsy is a cause of a random electrical discharge in the brain that could be followed by loss of consciousness and temporary amnesia. Seizures which are an aberration in the brain's neurological activity can be detected through the analysis of electroencephalogram (EEG) signals. It works by attaching small metal rods (electrodes) to the scalp of the skull. The billions of neurons in the brain produce very small electrical pulses that form a pattern of brain waves. During the test, the EEG detects these brain waves and amplifies these signals. The test is used to discern occurrences of seizures, tumors, insomnia, dizziness, head

---

C. Valliyammai (✉) · S. Surya Siddarthan · V. Aravind Priyamvadan · G. Dhilip Kumar
Madras Institute of Technology Campus, Anna University, Chennai, Tamil Nadu, India
e-mail: cva@mitindia.edu

B. Pati et al. (eds.), *Proceedings of the 6th International Conference on Advance Computing and Intelligent Engineering*, Lecture Notes in Networks and Systems 428,
https://doi.org/10.1007/978-981-19-2225-1_3

injuries, etc. The detection of epilepsy using EEG data with deep learning has been proposed in the neurological and biomedical fields for the effective use of such techniques which is bound to rapidly evolve. Two different time series classification (TSC) models, namely the echo state networks and the InceptionTime are used in the proposed epilepsy prediction model.

1.1 Motivation and Contribution

Epilepsy is fourth in the most common neurological disorders, and there is a growing need to detect accurately the onset of epileptic seizures, to ensure that these seizures can be mitigated at the earliest. Our methodology intends to introduce a novel approach to predict epilepsy while also maintaining a reasonable high accuracy.

The rest of this paper is organized as follows: Sect. 2 explores the literature survey. Section 3 presents the proposed methodology. The experimental results are discussed in Sect. 4. Section 5 concludes with future work.

2 Literature Survey

The problem dealing with epilepsy detection is a broad topic for which investigations have spanned decades. Epilepsy seizure detection is predominantly a classification problem. The workflow often involves extracting necessary features, and then classification is performed. An end-to-end LSTM was employed with segmentation of the time series is the initial step, ensuring these segments are nonoverlapping was paramount, next the nonoverlapping segments are fed to a LSTM network, the outputs of which are fed to a distributed average pooling layer to find the most useful EEG features, these features are provided to a soft max layer to form accurate labels [1]. The methodology achieved maximum accuracy of 100% for both A–E (healthy vs. epileptic) and ABCD-E (any non-seizure vs. seizure). A pyramidal 1-dimensional CNN model is used on the same dataset which achieves maximum accuracy of 99.8, 99.95 and 99.9% in cases of a three-class problem (AB vs. CD vs. E), two-class problem (AB vs. CDE) and two-class healthy versus epileptic (A vs. E) [2]. Bonn dataset was used with an independent component analysis (ICA) for preprocessing the data which was followed by spectral analysis, and linear decomposition was further used to clean the data. K-NN and Naive Bayes were used as classifiers, results achieving a maximum accuracy 100% for A versus E (healthy vs. epileptic) [3].

The Bonn dataset provides a distinct new way to classify the predictions by using a weighted visibility graph to predict the seizure presence in the intracranial stage [4]. The nodes for the WVG are found by using each sample point of the EEG time series data. After construction of the nodes, the edges between these nodes are calculated, with each edge being assigned weights. The model then uses a support vector machine (SVM) with an RBF kernel for classification, followed by validation performed using tenfold cross validation. Maximum classification accuracy for EEG data at normal versus interictal stages was 94.125%, whereas a maximum accuracy of 100% was achieved for the normal versus ictal stages. A novel technique of LS-SVM using a RBF kernel was used to classify the data, before which noise was removed using a bandpass filter, feature extraction was executed by computing the fractal dimension, and the experiments was done on the same Bonn dataset achieving the highest accuracy of 98.20% in discerning non-ictal and ictal EEG signals [5].

The shallow CNN with 2 blocks each consisting of a convolutional unit and a leaky ReLU features were extracted using a common spatial pattern, and these features were extracted from 9 frequency bands and 8 sub-bands, and this study achieved a maximum average accuracy of 92.2% [6]. A novel RNN model was built with stacking multiple 1D convolutional layer, and it was implemented on TUH corpus which consists of EEG data classified as either abnormal or normal, achieving a maximum testing accuracy of 86.57% [7]. An autoencoder was proposed to detect seizures and the model was trained and tested on the EEG data from the NTT Medical Centre Tokyo and University Hospital Tokyo containing 8 and 16 subjects, respectively [8]. An exhaustive review into the different methodologies used in predicting epileptic seizures during the interictal period [9]. The IED-based feature extraction was used with a CNN model followed by a sigmoid hidden layer and a logistic regression model to perform classification [10]. The performance was measured using precision and f-score results were favorably positive with 70% and 0.68, respectively.

3 Proposed Work

The proposed work compares two different deep learning networks that perform well with time series data with echo state network and InceptionTime network. The data is preprocessed to remove external noise, and the models are trained and validated using the filtered data.

3.1 Workflow

The workflow for the proposed epilepsy prediction model is shown in Fig. 1. Data such as EEG data are non-static. The purpose of this initial step is to remove any aberration in the frequencies of data that may not be adherent to the proposed model which is achieved by using a notch filter. Oftentimes these data are random and uneven in distribution, shuffling is necessary to ensure that data is of even distribution.

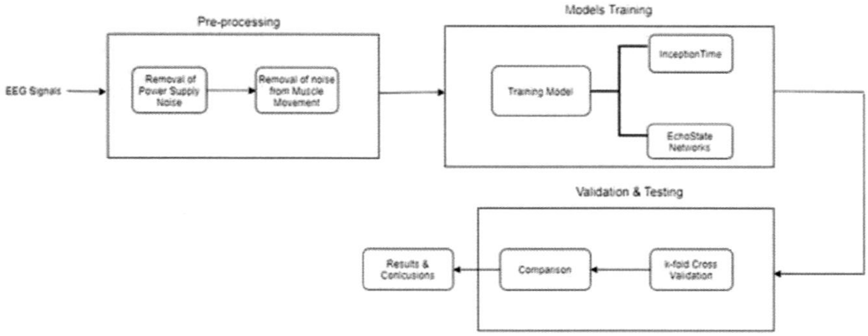

Fig. 1 Proposed epilepsy prediction model

The proposed method uses two TSC models to solve a two-class classification problem. The echo state network employs a reservoir computing methodology, where weights to the hidden layer are randomly assigned and untrainable, ESN also features sparse connectivity (often < 10%). The InceptionTime is a CNN ensemble model consisting of mainly the input layer, the inception network, and the inception module which is the major building block. The inception network itself consists of a series of inception modules. The main cause for comparison comes from the difference in the workings of both models. Accuracy of these models is compared with each other and with other state-of-the-art models.

3.2 Dataset Description

The dataset used is the well-known open-access dataset from the University of Bonn, German, by far, the most prevalently used dataset for detecting epileptic seizures. The Bonn dataset consists of five datasets labeled A through E with each label/set consisting of EEG data sampled with the duration of 23.6 s. These segments were selected and cut out from continuous multichannel EEG recordings after visual inspection for artifacts, e.g., due to muscle activity or eye movements. The dataset was sampled at 173.61 Hz and the specifics of each set A–E is provided in Table 1. The Bonn dataset does not contain any inherent artifacts.

Table 1 Overview of the EEG dataset

Set	Patients	Phase
A	Healthy	Open eyes
B	Healthy	Closed eyes
C	Epilepsy	Interictal
D	Epilepsy	Interictal
E	Epilepsy	Seizure

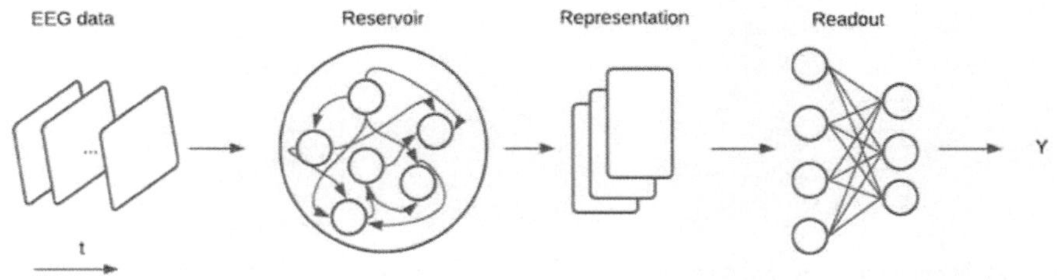

Fig. 2 Architecture of ESN

3.3 Echo State Networks

The echo state network is a type of reservoir computer that uses a set of recurrent neural networks; the weights in the input–hidden and the recurrent layer are randomly initialized and fixed throughout the learning stages. The echo state network consists of an input layer and a hidden layer (or "reservoir"), the hidden layer is sparsely connected, usually less than 10% connectivity. The outputs of the reservoir layer are connected to the readout layer which can be any predictive models such as SVM or MLP whose weights are trainable. Figure 2 gives a brief overview of the structure of the ESN.

```
Algorithm 1: Training and Testing the ESN Model
Input: EEG data of the patient
Output: Condition of the patient

  1. Import RC_Model, numpy, panda, sklearn, scipy
  2. Configure parameters for the reservoir layer
  3. Specify the readout layer and its parameters
  4. Load the txt data set and convert them into numpy arrays
  5. Filter the data set
  6. Split the data set for training and validation
```

3.4 InceptionTime Networks

InceptionTime network is an ensemble of inception networks each of which consists of 1D CNN networks whose outputs are concatenated by the depth concatenation layer. The input is also passed through a MaxPooling layer whose output is also concatenated at the end. There is also a shortcut layer for every *n*th inception networks which allows for gradients to flow to the earlier layers. The earlier layers are expected to capture the small patterns while the layers that come at the end are used to make

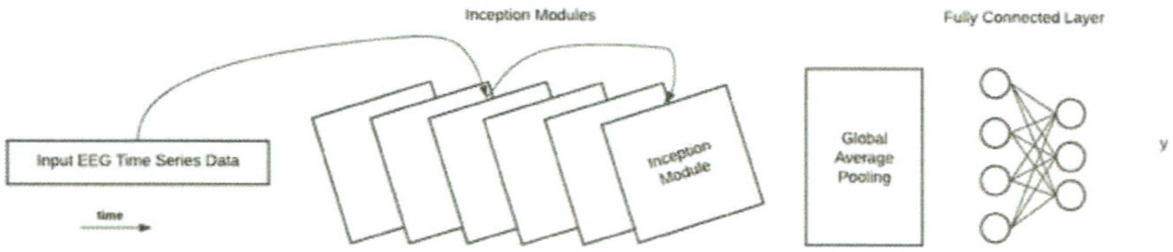

Fig. 3 Architecture of inception time network

predictions based on the details provided by the earlier layers. Figure 3 shows the architecture of simple InceptionTime network.

```
Algorithm 2: Training and Testing the InceptionTime Model
Input: EEG data of the patient
Output: Condition of the patient

   1. Import tsai, numpy, panda, scipy, sklearn
   2. Configure parameters for the Inception Network
   3. Load the txt data set and convert them into torch tensors
   4. Normalize the data set
   5. Filter the data set
   6. Split the data set for training and validation
   7. Train and test the model
```

4 Results

The expected results from the ESN model are to predict whether the person is epileptic or not. The parameters used for the reservoir are shown in Table 2, and parameters for the MLP layer are shown in Table 3. The proposed model is tested with the validation dataset after training. Based on the tests conducted, the ESN model achieves a maximum accuracy of 94.6%.

Table 2 Hyperparameters for the reservoir layer

Parameter	Value/function used
Processing units in the reservoir	450
Spectral radius	0.59
Leakage	0.6
Connectivity percent	0.25
Bi-directional reservoir	True

Table 3 Hyperparameters for the readout layer

Parameter	Value/function used
No. of layers	2
Units in each layer	20, 20
Activation function	ReLU

Table 4 Hyperparameters for the 3 1D CNN layers

Parameter	Value/function used
Kernel size	39, 19, 9
Stride	1, 1, 1
Padding	19, 9, 4

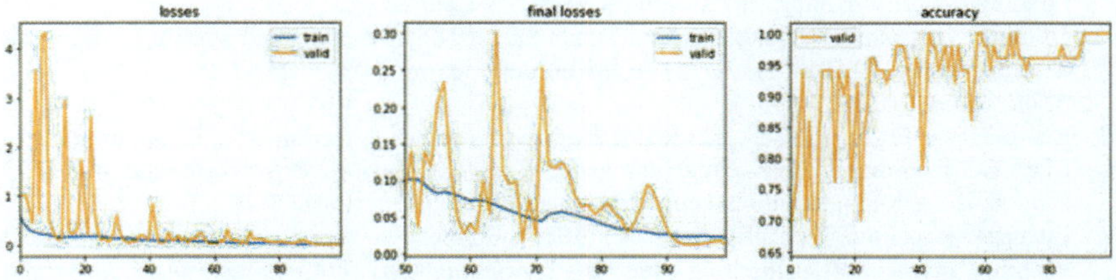

Fig. 4 This figure shows the loss and accuracy against the number of epochs. The analysis of the graphs indicates a pattern of a gradual decrease in loss as the number of epochs increased. Test results indicate that the model achieves a maximum accuracy of 100%

The expected results from the InceptionTime model are to predict whether the person is epileptic or not. The parameters used for the three 1D convolution layers of a single inception module are shown in Table 4. Figure 4 shows the loss and accuracy for the training and validation sets.

5 Conclusion and Future Work

The echo state network's readout layer was trained for 2000 epochs and the InceptionTime network was trained for 1000 epochs. Both networks are optimized to produce efficient outputs. After training and validating both the echo state network and InceptionTime network, test results indicate that the echo state network is much faster in training and significantly faster in testing consequently the model incurs a small but meaningful drop off in accuracy, achieving a maximum of 94.6% in the tests conducted, the InceptionTime network, however, takes more time in both the training and testing phases, this evidently large time complexity for train and test is compensated by its exceptionally high accuracy achieving a maximum of 100%.

The future work for this study could be twofold; the first would be to optimize both models, the ESN for its accuracy drop-off and the InceptionTime network for its invariably high training and testing time. Secondly, current wearable devices only detect a fraction of the seizures, while the EEG is the yardstick method by which all types of epileptic seizures can be detected. The proposed work can be extended to optimize the models and making them accessible for wearable applications to monitor brain waves and to detect seizures in real time.

References

1. Hussein, R., Palangi, H., Ward, R., Jane Wang, Z. (2018). Epileptic Seizure Detection: A Deep Learning Approach. https://arxiv.org/pdf/1803.09848.pdf (2018)
2. Ullah, I., Hussain, M., Qazi, E., Aboalsamh, H. (2018). An automated system for epilepsy detection using EEG brain signals based on deep learning approach. *Expert Systems with Applications, 2018*, 61–71.
3. Buettner, R., Frick, J., Rieg, T. (2019). High-performance detection of epilepsy in seizure-free EEG recordings: a novel machine learning approach using very specific epileptic EEG sub-bands. In 40th international conference on information systems.
4. Ebenezer Rajadurai, T., Valliyammai, C. (2018). Epileptic seizure prediction using weighted visibility graph. Soft computing systems. *Communications in Computer and Information Science, 2018*, 453–461.
5. Sharma, M., Pachori, R. B., Rajendra, U., Acharya (2017). A new approach to characterize epileptic seizures using analytic time-frequency flexible wavelet transform and fractal dimension. *Pattern Recognition Letters, 2017*, 172–179.
6. Zhang, Y., Guo, Y., Yang, P., Chen, W., Lo, B. (2020). Epilepsy seizure prediction on EEG using common spatial pattern and convolutional neural network. *IEEE Journal of Biomedical Health Informatics, 2020*, 465–474.
7. Mohanty, A., Parida, S., Nayak, S. C., Pati, B., Panigrahi, C. R. (2022). Study and impact analysis of machine learning approaches for smart healthcare in predicting mellitus diabetes on clinical data. In Smart healthcare analytics: State of the art. Intelligent systems reference library (Vol. 213). Springer.
8. Emami, A., Naoto, K., Matsuo, T., Shinozaki, T., Kawai, K., & Takahashi, H. (2019). Autoencoding of long-term scalp electroencephalogram to detect epileptic seizure for diagnosis support system. *Computers in Biology and Medicine, 2019*, 227–233.
9. Shoeibi, A., Khodatars, M., Ghassemi, N., Jafari, M., Moridian, P., Alizadehsani, R., Panahiazar, M., Khozeimeh, F., Zare, A., Hosseini-Nejad, H., Khosravi, A., Atiya, A. F., Aminshahidi, D., Hussain, S., Rouhani, M., Nahavandi, S., & Acharya, U. R. (2021). Epileptic seizures detection using deep learning techniques: A review. *International Journal of Environmental Research and Public Health, 18*(11), 5780.
10. Antoniades, A., Spyrou, L., Martin-Lopez, D., Valentin, A., Alarcon, G., Sanei, S., & Took, C. C. (2018). Deep neural architectures for mapping scalp to intracranial EEG. *International Journal of Neural Systems, 28*(8), 1850009.

YUDH: Multi-utility Tool to Reassure Serenity in the COVID-19 Pandemic

K. S. Krishna Das, Ananthu Vasudevan, Arun K. Nair, and S. Siji Rani

Abstract COVID-19, the virus that has affected current living standards, has undergone mutation, and the second wave has caused a much more devastating situation in India. In such a scenario, the alert of a third wave by the authorities has alarmingly increased concern in the nation. After being declared as an international emergency, the development of its vaccine has been conducted by different countries. India among other countries is also pursuing to develop much more efficient variants of the vaccine. The situation still persists to be hostile and maintaining the current precaution measures and maximizing the distribution of the vaccines is the only solution in hand. A necessity arises for a user-friendly app to reduce social interaction while assisting in medical support. In this paper, we have proposed an android application named YUDH, which focuses on the overall service that an individual requires from booking test centers, vaccine slot notification to home sanitization. The user can book COVID-19 testing centers and can arrange sanitization service after recovery with the provision of place, date, and time. In addition to booking test centers, swab testing at the doorstep is also available. The user also gets regular notifications on COVID vaccine slot availability in accordance with CoWIN portal and users' preferences. This deployment is aimed at the safety of the user and their privacy safeguard. The application also assists the government to maintain a database more efficiently.

Keywords COVID-19 · Sanitization · Third wave · COVID vaccine slot · Test centers

K. S. Krishna Das (✉) · A. Vasudevan · A. K. Nair · S. Siji Rani
Department of Computer Science and Engineering, Amrita School of Engineering, Amrita Vishwa Vidyapeetham, Amritapuri, India
e-mail: krishnadaskalathumkadavil@gmail.com

S. Siji Rani
e-mail: sijiranis@am.amrita.edu

B. Pati et al. (eds.), *Proceedings of the 6th International Conference on Advance Computing and Intelligent Engineering*, Lecture Notes in Networks and Systems 428,
https://doi.org/10.1007/978-981-19-2225-1_4

1 Introduction

SARS-CoV-2 is a pathogenic virus that adversely affects the lungs. This virus can be passed on to another person through direct or indirect contact. When a virus passes on to another person through one's respiratory secretions, this is termed direct contact. When a virus spreads onto a person through surfaces it is termed as indirect contact.

As per the current situation even though the vaccines are developed, the availability in India is still sparse. So, the proper distribution must be ensured under social distancing norms to avoid gathering. COVID-19 in this twenty-first century is the most devastating pandemic the human race has faced. There have been 181,722,790 COVID-19 positive cases and 3,942,233 deaths worldwide as per WHO dashboard records-July 2021. Analyzing the data, the World daily reported case remains the same (nearly 1 lakh) but while examining the report cases v/s death graph, the results are consolidating. The decrease in the average deaths and cases is a reason for the undertakings by individual countries in effectively providing the care and efficiency of the vaccine. In India, the cumulative count went up to 3,00,00,000 by the end of June, and the total deaths came around 4 lakh (396,761 deaths till July 1). Recent reports show an average count of 46,000 + in India alone, which adds up to nearly 40% of the global count. In accordance with the policy of Kerala, among other states of India having reportedly high rates of cases. It is zoned based on of local self-governments as per the test positivity rate (TPR) into four categories A, B, C, and D.

- Category A has 0–6% TPR, here there are no specific restrictions.
- Category B has 6–12% TPR, here minimum restrictions are implemented.
- Category C has 12–18% TPR, here half lockdown is enforced.
- Category D has 18% and above, here triple lockdown is enforced.

Special containment zones are enlisted in category C and D so that antigen tests are taken to those on a regular basis. The world economy was shaken by the novel coronavirus in 2020, and the ripples are still visible in 2021. In the three months since the last update of the world economic outlook in January 2021, the world has changed significantly. India after failing completely to the second wave of the coronavirus infection. Now on the verge of another major lockdown as an official third wave warning has come. The term wave defines a growing trend of infection over a long period of time. The variation in the curve and increase in the number of cases are indeed a premonition to the third wave. Periodic leaps and relative slowness as per the varying geography are signs. In India, the huge jump in the curve was observed during the second wave from the first wave, so the alarming rate of increase in the nation's number of reported cases certifies the coming of the third wave. But a decline in the curve has been observed since May.

So, the pertaining question is, Will the third wave cause a bigger impact than the second in India, mainly to the younger generations. Some speculations claim that the third wave is stronger than the second. Nevertheless, we cannot predict this. But in the case of a virus, it is proven to have less impact compared to a previous wave. Because some of them will have acquired immunity as it already affected a majority of the population. A concern arises as a newer version of COVID-19 viruses are being reported, which proves to be showing diverse symptoms to the vaccines. The virus can be mutated in a way that helps to escape the immune responses that have already developed in those who have been infected or who have been vaccinated. An estimate of 362,000,000 people has been vaccinated in India.

The main contribution of our research work is to provide a platform to insist a feeling of serenity in people. The lack of availability of services, medical supports including the vaccine, and the common aversion in people toward precursive undertakings show a trend. We utilize the current shortfall to introduce a platform for all in all services concerning the pandemic. This effortless interaction with the user and the health authority proves to be effective.

There are 15 approved vaccines by different laboratories from different parts of the world as of now. RNA vaccines which include *Pfizer-BioNTech and Moderna,* Adenovirus vector vaccines, namely *Oxford-AstraZeneca, Sputnik V, Sputnik Light, Convidecia and Johnson, and Johnson.*

Some Inactivated virus vaccines like *Sinopharm (BBIBP), CoronaVac, Covaxin, Sinopharm (WIBP), CoviVac, and QazCOVID-in, and finally,* protein subunit vaccines including *EpiVacCorona and ZF2001.* In India, the list of vaccines available is *Covishield (oxford AstraZenecayude version), Covaxin (Bharat Biotech developed), Sputnik V3 (from Russia), and Moderna (USA-based vaccine).* Some other vaccines under trial are *ZyCoV-D, Bio E COVID-19, HGC019, BBV154, and Covovax.* There are 15 approved vaccines by different laboratories from different parts of the world as of now. RNA vaccines which include *Pfizer-BioNTech and Moderna,* Adenovirus vector vaccines namely *Oxford-AstraZeneca, Sputnik V, Sputnik Light, Convidecia and Johnson and Johnson,* Some Inactivated virus vaccines like *Sinopharm (BBIBP), CoronaVac, Covaxin, Sinopharm (WIBP), CoviVac and QazCOVID-in, and finally* Protein subunit vaccines including *EpiVac-Corona and ZF2001.* In India, the list of vaccines available is *Covishield (oxford AstraZenecayude version), Covaxin (Bharat Biotech developed), Sputnik V3 (from Russia), and Moderna (USA based vaccine).* Some other vaccines under trial are *ZyCoV-D, Bio E COVID-19, HGC019, BBV154, and Covovax.* All the related works will be discussed in the next section.

2 Related Works

Nowadays studies are going on focusing COVID-19 pandemic. The paper [1] utilizes the topic, surveys on app stores like Google Play and Apple's App Store and some search engines, conducted on students and Web sites alike. They also conducted a detailed survey of COVID-19 existing applications. An application is developed with a focus on medical support making use of networking and its improvement. People in the primary contact, as well as the patients, can enroll and identify symptoms, and let the authorities keep track of their condition. A survey conducted in Jordanian showed light on some factors that cause people to show aversion to the COVID-19 contact-tracking [2]. Privacy, voluntariness, data beneficence, and a lack of adequate money were the issues. Introducing an application that makes use of the growing technology of artificial intelligence (AI) for COVID testing. The development is to find the geospatial location by making use of W3C with HTML5. The W3C geolocation APIs locate the position of the WebGIS on the map. This can be scaled from the model by including social networking nodes [3].

The workload handling is the main focus, so as the consumers can access the resources provided by the Amazon EC2. Marginal reduction as compared to the huge customer increase is controlled by analyzing using the heuristic algorithm. The global span of the purchasing adds to the workload, which can be optimized to a much greater extend [4]. The direct focus is on auto-scaling in cloud computing as it is in the initial baby step phase. The survey was conducted to identify the factors that affect its taxonomy such as auto-scaling techniques, its approaches, tools for monitoring, experiment, amount of work, and metric, etc., and to find articles, methods, and requirements for auto-scaling [5]. Almost, all the navigation systems including Google Maps are based on Dijkstra's algorithm. The paper focuses on the applications of this algorithm to find the shortest path on modern-day technologies and shows some light on the pseudocode of the algorithm [6]. They worked on conceptual algorithms used in Google Maps to get the shortest distance between source and destination [7]. Taking the state of Kerala into focus for other states to adapt and idealize in the country, while considering the pandemic conditions. To analyze the strategies taken by the state government like testing strategies, uninterrupted treatment services, community participation, surveillance, good quality quarantine, and special care for the elderly. Even factors like education and social mobilization all of which contribute to the situation and its control in Kerala [8]. They did proper studies to communicate with the public, and measures to reduce the rate of deaths and change in present conditions. The primary focus is to influence people into taking the vaccine [9].

The primary focus is on the pandemic and effective measures to prevent COVID-19. Social distancing being the main preventive mechanism, other situations and results of previous pandemics are also under observation on a global scale [10]. X-ray analysis is essential for COVID-19 tests during in the pandemic, which necessitates a lot of labor. Here the radiographic image of the lungs is taken under processing using a neural network-based deep learning approach known as nCOVnet [11]. They discussed ethical issues such as solidarity, equal moral respect, equity, autonomy, vulnerability, and faith, all based on the World Health Organization-Globid COVID-19 Global Research and Innovation Forum meeting in Geneva [12]. The main focus is on an alternative measure for COVID-19 patients with symptomatic hypoxia. Since the lack of proper treatment facilities and certified doctors in this pandemic, Ayurvedic care can be a perfect alternative and support. The patient who volunteered Ayurvedic treatment showed improvement on Ayurvedic medicines and was able to talk, eat, and sit on the bed without any breathing difficulty [13]. Their work was based on new studies proven to be an important factor. The new variants leading to the second wave of the pandemic prove to be challenging and demand the dynamic nature of the hospitals [14]. They did their research study to find out who is turning to

Ayurveda as the most backward people in metropolitan cities due to lack of oxygen cylinders [15]. The pandemic has proven to be adversely threatening as the care and medical support are limited and are the most effective way of prevention. Hence, a need for conducting studies and researches along with medical care and support is also needed. This calls for branching the research in different fields as well, especially Ayurveda [16]. Ongoing studies in India on the effect of COVID-19 pandemic on epilepsy care. The survey was conducted by neurologists and PWEs on patients inside and outside India. Although the COVID-19 pandemic does not seem to affect conditions, hospital visits have been declining, and the risk in India is similar to other parts of the world [17]. Some social relevant android apps described in [18, 19] are used in this proposed app development. By going through so many research papers, we realized that the growth in technology has a huge impact in aiding humanity during unfortunate situations. More than commercial growth, the involvement and the contributions of researchers had an immense role in the development of such technology. Hailing from several critical analysis and implementations, the COVID-19 pandemic was able to be identified, and measures to protect from it have been executed to a good extend. The recovery of many countries is the result of such measures as COVID protocols and vaccinations.

3 Proposed Methodologies

We intent to develop an android app as a multi-utility tool in the COVID-19 scenario featuring test center booking, sanitization service for the patient, and vaccine slot notification. The detailed work flow of the proposed system is shown in Fig. 1. Users can request COVID test and sanitization at doorstep. According to the availability of vaccine slots, user gets notified prioritizing date, time, and distance. The user can login or create an account using personal details. An account must be created prior to booking; otherwise, the features will not be accessible. Following, the location must be selected, a preferable location can be opted. After which the user can request the mode of service as required. Sanitization service, test booking, and vaccine slot booking, each taking to an individual page. Every add-on service is provided by their respective organizations. Sanitization service is provided by the LSGD, while the vaccine slot booking is directly under the CoWIN portal, hence the amenities vary for different services. For sanitization and testing services, the next step is to pick a suitable slot according to date and time. The approval of the request is solely upon the organization or the test center. After determining the method of payment, the status of the request will be displayed on the dashboard and regular alerts via message and mail. The vaccine slot booking has a discrete path, due to its scarce availability, the slot is searched district vice or by pin code. Varies filters are also provided, like age, dose, payment, and most importantly type of vaccine. Since the

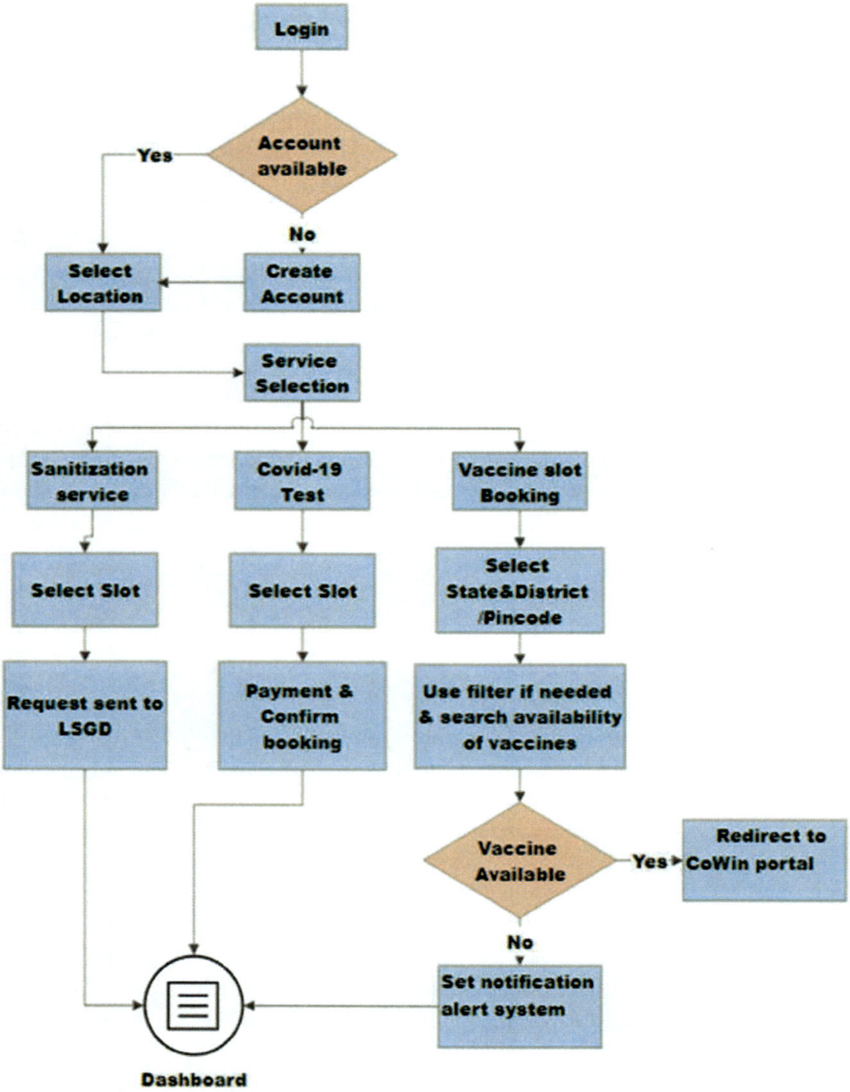

Fig. 1 Simple flow diagram of YUDH

availability and conformation of slots directly depend on the CoWIN portal, the user can set vaccine availability alerts and when available the user can access the CoWIN portal via the link attached with the alert messages. After taking or rescheduling the test, the slots will be empty. The test results are visible on the dashboard as well as email or SMS messaged personally.

A. **Proposed System Features**

The versatility of the proposed app includes the features as follows:

- Timely alerts for all the notifications.
- Durability in the transaction.
- Data isolation along with reallocation feature for individual appointments.
- Compatible with any network.
- Application requires less storage.

B. **System Configuration**

The Web application is hosted in Amazon Web Services (AWS) and configured as an EC2 instance. Amazon AWS is a cloud platform to host and manage Web applications using fewer resources and high-cost efficiency. For a better user experience, we offered 8 GB RAM and 100 GB SSD storage. This enhances the working of the application in limited network data and improves its speed. PHP framework is implemented (hypertext preprocessor) to reduce the complexity, increase the compatibility, and integration and as a cost reduction measure. We used MySQL for the construction of a relational database on Amazon (Amazon Web Services RDS) and the PHP framework connected to it.

A separate database is also included to add tables and scripts. The previously mentioned features are implemented as different modules. The booking module being the first has three sub-modules named COVID-19 test, home sanitization service, and COVID-19 vaccine booking service, which further articulated as hospitals, location, customer, or patient login module. To find the nearby testing centers for the user, we implemented two methods: distance priority algorithm and Geo-API using Google maps. To connect the application server online, we use the Web view module in Android Studio by embedding a public URL. For the third-party software application, Ministry of Health and Family Welfare (MOHFW) provides API-based access to the CoWIN portal to layout COVID-19 vaccine-related services. We are using a protected API to integrate it with CoWIN portal, because the protected API will give permission to the third-party application to admit onto CoWIN portal and update CoWIN database. In order to synchronize with CoWIN using API, we must first register as an application service provider (ASP). Then follow the guidelines provided by the MOHFW. Authentication API, metadata API, vaccine appointment API will be used for the integration. Corroboration will be done in each panel, and the response will be received in JSON format and displayed by converting it into tables. All API keys will be managed using the CoWIN API console.

C. **Locating Test Center and LSG-Algorithms**

We have adopted the existing Google map's Dijkstra's algorithm [15] to find the nearest testing center. The algorithm has a source node and a destination node. The lab center location is the source node, and the user's location is the destination node. Depending on the location of the user, we were able to provide information on nearest testing centers using the mentioned algorithm and with the assistance of Geo-API. In accordance of Geo-mapping, we can assist the admin holder (in this case, testing center authority or the LSG authorized health organization) to find the shortest path to the individual's location. This is because Dijkstra's is a greedy algorithm, so the shortest path gets the highest priority. Dijkstra's algorithm has a time complexity of $O(|V2V2|)$ where V is a node that represents the individual's house, city, or an intersection in real life. But this complexity can be reduced because we use a minimum-priority queue then the time complexity reduces to $O(|E|+ |V|\log|V|)$. For improving the efficiency of gathering and locating, the user another algorithm is appended to the Dijkstra's algorithm. It also helps while a greater number of nodes access, the increase in the amount of data may cause the latter algorithm to fail. Another algorithm incorporated is the A* algorithm which identifies the shortest distance between the source and destination. A* Algorithm is a cutting-edge algorithm similar to Dijkstra's algorithm. It uses a heuristic function to navigate, while Dijkstra explores all the nodes. This algorithm checks for parameters such as distance, time, and date as required by the user to improve the performance by prioritizing the nodes. The nodes will be prioritized in a hierarchy from higher to lower, this proves to be more of an improvement compared to others.

D. COVID Test Slot Allocation and Sanitization Request

The slot availability is computed as follows:

$$n = t/p \tag{1}$$

where

- n number of time slots available in a hospital.
- t total working hours.
- p maximum time for an individual test (15 min).

The slot is available when $n > 0$.

After choosing and the confirmation (payment) being complete, the allotted slot will be taken out. Thereby the slots will be reduced $(n - 1)$. If a patient cancels the appointment, the slots would be incremented $(n + 1)$.

The number of sanitization requests has no limit, but at a time only one request per account can be sent, and a second request can be sent if no preliminary response from the LSG is received. There is no limit on the sanitization units available in an area, since LSG has the complete privilege of assigning organizations to the task.

E. Steps for Various Booking Services

The choice of selecting the type of test antigen or RTPCR is provided as per the user's selection of location. After that only, we can schedule the time slots and convenient dates. The reallocation feature is added such that if a person reschedules his appointment the initial reservation must be retained. The reservation and reallocation feature can be done concerning time as well as the date. An option to book according to a hierarchy of preferences is also available but a refreshing buffer time is allocated, so that the slots are effectively utilized. All these features are in real time. The user's current status will be displayed on the dashboard as a My Appointments tab. The user will be required to book a new slot if the reservation is still pending or yet to be approved. Each individual must create an account with personal details to use all the features of YUDH. A unique mobile number or email ID must be provided to create the account and for its verification. After creating, the services will be provided. The slots will be available for selection from the nearby testing centers, which will be enlisted by the application. The reallocation facility is not available after the approval of the payment. Double verification of the personal details is done during the payment time and changes cannot be made after that. The confirmation message will be sent to the mobile number and the mail ID provided by the user, which contains a four-digit security code. The code will be verified by the authority while attending to the COVID-19 test (including test at doorstep). This increases the efficiency as well as security of the process. The test results will be available in the dashboard and forward via mail and message immediately after they are obtained. The hospital will provide the follow-up procedures if the results are positive. The feature to select quarantine service is also included. The user can opt for home quarantine or get hospitalized according to the doctor's diagnosis. In case of emergency, the user can request hospital service and counseling sessions with the doctor.

YUDH provides a feature to view the vaccine slots that are available in the CoWIN portal. Along with an addition of filtering and according to the user's relevance, the direct selection of slots with prioritizing location (District and pin code), age group, and the type of vaccine available. The filters include dose number, age category, and available vaccine type, which can be selected as per the desire. The primarily registered mobile number of the CoWIN portal must be used here during the login process, since the OTP verification of the individual's CoWIN portal ID must be carried out. During the slot selection, the user can identify the vaccines to be paid or free as in the portal. To increase the rate of efficiency and alertness, a dedicated notification service is also enabled. A pre-fixed filter can be set by the user, so that the availability of slots in the predefined filter will be notified accordingly when the database gets updated by the respective centers (hospitals). The notification and viewing of the slot availability pertain even after a slot is chosen. This assists the user to reschedule and chose an even better slot if one gets available at a much more desirable location, while keeping the former. When the slot gets successfully booked, a slip will be available for download similar to the CoWIN portal. The total feature is available for booking only after having a CoWIN portal account. While a single CoWIN portal account can handle four individuals.

The primary focus is based on the fact that the government hospitals have constraints to facilitate a large number of patients. The lack of proper quarantine areas and hospital beds all of which contribute to this fact. Hence, people tend to adopt home quarantine especially lower- and middle-class people. Even with facilities available in private hospitals most of the population avoid private treatment because of the huge expense, which they cannot afford in the current scenario. The state government also supports home quarantine and as of now hospitalization is recommended only in emergencies. This undertaking by the government protocols proves to be helpful for the public.

The sanitization of the place where a person has undergone isolation or quarantine after being test negative is mandatory. Sanitization is as important as social distancing in case of a virus outbreak, hence improving its efficiency is a requirement. Currently, it is being taken care of by the social workers, charity organizations, and political federations, but this is still not practiced in some parts of the state. Our implementation focuses on this situation. The process is to split regions in panchayat, corporation, and municipality as an undertaking of the local self-government (LSG) and to increase participation of maximum political and charity organizations so that uniform distribution of the service is ensured. The procedure starts when a user requests a service, and the selected LSG determines a particular organization available in that area and assigns the sanitization duty. When the LSG approves a sanitization request, the assigned organization will contact the user and will verify the location in real time. The request and its status will be available in the dashboard and a request number will be assigned to the user as a generated token. The social workers and such organizations are carrying out the service as a charity, and the adaptation aids to be a medium to continue the work.

The protocols taken by the Kerala state government is that a patient must undergo seven days of compulsory quarantine if he is tested to be positive and again seven days of reverse quarantine after being tested negative. The protocol may change according to ICMR circular. The sanitization process can be requested only after the patient being tested negative. This condition is put forward to ensure the safety of the social workers. The self-verification of COVID test results must be overseen before the sanitization procedures, to make sure that the patient is COVID negative.

F. **Support of Admin Panel**

The admin app version of YUDH and its out-and-out control is fully entitled to the authority. The COVID-19 testing centers have the provision to approve slots. And since the sanitization requisites are provided by the local self-government (LSG), the complete authority of the home sanitation falls under the Kerala LSG COVID management team. Even though the data is transferred to different authorities, the user's personal data safekeeping is ensured. All the requests are approved as well as declined by only the above-mentioned authorities, and the information and details are notified by the application. The approval message as well as the decline message in case of any difficulty will be visible on the dashboard. The admin panel also has the authority to reschedule services. For further services, the COVID test details are transferred to corresponding COVID data collectors (health authorities), and this competence also lies under the admin. This feature is necessary to locate the COVID positive individuals and provides them with medical and sanitary assistance via informing corresponding hospital and government health officials. Figure 2 represents the user-admin chat window and Fig. 3 represents the dashboard view of the admin panel.

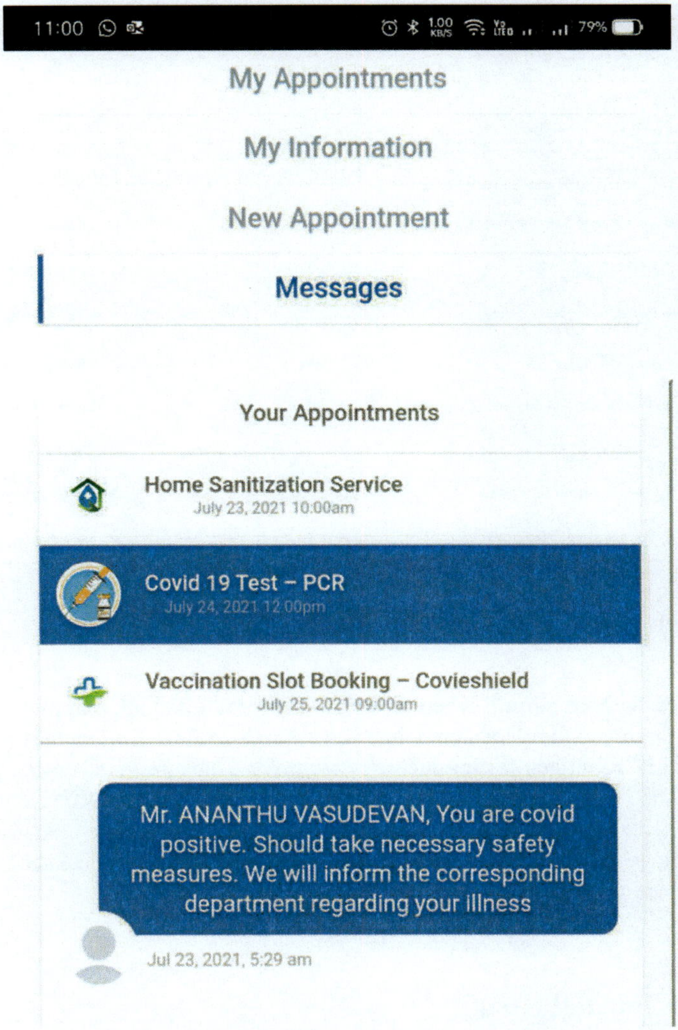

Fig. 2 User-admin chat window

G. User Privileges

The user can reschedule the services irrespective of sanitization request or COVID test booking, to change the date, time, or even place. The request status will be changed to a rescheduled status if the reschedule is requested after the latter being approved. And for each service, the reschedule option is limited to two to reduce the complexity. The request after being approved, reschedule option will be disabled two hours before the actually approved time slot to reduce the confusion and collisions. The search option while COVID vaccine slot booking to check availability of slots

Fig. 3 Admin panel dashboard

can be used without a limit; whereas, in CoWIN portal and other applications it is limited to twenty while also having time constrain to limit the load on the server. The COVID RTPCR/antigen testing facilities are private labs provided so the payment feature is also including with an option to switch from online to offline payment. The test results will be visible in the dashboard as soon as they are uploaded by the centers. Figure 4 shows the types of services provided. Figures 5 and 6 show the user dashboard view.

4 Experimental Result

Overall, it assists trouble-free sanitization and the existing social distancing norms by reducing the number of hospital visits. This prevents the virus from further spreading. During the time of lockdown, community gathering is not allowed, and most of the people are not willing for taking tests due to the leakage of privacy. The reallocation feature ensures effective slot booking and shows more affinity toward the user. The online and offline choice of payment for the COVID testing proves to be helpful. The separate database maintained for COVID details manages to avoid errors that are presently occurring. The single application that maintains both COVID testing and sanitization is an advantage as to communicate the test results with the health

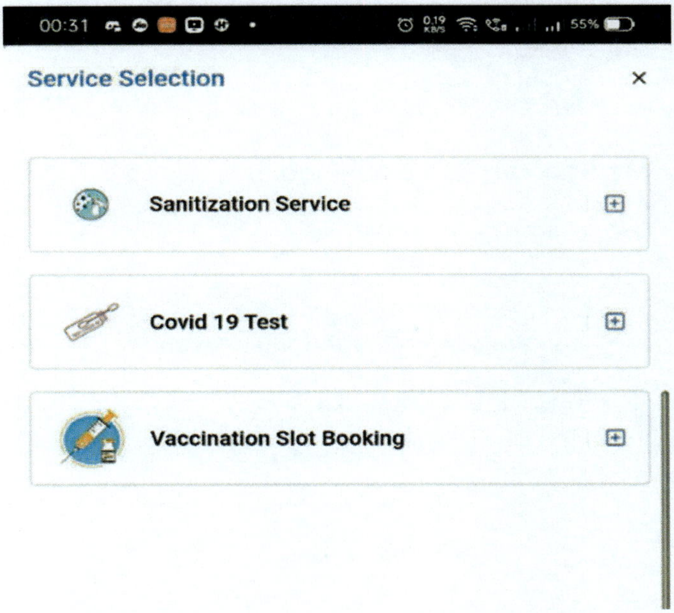

Fig. 4 Types of services provided

officials. The dashboard facility also seems to ease the user's procedures. Figures 7 and 8 show the details regarding the vaccination recorded daily as per WHO.

In the beginning, many were in fear of taking the vaccine, it is because of the misconceptions. At the commencement of various awareness systems, there has been a change of tides. YUDH being an active part of this change. At least 23.1% of the population has taken the first dose of vaccination; whereas, 8.1 CR has fully vaccinated so far. The total percentile of the population that has fully vaccinated is 5.9%. The statistics are gradually improving.

5 Conclusion

COVID-19, the virus that has affected current living standards, has a drastic increase rate since it has emerged in 2019.The virus spreads mainly by direct contact, and the primary measure still under practice is social distancing along with usage of double and N95 masks and regular sanitization. As the release of various vaccines are yet to been proven affective, proper vaccination measures are undergoing to vaccinate the whole country. In India, the emergence of the mutated coronavirus and the second wave had a huge impact; whereas, predictions of the third wave has raised an even larger concern. In consideration of all these factors, the application focuses on each stage of the disease, before and after being tested positive. All the services provided

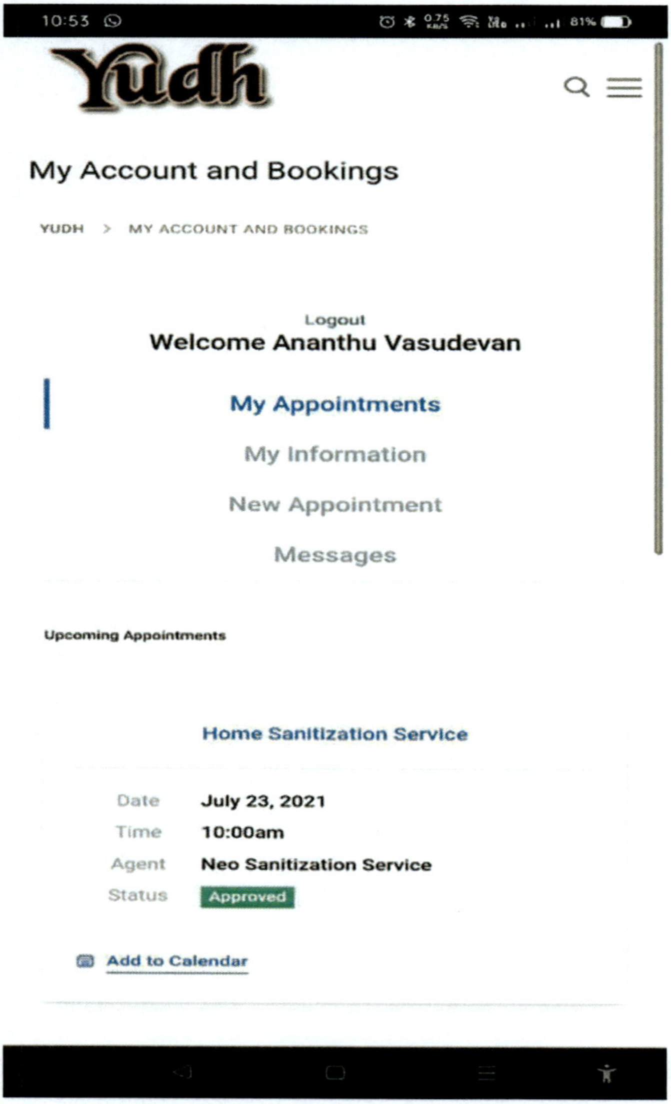

Fig. 5 User's dashboard

by YUDH make sure the safety of the user is to the fullest. From slot booking for swab test at doorsteps and vaccine notification to home sanitization after recovery, all of which without physical effort and social interaction and minimizing hospital visits. Community cohesion is now seen in hospitals and many medical centers. The government also needs to include active cases so they can take the necessary action. The proposed app supports to achieve both. In the long run, if a new variant of the

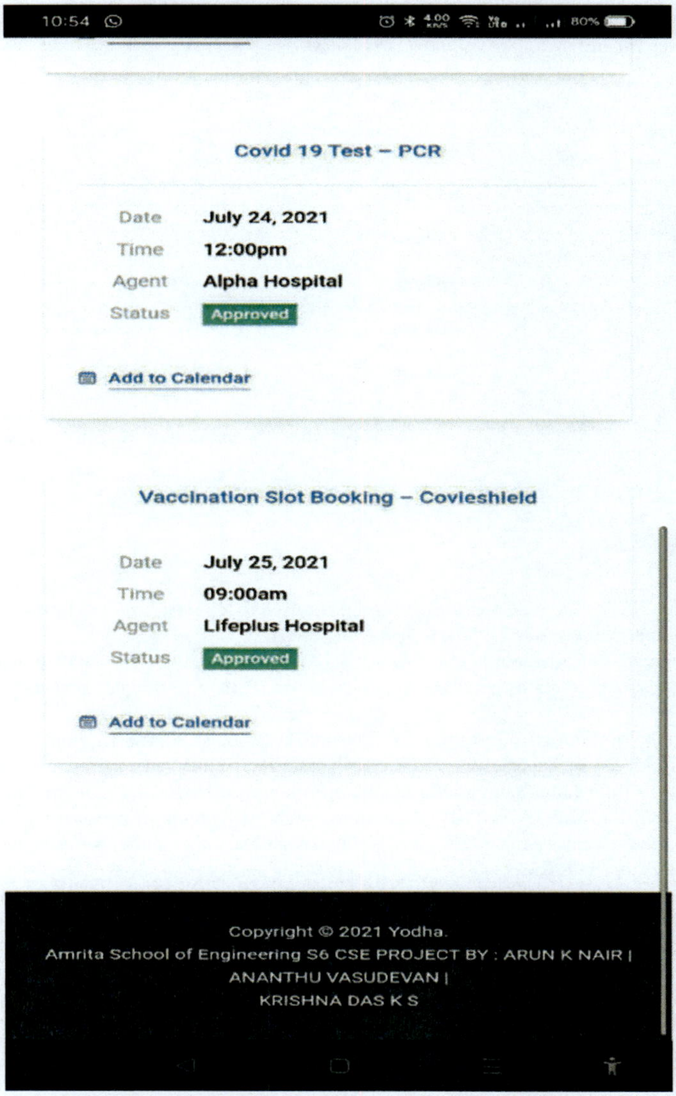

Fig. 6 User's dashboard

current COVID-19 or any other outbreak deems to occur, the app is designed in such a way that it provides service in a one-touch format. Since, we have already created an environment it will be easier for taking the right action and thereby maintaining serenity and avoiding panicking in case of an emergency in the society.

Fig. 7 Average number of vaccines given monthly

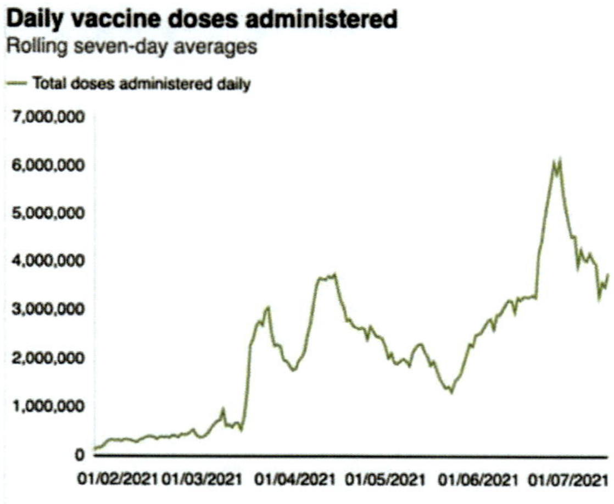

References

1. Noronha, N., D'Elia, A., Coletta, G., Wagner, N., Archer, N., Navarro, T., & Lokker, C. (2020). *Mobile applications for COVID-19: A scoping review.*
2. Abuhammad, S., Khabour, O. F., & Alzoubi, K. H. (2020). COVID-19 contact-tracing technology: Acceptability and ethical issues of use. *Patient Preference and Adherence, 14*, 1639.
3. Imran, A., Posokhova, I., Qureshi, H. N., Masood, U., Riaz, M. S., Ali, K., John, C. N., Iftikhar Hussain, M. D., & Nabeel, M. (2020). AI4COVID-19: AI enabled preliminary diagnosis for COVID-19 from cough samples via an app. *Informatics in Medicine Unlocked, 20*, 100378.
4. Bhise, V. K., & Mali, A. S. (2013). Cloud resource provisioning for Amazon EC2. In *2013 fourth international conference on computing, communications and networking technologies (ICCCNT)*, pp. 1–7. IEEE.
5. Singh, P., Gupta, P., Jyoti, K., & Nayyar, A. (2019). Research on auto-scaling of web applications in cloud: Survey, trends and future directions. *Scalable Computing: Practice and Experience, 20*(2), 399–432.
6. Parida, S., Mohanty, A., Nayak, S. C., Pati, B., & Panigrahi, C. R. (2021). Study and impact analysis of COVID-19 pandemic clinical data on infection spreading (p. 225). In *Data science for COVID-19: Volume 2: Societal and medical perspectives*
7. Mehta, H., Kanani, P., & Lande., P. (2019). Google maps. *International Journal of Computer Application, 178*, 41–46.
8. Menon, J. C., Rakesh, P. S., John, D., Thachathodiyl, R., & Banerjee, A. (2020). What was right about Kerala's response to the COVID-19 pandemic? *BMJ Global Health, 5*(7), e003212.
9. Malik, A. A., McFadden, S. M., Elharake, J., & Omer, S. B. (2020). Determinants of COVID-19 vaccine acceptance in the US. *EClinicalMedicine, 26*, 100495.
10. Sarkar, J. L., Ramasamy, V., Majumder, A., Panigrahi, C. R., Gomathy, B., Pati, B., & Saha, A. K. (2021). SensMask: An intelligent mask for assisting patients during COVID-19 emergencies. *Computación y Sistemas, 25*(3).
11. Panwar, H., Gupta, P. K., Siddiqui, M. K., Morales-Menendez, R., & Singh, V. (2020). Application of deep learning for fast detection of COVID-19 in X-Rays using nCOVnet. *Chaos, Solitons & Fractals, 138*, 109944.
12. Dawson, A., Emanuel, E. J., Parker, M., Smith, M. J., & Voo, T. C. (2020). Key ethical concepts and their application to COVID-19 research. *Public Health Ethics, 13*(2), 127–132.

13. Das Mohapatra, S., Nayak, S. C., Parida, S., Panigrahi, C. R., & Pati, B. (2021). *COVTrac*: COVID-19 tracker and social distancing app. In C. R. Panigrahi, B. Pati, B. K. Pattanayak, S. Amic & K. C. Li (Eds.), *Progress in advanced computing and intelligent engineering. Advances in intelligent systems and computing* (Vol. 1299). Springer. https://doi.org/10.1007/978-981-33-4299-6_50
14. Joshi, J. A., & Puthiyedath, R. (2020). Outcomes of ayurvedic care in a COVID-19 patient with hypoxia—A case report. *Journal of Ayurveda and Integrative Medicine.*
15. Suresh, M., Roobaswathiny, A., & Lakshmi Priyadarsini, S. (2021). A study on the factors that influence the agility of COVID-19 hospitals. *International Journal of Healthcare Management, 14*(1), 290–299.
16. Joshi, J., Payyappalimana, U., & Puthiyedath, R. (2021). Potential for supportive ayurvedic care in hypoxemic COVID-19 patients. *Journal of Ayurveda and Integrative Medicine.*
17. Payyappallimana, U., Patwardhan, K., Mangalath, P., Kessler, C. S., Jayasundar, R., Kizhakkeveettil, A., Morandi, A., & Puthiyedath, R. (2020). The COVID-19 pandemic and the relevance of ayurveda's whole systems approach to health and disease management. *The Journal of Alternative and Complementary Medicine, 26*(12), 1089–1092.
18. Siji Rani, S., & Krishnanunni, S. (2020). Educational app for android-specific users—EA-ASU. In *Ambient Communications and Computer Systems* (pp. 325–335). Springer.
19. Siji, R. S., & Rajeev, R. (2016). Colour transformation for deuteranopic viewers. *International Journal of Control Theory and Applications, 9*(10), 4527–4535.

Easy Detect—Web Application for Symptom Identification and Doctor Recommendation

Abhinay Krishna Chitirala, Chunduri Sai Akhilesh, Garimella Venkasatyasai Phani Ramaaditya, Kollipara Bala Vineel, Krosuri Sri Chandan, and M. Rajesh

Abstract It is not very safe to go to hospitals for regular check-ups as we are used to. Patients can get medical advice from the comfort of their homes from specialized medical professionals. Patient data can also be digitalized, so that it can be used at any hospital. Symptoms are taken as input, and our deep learning model will give the possible disease, which the person might be suffering from, and the doctor he/she needs to consult for further medication.

Keywords Deep learning · Medical · Patients · Doctors · Symptoms · Consultation · Medication

A. Krishna Chitirala · C. Sai Akhilesh · G. Venkasatyasai Phani Ramaaditya · K. Bala Vineel · K. Sri Chandan · M. Rajesh (✉)
Department of Computer Science and Engineering, Amrita School of Engineering, Amrita Vishwa Vidyapeetham, Bengaluru, India
e-mail: m_rajesh@blr.amrita.edu

A. Krishna Chitirala
e-mail: blenu4cse17001@bl.students.amrita.edu

C. Sai Akhilesh
e-mail: blenu4cse17019@bl.students.amrita.edu

G. Venkasatyasai Phani Ramaaditya
e-mail: blenu4cse17031@bl.students.amrita.edu

K. Bala Vineel
e-mail: blenu4cse17062@bl.students.amrita.edu

K. Sri Chandan
e-mail: blenu4cse17065@bl.students.amrita.edu

B. Pati et al. (eds.), *Proceedings of the 6th International Conference on Advance Computing and Intelligent Engineering*, Lecture Notes in Networks and Systems 428,
https://doi.org/10.1007/978-981-19-2225-1_5

1 Introduction

1.1 Motivation and Contributions of the Research

COVID-19 has had an impact on all of us. The pandemic's implications and consequences, on the other hand, are felt differently in different sectors. This pandemic has had the most significant impact on the health care industry. As the number of cases of COVID increased, hospitals converted various wards into COVID units. This increase in COVID wards, and the need to avoid overcrowding make it more difficult for people with other diseases to see doctors and get to laboratories for consultation and testing. As a result, we need an app that provides an essential diagnosis based on the patient's symptoms and recommendations for which doctor to visit. The patient can even book an appointment in advance at a particular time so that he/she can avoid waiting in the hospitals in case of emergencies.

This study focuses on the design features of the prediction system for medical conditions to detect many typical diseases. Techniques like neural networks, decision making and logistic regression are used for the topic implementation. We have acquired the required data set. The algorithm also proposes physicians that are applicable for the pattern detected disease(s).

1.2 Introduction to the Easy Detect Application

We have developed a Web application that can detect diseases in patients based on their symptoms. They are connected with a specialized doctor for further consultation based on the application's results. We can schedule appointments ahead of time to avoid waiting in hospitals. Determine which specialized doctor to consult ahead of time. We cannot risk the patient's health during his regular check-ups in the given circumstances, where social distancing is critical. As a result, this application will keep track of its clients' health regularly and with the help of doctors.

So, we would like to propose an intelligent system trained based on past medical records (symptoms for specific diseases).

The proposed system is designed to support the decisions of doctors and is not designed for a patient without supervision by a medical practitioner for individual usage. The remaining part of this article has the following structure: The literature review conducted in the medical field is described in Sect. 2.

2 Literature Review

Intelligently analysed data becomes a corporate requirement to find effective and trustworthy detections of disease as quickly as possible to ensure the best possible treatment for patients. This detection has been conducted in recent decades by finding remarkable patterns in databases. The technique of retrieving information from the database is known as data mining. Finding these patterns, however, is a challenging process. This has led to the development of various artificial intelligence approaches, including machine learning as a tool for providing intelligent data processing. Medical data sets are usually multidimensional, on the other hand. The use of big data technology is necessary, in certain situations, when machine learning techniques fail. Deep learning has, therefore, developed into a subset of machine learning, which allows us to work with such data sets.

Caballé et al. [1] gave a comprehensive overview of smart data analysis tools from the medical area. They also include examples of algorithms used in different medical fields as well as an overview of probable trends depending on the objective, process employed and the application field. The benefits and cons of each approach were also overcome.

In all shown fields of application, the author states that the categorization is the most usual action in the medical profession. In the realm of infectious diseases, regression, on the other hand, is a regular task. In illnesses like Alzheimer's or Parkinson's diseases, this duty is rare to be employed. In addition, the task of clustering in liver and cardiovascular diseases is briefly studied, but is used extensively in Alzheimer and Parkinson diseases. In the case of cancer, Alzheimer's, Parkinson's and renal disease studies, neural networks and other supervised algorithms are commonly employed in study into metabolism, hepatic, infectious and heart illness.

The author chose the technique based on the advantages and disadvantages of each tool in the specific application area and under his or her experimental conditions.

Traditional approaches can be used with large volumes of data and powerful hardware architectures to represent more complex statistical phenomena, while ML enables previously hidden patterns to be identified and trends extrapolated and the result to be predicted in the absence of trace problems as well. Currently, machine learning algorithms are employed in clinical practice in medical records, for instance, to forecast which patients would most likely be hospitalized or who are less susceptible to a prescription of therapies. Diagnostic, research, drug development and clinical trials have unlimited possibilities. Although there are large numbers of digital data, predictive medical record models are typically based on basic linear models and seldom take into account more than 20 or 30 parameters.

Dhomse Kanchan et al. [2] used SVM, Naive Bayes and decision tree with and without PCA on the data set to predict heart disease. The principal component analysis (PCA) approach is used to reduce the number of characteristics in a data set. When the data set size is decreased, SVM beats, Naive Bayes and decision tree. SVM may potentially be used to forecast the start of cardiovascular illness. Their algorithms were developed using the WEKA data mining approach, which was utilized to evaluate algorithm accuracy after executing them in the output window.

These techniques evaluate classifier accuracy relying on properly identified examples, the time required to create a model, mean absolute error and ROC area. As a consequence, they concluded that, when compared to other methods, the maximum ROC area indicates outstanding prediction performance.

The methods are rated based on how long it takes to create a model, how many cases are properly categorized, the error rate and the ROC area. The algorithm's accuracy is displayed in Naive Bayes 34.8958 per cent correctly instances accuracy with a minimum Naive Bayes mean absolute error = 0.2841 and a maximum Naive Bayes ROC = 0.819 times needed to construct the model = 0.02 s. Based on the explorer interface data mining approach, we can infer that Naive Bayes has the greatest accuracy, the lowest error, the shortest time to develop and the maximum ROC.

Human illness diagnosis is a tough process that requires a high level of skill. Any attempt to develop a Web-based expert system for human illness diagnosis must overcome a number of obstacles.

This project's [3] objective is to develop a Web-based fuzzy expert system for detecting human illnesses. Fuzzy systems, which portray systems utilizing linguistic principles are currently being employed successfully in a growing variety of application domains. Hasan et al. [3] are investigating and developing a Web-based clinical tool to increase the quality of health information sharing between physicians and patients. This Web-based tool can also be used by practitioners to confirm diagnoses. To assess its performance, the proposed system is tested in a variety of scenarios. The proposed system achieves satisfactory results in all cases.

A control programme is created by gathering, encoding and storing knowledge. To diagnose the fuzzy expert system, a uniform structure was developed, and mathematical equivalence will be employed. The likelihood of illnesses was calculated using that equation, the value of which was determined via feedback during diagnosis. In this case, a catalytic factor is employed in the form of a question about prior results, which is also taken into consideration during the probability calculation.

The addition of catalyst after evaluation increases the accuracy of the system as past results play a significant role in illness prediction. The following system increases the accuracy and it works with real-time diagnosis. It was even found that the confidence level of this system after observing past pathological tests was far better than otherwise.

Laxmi et al. [4] The usage of Bayesian networks is presented in the creation of a system of clinical decision support. Infer network parameters, which offer the idea of learning were used to the Bayes ML technique. The study is unique in that, in addition to identifying diseases, it attempts to propose laboratory testing, infers diseases from laboratory test data and offer age-based therapeutic prescriptions for regularly occurring diseases in India. For simulating laboratory testing and medical prescriptions, a rule-based technique is employed.

Mohanty et al. [5] deal with the problem of the symptoms and seriousness of the most likely sickness in the physician. ANFIS benefits from the classic fuzzy models by being extremely flexible and easily learned. The patient and the diagnostic information will be the learning and testing of the system when the system is deployed to a clinic.

Based on the above citations, we have used a filter to reduce the number of features based on their importance in finding the result [6, 7]. The importance of each feature is determined using a coefficient matrix. We inferred that SVM can be more effective while dealing with cardiovascular illness but with an overall data set of large size, we concluded that Naive Bayes is better.

3 Design and Implementation

In this section, we are going to discuss the design and implementation of the modules used in the application. We have used different modules such as the data collection module, logistic regression module, decision tree module, neural network module and a disease prediction module.

3.1 Design

This diagram depicts the operation of our application. The data will be separated into training, testing and model training once the deep learning model is pre-processed. The model will then be loaded into the Web application to forecast the ailment that the patient is suffering from (Fig. 1).

3.2 Implementation

a. **Data Collection Module**: The data collecting module is used to build a knowledge base for a medical illness prediction system. The collection of disease-related symptoms is the first step in the data collection process. 41 disorders and 132 symptoms were picked for the initial deployment. The symptoms considered were a wide range of common symptoms that a patient might experience. Later this data set is processed for feature extraction using the coefficient matrix as shown in the below figure. Among the coefficients, a 0.4 quantile of symptoms is removed which brings down the data set to make the data set more feasible for the model to be used. We use a pre-processing input function to pre-process the data, i.e. labelling the data and splitting it into a 70:30 ratios [8, 9].
b. **Logistic Regression Module**: Training and testing are the two phases of the logistic regression module. The first phase is designing the model and training it

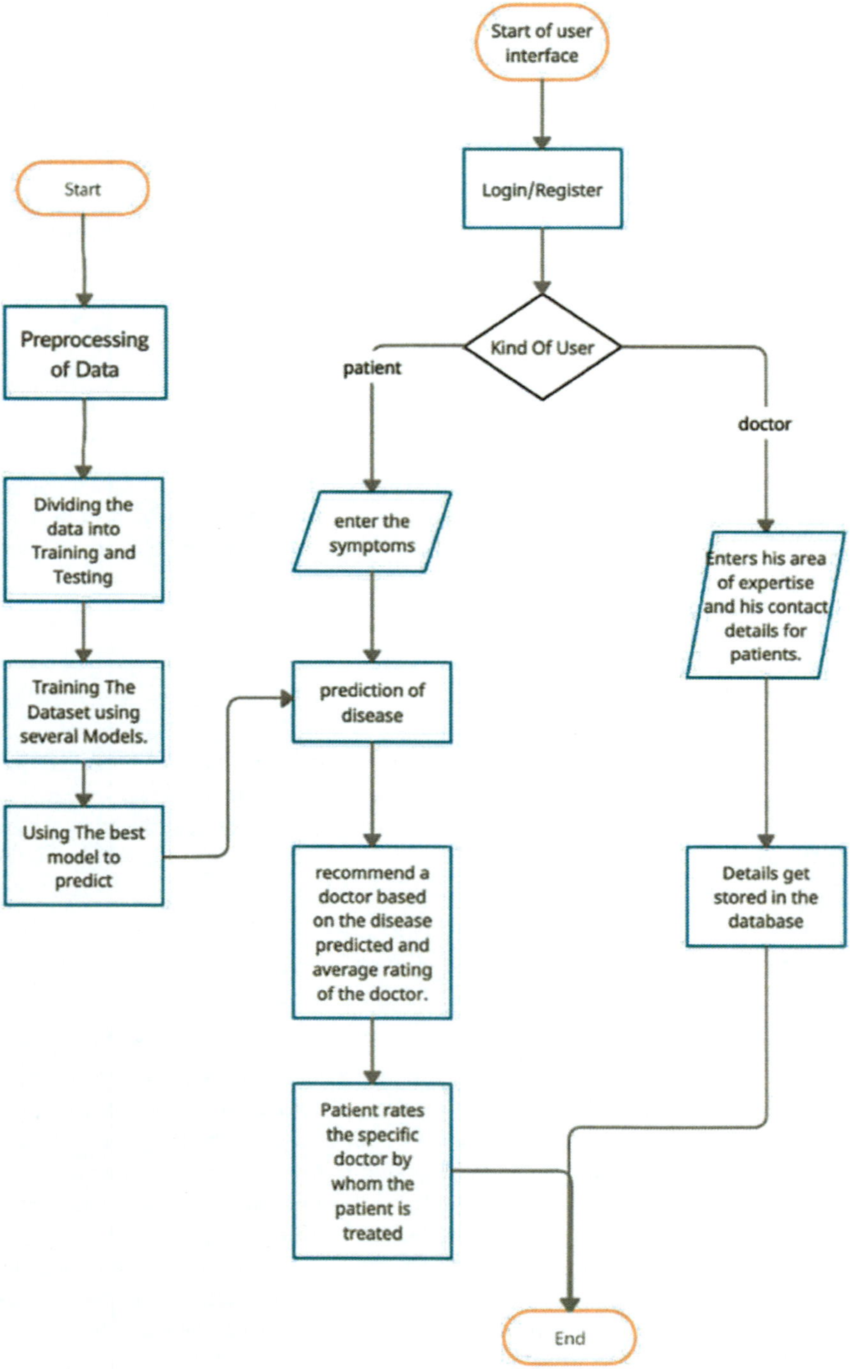

Fig. 1 Flow diagram of the application

with data gathered from the data collecting module; whereas, the second phase involves testing the model and finding accuracy [10].

- Logistic Regression Model Creation: The data set created in the above module is used to create logistic regression. The multinomial class detection and solver as limited-memory Broyden–Fletcher–Goldfarb–Shanno (LBFGS) [11–13].
- Logistic Regression Testing: Testing data is inputted into the trained model, which involves creation of probability for diseases using Gaussian algorithm [14, 15].

c. **Decision Tree Module**: It is constructed with a data set from the data collection module. The module is separated into two phases: training and testing.

- Decision Tree Model Creation: The model is built using the data set from the data collection model. It uses the information gain algorithm to build the decision tree in which internal nodes represent the symptoms and leaf nodes represent diseases.
- Decision Tree Testing: Testing data is inputted into the trained model, which involves traversing the tree through the symptoms to find the disease.

d. **Neural Network Module**: The neural network model is a sequential model, which is built using different layers containing a different number of nodes or neurons. The model consists of three dense layers and three activation layers in the following order:

1. Dense layer (32-nodes)
2. Activation layer (ReLU)
3. Dense layer (16-nodes)
4. Activation layer (ReLU)
5. Dense layer (41-nodes): Output layer
6. Activation layer (Softmax).

The training and testing phases of this model are divided into two parts:

- Neural Network Model Creation: The model is built using the data set from the data collection module. Each neuron has a weight associated with it. Activation functions are applied to a whole layer of neurons. These provide nonlinearity, without which the neural network reduces to a mere logistic regression model. After every epoch, the parameters and hyper-parameters of the model are modified such that the cost function is reduced till it reaches the point of global minima. The ReLU is the activation function employed here (rectified linear unit). The output layer, also known as the last layer, is made up of 41 neurons whose outputs are passed through the final activation layer containing Softmax activation function, which returns the probability of occurrence of the corresponding diseases. This model is compiled with categorical cross-entropy as loss (as we are performing a classification), validation accuracy as a metric and Adam as optimizer. An early stopping mechanism is also added with the patience of two epochs to prevent the overfitting of the model on training data, i.e. if the validation accuracy is either decreasing or is constant, the training would end there …

- Neural Network Model Testing: Testing data is passed to the model along with another data called validation data, which validates or verifies the performance of the model on testing data.
- **Disease Prediction Module**: The disease prediction module is designed on the trained model. The symptoms data is gathered from the UI provided for the user. The symptoms thus gathered are made into a NumPy list where the symptoms, which are marked by the user are given the value "1" and others have default value of "0" and this list is passed using the trained model to forecast the likelihood of each disease's occurrence

4 Results and Analysis

A sample testing set of roughly 42 records was used to evaluate the decision tree approach for the current paper's implementation.

- Accuracy for the Decision Tree Model is: 97.62
- Accuracy for the Logistic Regression is: 94.93
- Accuracy for the Neural Network is: 94.3.

Figure 2 illustrates different values of (accuracy and validation accuracy) versus (epochs), i.e. graph on left, the graph on the right illustrates the distribution of (loss and validation loss) versus (epochs). After the twentieth epoch, accuracy and validation accuracy remain almost constant.

Figure 3 shows the list of symptoms from which the patient can select particular symptoms, which he is suffering from and submit to generate a report.

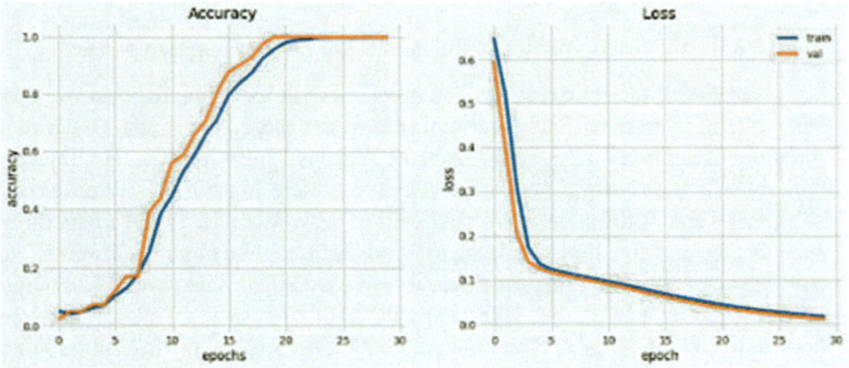

Fig. 2 Values of accuracy and loss for validation and training versus epochs

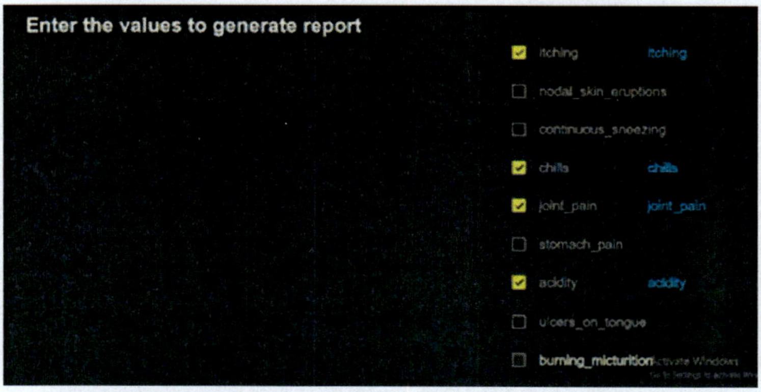

Fig. 3 User interface for patients to enter data

Figure 4 depicts the output generated by the model. This output consists of different values, which range from 0 to 1 multiplied by 100, that represent the probability of occurrence of the list of diseases, and the result of the record illustrated by figure informs us that there is a high chance the patient or the details related to a person is suffering from "urinary tract infection".

Figure 5 shows the details of the doctors recommended for the respective diseases.

In this page, the user can decide on a doctor and move ahead as they will be redirected to booking appointments.

5 Conclusion

A linear regression model to predict a most likely disease from a particular set of symptoms is developed. As a result, the number of symptoms reduced from 133 to 79 symptoms using a coefficient matrix and took 0.4 quantile out of it, and trained the model which gave the accuracy of 95.93%. A decision tree model to predict disease using all the symptoms which gave us an accuracy of 97.6% is also developed. A neural networks model is developed with two hidden layers and an output layer with 21 epochs, which gives the accuracy of 95.3%. From these results, it can conclude that decision tree is the best model for the given data set. It is able to provide a user interface for the disease prediction, mapped the respective diseases with a specialization, so that a doctor with the required specialization can be recommended to the patient. Provided an option of rating the doctor after the respective appointment based on which the doctors are recommended later on.

There is a possibility of advancement in the machine learning part, where we can improve or add new models such as neural networks with different activation functions. We can also try out different models such as SVM, and also, we can include feature reduction methods like PCA. Regarding the Web application, we can

```
'Allergy': 0.29797658841897195,
'Arthritis': 0.10202953011841367,
'Bronchial Asthma': 0.08577025658186797,
'Cervical spondylosis': 0.374924267671638,
'Chicken pox': 0.031507094918759394,
'Chronic cholestasis': 0.0594407229662809,
'Common Cold': 0.021243515902439482,
'Dengue': 0.015267691518277605,
'Diabetes ': 0.08082729040338064,
'Dimorphic hemmorhoids(piles)': 0.08282832122938537,
'Drug Reaction': 2.348780216413582,
'Fungal infection': 0.12382000725163839,
'GERD': 0.07026928262853761,
'Gastroenteritis': 0.15664442618135044,
'Heart attack': 0.15147591201697141,
'Hepatitis B': 0.022172753797773512,
'Hepatitis C': 0.035939072727777648,
'Hepatitis D': 0.024946006896839014,
'Hepatitis E': 0.022878607829241210,
'Hypertension ': 0.17533155907786818,
'Hyperthyroidism': 0.0539436252361747,
'Hypoglycemia': 0.025092467487966483,
'Hypothyroidism': 0.043682263052353856,
'Impetigo': 0.1441766106480116,
'Jaundice': 0.055863802668430385,
'Malaria': 0.04184641012666852,
'Migraine': 0.045926798979937095,
'Osteoarthristis': 0.05848399011768744,
'Paralysis (brain hemorrhage)': 0.41466098140482055,
'Peptic ulcer diseae': 0.10328803030986862,
'Pneumonia': 0.024423020142999012,
'Psoriasis': 0.08096796127646837,
'Tuberculosis': 0.021387069013669272,
'Typhoid': 0.029587648713441043,
'Urinary tract infection': 90.42662461719962,
'Varicose veins': 0.3651929313327923,
'hepatitis A': 0.030051155196802487}
```


Fig. 4 Probabilities of each disease as predicted by the model

include the exact time limit for the appointment booking. The location of the doctor can be known to the patient using Google Maps API. We can also include payment methods like UPI, credit card billing, etc. We can also provide an electronic health record (EHR) facility for large type organizations.

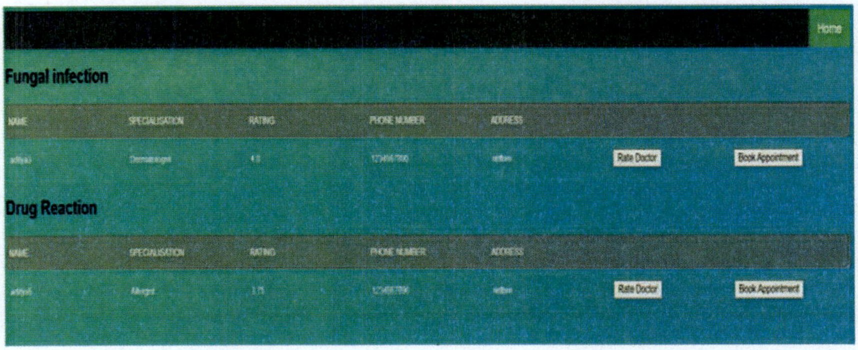

Fig. 5 Recommended doctors

References

1. Caballé, N. C., Castillo-Sequera, J. L., Gómez-Pulido, J. A., Gómez-Pulido, J. M., & Polo-Luque, M. L. Machine learning applied to diagnosis of human diseases: A systematic review. *Journal of Applied Sciences.*
2. Dhomse Kanchan, B., & Mahale Kishor, M. Study of machine learning algorithms for special disease prediction using principal of component analysis. In *2016 international conference on global trends in signal processing, information computing and communication.*
3. Hasan, M. A., Sher-E-Alam, K. M., & Chowdhury, A. R. (2010). Human disease diagnosis using a fuzzy ExpertSystem. *Journal of Computing, 2*(6).
4. Laxmi, P., Gupta, D., Radhakrishnan, G., Amudha, J., & Sharma, K. (2021). Automatic multi-disease diagnosis and prescription system using bayesian network approach for clinical decision making. In Advances in artificial intelligence and data engineering (Vol. 1133, pp. 393–409). Springer.
5. Mohanty, A., Parida, S., Nayak, S. C., Pati, B., Panigrahi, C. R. (2022). *Study and impact analysis of machine learning approaches for smart healthcare in predicting mellitus diabetes on clinical data. Smart healthcare analytics: State of the art. Intelligent systems reference library* (Vol. 213). Springer. https://doi.org/10.1007/978-981-16-5304-9_7
6. Sasikala, T., Rajesh, M., & Sreevidya, B. (2020). Prediction of academic performance of alcoholic students using data mining techniques. In *Cognitive informatics and soft computing, advances in intelligent systems and computing* (Vol. 1040). Springer.
7. Sreevidya, B., Rajesh, M., & Sasikala, T. (2019). Performance analysis of various anonymization techniques for privacy preservation of sensitive data. In *International conference on intelligent data communication technologies and internet of things (ICICI) 2018, lecture notes on data engineering and communications technologies* (Vol. 26). Springer.
8. Das Mohapatra, S., Nayak, S. C., Parida, S., Panigrahi, C. R., Pati, B. (2021). *COVTrac*: Covid-19 tracker and social distancing app. In C. R. Panigrahi, B. Pati, B. K. Pattanayak, S. Amic & K. C. Li (Eds.), *Progress in advanced computing and intelligent engineering. Advances in intelligent systems and computing* (Vol. 1299). Springer. https://doi.org/10.1007/978-981-33-4299-6_50
9. Krishna, C. S., & Sasikala, T. (2019). Healthcare monitoring system based on IoT using AMQP protocol. In *International conference on computer networks and communication technologies, lecture notes on data engineering and communications technologies* (Vol. 15). Springer.
10. Zadeh, L. A. (1983). The role of fuzzy logic in the management of uncertainty in expert systems. *Fuzzy Sets and Systems On Elsevier, 11*(1–3), 197–198.
11. National Research Council Canada. Available http://www.nrccnrc.gc.ca/eng/index.html

12. Yourdiagnosis, an online diagnosis tool. Available http://www.yourdiagnosis.com/
13. Easydiagnosis, an online diagnosis tool. Available http://easydiagnosis.com/
14. Symptoms of different diseases. Available http://www.wrongdiagnosis.com/
15. Symptoms of different diseases. Available http://www.webmd.com/

Role of Machine Learning Approaches in Predicting COVID-19 New Active Cases Using Multiple Models

Ritesh Kumar Sinha, Sukant Kishoro Bisoy, Bibudhendu Pati, Rasmi Ranjan Khansama, Chhabi Rani Panigrahi, and Saurabh Kumar

Abstract The coronavirus epidemic began in Wuhan and has already spread to practically every country on the planet. Conravirus has a big population in India, and people are becoming infected at an alarming rate. Machine learning algorithms have been utilized to find trends in the number of active cases owing to COIVD in India and the state of Odisha in this study. The data was gathered from the WHO and studied to see if there was a link between the number of current cases, those who died, and those who recovered. The model was entirely based on multiple regression, support vector machine, and random forest which fits as an effective tool for prediction and error reduction. Based on the dataset taken from March 16, 2020, to August 20, 2020, from the ICMR website, the mean absolute error (MAE) of SVM is less for Odisha and multiple linear regression is less for India. The multiple learner regression model is able to predict number of active cases properly as its R^2 score value are 1 and 0.999 for Odisha and India, respectively. Machine leaning model helps us to find trends of effected cases accurately. The model is able to predict what extent the COVID cases will grow or fall in the next 30 days which enables us to be prepared in advance and take some preventive measures to fight against this deadly COVID virus. It is observed that features are positively correlated with each other.

Keywords Multiple regression · Support vector machine · Random forest · Machine learning · Coronavirus

R. K. Sinha (✉) · S. K. Bisoy · R. R. Khansama · S. Kumar
C.V. Raman Global University, Bhubaneswar, Odisha, India
e-mail: sinharitesh567@gmail.com

B. Pati · C. R. Panigrahi
Department of Computer Science, Rama Devi Women's University, Bhubaneswar, Odisha, India

B. Pati et al. (eds.), *Proceedings of the 6th International Conference on Advance Computing and Intelligent Engineering*, Lecture Notes in Networks and Systems 428,
https://doi.org/10.1007/978-981-19-2225-1_6

1 Introduction

A pandemic or epidemic is a disease that spreads over a long period of time. Coronaviruses are RNA virus families found in animals that cause respiratory tract infections in people and birds ranging from moderate to fatal, as well as loose bowels, hepatitis, and encephalomyelitis in mice. SARS-COV2 produced the COVID19 pandemic, which is the most dangerous public health emergency because it spreads quickly [1]. The spread trends shows transmission through human interaction through tiny droplets formed through coughing, sneezing, and talking whereas being contact with an affected exterior is becoming one of primary reasons for the transmission. The world is facing a great crisis since after World War Two leading to 792,457 deaths and more than 22,589,017 confirmed cases as of August 21, 2020, and its effect would be felt for many upcoming years. According to the sources [1], the epidemic started from Wuhan, China, in December, 2019 as a pneumonia outbreak which was further found out to be a novel strain of coronavirus and was called 2019-nCoV by the WHO, at that point changed to SARS-COV2 by the International Committee on Taxonomy of Viruses (ICTV) [2]. On March 1, WHO declared the pandemic which has spread over through the world [1, 3–6]. This pandemic has brought about movement limitations and lockdowns in numerous nations. There have been 2,973,317 confirmed cases in over 29 states and 5 union territories with a total of 55,928 deaths in India as of August 21, 2020 [1]. India's index case was reported on January 30, 2020, in Thrissur, Kerala (10° 31′ 39″ N 76° 12′ 52″ E), and the country is facing the pandemic since six months, three weeks, and one day as on August 21, 2020. The country underwent a series of lockdowns since March 22, 2020. As of August 21, 2020, 33,467,237 tests were conducted, out of which 2,904,340 individuals were confirmed positive, 690,894 are active, whereas 2,157,941 have recovered with fatalities of 54,975 individuals. India's fatality rate stands at 2.41% with a steady decline and recovery rate of 63.18% as of July 23, 2020 [1]. The lockdown in India has started the month of March 2020 to stop the number of active cases looking the high speed growth rate of Italy [7]. As per the report given by Indian Government, Maharastra state of India is severely affected by cornoavirus [8].

Different models have been utilized for foreseeing the lifecycle of COVID-19 for determining the trends with data for outbreak severity, observing the agitation, correlating susceptible-infected-removed (SIR) and SIER, predicting the spread and death rate and its impact. Author in [9] developed a theoretical approach to investigate the spreading of corona using SIR model. Daily expected COVID cases are forecasted in Saudi Arabi for four weeks using Autoregressive Integrated Moving Average (ARIMA) model [10]. Malware detection [11, 12], mobile malware detection [13, 14], medicine [15–17], and information retrieval [18, 19] are just a few of the sectors where machine learning (ML) is being used more frequently. Statistics and time series models have recently been utilized to develop models for estimating the duration of the global problem caused by coronavirus. The authors used data from confirmed cases in Nigeria in bent quantifiable estimating techniques as well as a linear regression model to predict the number of deaths caused by coronavirus in

India. Due to the rapid increase in the number of current cases, doctors are finding it difficult to diagnose in a timely manner. Techniques to predict the number of current cases, recovered cases, and deceased cases are needed. Many machine tools have made it feasible to evaluate enormous amounts of data and forecast cases. Our goal is to use multiple regression, support vector machine, and random forest to forecast the number of active cases in the following 30 days so that we may be prepared ahead of time to take preventive steps.

The remainder of the paper is organized in the following manner. The material and method employed in Sect. 2 will be presented, and the outcome and analysis will be explained in Sect. 3. Finally, in part 4, we bring our work to a close.

2 Material and Method Used

Detecting COVID-19 instances around the world is currently one of the most difficult issues, resulting in the disease's rapid spread. According to the figures, the number of patients affected with COVID-19 is geometrically expanding, with over confirmed cases 1.6 million; it has spreaded to many nations throughout the world. We gathered daily data from the WHO site on the number of active, recovering, and deceased people. Data is collected from the 16th of March to the 20th of August 2020. The most affected countries in the globe are Odisha and India. India has taken the lead and has implemented a lockdown on the country beginning March 23, 2020. Figures 1 and 2 indicate the predicted number of active, recovered, and deceased cases for Odisha and India, respectively. In both Odisha and India, the number of active, recovered, and deceased people has been climbing linearly since July. However, as contrasted to Odisha, the number of deaths has been steadily rising since June.

In recent months, there has been a lot of research into the analysis and detection of COVID-19. Because exposed people do not develop sickness signs right once, it is difficult to detect them. We need a system which can estimate the number of potentially affected people on a frequent basis to take the essential precautions. As a result, the next step is to look at the correlation and distribution of the number of active cases. Following the collection of data, the data cleaning is done, and a correlation analysis is performed. The correlation analysis depicts the relationship between two variables using a scale of 0 to 1. Higher value represents higher correlation. The correlation among different features is shown in Fig. 3 for Odisha and India dataset. From the figure, it is observed that features are positively correlated with each other. The distribution of active cases in both India and Odisha dataset is shown in Fig. 4.

Because of rapid spread of COVID-19, it is critical to develop a plan for estimating the possibly infected cases regularly so that necessary steps may be taken. India and Odisha currently rely on specific decision-making indicators such as patterns and seasonality. The trend and seasonality of the active cases for India and Odisha dataset are shown Figs. 5 and 6. From these figure, we can say that there is a growing trend of active cases and also less impact of seasonality in active cases. In [20], author

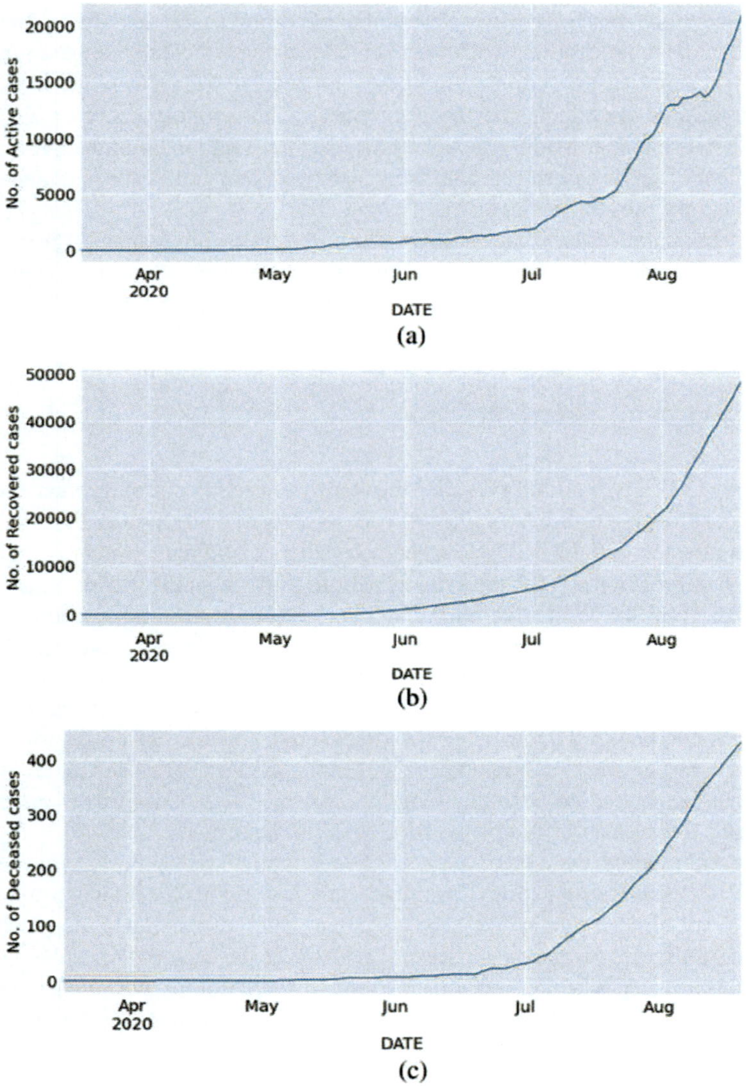

Fig. 1 No. of cases in Odisha **a** active, **b** recovered, and **c** deceased

analyzed the association of census feature and weather data on number of active cases and found that mortality cases depends on weather data.

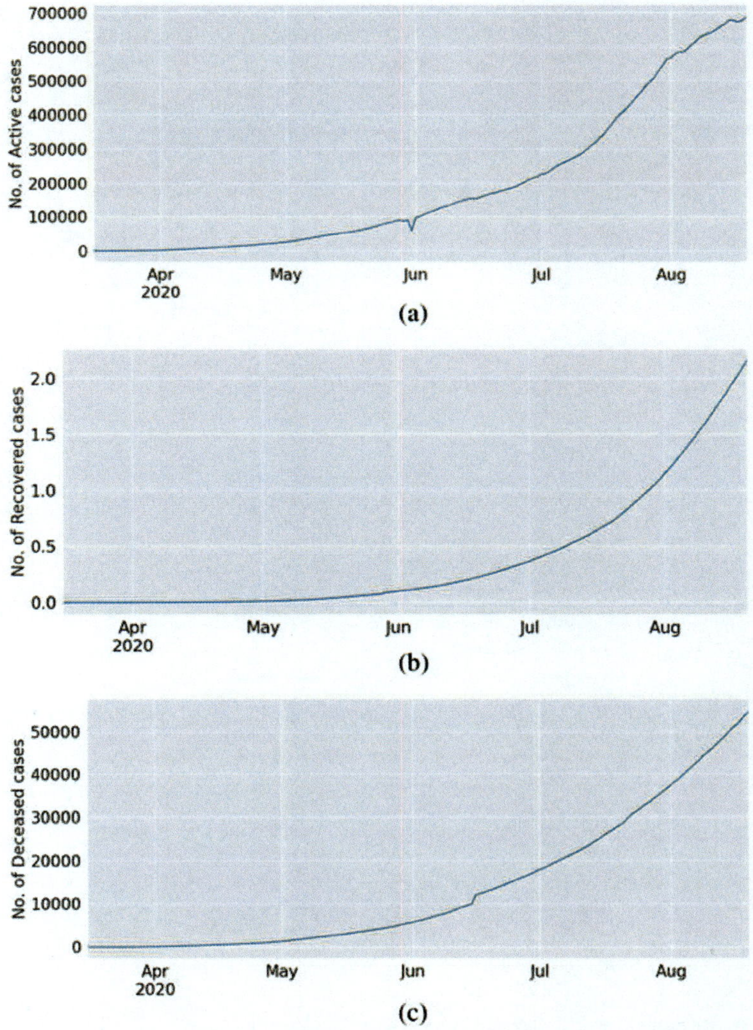

Fig. 2 No. of cases in India **a** active, **b** recovered, and **c** deceased

3 Result Analysis

Many state of art analysis is done to predict number of coronavirus cases using machine learning and deep learning models [21]. To analyze the number of infected deceases, statistical plays an import role. The effect of lockdown was studied in [22], and the author used the susceptible-exposed-infectious-recovered (SEIR) model to anticipate the number of active cases in India. To stop the disease from spreading, machine learning models can be used to analyze the behaviors of active cases. For

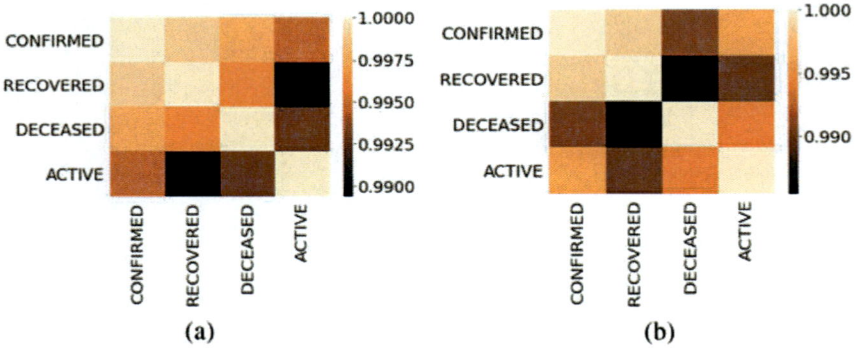

Fig. 3 Correlation table of **a** Odisha, **b** India

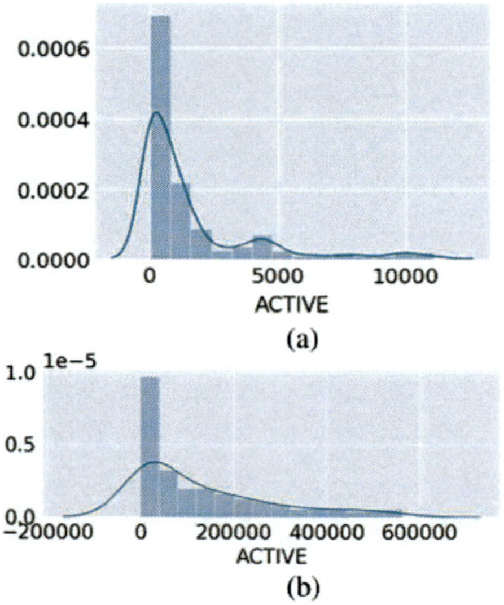

Fig. 4 Active cases showing peak and average cases in **a** Odisha, **b** India

COVID case prediction, we used three different machine learning models: Random Forest, multiple linear regression, and support vector machine. As previously stated, we considered data from the 16th of March 2020 to the 20th of August 2020 and divided it into two datasets. Data from 80% of these was used for training and 20% was used for testing. We looked over the case of Odisha.

Case-1: In this section, we have provided the performances of all three models for the prediction of active cases using number of confirmed, recovered and deceased cases and recovered cases. First, the dataset for Odisha is divided into training set and testing set. All the models have been trained on first 138 instances of the dataset that is from March 16, 2020, to July 31, 2020. Then, the number of active cases

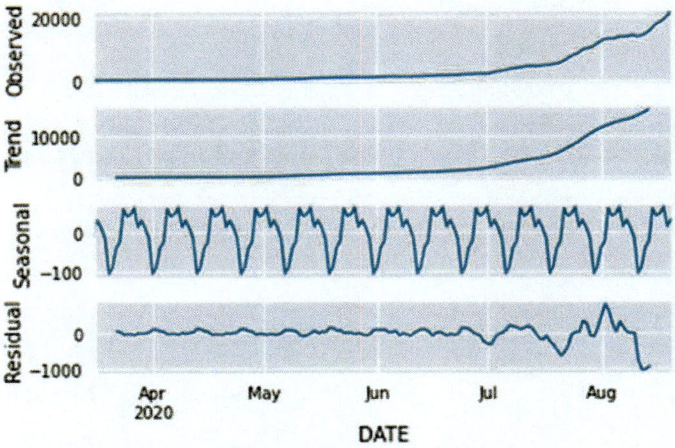

Fig. 5 Active cases of Odisha

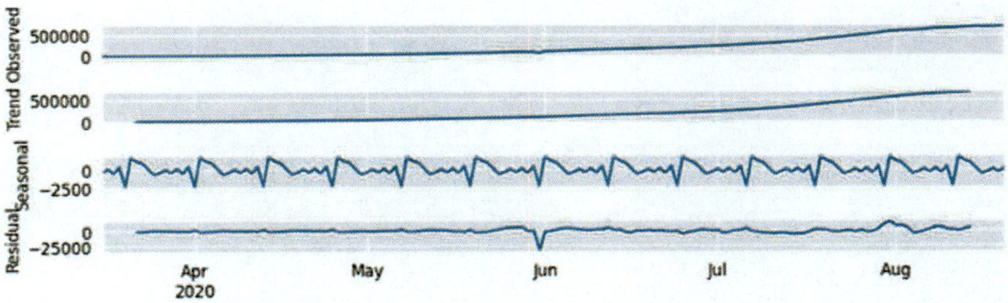

Fig. 6 Active cases of India

is predicted for 20 instances that are from August 1, 2020, to August 20, 2020, to evaluate and analyze the performance of the different model. The values of parameters for multiple linear regressions after fitting on training set are shown in Table 1. Here, the value of R^2 indicates the multiple regression models as the strong fitted model. Random forest and support vector machine are both trained and assessed on the same training and testing sets. Figures 7, 8, and 9 provide a comparison of actual and expected values of random forest, multiple linear regression, and support vector machines for the next 20 days on the Odisha dataset. All of these models' prediction performance is assessed using a variety of error metrics. Table 2 shows the results of

Table 1 Values of parameter after training with multiple linear regression

Dataset	Intercept	Co-efficient ($B1$)	Co-efficient ($B2$)	Co-efficient ($B3$)	R^2
India	36.3703774651658	0.97167467	−0.64740431	−0.97014202	0.999
Odisha	0.38181944201880924	0.98538383	0.62198826	−0.98967798	1.000

Fig. 7 No. of active cases of Odisha using random forest

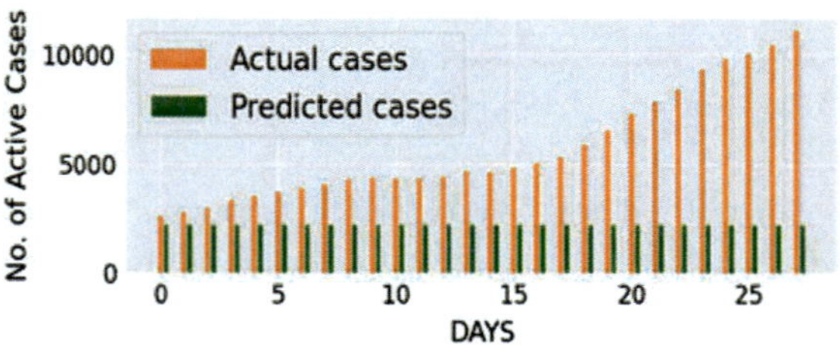

Fig. 8 No. of active cases of Odisha using multiple linear regression

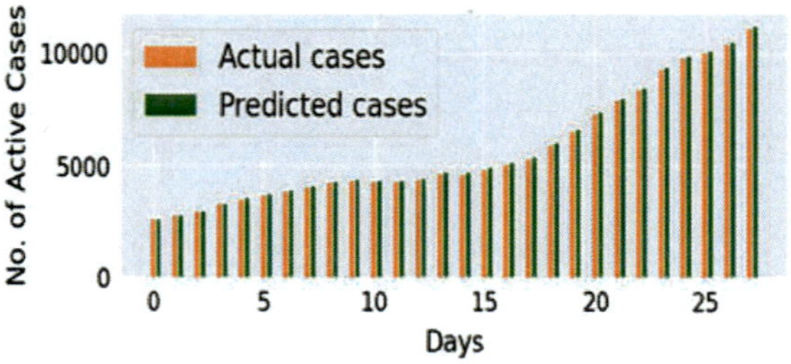

Fig. 9 No. of active cases of Odisha using support vector machine

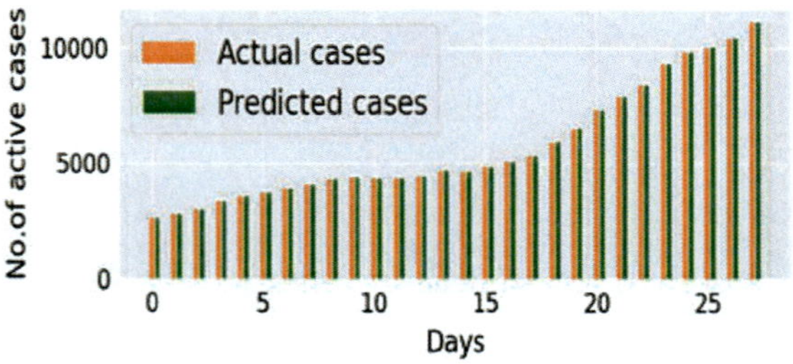

Table 2 Values of performance metrics on Odisha test data for all three models

Odisha	Linear regression	SVM	Random forest
Mean absolute percentage error	0.9094964300984387	**0.14417560953898045**	54.38628892838645
Mean squared error	3774.501697282683	**100.77708155781197**	18,631,195.07867186
Mean absolute error	54.40728328005709	**8.793622275305292**	3519.773
Mean error	−54.40728328005709	**−8.793622275305292**	3519.773
Root mean squared error	61.436973373390416	**10.038778887783712**	4316.386808277482

these performance metrics. Table 2 shows that the support vector machine beat all other models for predicting active cases in Odisha across all performance criteria.

Case-2: The actual and predicted values of random forest, multiple linear regression, and support vector machines on India test dataset are shown in Figs. 10, 11 and 12. All of these models' prediction performance is assessed using a variety of error metrics for prediction of active cases from August 1, 2020, to August 20, 2020. The values of these performance metrics are provided in Table 3. From Table 2, we can conclude that the multiple linear regression has outperformed all the models in all the performance metrics for prediction active cases in India.

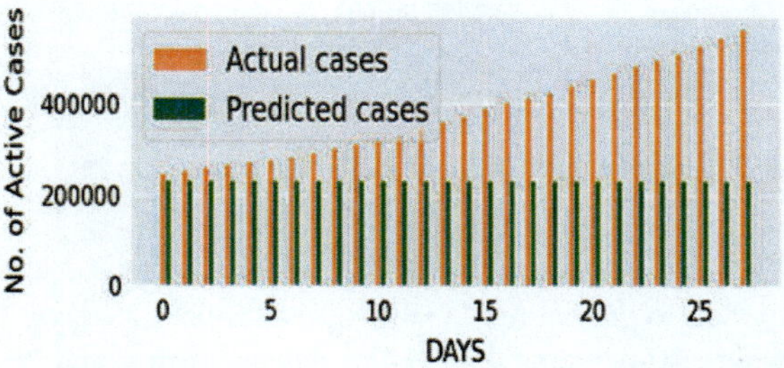

Fig. 10 No. of active cases of India using random forest

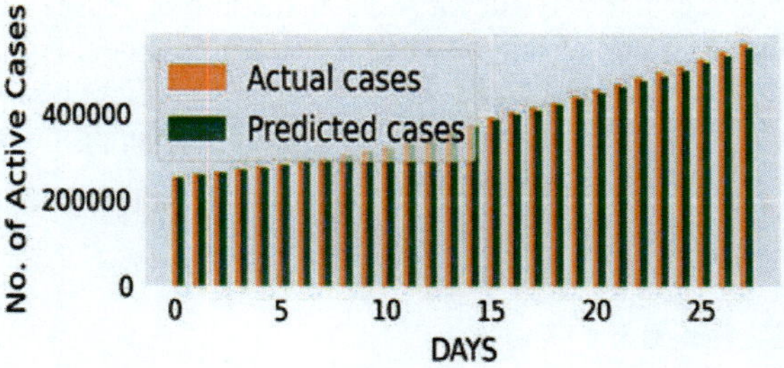

Fig. 11 No. of active cases of India using multiple linear regression

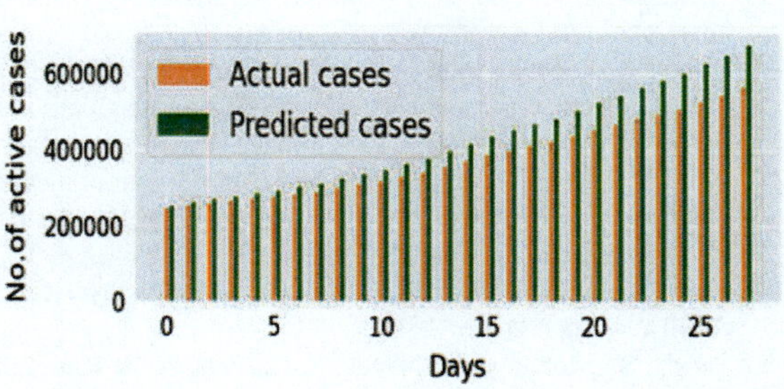

Fig. 12 No. of active cases of India using support vector machine

Table 3 Values of performance metrics on India test data for all three models

India	Linear regression	SVM	Random forest
Mean absolute percentage error	**0.9097190390079295**	11.84211003401615	34.34598199650979
Mean squared error	**21,479,699.96355724**	3,421,566,330.222326	30,691,463,380.963783
Mean absolute error	**3848.2230178134155**	49,546.180282397196	145,814.60914285714
Mean error	**3848.2230178134155**	−49,546.180282397196	145,814.60914285714
Root mean squared error	**4634.6197215691**	58,494.15637670421	175,189.79245653492

4 Conclusions

We analyzed the number of active cases of Odisha and India and used three different models to predict the COVID cased for next 25 days. The data collected from WHO where 80 percentage of data used for testing and 20 percentages for testing. It is observed that SVM gives better result for Odisha and multiple linear regression provides better result for India as the number cases are very high. Our result shows that R^2 value of multiple linear regression model is 1 for Odisha and keeping close to 1 (0.9999) for India which indicates prediction is robust. Our prediction model could help the medical professional to take appropriate plan so that future cases in COVID-19 can be minimized.

References

1. World Health Organization, Coronavirus Disease (COVID-19) outbreak. https://www.who.int/emergencies/diseases/novel-coronavirus-2019
2. Banerjee, A., Mukherjee, S., Panigrahi, C. R., Pati, B., & Mall, R. (2021). Analysis of COVID-19 data using consensus clustering technique. In *Computational modeling and data analysis in COVID-19 research* (pp. 17–28). CRC Press.
3. Wu, J. T., Leung, K., & Leung, G. M. (2020). Nowcasting and forecasting the potential domestic and international spread of the 2019-ncov outbreak originating in Wuhan, china: A modelling study. *The Lancet,395*(10225), 689–697.
4. Novel coronavirus (COVID-19) cases, provided by JHU CSSE. https://github.com/CSSEGISandData/COVID-19
5. Parida, S., Mohanty, A., Nayak, S. C., Pati, B., & Panigrahi, C. R. (2021). Study and impact analysis of COVID-19 pandemic clinical data on infection spreading (p. 225). *Data Science for COVID-19: Volume 2: Societal and medical perspectives.*
6. Kraemer, M. U., Yang, C. H., Gutierrez, B., Wu, C. H., Klein, B., Pigott, D. M., & Brownstein, J. S. (2020). The effect of human mobility and control measures on the COVID-19 epidemic in china. *Science,368*(6490), 493–497.

7. Sarkar, J. L., Ramasamy, V., Majumder, A., Panigrahi, C. R., Gomathy, B., Pati, B., & Saha, A. K. (2021). SensMask: An intelligent mask for assisting patients during COVID-19 emergencies. *Computación y Sistemas, 25*(3).
8. India COVID-19 TRACKER. 2020 [online]. Available at https://www.covid19india.org/. Accessed on 11th September 2020.
9. Biswal, A., Nanda, S., Panigrahi, C. R., Cowlessur, S. K., & Pati, B. (2021). Human activity recognition using machine learning: A review. In C. R. Panigrahi, B. Pati, B. K. Pattanayak, S. Amic & K. C. Li (Eds.), *Progress in advanced computing and intelligent engineering. Advances in intelligent systems and computing* (Vol. 1299). Springer. https://doi.org/10.1007/978-981-33-4299-6_27
10. Alzahrani, S. I., Aljamaan, I. A., & Al-Fakih, E. A. (2020). Forecasting the spread of the COVID-19 pandemic in Saudi Arabia using ARIMA prediction model under current public health interventions. *Journal of Infection and Public Health, 13*, 914–919.
11. Alazab, A., Hobbs, M., Abawajy, J., & Alazab, M. (2012). Using feature selection for intrusion detection system. In *2012 international symposium on communications and information technologies (ISCIT)*, pp. 296–301.
12. Alazab, A., Alazab, M., Abawajy, J., Hobbs, M. (2011). Web application protection against SQL injection attack. In *Proceedings of the 7th international conference on information technology and applications*, pp. 1–7.
13. Nanda, S., Panigrahi, C. R., Pati, B., & Mishra, A. (2021). A novel approach to detect emergency using machine learning. In C. R. Panigrahi, B. Pati, P. Mohapatra, R. Buyya & K. C. Li (Eds.), *Progress in advanced computing and intelligent engineering. Advances in intelligent systems and computing* (Vol. 1199). Springer. https://doi.org/10.1007/978-981-15-6353-9_17
14. Alazab, M., Monsamy, V., Batten, L., Lantz, P., & Tian, R. (2012). Analysis of malicious and benign android applications. In *2012 32nd international conference on distributed computing systems workshops (ICDCSW)*, pp. 608–616.
15. Xu, Y., Wang, Y., Yuan, J., Cheng, Q., Wang, X., & Carson, P. L. (2019). Medical breast ultrasound image segmentation by machine learning. *Ultrasonics,91*, 1–9.
16. Das Mohapatra, S., Nayak, S. C., Parida, S., Panigrahi, C. R., & Pati, B. (2021). *COVTrac*: Covid-19 tracker and social distancing app. In C. R. Panigrahi, B. Pati, B. K. Pattanayak, A. Amic & K. C. Li (Eds.), *Progress in advanced computing and intelligent engineering. Advances in intelligent systems and computing* (Vol. 1299). Springer. https://doi.org/10.1007/978-981-33-4299-6_50
17. Mesleh, A., Skopin, D., Baglikov, S., & Quteishat, A. (2012). Heart rate extraction from vowel speech signals. *Journal of Computer Science and Technology,27*, 1243–1251.
18. Nanda, S., Panigrahi, C. R., Pati, B., Rath, M., & Weng, T. H. (2021). COVID-19 risk assessment using the C4. 5 algorithm. *Computational Intelligence Techniques for Combating COVID-19, 61*.
19. Mohanty, A., Parida, S., Nayak, S. C., Pati, B., & Panigrahi, C. R. (2022). Study and impact analysis of machine learning approaches for smart healthcare in predicting mellitus diabetes on clinical data. In P. K. Pattnaik, A. Vaidya, S. Mohanty, S. Mohanty & A. Hol (Eds.), *Smart healthcare analytics: State of the art. Intelligent systems reference library* (Vol. 213). Springer. https://doi.org/10.1007/978-981-16-5304-9_7
20. Malki, Z., Atlam, E. S., Hassanien, A. E., Dagnew, G., Elhosseini, M. A., & Gad, A. (2020). Association between weather data and COVID-19 pandemic predicting mortality rate: Machine learning approaches. *Chaos, Solitons and Fractals,138*, 110137.
21. Swapnarekha, H., Behera, H. S., Nayak, J., & Naik, B. (2020). Role of intelligent computing in COVID-19 prognosis: A state-of-the-art review. *Chaos, Solitons and Fractals,138*, 109947.
22. Pai, P., Bhaskar, A., & Rawoot, V. (2020). Investigating the dynamics of COVID-19 pandemic in India under lockdown. *Chaos, Solitons and Fractals,138*, 109988.

An Overview of Applications of Machine Learning During COVID-19

Harsh Panchal and Ankit K. Sharma

Abstract Despite the recent global concern, healthcare specialists, doctors, and scientists around the globe are still looking for a breakthrough solution to help fight the COVID-19 outburst. By use of artificial intelligence (AI) and machine learning (ML) in past, outbreaks have intrigued scientists, suggesting a particular methodology to tackling the existing coronavirus pandemic. In terms of the outbreak that followed after coronavirus, widely recognized as SARS-CoV-2. This paper provides an in-depth analysis of appraisal of AI and ML as one good approach for monitoring for contact tracking, prediction, forecasting, and therapeutic development.

Keywords Application of machine learning · Diagnostics · Treatment · Forecasting and prediction · Tracking of contacts · COVID-19

1 Introduction

The outbreak has started to spread throughout the world since the very first instance of COVID-19 (coronavirus) infection was discovered in China's Wuhan District in December 2019, with the WHO declaring the pandemic a serious global concern on January 30, 2020. According to recent studies, AI and ML are the most advanced technologies that are being used by a variety of healthcare providers because they allow faster processing power, better scale-up, and even outperform humans in certain healthcare jobs. As a consequence, health systems and doctors across the whole world adopted a myriad of ML and AI solutions to tackle the COVID breakout and tackle the problems that resulted. This study examines the unique COVID epidemic and how modern AI and machine learning technology were recently used to address the issues that arose during the outbreak.

H. Panchal (✉) · A. K. Sharma (✉)
Instrumentation and Control Engineering Department Institute of Technology, Nirma University, Ahmedabad, Gujarat, India
e-mail: harshpanchal874@gmail.com

A. K. Sharma
e-mail: ankit.sharma@nirmauni.ac.in

B. Pati et al. (eds.), *Proceedings of the 6th International Conference on Advance Computing and Intelligent Engineering*, Lecture Notes in Networks and Systems 428,
https://doi.org/10.1007/978-981-19-2225-1_7

Section 2 discusses the use of ML and AI in diagnosis and treatment, Sect. 3 discusses the use of ML and AI in forecasting and prediction, Sect. 4 discusses the use of ML and AI in tracing of contacts, Sect. 5 discusses the use of ML and AI in vaccines and pharmaceuticals, and finally, Sect. 6 discusses the study's generated conclusions.

2 Diagnostic and Treatment

Recognition of any disease, noninfectious or infectious, is vital for effective treatment. With technology like X-ray, computed tomography (CT), and blood sample data, AI and ML are used to assist the treatment and diagnosis method of the patient. In this regard, this section provides selective information on the coronavirus diseases diagnosis and treatment protocols.

Ardakani et al. [1]: COVID-19 infection has been differentiated from other unusual and viral respiratory diseases using a potential CAD methodology based on CT imaging. In this study, the ResNet-101 was reported to be an optimistic model for predicting and assessing COVID illness. This model is economical which can be used as an augment procedure in radiology departments as CT imaging. AlexNet, SqueezeNet, and 101, VGG-16, 19, GoogleNet, MobileNet-V2, along with Xception are among the ten possible best convolutional neural networks (CNNs) applied to discriminate non-COVID infection against COVID infection. Every one of the networks performed the best apart from Xception and ResNet-10. ResNet-101 was able to tell the difference with an accuracy of 99.51%, specificity of 99.02%, and Sensitivity of 100%. Xception had a specificity of 100%, accuracy was 99.02%, and sensitivity was 98.04%.

Ozturk et al. [2]: For binary and multiclass categorization, the proposed approach is meant to offer useful diagnostics. Their method accurately recognized multiclass instances with 87.02% efficacy and binary groups with 98.08% efficacy. They employed the DarkNet system as a classifier in their research on the YOLO real-time object categorization system. They used 17 layers of CNNs each of which has its filter.

Sun et al. [3]: For this investigation, all 336 cases of COVID transmission in Shanghai from the March 12 were sequentially enrolled and separated into test and train data. There were also 220 clinic pathological reports acquired. A technique for measuring the intensity of severe cases was constructed, and diagnostic or therapeutic signals for obvious illnesses were recognized. There would have been 36 medical indications that were linked to major illnesses in a meaningful way. Thyroxine, immune system cells, and metabolites seem to be the most frequent surgical indications. It has been shown that using a support vector system and a proper mix of longevity, GSH, protein content, and CD3 ratio to distinguish among moderate—to—severe diseases is effective. In the test and train info, the area beneath the receiver operating characteristic (ROC) was 0.9757 and 0.9996, respectively. When the rate of recall was 93.33%, the value of cut-off was 0.0667 in the test dataset and for the training dataset it was 100%. As per Cox multivariate regression and longevity studies, the method significantly distinguished crucial cases and incorporated material from the relevant clinical features. The system was indeed accurate and effective in forecasting extreme COVID instances.

Wu et al. [4]: To address the final aid categorization system, 11 essential blood variables were gathered from 49 clinically important blood test findings available by commercialized blood test machinery using just a random forest system. With the efficacy of 0.9697 and 0.9795 for the validation sets and cross-validation, separately, the methodology was able to successfully recognize COVID from a significant number of potential users with indistinguishable evidence of CT or characteristics. After several verifications, the tool's reliability and repeatability have also been thoroughly examined, and it tends to advance into technological innovation for identifying COVID-19 and reducing the worldwide public health strain. The developed technology is useful for conducting experimental assessments of individuals recognized and supporting individuals in acquiring quick medication and quarantine suggestions.

Brinati et al. [5]: They designed two machine learning algorithms models based on hematological values from taking a blood sample from 279 sick people who were examined with RRT-PCR also known as real-time reverse transcription polymerase chain reaction tests after being admitted in Italy to the San Raffaele Health clinic with COVID signs. About 177 people had a positive response, whereas 102 people had a re-response that was negative. They developed two models based on ML to discern among patients who are negative or positive for coronavirus: Their sensitivity and accuracy are equivalent to the benchmark, at 92 and 95% and 82 and 86%, correspondingly. They also created an interpretable decision tree model as a simple aid to take decisions for clinicians assessing COVID suspicious blood tests (even offline). This analysis demonstrated that using ML and blood test analysis to determine COVID-19 positive patients as an adjunct to RRT-PCR is doable and therapeutically sound.

Alakus and Turkoglu, [6]: In this study, Alakus and Turkoglu employed deep learning and laboratory actionable insights to generate clinical forecasting algorithms that predict which people are supposed to get COVID-19 sickness. To assess the forecasting accuracy of their systems, they calculated AUC, precision, accuracy, F1-score, and recall. The hypotheses were evaluated on 18 laboratory results from 600 individuals employing tenfold cross-validation and a strategy of train-test splitting. According to empirical observations, their prediction models properly diagnosed COVID-19 illness patients with a recall of 99.42%, F1-score of 91.89%, an AUC of 62.50%, 86.75% of precision, and 86.66% accuracy. Neural networks relied on experimental information have been established which could be utilized to recognize COVID-19 transmission, which can aid legal professionals inadequately targeting resources.

Meza et al. [7]: Their goal was to create and test an ML framework for recognizing COVID in patients that are in the hospital. This algorithm was created to act as a tool for screening in hospitals, where testing is absent or poor. It relied on rudimentary demographic and laboratory parameters. They experimented with seven ML systems before combining them to create the final diagnostic categorization. In the validation set, our training set had a region underneath the ROC arc of 0.91 (95% confidence interval 0.87–0.96). The system had a specificity of 0.64 (interval of 0.58–0.69) and a 0.93 sensitivity (interval of 0.85–0.98). When compared to COVID-19 PCR, they discovered that their machine learning system had superior diagnostic metrics.

3 Forecasting and Prediction

This section describes how ML and AI are used to forecast and predict the new epidemic. Ribeiro et al. [8]: The results of random forest, autoregressive integrated moving average, support vector regression, cubist, stacking ensemble learning, and ridge regression in time series data with 1, 2, and 6 days before the COVID total combined verified incidents in ten areas of Brazil with greater frequency of occurrence are presented in this paper. The symmetric mean absolute percentage error, mean absolute error, and improvement index parameters are being used to evaluate the algorithms' efficacy. Stacking ensemble learning and SVR outperform comparison models in the majority of cases in terms of adopted criteria. In general, the created models can produce reliable forecasts with errors ranging from 0.87 to 3.51% in one day, 1.02% to 5.63% in three days, and 0.95% to 6.90% in six days. Once these models can assist managers in making decisions, decision-making assistance systems will be implemented, it is advised that they be used to forecast and monitor the continued rise of COVID-19 instances.

Yan et al. [9]: This study identified three parameters (hs-CRP, LDH, and lymphocytes) as well as a clinical approach for COVID-19 clinical prediction. They created an XGBoost machine learning-based model for predicting patient mortality rates with better than 90% accuracy longer than ten days before the date, allowing for early diagnosis, intervention, and maybe a reduction in COVID-19 patient mortality. Overall, this article proposes a simple and easy-to-implement decision rule for swiftly identifying high-risk patients, allowing for prioritization and perhaps lowering rates of mortality.

Chimmula et al. [10]: In this innovative work, they analyzed the critical elements for predicting the trends and likely stopping time of the current COVID outbreak in Canada and around the world. They anticipated that the outbreak would cease around June 2020 using the LSTM model known as long short-term memory which is a method of deep learning for anticipating future cases. They also compared transmission rates in Canada to those in Italy and the United States. They also correctly predicted the 2, 4, 6, 8, 10, 12, and 14th days for two days in a row.

Booth et al. [11]: They use five serum chemistry laboratory parameters from 398 people to construct a model based on ML that forecasts patient death up to 48 h before it happens. The constructed SVM model exhibits 91% sensitivity and specificity for predicting the status of a patient's expiry based on data from previously held-out diagnostics. Finally, they look at how each feature combination and feature affects a variety of model predictions, with a focus on important laboratory value patterns to establish the infection's characteristics and its effects.

Yao et al. [12]: The COVID-19 severeness detection model was created using machine learning methods in this research. After identifying 32 parameters that were strongly linked with COVID severity, in identifying them, an SVM algorithm showed great accuracy. These 32 characteristics were also examined for Redundancy between features. The SVM model has a 0.8148 precision and was trained with 28 features. This research could reveal whether COVID patients would experience severe signs and symptoms. The underlying mechanisms of action of the 28 COVID biomarkers linked to severity could potentially be examined in COVID infections.

4 Tracking of Contacts

After a person has been diagnosed and confirmed with COVID-19, the next important step is to avoid tracing of contact. This technique, if used correctly, can break the present new coronavirus transmission chain and reduce the epidemic by increasing the chances of sufficient controls.

Generally speaking, the method identifies the sick person after a 14-day follow-up after the exposure. This technique, if used correctly, can break the transmission chain of the current new coronavirus, suppress the epidemic, and assist lower the scale of the recent pandemic by increasing the chances of proper controls. Contact tracking applications in different countries are listed in Table 1 [13, 14].

5 Vaccines and Pharmaceuticals

Since the arrival of the COVID, specialists in health and researchers have been urging for a reasonable solution to approach the development of vaccines for the COVID pandemic, and ML and AI technology had also been shown to be an exhilarating pathway to go. This section describes how ML and AI are used in vaccines and pharmaceuticals.

Beck et al. [15]: In this experiment, they employed their MT-DTI model that has been pre-trained to look for antiviral medicines that potentially alter viral components that cause COVID. Their method was built on an MT-DTI a model that has already been conditioned that comprehends interactions between drugs and their targets without domain knowledge. In fact, in previous research, among the 1794 chemical compounds recorded in the database of DrugBank, MT-DTI effectively identified EGFR also known as epidermal growth factor receptor—pharmaceuticals used in clinics, implying that three-dimensional structure knowledge of molecules or proteins is not necessary for drug-target interactions prediction.

Ke et al. [16]: A platform for artificial intelligence was developed using two separate learning datasets to identify possible anti-coronavirus activities in outdated medications. One database contained compounds proven or reported active against human immunodeficiency virus, influenza virus, COVID, and the other database

Table 1 Contact tracking applications

Application	Country	Technology used
AMAN app	Jordan	Global positioning system (GPS)
Aarogya setu	India	GPS and Bluetooth
Apturi covid	Latvia	Bluetooth
BeAware Bahrain	Bahrain	Global system for mobile communication (GSM) and Bluetooth
Beat Covid Gibraltar	Gibraltar	Bluetooth
BlueZone	Vietnam	Bluetooth
COCOA	Japan	Apple/Google and Bluetooth
COVID Alert	Canada	Bluetooth
COVID Alert SA	South Africa	Apple/Google and Bluetooth
COVIDSafe	Australia	BlueTrace protocol
CareFIJI	Fiji	BlueTrace protocol
CoronApp	Columbia	GPS
Corona Tracer BD	Bangladesh	GPS and Bluetooth
Corona-Warn-App	Germany	Apple/Google and Bluetooth
Coronalert	Belgium	Decentralized privacy-preserving proximity tracing (DP3T), Apple/Google and Bluetooth
CovTracer	Cyprus	GSM and GPS
CovidRadar	Mexica	Bluetooth
E7mi	Tunisia	Bluetooth
Ehteraz	Qatar	GSM and Bluetooth
GH Covid-19 tracker	Ghana	GPS
HOIA	Estonia	DP3T, Apple/Google and Bluetooth
HSE Covid-19 App	Ireland	Apple/Google and Bluetooth
HaMagen	Israel	Standard location API
Hayat Eve Sigar	Turkey	GSM and Bluetooth
Immuni	Italy	Apple/Google and Bluetooth
Ketju	Finland	Bluetooth and DP3T
Mask.ir	Iran	GSM
MorChana	Thailand	Location and Bluetooth
MyTrace	Malaysia	Bluetooth
NHS Covid-19 App	UK	Bluetooth
NZ COVID Tracer	New Zealand	QR codes and Bluetooth
Non-app-based	South Korea	Data from card transactions and mobile devices
PeduliLindungi	Indonesia	GSM and Bluetooth
ProteGO	Poland	Bluetooth

(continued)

Table 1 (continued)

CovidRadar	Mexica	Bluetooth
Radar COVID	Spain	DP3T
Rakning C-19	Iceland	GPS
Shlonik	Kuwait	GSM and GPS
Smittestopp	Norway	GSM and Bluetooth
Smittelstop	Denmark	Apple/Google and Bluetooth
StaySafe	Philippines	Bluetooth
StopKorona	North Macedonia	Bluetooth
Stopp Corona	Austria	Bluetooth
SwissCovid	Switzerland	Bluetooth and DP3T
Tawakkalna	Saudi Arabia	Bluetooth
TousAntiCovid	France	Bluetooth
TraceTogether	Singapore	BlueTrace protocol
ViruSafe	Bulgaria	GSM
VírusRadar	Hungary	Bluetooth
Conjunction with Alipay	Chine	Credit card transaction, GSM, and GPS
eRouška (eFacemask)	Czech Republic	BlueTrace protocol

3C-like protease known as 3CLpro inhibitors were found in the mixture [17]. All drugs based on an AI prediction were tested for activity against a feline COVID in an in vitro cell-based assay. The AI-based system received feedback from the assay results, allowing it to retrain and construct a new model based on AI to search for expired drugs. The AI system discovered 80 marketable medications with potential after a few cycles of AI prediction and learning operations.

6 Conclusion

This study examines recent studies that use advanced technology to assist researchers in a wide range of ways, addressing the difficulties and obstacles that arise when utilizing such algorithms to assist medical experts in problems of the real world. This study also includes recommendations from ML and AI-based model designers, policymakers, and medical specialists on a few faults committed in the face of the epidemic in the current situation. The application of modern technologies, such as AI and ML, greatly enhances screening, forecasting, prediction, and vaccine and drug development with extraordinary dependability, according to this review. Finally, AI and machine learning may dramatically improve COVID-19 pandemic therapy, prescription, screening and prediction, forecasting, and drug/vaccine research while reducing human engagement in medical practice. The majority of the models, but for the other part, have yet to be put to the test in real-life situations, but they are still capable of combatting the epidemic.

References

1. Ardakani, A. A., Kanafi, A. R., Acharya, U. R., Khadem, N., & Mohammadi, A. (2020). Application of deep learning technique to manage COVID-19 in routine clinical practice using CT images: Results of 10 convolutional neural networks. *Computers in Biology Medicine, 121*, 103795.
2. Ozturk, T., Talo, M., Yildirim, E. A., Baloglu, U. B., Yildirim, O., & Acharya, U. R. (2020). Automated detection of COVID-19 cases using deep neural networks with X-ray images. *Computers in Biology and Medicine, 121*, 103792.
3. Sun, L., Song, F., Shi, N., Liu, F., Li, S., Li, P., Zhang, W., et al. (2020). Combination of four clinical indicators predicts the severe/critical symptom of patients infected COVID-19. *Journal of Clinical Virology, 128*, 104431.
4. Wu, J., Zhang, P., Zhang, L., Meng, W., Li, J., Tong, C., Li, Y., et al. (2020) Rapid and accurate identification of COVID-19 infection through machine learning based on clinical available blood test results. *MedRxiv*.
5. Brinati, D., Campagner, A., Ferrari, D., Locatelli, M., Banfi, G., & Cabitza, F. (2020). Detection of COVID-19 infection from routine blood exams with machine learning: A feasibility study. *Journal of Medical Systems, 44*(8), 1–12.
6. Alakus, T. B., & Turkoglu, I. (2020). Comparison of deep learning approaches to predict COVID-19 infection. *Chaos, Solitons and Fractals, 140*, 110120.
7. Goodman-Meza, D., Rudas, A., Chiang, J. N., Adamson, P. C., Ebinger, J., Sun, N., Botting, P., et al. (2020). A machine learning algorithm to increase COVID-19 inpatient diagnostic capacity. *PLoS ONE, 15*(9), e0239474.
8. Ribeiro, M.H.D.M., da Silva, R. G., Mariani, V. C., & dos Santos Coelho, L. (2020) Short-term forecasting COVID-19 cumulative confirmed cases: Perspectives for Brazil. *Chaos, Solitons and Fractals* 135:109853
9. Yan, L., Zhang, H.-T., Goncalves, J., Xiao, Y., Wang, M., Guo, Y., Sun, C., et al. (2020). An interpretable mortality prediction model for COVID-19 patients. *Nature Machine Intelligence, 2*(5), 283–288.
10. Chimmula, V. K. R., & Zhang, L. (2020). Time series forecasting of COVID-19 transmission in Canada using LSTM networks. *Chaos, Solitons and Fractals, 135*, 109864.
11. Booth, A. L., Abels, E., & McCaffrey, P. (2021). Development of a prognostic model for mortality in COVID-19 infection using machine learning. *Modern Pathology, 34*(3), 522–531.
12. Yao, H., Zhang, N., Zhang, R., Duan, M., Xie, T., Pan, J., Peng, E., et al. (2020). Severity detection for the coronavirus disease 2019 (COVID-19) patients using a machine learning model based on the blood and urine tests. *Frontiers in Cell and Developmental Biology, 8*, 683.
13. COVID-19 Apps. Wikipedia. October 20, 2021. https://en.wikipedia.org/wiki/COVID-19_apps
14. O'Neill, P., Ryan-Mosley, T., & Johnson, B. (2020). A flood of coronavirus apps are tracking us. *Now it's Time to Keep Track of Them*.
15. Beck, B. R., Shin, B., Choi, Y., Park, S., & Kang, K. (2020). Predicting commercially available antiviral drugs that may act on the novel coronavirus (SARS-CoV-2) through a drug-target interaction deep learning model. *Computational and Structural Biotechnology Journal, 18*, 784–790.

16. Ke, Y.-Y., Peng, T.-T., Yeh, T.-K., Huang, W.-Z., Chang, S.-E., Wu, S.-H., Hung, H.-C., et al. (2020). Artificial intelligence approach fighting COVID-19 with repurposing drugs. *Biomedical Journal, 43*(4), 355–362.
17. Banerjee, A., Mukherjee, S., Panigrahi, C. R., Pati, B., & Mall, R. (2021). Analysis of COVID-19 data using consensus clustering technique. In *Computational modeling and data analysis in COVID-19 research* (pp. 17–28). CRC Press.

Design and Implementation of Wearable Cap for Parkinson's Population

M. Gokul, S. Surekha, R. Monisha, A. N. Nithya, and M. Anisha

Abstract Parkinson's disease is a progressive neurodegenerative illness that causes movement abnormalities. Because of their dread of retropulsion, many with Parkinson's disease refuse to leave their rooms and remain immobile. Injury-related permanent impairment exacerbates the issue. Deep brain stimulation is now the sole therapy option for the illness, but it is not accessible for everyone because it is more expensive, intrusive, and requires the installation of electrodes and a pacemaker. While existing methods fail to give long-term relief at a high cost, our discovery helps to slow the course of Parkinson's disease non-invasively and also provides better therapy to the majority of the senior population with motor problems at a lower cost. Our concept is to create a wearable head cap with motors and drivers that would provide mechanical stimulation in the manner of the ancient Varma medical technique. As it has Bluetooth interference it can be easily connected to android and make it work accordingly. It can give care at home, making therapy outside of hospitals more convenient. We believe that our project's originality and creativity will help us reach our aim.

Keywords Mechanical stimulus · Parkinson's · Retropulsion · Varma · Wearable

1 Introduction

Parkinson's Disease (PD) is a neurological chronic condition that results in increasing impairment causing motor and non-motor abnormalities [1, 2]. The disease has substantial repercussions on many aspects of patients' individual intellectual lives, including cognitive processing, emotions, interaction, psychiatric conditions, communications, and livelihood. Although a custodian is supplied to them. They

---

M. Gokul (✉) · S. Surekha · R. Monisha
Department of Biomedical Engineering, Kalasalingam Academy of Research and Education, Krishnankoil, Srivilliputhur, India
e-mail: mrgokul3@gmail.com

R. Monisha · A. N. Nithya · M. Anisha
Department of Biomedical Engineering, Rajalakshmi Engineering College, Thandalam, India

B. Pati et al. (eds.), *Proceedings of the 6th International Conference on Advance Computing and Intelligent Engineering*, Lecture Notes in Networks and Systems 428,
https://doi.org/10.1007/978-981-19-2225-1_8

cannot have them around every time. Geriatrics believe healthy seniors can look for themselves up to 75 years of age, but people aged 80 need some help [2–4]. Problems arise when falls, minor injuries or illnesses are present. Injury-induced permanent impairment aggravates the issue. Trouble arises when they become dependent, as they are frequently regarded as a liability [2, 4, 5]. While the state government has taken the first step by announcing that senior persons would be given priority care in its hospitals, social workers, geriatricians, and senior citizens advocate for the establishment of a geriatric ward in every hospital, private or public. It is critical to educate the family on the changes that occur in the elders who live in their house, particularly regarding illnesses linked with old age. People in the later stages of Parkinson's disease, for example, require particular treatment [2, 6].

1.1 Statistics and Analysis

Neurological diseases result in considerable morbidity, death, disability, economical losses, and a decrease in life quality. The majority of Indian epidemiological data on movement disorders comes from research of neurological diseases rather than investigations of Parkinson's disease or essential tremors [7]. People of various ethnicities and cultures are affected by Parkinson's disease. The disease affects around 10 million individuals globally, accounting for less than 1% of the overall population. The majority of Parkinson's patients are above the age of 60, while one in every ten is under the age of 50. Men are impacted slightly more than women. Movement problems made up 20% of the neurological patients in a hospital-based study. The most prevalent movement disorders were Parkinsonism (24%) and essential tremors (4.5%), with other movement disorders being less common.

1.2 Motivation and Contribution

According to ancient and modern medicine, our body is considered to be a closed electrical circuit. When we talk about ancient medicine, it is all about Ayurveda. Though there's no cure for Parkinson's, therapies with Ayurveda can assist in stopping brain cell degradation, which enhances fine and gross motor traffic. The VARMA is one such method [8]. Which integrates different combinations of traditional massages and yoga, which are all used to manipulate the body's pressure points and cure the body. The currently available stimulation techniques are electrical-based, minimally invasive, and unsafe at times [9]. As a result, we developed an invention that helps to slow the course of Parkinson's disease while simultaneously providing improved therapy non-invasively and enhancing their quality of life [10, 11]. Our concept is to create a wearable head cap with motors and drivers that would provide regulated mechanical stimulation at 3 fingers above each ear lobe, similar to the ancient Varma medical technique. A mobile application using Bluetooth may pre-set or change the pressure and time restriction. To make our product really user-friendly, a specific emphasis will be placed on informing the user in the event of an emergency [12, 13]. It can offer care at home, making treatment outside of health institutions more convenient.

2 Materials and Methods

2.1 Methods

Based on the old medical method, Varma, we chose two key stimulation sites at three fingers spacing above each ear lobe. Our concept is to create a wearable head cap equipped with motors and drivers to provide mechanical stimulation at different locations [14]. The pressure and duration limit may be fixed or adjusted wirelessly using a Bluetooth smartphone application [15]. The most important aspect of our topic is the provision of mechanical motion which is explained in (Fig. 1). Adjusting the number of steps provided by the stepper motor yields the appropriate pressure.

(a) *PIC16F877A Microcontroller*

The PIC16F877A microcontroller controls the forward and reverses motions of the stepper motor as well as the delay. PIC16F877A is the control unit of all other components in the prototype. It commands the drivers and motors to operate at the appropriate level. The PIC controller was chosen because it is simple to construct and can be operated easily.

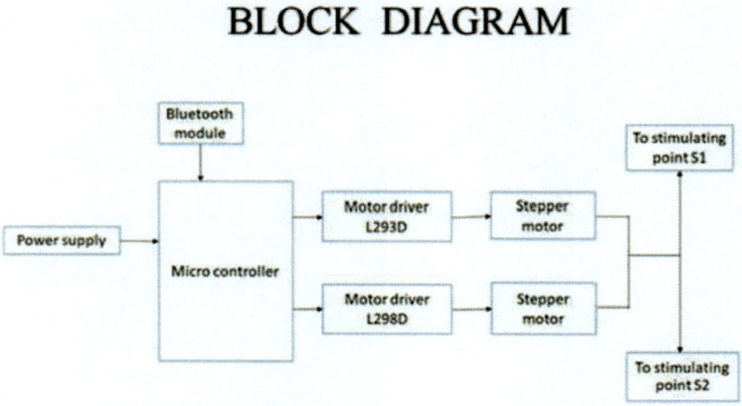

Fig. 1 Block diagram of proposed model

(b) *Driving Unit L293D*

It is a typical motor driver which allows the motor to drive in either direction. Since it works on the concept of H-Bridge, it allows the voltage to be flown in either direction, so this 16-pin IC L293 Driver can control a set of two motors simultaneously in any direction, hence it is used to controls the forward, reverse, and sliding movement of the motor [16].

(c) *DC Motor*

The exterior construction of a DC transducing engine seems at the first glance like a linear extension across the basic DCs. Figure 2a depicts the side view of the engine reveals the outside of the gear head. In Fig. 2b, a nut is positioned near the motor shaft to assemble the other elements of the assembly. An inner threaded hole on the shaft is also there to allow the motor to be connected to attachments or extensions such as wheel.

(d) *Android Application*

It addresses Parkinson's requirements by turning it into a portable wireless device. In the wireless connection between the device and the smartphone, the Android

Fig. 2 **a** Geared DC motor, **b** DC motor with rotatory shaft and nut

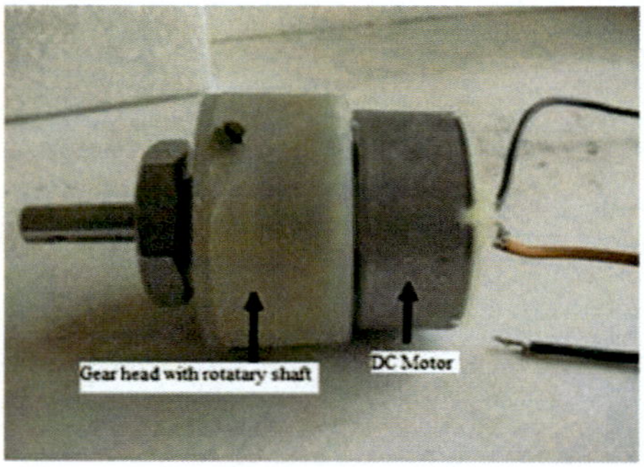

(a) Geared DC Motor

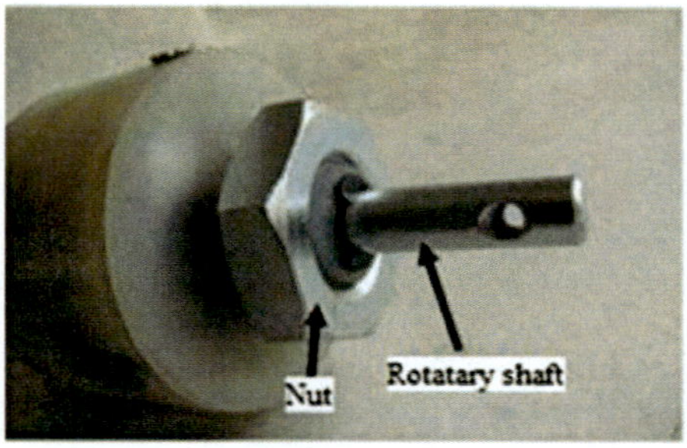

(b) DC Motor with rotatory shaft and nut

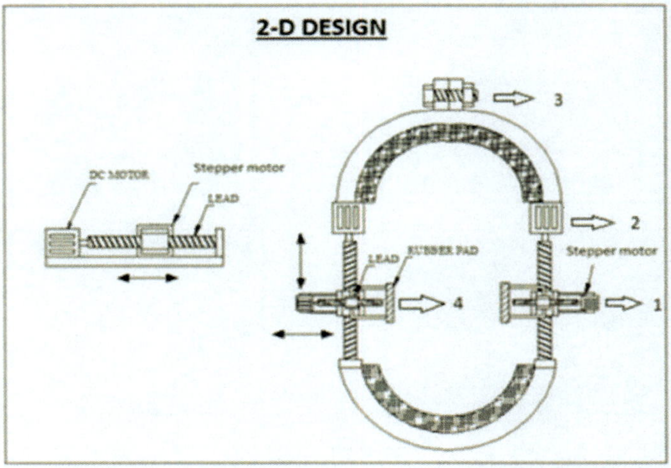

Fig. 3 2-D design

app plays a key part. Since all mobile devices have the Bluetooth option accessible regardless of brand or price, connecting the item to its cell phone will be convenient for consumers. It also offers simplified communication for improved performance [17].

2.2 2-D Design of Equipment

A product 2-D design has been developed (Fig. 3). The wearable cap stepper motor is counted as 1. Applying the proper number of rotations to the step motor can set the pressure to be applied. The progress of the motor to the stimulating point may be seen in the arrow here. The DC engine is number 2. It creates a temporal delay in the sliding movement. The adjustable screw is indicated as number 3 at the rear of the frame. By this function, the gadget may be adapted to the patient's head sizes. The rubber pad is set to number 4 in the frame. It gives an easy and user-pleasing feel to the stimulating spots. The sponge in the frame gives a comfortable feeling throughout the therapy.

2.3 Expected Outlook of Device

This will be the final view of the product (Fig. 4). It shows the product from various angles. The numbered portion as 1 is the cap where the engines and shaft are attached. In the prototype, the hat has a circumference of around 57 cm. In the side and front

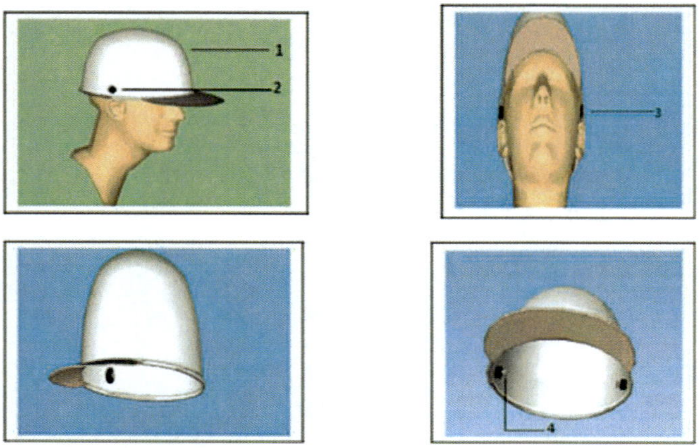

Fig. 4 Expected outlook of device

view, the section depicted in numbers 2 and 3 is motorized. The engines have around 3 finger spacious spaces above every earring. The shaft is indicated as 4 that is the silver-colored motor extensions.

Thus this chapter completely described the hardware along with interfacing of components effectively. The skeletal structure of the prototype is also portrayed distinctly. In essence, the technical part of the concept is clearly illustrated in the above text.

3 Results and Discussion

The force sensor was used to determine the pressure necessary for mechanical stimulation. The sensor was positioned roughly over the stimulation sites on the skull of an elderly Parkinson's patient (Fig. 5). Manual stimulation was provided by the Varma specialist over the stimulation sites where the sensor was inserted.

The sensor results revealed that the needed pressure for optimal stimulation might range from 1 to 4 Nm depending on the severity of the condition. The prototype was designed based on the concept and the pressure level derived from the manual stimulation is produced (Fig. 6). It cannot be tested over the patient since it is at the preliminary phase. The force sensor has therefore been tested to check the pressure range provided by the engines. The sensor was mounted on about the stimulation locations over the head of the phantom. The prototype was placed over the phantom then it was made to run. The prototype pressures were found to be somewhat different from the predicted results from sensor readings.

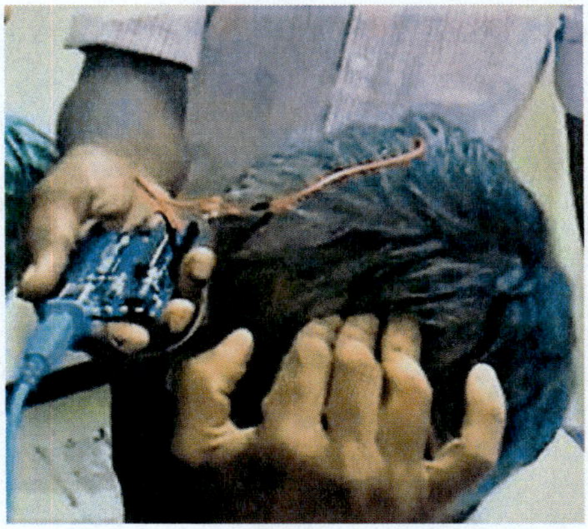

Fig. 5 Pressure value analysis

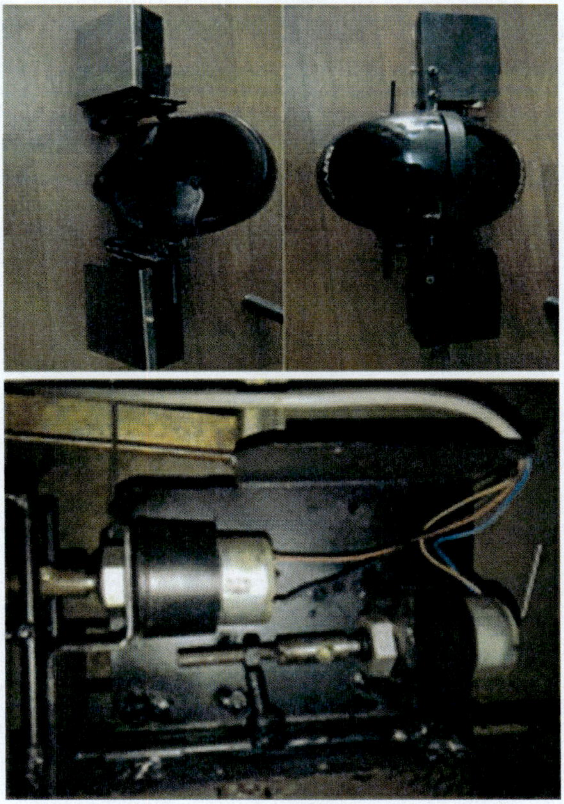

Fig. 6 Product prototype

4 Conclusion and Future Work

The Project provides an effective treatment for Parkinson's. It will non-invasively slow the course of Parkinson's disease while also providing better therapy to the majority of the senior population suffering from motor problems at a lower cost. Its distinguishing characteristic is that it addresses the limitations of existing treatments by using non-invasive mechanical stimulation rather than electrical stimulation.

Mechanical communications are more diverse than electrical and chemical interactions. Stimulation has therefore become an improved therapy process, notably for mechanical stimulation in treating neurological diseases, because the currently available stimulation techniques are electric, less intrusive, and insecure at times. Another noteworthy result is that when the stimulus is mechanical, the cell pace is longer, which indicates that mechanical communication leads to cell-induced long-term changes. The user will be notified in case of an emergency in a particular way, to make our thing a completely user-pleasant product.

References

1. Gokul, M., Durgadevi, N., Bhuvaneshwari, R., Ramya, V., & Kumar, C. R. (2018). Rehabilitation tool for gastroparesis by the analysis of interstitial cells of Cajal (The external gastric pacmaker with a feedback of gastric potential). *Journal of Gastrointestinal and Digestive System. 8*. https://doi.org/10.4172/2161-069X.1000557
2. De Lau, L. M., & Breteler, M. M. (2006). Epidemiology of Parkinson's disease. *The Lancet Neurology, 5*(6), 525–535.
3. Russmann, H., Ghika, J., Villemure, J. G., Robert, B., Bogousslavsky, J., Burkhard, P. R., & Vingerhoets, F. J. (2004). Subthalamic nucleus deep brain stimulation in Parkinson disease patients over age 70 years. *Neurology, 63*(10), 1952–1954.
4. Pils, K. (2016). Aspects of physical medicine and rehabilitation in geriatrics. *Wiener Medizinische Wochenschrift, 166*(1–2), 44–47.
5. Colombo, M., Guaita, A., Cottino, M., Previdere, G., Ferrari, D., & Vitali, S. (2004). The impact of cognitive impairment on the rehabilitation process in geriatrics. *Archives of Gerontology and Geriatrics, 38*, 85–92.
6. Opara, J. A., Brola, W., Leonardi, M., & Błaszczyk, B. (2012). Quality of life in Parkinsons disease. *Journal of Medicine and Life, 5*(4), 375.
7. Pelicioni, P. H., Lord, S. R., Okubo, Y., Sturnieks, D. L., & Menant, J. C. (2020). People with Parkinson's disease exhibit reduced cognitive and motor cortical activity when undertaking complex stepping tasks requiring inhibitory control. *Neurorehabilitation and Neural Repair, 34*(12), 1088–1098.
8. Senthilvel, G., & Amuthan, A. (2019). Application of Varmam (Physical manipulation therapy of traditional siddha medicine) for contemporary health issues: An update. *Journal of Ayurveda Medical Sciences, 4*(1).
9. O'Callaghan, C., & Lewis, S. J. (2017). Cognition in Parkinson's disease. *International Review of Neurobiology, 133*, 557–583.
10. Benabid, A. L. (2003). Deep brain stimulation for Parkinson's disease. *Current Opinion in Neurobiology, 13*(6), 696–706.
11. Abbruzzese, G., Marchese, R., Avanzino, L., & Pelosin, E. (2016). Rehabilitation for Parkinson's disease: Current outlook and future challenges. *Parkinsonism and Related Disorders, 22*, S60–S64.

12. Mazilu, S., Blanke, U., Hardegger, M., Tröster, G., Gazit, E., Dorfman, M., & Hausdorff, J. M. (March 2014). GaitAssist: A wearable assistant for gait training and rehabilitation in Parkinson's disease. In *2014 IEEE international conference on pervasive computing and communication workshops (PERCOM WORKSHOPS)* (pp. 135–137). IEEE.
13. Tous, F., Ferriol, P., Alcalde, M. A., Melià, M., Milosevic, B., Hardegger, M., & Roggen, D. (2014). Incorporating the rehabilitation of Parkinson's disease in the play for health platform using a body area network. In *XIII mediterranean conference on medical and biological engineering and computing 2013* (pp. 1481–1484). Springer.
14. Nieuwboer, A., Rochester, L., Müncks, L., & Swinnen, S. P. (2009). Motor learning in Parkinson's disease: Limitations and potential for rehabilitation. *Parkinsonism and Related Disorders, 15*, S53–S58.
15. Barry, G., Galna, B., & Rochester, L. (2014). The role of exergaming in Parkinson's disease rehabilitation: A systematic review of the evidence. *Journal of Neuroengineering and Rehabilitation, 11*(1), 1–10.
16. Delrobaei, M., Baktash, N., Gilmore, G., McIsaac, K., & Jog, M. (2017). Using wearable technology to generate objective Parkinson's disease dyskinesia severity score: Possibilities for home monitoring. *IEEE Transactions on Neural Systems and Rehabilitation Engineering, 25*(10), 1853–1863.
17. Rodgers, M. M., Pai, V. M., & Conroy, R. S. (2014). Recent advances in wearable sensors for health monitoring. *IEEE Sensors Journal, 15*(6), 3119–3126.

RETRACTED CHAPTER: A Survey on Applications of Machine Learning Algorithms in Health care

Deep Rahul Shah, Samit Nikesh Shah, and Seema Shah

Abstract As one of the main early adopters of innovative advances, the medical care industry has delighted in much accomplishment accordingly. In various wellbeing-related fields like new operations, patient information the executives, and ongoing infection treatment, artificial insight subset machine learning is assuming a key part. Inside the medical services industry, machine learning is gradually acquiring a foothold. An assortment of medical services circumstances is now being affected by machine learning (ML). With machine learning (ML) applied to the medical services industry, a great many different datasets can be investigated to make expectations about results, just as given opportune danger scores and exact asset designation. This exploration prompted the making of a more effective choice organization for clinical applications.

Keywords Machine learning · Artificial intelligence · Disease prediction · Health care

1 Introduction

Lately, machine learning has turned into a significant pattern in the business. There are various fields inside the field of machine learning, including measurements, polynomial math, information handling, and information examination, that make it hard to think of another definition [1]. There is no deficiency of uses for machine learning,

---

The original version of this chapter was retracted: The retraction note to this chapter is available at https://doi.org/10.1007/978-981-19-2225-1_61

D. Rahul Shah (✉) · S. Nikesh Shah · S. Shah
Department of Computer Engineering, NMIMS Mukesh Patel School of Technology Management and Engineering, Mumbai, Maharashtra, India
e-mail: deepshah3110@gmail.com

S. Shah
e-mail: seema.shah@nmims.edu

B. Pati et al. (eds.), *Proceedings of the 6th International Conference on Advance Computing and Intelligent Engineering*, Lecture Notes in Networks and Systems 428,
https://doi.org/10.1007/978-981-19-2225-1_9

and it is turning out to be increasingly normal. Various ventures like money, medication, and security depend on it. The appearance of advanced innovation in the medical care region has been set apart by continuous obstacles in both application and reasonableness. The mix of different medical services frameworks has been slow, and most nations of the world presently cannot seem to accept a totally coordinated medical care framework. The intrinsic person and intricacy of human science, just as the fluctuation between individual patients, have reliably exhibited the need for the human factor in illness conclusion and treatment. Notwithstanding, upgrades in advanced innovations are indeed becoming indispensable devices for medical care laborers in giving the best therapy to patients. Utilizing machine learning, clinical information sources can be investigated to discover designs that can assist with expecting illness. Computerizing clinic regulatory cycles, planning irresistible illnesses, and customizing clinical medicines are completely made conceivable by machine learning today. A huge number of individuals depend on medical care benefits that depend on esteem. This is one of many nations' top-income workers. There is a lot of interest in the medical care area to offer quality therapy and medical care administrations on account of the world's steadily developing populace development. Medical care administrations, wearables, and applications that assist individuals with living longer; better lives are sought after presently like never before. The progression of information advances, for example, stockpiling size, preparing force, and information transport speeds have empowered the expansive utilization of machine learning in an assortment of disciplines, including medical services. Since giving ideal medical care to an individual is complex, late clinical improvements have underlined the need for a tweaked medication or "accuracy medication" way to deal with medical care.

1.1 Motivation and Contribution

The motivation behind the customized medication is to utilize gigantic measures of medical care information to reveal, expect, and dissect indicative choices that clinicians may then apply to every individual patient. Machine learning applications underway incorporate an indicative device for diabetic retinopathy and prescient investigation to estimate bosom malignant growth repeat utilizing clinical data and photographs. Various machine learning methods and their applications utilized in healthcare decision assistance are discussed in this study. As a result of this research, a more efficient decision support system for medical applications may now be built more efficiently.

2 Related Work

Zhang et al. [2] used SVM with RBF piece work which relies upon the measurable learning hypothesis for the visualization of heart illness. To pick the magnificent boundaries of part capacity and fine fundamental component, the grid search technique for making upgrades to rules is used, to achieve likely the most raised order exactness.

Shariati et al. [3] used fuzzy NN with SVM and ANN to find and guess hepatitis and thyroid ailments. Further, examination of illness, they recognized the assortment and the phase of infirmity which consolidate six classes for hepatitis infirmity, for example, Hep B (two stages) Hep C (two phases), hepatitis and non-hepatitis, and for thyroid illness, five classes were named, explicitly: hypothyroid, thyrotoxicosis, subclinical hypothyroid, subclinical thyrotoxicosis; furthermore, non-presence of thyroid. For hepatitis illness, the phenomenal correctness limits to 98% and for thyroid illness to close to 100%.

Vassis et al. [4] made a total outline concerning the use of neural organizations in the computerized clinical forecast, with an assigned priority in support vector machines (SVMs), which are precise assortments of neural capacities. Over the assessment, in proliferating cases, side effects and afflictions may likewise be all around anticipated via neural programs, while SVMs are bit-by-bit utilized in the clinical forecast due to their unique order components.

Elshazly et al. [5] proposed a genetic calculation settled help vector machine classifier for lymph illnesses investigation. In the premier stage, the components of the lymph illnesses dataset have 18 provisions, and they are reduced to six viewpoints by method for using GA. A help vector machine with an excess of a couple of portion capacities including direct, quadratic, and Gaussian was utilized as a classifier in the second stage. The SVM classifier with each portion work is utilized to survey the exhibition by utilizing proficiency records like exactness, affectability, particularity, AUC/ROC, Matthew's correlation coefficient, and F-measure. Straight bit work purchased a most further developed impact which approves the skill of GA-direct framework.

Saiti et al. [6] suggested SVM and probabilistic NN for the characterization of two thyroid sicknesses: hypothyroidism and hyperthyroidism from the thyroid issue information base. These calculations depend routinely reasonable on successful arrangement calculations to deal with unnecessary and irrelevant viewpoints. Hereditary algorithm tried a helpful, furthermore, solid structure for choosing reasonable subsets of viewpoints that outcome in braces forecast rates. Elsayad et al. [7] assessed the affectivity of the Bayesian classifier in diagnosing the risk of cardiovascular sickness. Two Bayesian network classifiers: tree augmented Naïve Bayes and the Markov cover estimation are executed, and their forecast sureness is the reference guides in real life toward the support vector machine. The exploratory results display that Bayesian organizations with MBE have the characterization accuracy of 97%, insofar as TAN and SVM units have 88 and 71 rates.

Hongzong et al. [8] excited about the use of SVM on the coronary heart infirmity and non-coronary heart problem characterization. Direct discriminant assessment and SVM with an outspread premise work portion are analyzed. The expectation precisions of preparing and assessment units of SVM were 96% and 78% in this way, and for LDA it was 90 also, 72 rates in this way. The cross-approved accuracy of SVM and LDA was 92 and 85 rates. Çomak et al. [9] introduced a DSS that orders the Doppler signs of a heart valve to two examples (conventional and unusual) to help the cardiologist. LS-SVM and ANN with back propagation are executed to group the somewhat long components. Further, the ROC bend is used to investigate the sensitivities and specificities of those classifiers and gauge the AUC. Eventually, two classifiers are assessed in all aspects.

Sartakhti et al. [10] proposed an inventive machine learning framework that injects support vector machine and reenacted toughening for the examination of the hepatitis issue. The precision of the arrangement is gotten by utilizing ten-overlay cross-approval. The surveyed order accuracy of the proposed interaction was 96.25 rates. Studying every one of the top algorithms in one analysis is challenging. Considering this, I have decided to distribute an overview section in which each of the top algorithms have been assembled for faster examination and study.

3 Machine Learning Algorithms

Patients' "traits" and medical outcomes of interest are some of the most common features extracted by ML algorithms. When it comes to classifying things, the logistic regression algorithm has dominated AI in health care for quite some time, was easy to use and finish, and straightforward to understand, and would definitely recommend. The situation has altered in recent years, with SVM and neural networks taking the lead. A few of the leading algorithms are explained below (Fig. 1).

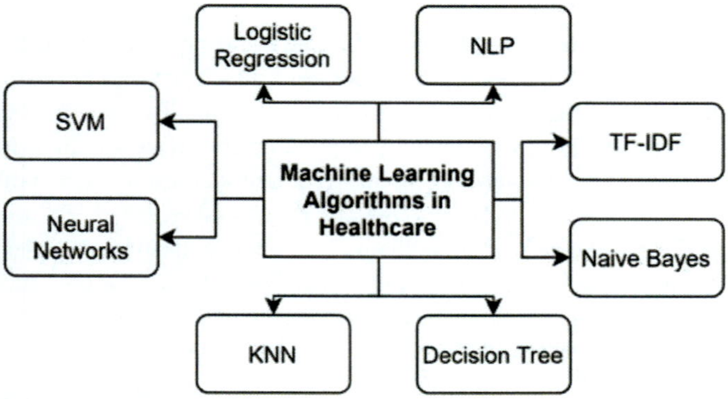

Fig. 1 Various machine learning algorithms in health care

3.1 *Support Vector Machine*

The main use of SVM is in classification tasks that require dividing a dataset into separate classes through a hyperplane. The best way to classify new data is to choose a hyperplane with the maximum possible margin, or distance, between the hyperplane and the points in the training set. The hyperplane's position would be affected if these points were removed. SVM is an optimization problem, so the solution is always global. SVMs are extensively employed in clinical research, for example, to identify imaging biomarkers, to diagnose cancer or neurological diseases, and in general to classify data from imbalanced datasets and datasets with missing values. In machine learning (ML) tasks, support vector machines (SVM) are commonly used. Each training sample in this technique is divided into several categories. Support vector machines (SVM) are mainly used for classification and regression [11].

3.2 *Neural Networks*

In neural networks, the associations between output and input variables are represented by hidden layer combinations of predefined functions. In order to achieve the most accurate weight estimation, input and outcome data must be combined in such a way as to minimize the average difference between predictions and the actual outcome. A textbook example of neural networks being applied to health care is the detection of Breast cancer from mammographic images. But neural networks have also been used successfully in diagnostic systems, biochemical analysis, and image analysis.

3.3 *Logistic Regression*

Logistic regression is one of the most widely used multivariable algorithms for modeling dichotomous outcomes. The odds ratio is calculated using logistic regression for several explanatory variables. The response variable is a binomial response variable, similar to multiple linear regression. In the graph, the effect of each variable on the odds ratio of an observed event is represented. In comparison with linear regression, it does not have confounding effects with all variables analyzed jointly. As a tool for disease risk assessment and enhancing medical decision-making, logistic regression can be used to solve classification problems and estimate the likelihood of an event occurring in healthcare settings.

3.4 Natural Language Processing

Unstructured and unintelligible narrative prose makes up a major amount of clinical information in health care. On the basis of historical databases, natural language processing (NLP) finds disease-relevant keywords in medical records, which are then entered and enhanced in structured data to improve clinical decision-making. NLPs use in health care has grown as its recognized promise for analyzing and interpreting large patient datasets. NLP technology services, including medical algorithms, machine learning in health care, and NLP technology can now provide insight into unstructured data, providing insight into the understanding quality and refining techniques, as well as better results for patients.

3.5 Tf-Idf

The TF-IDF keyword extraction technique uses the inverse frequency of terms and the document frequency. In terms of statistical significance, this reveals how often a term appears in a corpus of documents. As the frequency of a term increases in the corpus, its relevance increases, but it is offset by its increasing frequency in a document. By using the TF-IDF algorithm, healthcare professionals can detect similar patients in observational studies, study medical reports for disease correlations, or search databases to find patterns.

3.6 Naïve Bayes

The Naïve Bayesian classifier is used to classify documents in text classification, a challenge in categorizing documents. In Bayesian classification, a given feature is not related to other features in the class. In fact, this is not the case. Even though they are interconnected, they each contribute to their likelihood of being assigned to each group independently. With one of the most effective and efficient classification algorithms, it has been successfully applied to a wide range of medical challenges, including the classification of medical reports and journal articles. Statistical classifiers are the case of Bayesian classifiers. Naïve Bayes identify probabilities of class membership based on a given class label [12]. It does a data sweep, and therefore, the classification is easy.

3.7 *Decision Tree*

There are other supervised learning algorithms, such as the decision tree algorithm (DTA). The decision tree algorithm, unlike other supervised learning algorithms, may also be used to solve regression and classification issues, unlike other supervised learning techniques. Learning simple decision rules from prior data allows for the usage of decision trees to develop a model that can be used to predict the class or value of the target variable (training data). An internal node plus a leaf node with a class designation form a decision tree (DT). The root nodes are the nodes at the top of the decision tree. The decision tree is popular because it is easy to design and does not require any parameters [13].

3.8 *K-nearest Neighbor*

One of the simplest machine learning algorithms, K-nearest neighbor employs the supervised learning technique to find the neighbor closest to you. For example, the ANN method assumes a similarity between a new case/data and the existing cases and places them in a category that is more similar to the existing cases. Using the K-NN method, all the available data is stored, and a new data point is classified based on its resemblance to previous data points. As a result, new data can be simply categorized using the K-NN method. We often utilize K-nearest neighbor to classify samples. With this technique, we might also additionally calculate distance from N schooling samples with the aid of using making use of a distance measure [14].

4 Applications of Machine Learning in Health care

Making machines more proficient and solid is one of the objectives of machine learning. In case you are hoping to find out with regards to machine learning in medical services, your best asset is your PCP. Up to a patient is alive, they will consistently require human touch and care. This cannot be supplanted by machine learning or some other innovation. The help can be better given by a mechanized machine. The best ten employments of machine learning in medical care are introduced in the accompanying passages below.

4.1 *Heart Disease Diagnosis*

As one of our body's most imperative organs, the guts might be an indispensable part. Cardiovascular problems including arteria coronaria infection (CAD), coronary heart condition (CHD), then, at that point, are normal. Because of this, numerous scientists are performing machine learning calculations to analyze heart infections and Naive Bayes are tests of administered machine learning calculations that analysts are that work in for heart conditions distinguishing prediction. A machine learning procedure created by Parthiban and Srivatsa [15] utilizes the Naive Bayes calculation and support vector machine to distinguish and examine heart sickness. The Naïve Bayes technique yields 74% precision, though SVM offers 94.60% exactness. To estimate coronary illness, Otoom has utilized support vector machine and Naïve Bayes [16]. As far as exactness or accuracy, SVM conveys 88%, while Naïve Bayes gives 84%.

4.2 Predicting Diabetes

Diabetes is a typical and destructive infection. As well as causing other genuine diseases and passing, this condition is one of the most well-known reasons for other genuine ailments and mortality. This infection can make harm our kidneys, heart, and nerves, among different parts of our bodies. An early identification of diabetes is the objective of utilizing machine learning in this field. To build a diabetes forecast framework, random forest, K-NN, decision tree, or Naive Bayes can be utilized. Guileless Bayes beats different calculations with regards to precision among these calculations due to its exceptional presentation and diminished preparing time. With the utilization of Naive Bayes and decision trees, Iyer fostered a machine learning framework to foresee diabetic inconveniences. Both Bayesian surmising and choice tree induction are profoundly precise at 79.56% [17]. Utilizing machine learning techniques, Dash and Sen had the option to analyze diabetes. Utilization of 77.479% accuracy Logitboost and CART algorithms [18].

4.3 Disclosure of Breast Cancer

Breast cancer chances in the USA were distinguished utilizing the j48 Naive Bayes model by Williams et al. WEKA is utilized to lead the trial. As per their discoveries, j48 has a 94.2% precision rate and Naive Bayes has an 82.6% exactness rate [19]. In Breast disease prediction, there are a few distinctive grouping models that Senturk et al. used, for example, support vector machines, credulous Bayes, K-closest neighbor, and choice trees (DT). K-NN has an exactness of 95.15%, while SVM has a precision of 96.40% [20].

4.4 *Prediction of Liver Disease*

The liver is our body's second most significant inner organ. Metabolic cycles depend on it for energy. On the off chance that you have Cirrhosis, Chronic Hepatitis, or Liver cancer you can treat them. Hepatitis has been anticipated with a lot of accomplishment by utilizing machine learning and information mining strategies that were grown as of late. Utilizing a lot of clinical information to anticipate sickness is a troublesome errand. Subsequently, scholastics are buckling down on machine learning standards like order, bunching, and a lot more to resolve these issues. One way of utilizing ILPD is to make a liver illness expectation framework.

4.5 *Personalized Treatment*

Machine learning for individualized treatment is a fervently discussed subject in the exploration local area. Individual wellbeing information and prescient examination will be utilized to further develop benefits around here. A customized treatment framework dependent on patients' indications and hereditary data is created utilizing machine learning computational and factual techniques. Administered machine learning algorithm is used to develop the personalization framework. Patients' clinical records were utilized in the advancement of this framework. An illustration of individualized treatment is SkinVision. On his/her telephone, an individual can check for skin malignant growth by using the skin cancer checker application. Because of custom-made treatment, the expense of medical services can be diminished.

4.6 *Robotic Surgery*

Machine learning applications in medical care incorporate mechanical medical procedures. Computerized stitching, careful expertise assessment, and improvement of automated medical procedure materials are part of the class. Neurosurgery is another field where robots cannot perform successfully. The Raven Surgical Robot (Raven) is being developed at the University of California, San Diego. Neurosurgery is another field where robots cannot perform successfully. Hand-worked a medical procedure is tedious, and it does not give programmed input. Machine learning methods can be utilized to accelerate the cycle.

4.7 *Machine Learning in Radiology*

Machine learning and computerized reasoning have been the focal point of late examination in radiology. Utilizing machine learning procedures, Aidoc's product assists radiologists with accelerating the identification process. A clinical picture should be examined so irregularities can be recognized all through the body. Machine learning calculations that are managed are most usually used. Machine learning methods are used for clinical picture division, which is a complicated assignment. Designs in a picture are distinguished by utilizing segmentation. Images are typically separated into fragments utilizing a system called chart cut. Text reports in radiography are broken down utilizing natural language processing (NLP). Thus, the utilization of machine learning in radiology can work on understanding consideration.

4.8 Clinical Trial and Research

In a clinical preliminary, questions are approached to decide the viability and security of a specific organic or pharmacological item. In this preliminary, the analysts are zeroing in on creating novel drugs. Clinical preliminaries are costly and tedious. Machine learning has an enormous effect in this industry. Ongoing observing and strong assistance are conceivable with a machine learning (ML)-based framework clinical preliminaries, and examination utilizing machine learning strategies enjoy the benefit of being distantly managed. Besides that, machine learning establishes a safe clinical climate for patients to work in. Directed machine learning can work on the proficiency of clinical preliminaries.

5 Conclusion and Future Work

As of today, there are various machine learning applications in the real world that are not widely acclaimed because of their hackneyed nature (essentially in the information science local area). Each persistent in turn may not profit from this innovation, yet it is now working on the existences of others. When the foundation is set up, we will see extra clinical preliminaries joining machine learning methods in the coming months. At present, machine learning is a piece of our regular routines. Climate determining, promoting applications, deals expectation, and numerous different fields depend on this strategy. Because of clinical multifaceted design and an absence of information, the utilization of machine learning in medical services is as yet restricted. In case you are keen on learning more with regards to machine learning, this article is an extraordinary spot to begin. In the future, we will investigate utilizing a few more machine algorithms such as CNN, ANN, and deep learning to overview different illnesses and their precision.

References

1. Mitchell, T. (1997). *Machine learning*, (p. 2). McGraw Hill.
2. Zhang, Y., Liu, F., Zhao, Z., Li, D., Zhou, X., & Wang, J. (2012). Studies on application of support vector machine in diagnose of coronary heart disease. *Sixth international conference*

on electromagnetic field problems and applications (ICEF), (pp 1–4).
3. Shariati, S., & Haghighi, M. M. (2010). Comparison of ANFIS neural network with several other ANNs and support vector machine for diagnosing hepatitis and thyroid diseases. *International conference on computer information systems and industrial management applications (CISIM)*, (pp. 596–599).
4. Vassis, D., Kampouraki, B. A., Belsis, P., Zafeiris, V., Vassilas, N., Galiotou, E., Karanikolas, N. N., Fragos, K., Kaburlasos, V. G., Papadakis, S. E., Tsoukalas, V., & Skourlas, C. (2015). Using neural networks and SVMs for automatic medical diagnosis: A comprehensive review. International conference on integrated information (ICININFO 2014) AIP conference proceedings, (vol. 1644, pp. 32–36). https://doi.org/10.1063/1.4907814
5. Elshazly, H. I., Elkorany, A. M., & Hassanien, A. E. (2014). Lymph diseases diagnosis approach based on support vector machines with different kernel functions. 9th *International conference on computer engineering and systems (ICCES)*, (pp. 198–203).
6. Saiti, F., Naini, A. A., Shoorehdeli, M. A., & Teshnehlab, M. (11–13 June 2009). Thyroid disease diagnosis based on genetic algorithms using PNN and SVM. *The international bioinformatics and biomedical engineering (ICBBE)*, (pp. 1–4). Beijing
7. Elsayad, A., & Fakhr, M. (2015). Diagnosis of cardiovascular diseases with bayesian classifiers. *Journal of Computer Sciences, 11*(2), 274–282. https://doi.org/10.3844/jcssp.2015.274.282
8. Hongzong, S., Tao, W., Xiaojun, Y., Huanxiang, L., Zhide, H., Mancang, L., & BoTao, F. (2007). Support vector machines classification for discriminating coronary heart disease patients from non-coronary heart disease. *West Indian Medical Journal, 56*(5), 451–457. PMID: 18303759.
9. Çomak, E., Arslan, A., & Türkoğlu, I. (2007). A decision support system based on support vector machines for diagnosis of the heart valve diseases. *Computers in Biology and Medicine, 37*:21–27.
10. Sartakhti, J. S., Zangooei, M. H., & Mozafari, K. (2012). Hepatitis disease diagnosis using a novel hybrid method based on support vector machine and simulated annealing (SVM-SA). *Computer Methods and Programs in Biomedicine, 108*, 570–579.
11. Moein, M., Davarpanah, M., Montazeri, M. A., & Ataei, M. (2010). Classifying ear disorders using support vector machines. In *2010 second international conference on computational intelligence and natural computing*, (vol. 1, pp. 321–324). IEEE.
12. Animesh, et al. (2016). *Study and analysis of breast cancer cell detection using Naïve Bayes, SVM and ensemble algorithms. International Journal of Computer Applications, 2.*
13. Jalpaiguri, et al. (2017). Heart disease diagnosis and prediction using machine learning and data mining techniques: A review. *Advances in Computational Sciences and Technology, 10*, 2137–2159.
14. Bishop, C. M. (1995). *Neural networks for pattern recognition.* Oxford University.
15. Parthiban, G., & Srivatsa, S. K. (2012). Applying machine learning methods in diagnosing heart disease for diabetic patients. *International Journal of Applied Information Systems, 3*, 25–30.
16. Otoom, A. F., Abdallah, E. E., Kilani, Y., & Kefaye, A. (2015). Effective diagnosis and monitoring of heart disease. *International Journal of Software Engineering and Its Applications, 9*, 143–156.
17. Iyer, A., Jeyalatha, S., & Sumbaly, R. (2015). Diagnosis of diabetes using classification mining techniques. *International Journal of Data Mining and Knowledge Management Process (IJDKP), 5*, 1–14.
18. Sen, S. K., & Dash, S. (2014). Application of meta learning algorithms for the prediction of diabetes disease. *International Journal of Advance Research in Computer Science and Management Studies, 2*, 396–401.
19. Williams, et al. (2015). Breast cancer risk prediction using data mining classification techniques. *Transactions on Networks and Communicatons, 2*, 1–11.
20. Senturk, et al. (2014). Breast cancer diagnosis via data mining: Performance analysis of seven different algorithms. *Computer Science and Engineering, 1*, 1–10.

Anomaly Detection of Unstructured Logs Generated from Complex Micro-service Based Architecture Using One-Class SVM

Anukampa Behera, Chhabi Rani Panigrahi, Sitesh Behera, and Bibudhendu Pati

Abstract With the increase of micro-service-based architecture, detection of anomalies in enormous, complex production infrastructure has become complicated. In this work, authors provide a solution for finding anomalies using the structured event objects in a complex micro-service-based productions environment, where the flow of data is assumed to be periodic, deterministic and predictable in nature. As these objects are multivariate and multidimensional in nature, the number of features and dimensions is reduced without losing the quality of data. Next, a method to find anomalies has been proposed with the obtained dataset. The proposed method uses an unsupervised anomaly detection model using Tax–Duin approach for one-class support vector machine (SVM) to classify outliers. The basic assumption of this model includes classifying the first occurrence of any event as anomaly by default. The experimental results obtained indicate an accuracy of 88% by applying one-class SVM on the considered dataset.

Keywords Micro-service · Unstructured logs · Unsupervised · Tax–Duin one-class SVM · Hyper-sphere · Multidimensional anomaly detection

A. Behera
ITER, S'O'A (Deemed to Be) University, Bhubaneswar, India
e-mail: anukampa1@gmail.com

C. R. Panigrahi (✉) · B. Pati
Department of Computer Science, Rama Devi Women's University, Bhubaneswar, India
e-mail: panigrahichhabi@gmail.com

B. Pati
e-mail: patibibudhendu@gmail.com

S. Behera
Plivo.Com, Bangalore, India
e-mail: sitesh.citzen@gmail.com

B. Pati et al. (eds.), *Proceedings of the 6th International Conference on Advance Computing and Intelligent Engineering*, Lecture Notes in Networks and Systems 428,
https://doi.org/10.1007/978-981-19-2225-1_10

1 Introduction

As we move towards deployments of complicated and complex large-scale micro-service architecture, the kind of data generated from all these systems is huge. In the service cluster, generally the data can be generated through user access, invoked service or from host-based intrusion detection system (HIDS) which particularly generates huge amount of data [1]. The access-related data includes the user details of who logged in, pseudo users, etc. and similarly service-related data is the details of the services those are getting restarted, or services leading to memory leaks, memory killer services, etc. HIDS [1] records the events such as when a file is manipulated, any changes are made to the file system, any root change is noticed, intrusion found, any antivirus information or network communication getting failed, etc.

Now, in such a scenario, it becomes very difficult to identify if anything goes wrong in the underlying system, i.e. it might have been affected by various attacks or there might be malfunctioning or sometimes it might be some genuine changes that has been incorporated [2]. These problems include change in network access control list (ACL), i.e. it might have been manipulated which may stop or allow communication between two systems. Similarly, a service may be deployed to be configured wrongly, it may keep restarting, there may be a service with memory leak or abusing the system and an unauthorized attempt to access the system. By carefully analysing these various use cases, one can see the scope of attacks start with operating system layer, covers network, deployed applications and can go till security [3].

Thus, these logs need to be analysed. The events between services must be related, so that the event can be traced to find out any anomaly available to be looked upon. In order to ensure safety, the huge data generated by these systems need to be monitored. The logs, system metrics those are used for this purpose, are mixed structured and unstructured data. Unstructured data includes logs, whose sources are auth logs, sys logs and HIDS generated logs that need monitoring. Then, these unstructured data need to be converted to structured data so that these can be used by any data analytics model.

Behera et al., proposed an approach [4] to collect the output generated by HIDS server to get a structured dataset for analysis. Authors have also proposed methods to reduce the high-dimensional data obtained to 3-dimensional dataset for further analysis without compromising the quality of data.

In this work, we have proposed an anomaly detection model based on one-class support vector machine (one-class SVM) to detect any deviation observed in the system logs. The remaining part of the paper is organized as follows. The related works in the field of anomaly detection on unstructured logs are briefly discussed in Sect. 2. Some prerequisites for our work and a brief insight on the working principle of one-class SVM are given in Sect. 3. In Sect. 4, our proposed model is described followed by the experimental set-up and result in Sect. 5. Sect. 6 concludes the paper along with the highlights of scope for future work.

2 Related Work

In recent years with the adaptation of micro-service-based architectures, anomaly detection for high-dimensional data has become a rudimentary need of the industry. Thus, it has grabbed the attention of researchers all over the world in the last decade. However, very few of the researchers are found to dually focus on both the area of anomaly detection as well as high-dimensional data handling [5].

For the reduction in time needed for log analysis, Breiber et al. used Hadoop technology [6]. The anomaly detection by the authors was done by the creation of anomaly profile based on the analysis of several log sources.

Recurrent neural network (RNN)-based network language model was used by Tuor et al. for the purpose of cyber anomaly detection [7]. A similarity check was run for the current events with those of the events occurred previously and any deviation found was reported [8].

Natural language processing (NLP) was found to be the most adapted technology for anomaly detection recently. Influenced by NLP, a deep neural network-based engine named as DeepLog was proposed by Du et al.[9]. It used long short-term memory (LSTM) to establish a pattern as well as syntax in the received log-sequences and raised an alarm over any deviations incurred.

Auto-LSTM, Auto-BLSTM and Auto-GRU-based models were used for the detection and classification of anomalies by Farzad et al.[10]. Processed data obtained from standard repositories was used for testing the models. Deep learning framework with LSTM as a key method has been used in some anomaly detection models proposed by various researchers [11–13].

All the above discussed models in the literature acted on the log messages directly by using some pre-processing methods such as log parsing, FT-Tree. To search the keywords, methods like Template2Vec, Counting Word Frequency, Template2Vec, FastText and TF-IDF were used.

3 Prerequisites

In this section, the background of the problem and one-class SVM mechanism is presented.

3.1 Background

Model is valid for production environment where activities are periodic in nature. As for the detection of anomaly, an environment is needed where activities are fixed in nature so that any deviation can be detected.

For tracking an activity, a footprint along with a particular event of the activity is required. For example, 'Source 'A' makes SSH on PORT 22 on example.com' is treated as one activity. Again the frequency of activity is single which is periodic in nature, i.e. say the event occurs every day at 9 p.m. once. Now it is found that on a day if such events occur eight times at different hours of a particular day, then it is treated as an anomaly.

So when the frequency of activities exceeds a set threshold value, then it is raised as an anomaly. At the same time suppose there is first time occurrence of an activity, then even though the model will not be able to show it as an outlier but it can be viewed as a single instance, hence marked as anomaly too.

3.2 One-Class SVM

The idea and working mechanism of one-class SVM is described in this section.

The property that thrives SVM from other algorithms is its capability to create a nonlinear decision boundary to a higher dimensional space using a nonlinear function ϕ. So in cases where the data points originally do not exist in the same plane, clustering cannot be applied on them. Using SVM, these data points can be elevated to a feature space F where these can be separated by a hyper plane to distinctly categorize them into different classes. It helps to classify the data into two different classes with the help of a nonlinear decision function. But if the requirement is just to label for a single class, then one-class SVM is used.

One-class SVM is categorized under the group of unsupervised algorithms. For the purpose of novelty detection, the model learns a decision function so that the input unlabelled data can be classified as similar or different in comparison with the dataset on which the model is trained. For a dataset having severely skewed distribution of class, one-class SVM is ideal. It is best suited to be used for the imbalanced classification datasets, where there are a few or no instances available for the minority class or in case of absence of coherent structure in dataset towards division of classes.

There are basically two types of approaches used for one-class SVM: one-class SVM according to Schölkopf [14] and one-class SVM according to Tax and Duin [15]. In this work, we have used one-class SVM as proposed by Tax and Duin.

One-Class SVM according to Tax and Duin

It is also known as support vector data description (SVDD). Unlike the planar approach by Schölkopf, SVDD takes a spherical approach. In feature space *F*, a spherical boundary is formed around the data by this algorithm. In order to keep the outlier's effect down, the hyper-sphere's volume is minimized.

The representation of the resultant hyper-sphere is done by centre c and radius $r > 0$ which denotes the distance of any support vector on the boundary from the centre where volume V is minimized as in Eq. (1) with constraints as stated in Eq. (2). Linear combination of all support vectors forms the centre. Though ideally the distance from

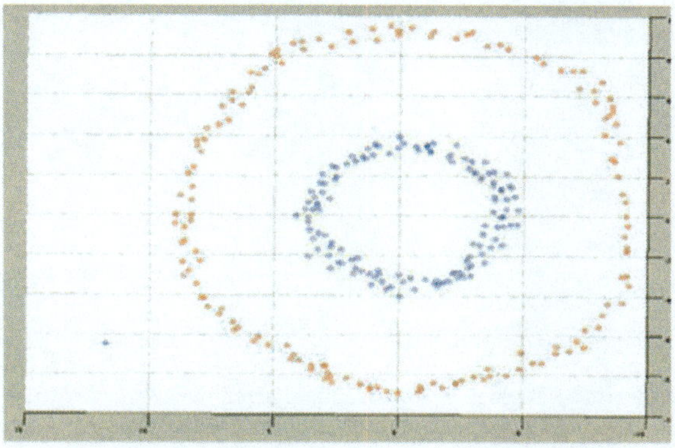

Fig. 1 Data inseparable linearly on a 2D plane [16]

any point a_i to the centre must be less than r, use of slack variable ξ_i with C as penalty parameter is used to create a soft margin. Equation (3) defines the values for the slack variable.

$$\min_{r,c} r^2 + C\sum_{i=1}^{n} \xi_i \tag{1}$$

$$\text{Subject to} : a_i - c^2 \leq r^2 + \xi_i \quad \text{for all } i = 1, \ldots, n \tag{2}$$

$$\xi_i \geq 0 \quad \text{for all } i = 1, \ldots, n \tag{3}$$

Testing for the detection of outlier of a new data point, χ can be done after Eq. (4) is introduced with Lagrange multipliers α_i.

$$\|\chi - A\|^2 = \sum_{i=1}^{n} \alpha_i \exp\left(\frac{-\chi - a_i^2}{\delta^2}\right) \geq -r^2/2 + C_r \tag{4}$$

Here, $\delta \in R$, is a kernel parameter.

In Fig. 1, the placement of the data is shown on a two-dimensional plane where it is not possible to separate them linearly. Figure 2 represents the projection of data to a 3-dimensional plane where data could be separated by using hyper-sphere.

4 Proposed Methodology

In this section, the dataset preparation steps along with the model used in the proposed method are discussed.

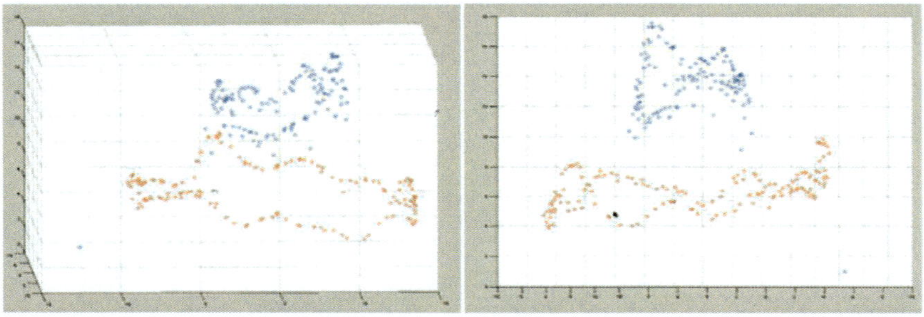

Fig. 2 Data projection on a 3D plane to separate them using hyper-sphere [16]

Dataset Design

The steps involved in the dataset design are described as follows.

Step 1: The initial structured and tagged logs are collected from HIDS server which is further converted into an N-dimensional dataset. The generation of these logs is based on any deviation found to the various rules defined by the HIDS server.

Step 2: Keeping the quality of data intact, the reduction of the obtained dataset to a 4-dimensional feature-set is done as follows:

The *source host* is treated as the first feature which represents the source from where events are generated. To handle the enormous volume of data generated, the logs generate by HIDS for 24 h are suggested to be divided into smaller time buckets represented by unique identifier. A single time bucket represents complete instance from dataset taken on a particular time-stamp for a specific user. This *time bucket identifier* is treated as the second feature. Next, the most contributing components need to be derived from the preprocessed and normalized dataset. By making a union of all unique principal components across rule IDs, a common format is produced indicating the fired alerts. All the alerts generated by a specific rule ID are merged and are represented as the frequency there by reducing volume of instances in the dataset. *Frequency* is used as the third feature. Each instance of this dataset represents a unique footprint of a user, i.e. the activities undertaken by the user in its each tenure in the application. So by grouping all the features related to a particular footprint, a unique signature is obtained which is represented by a *profile identifier* and serves as the fourth feature.

Hence, the dataset is prepared with four varying dimensions as source host, time bucket identifier, frequency of event and profile identifier.

Model used

One-class SVM is used for anomaly detection purpose. In our problem statement, the model needs to point out if something goes wrong in the existing system by analysing the log lines and finding the deviation in the behavioural pattern. Now, in a typical production scenario, even though instances for normal situation, i.e. the instances when system runs with no issues are amply available, and hence are easy to collect, but collection of instances where the system had turned out to be faulty is very expensive and not possible in real scenario. At the same time, simulation of a faulty system is not guaranteed simulate all faulty cases to make the dataset work full proof. So a traditional two-class problem solution will not be able to solve the cause. Hence, one-class classification approach is proposed to deal with our problem statement. The model using its in-built algorithm determines a representational model by evaluating the input data. Thus whenever any data is encountered which deviates from the existing pattern, it is labelled as an outlier or out-of-class. In this work, we have used one-class SVM proposed by Tax and Duin.

5 Experimental Set-up and Result Analysis

In this section, the detailed experimental set-up along with analysis of results is presented.

The experimental set-up is done on production environment where multiple micro services are implemented. The initial dataset has been created by collecting logs generated from the open source HIDS security server (OSSEC) through which all these micro services generated logs pass. The 24 h data is divided into slots of 15 min duration which is fixed as the time bucket for our experimentation.

Then required features were selected as ['t_bucket', 'hits', 'profile_hash', 'source_host'], and the data frame was created by extracting the selected features from the original dataset for model implementation.

One-class SVM model was defined by tuning the hyperparameters, i.e. the gamma value was passed in a range of 0.05 to 1 with 0.01 as the step value, and then the data frame was fitted to the model. Once the predicted values by the model were obtained, a result column 'category' was added to the dataset and the values were labelled as 'normal' for predicted value 1 and 'anomaly' for -1.

To test the efficiency of the model, the threshold value was set to 3 and one month's reference data was collected for any profile which is targeted for classification. This is because, if the duration is taken as a month, then for the events those occur on weekly basis, at least four events will be available for classification. Thus, we prepared a dataset with a reference column as 'category' by doing a manual verification of events generated over one month for testing purpose. Then this dataset of 7950 instances was obtained from six different rule IDs, i.e. 5715, 5501, 5100, 5201, 5275 and 5291, and was used for testing the mode. After dropping the category column, this dataset was passed to the model, and the predictions updated in category field were compared with the original prediction. When our proposed model was tested on six different rule IDs, the precision, recall and F1-score obtained by the classification-report are given in Table 1.

From the experimental results, it was found that the considered model has obtained an accuracy of 88%.

Even though the proposed solution is experimented and scoped to the logs collected in the above stated environment, the same principle can be successfully implemented across various log types those are defined under OSSEC.

Table 1 Precision, recall and F1-score for model tested on six different rule IDs

Rule Id	Precision	Recall	F1-score
5715	0.93	0.92	0.92
5501	0.99	0.16	0.27
5100	0.01	1.00	0.01
5201	1.00	0.17	0.29
5275	1.00	0.04	0.08
5291	0.85	0.86	0.86
Avg/total	**0.92**	**0.87**	**0.88**

6 Conclusion

In the recent IT scenario where data has to play a very big and important role, any standard approaches for detection of anomalies in big data are yet to be devised. So organizations dealing with high-dimensional big data are facing a lot of difficulty in tracking the faulty or compromised service underneath. With the increasing volume and complexity of data, identification of anomalous activities has become a big challenge for the researchers. In this work, we proposed an approach for anomaly detection from unstructured log generated from complex micro-service-based architecture, using one-class SVM. For this, profiling method was used on the multivariate dataset to reduce the dimensions without losing the accuracy and quality of data. The one-class SVM model was used to find out the anomalous data points, and our experimented results indicate an accuracy of 88% on the considered dataset. In future, the model can be checked with varying size of time buckets for improving the accuracy of the model.

References

1. Wang, Z., & Zhu, Y. (2017). A centralized HIDS framework for private cloud. *18th IEEE/ACIS international conference on software engineering, artificial intelligence, networking and parallel/distributed computing (SNPD)*, (pp. 115–120). Kanazawa.
2. Behera, A., Panigrahi, C. R., & Pati, B. (2021). Unstructured log analysis for system anomaly detection—A study. In *Proceedings of ICDSM 2021, advances in data science and management*, (vol 86). Springer, Singapore, [accepted]. https://doi.org/10.1007/978-981-16-5685-9
3. Kappagantula, S. What are microservices? An introduction to microservice architecture, May. 17, 18 Microservices Zone, https://dzone.com/articles/what-is-microservices-an-introduction-to-microserv. Last accessed on 03/10/2021.
4. Behera, A., Behera, S., Panigrahi, C. R. & Tien-Hsiung, W. (2021). Using unstructured logs generated in complex large scale micro-service-based architecture for data analysis. *International Journal of Business Intelligence and Data Mining*. [accepted]
5. Thudumu, S., Branch, P., Jin, J., & Singh, J. (Jack). (2020). A comprehensive survey of anomaly detection techniques for high dimensional big data. *Journal of Big Data, 7*(1). https://doi.org/10.1186/s40537-020-00320-x

6. Breier, J., & Branišová, J. (2015). Anomaly detection from log files using data mining techniques. In *Information science and applications*. Springer, 449–457.
7. Tuor, A., Baerwolf, R., Knowles, N., Hutchinson, B., Nichols, N., & Jasper, R. (2017). *Recurrent neural network language models for open vocabulary event-level cyber anomaly detection.* arXiv preprint arXiv:1712.00557.
8. Brown, A., Tuor, A., Hutchinson, B., & Nichols, N. (2018). Recurrent neural network attention mechanisms for interpretable system log anomaly detection. In *Proceedings of the first workshop on machine learning for computing systems*, (pp. 1–8).
9. Du, M., Li, F., Zheng, G., & Srikumar, V. (2017). DeepLog: Anomaly detection and diagnosis from system logs through deep learning. In *Proceedings of the 2017 ACM SIGSAC conference on computer and communications security (CCS '17). association for computing machinery*, (pp 1285–1298). New York, NY, USA. https://doi.org/10.1145/3133956.3134015
10. Farzad, A., & Gulliver, T. A. (2019). Log message anomaly detection and classification using auto-b/lstm and auto-gru. arXiv preprint arXiv:1911.08744
11. Zhang, S., Meng, W., Bu, J., Yang, S., Liu, Y., Pei, D., Xu, J., Chen, Y., Dong, H., & Qu, X. (2017). Syslog processing for switch failure diagnosis and prediction in datacenter networks. In *2017 IEEE/ACM 25th international symposium on quality of service (IWQoS)*, (pp. 1–10), IEEE.
12. Meng, W., Liu, Y., Zhu, Y., Zhang, S., Pei, D., Liu, Y., Chen, Y., Zhang, R., Tao, S., Sun, P. (2019). Loganomaly: Unsupervised detection of sequential and quantitative anomalies in unstructured logs. In Proceedings of the twenty-eighth international joint conference on artificial intelligence, IJCAI-19. *International joint conferences on artificial intelligence organization*, (vol. 7, pp. 4739–4745).
13. Wang, X., Wang, D., Zhang, Y., Jin, L., & Song, M. (2019). Unsupervised learning for log data analysis based on behavior and attribute features, In *Proceedings of the 2019 international conference on artificial intelligence and computer science*, (pp. 510–518).
14. Schölkopf, B., Williamson, R., Smola, A., Shawe-Taylor, J., & Platt, J. (1999). Support vector method for novelty detection. NIPS. *12*, 582–588.
15. Tax, D. M., & Duin, R. P. (2004). Support vector data description. *Machine Learning, 54*, 45–66. https://doi.org/10.1023/B:MACH.0000008084.60811.49
16. Vlasveld, R. Introduction to one-class support vector machines (2013) [online] Available at: http://rvlasveld.github.io/blog/2013/07/12/introduction-to-one-class-support-vector-machines/ Last accessed: 01 October 2021.

Advanced Computer Networks, IoT, and Security

Lightweight Model for Waifu Creation Using Deep Convolutional Generative Adversarial Network (DCGAN)

Bravish Ghosh and Manoranjan Parhi

Abstract The inceptions of generative adversarial networks have made it possible for machines to mimic creativity, one of the most unique and sophisticated human characteristics. Due to the rapid advancements in the field of generative adversarial models, lots of approaches have been proposed in the past. One of the most efficient GANs is deep convolutional generative adversarial network (DCGAN), which uses convolutional layers in the generator model to generate more realistic fake images. In this paper, we propose a lightweight implementation of the DCGAN that can be productive for the animation industry. Our model can be used by animators and designers to innovate ideas about creative anime avatars that have never existed before. This novel approach not only saves a lot of time on creative thinking but also provides brand new character designs for anime and manga avatars production.

Keywords Generative adversarial network (GAN) · Image generation · Deep convolutional generative adversarial network (DCGAN) · Deep learning (DL) · Generative adversarial model · Convolutional neural network (CNN) · PyTorch-Lightning

1 Introduction

GANs have been a topic of interest in the research domain since being proposed in 2014 by Goodfellow et al. [1]. They are an emergent class of deep learning (DL) algorithms that generate incredibly realistic images without an extensively annotated training dataset. Turing award laureate and a deep learning legend, Yann LeCun, had said, "GANs is the most interesting idea in the last 10 years in ML" [2]. They are an emerging technique for semi-supervised as well as unsupervised learning. Furthermore, hybrids GANs have been developed for specific applications, by combining them with other machine learning (ML) algorithms, such as transfer learning (TL)

B. Ghosh · M. Parhi (✉)
Department of Computer Science and Engineering, ITER Siksha 'O' Anusandhan (Deemed to Be University), Bhubaneswar, Odisha, India
e-mail: manoranjanparhi@soa.ac.in

B. Pati et al. (eds.), *Proceedings of the 6th International Conference on Advance Computing and Intelligent Engineering*, Lecture Notes in Networks and Systems 428,
https://doi.org/10.1007/978-981-19-2225-1_11

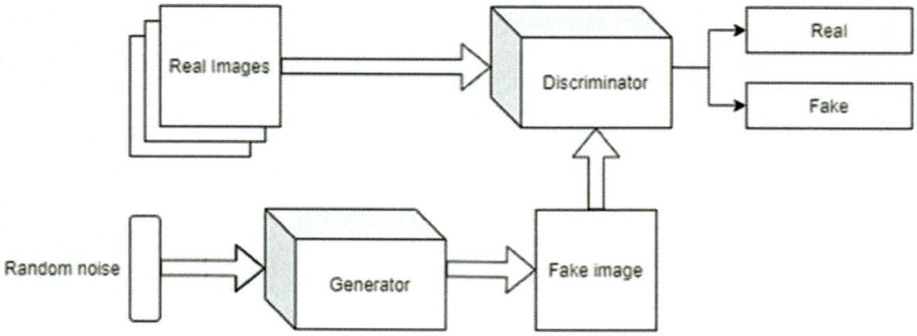

Fig. 1 Illustration of GAN framework

and reinforcement learning (RL). This is achieved through implicit modeling high-dimensional data distribution. In the image generation context, GANs are deep generative models composed of a pair of neural networks, which back-propagate signals through a competitive process [3].

1. Generator (*G*) is the neural network that generates a fake image (new data points) from input random noise (random uniform distribution).
2. Discriminator (*D*) is the other neural network that identifies fake images produced by *G* from real images.

We can think of *G* as the art forger that aims to create realistic images and *D* as the art inspector that distinguishes between the authentic and forgery images [3]. Both models are trained simultaneously, and in competition with one another (see Fig. 1). During training, *G* progressively becomes better at creating realistic images, while *D* becomes better at distinguishing them apart. The process reaches equilibrium when *D* can no longer differentiate real images from fake images.

1.1 Motivations and Contributions

GANs can be used to boost the creativity of artists and designers and generate new content in a shorter time period. Moreover, a special GAN can be used for image data, particularly deep convolutional GAN (DCGAN) [4] that uses convolution layers in the discriminator and transposes convolution layers in the generator. In this paper, we demonstrate a model; WaifuGAN Epitome inspired by the DCGAN architecture implemented using PyTorch-Lightning. The following are some of the features of the proposed model:

1. Create a new anime character, which is not a replica of any existing character
2. Improved image generation with the aid of deep convolutional GAN architecture
3. Optimize the coding complexity of neural networks using PyTorch-Lightning.

The rest of the paper has been organized in the following manner: Sect. 2 reflects on the spectrum of research work conducted in this field. Section 3 provides some information regarding methodologies and datasets used in the proposed model. Section 4 gives an understanding of the proposed work. Section 5 covers the performance evaluations in the field of analysis and a discussion on results. Section 6 concludes the paper with a summary of findings and future scope.

2 Related Work

GANs have become popular throughout the DL research areas because of the flexibility of the GAN architecture to combine with other ML algorithms for specific applications. Chen et al. [5] proposed CartoonGAN, a GAN framework for photo cartoonization. They used a network architecture that used a mapping function to transform real-life photos into cartoon manifold. Yeh et al. [6] proposed a GAN framework for semantic image inpainting with the help of deep generative models. IIzuka et al. [7] introduced a technique for consistent image completion with the help of deep convolutional neural networks (DCNN). Wang et al. [8] formulated an information retrieval (IR) framework called information retrieval GAN (IRGAN). It consisted of a generative retrieval model and a discriminative retrieval model. IRGAN is used in Web searching, item recommendation, and question answering. Qiao et al. [9] proposed a text-to-image generation model called MirrorGAN that uses a three generators system. Chen et al. [10] proposed a GAN model for image-to-text conversion (image captioning) using a deep compositional captioner (DCC) model as the baseline. GANs are widely utilized in medical fields such as Killoran et al. [11] proposed a model for DNA generation and designing; Benhenda et al. [12] proposed a technique for drug discovery; Choi et al. [13] introduced a method for generating patient labels in multiple labels, and Dai et al. [14] introduced a technique in medical image processing for organ segmentation in chest X-rays.

GANs have played a major role in creating AI-aided designs and images. Cartoons and animes are great examples of human creativity in design and animation. Zhang et al. [15] introduced a style transfer technique for anime sketches using U-net and auxiliary classifier GAN (AC-GAN). Jin et al. [16] proposed anime characters creation using GAN. The generator's architecture used super-resolution ResNet. Li et al. [17] created AniGAN, a style-guided GAN for unsupervised anime face generation using image-to-image translation. Wang et al. [18] proposed a method for anime sketch colorization using conditional GAN (C-GAN). GANs have proven to be a useful technology in the anime industry. Keras (open-source) [19], TensorFlow (Google) [20], and PyTorch (Facebook) [21] are some of the most popular deep learning frameworks for GAN implementations.

With the basic understanding of different GAN architectures, we propose a novel framework for non-existing anime face creation from random noise using the DCGAN model by Radford et al. [4] and PyTorch-Lightning by Falcon et al. [22].

3 Methodologies and Datasets

Our model uses the deep convolutional generative adversarial network technique and anime face dataset to create non-existing Waifu images. With the help of the PyTorch-Lightning module, we are able to create a lightweight model without any bulk of codes.

3.1 Dataset Used

There are various anime face datasets available online using Web scraping. In our model, we have used the anime face dataset available on Kaggle by Churchill et al. [23] based on Mckinsey666's anime face dataset in GitHub https://github.com/bchao1/Anime-Face-Dataset (see Fig. 4, real images). The dataset has 63,632 anime faces scraped from www.getchu.com, which is cropped based on the anime face detection algorithm.

3.2 Technologies Used

The code implementation is done on a Jupyter notebook environment called Google Colab using Python language and the PyTorch-Lightning module.

3.2.1 Platform Used: Jupyter Notebook

Jupyter notebook is user-friendly open-source Web-application software that provides programming services for interactive computing such as data cleaning, numerical simulation, statistical modeling, data visualization, machine learning, etc., across varieties of programming languages. It runs entirely in the cloud.

3.2.2 Language Used: Python

Here, we use the Python programming language as it is one of the most popular programming languages, and it enhances its open-source libraries regularly.

3.2.3 Library Used: PyTorch-Lightning

In this proposed approach, we use PyTorch-Lightning, a lightweight PyTorch wrapper for high-performance ML/AI research. PyTorch is an open-source python library for machine learning and deep learning applications. PyTorch-Lightning helps in scaling the model by writing less boilerplate code as compared to PyTorch. With the help of the Lightning Module and trainer, the code becomes simpler and flexible, hence making it easy to write and execute complex code.

3.3 *Deep Convolutional Generative Adversarial Networks*

DCGAN is an extension of vanilla GAN that uses convolution neural networks (CNN) in the generator for a stable architecture. This model proves to be befitting especially for image data sampling from a latent space to generate entirely new images.

3.3.1 Generative Adversarial Model

In GAN, *G* and *D* learn through an adversarial fashion. *G* takes random noise vector *z* as input and learns to output *G*(*z*) having similar distribution as the data samples *x*. D leans to discriminate between the real data x and the fake data *G*(*z*). *G* and *D* are trained alternatively to minimize the min–max loss [1]. GAN function is expressed as follows:

$$\min_{G}\max_{D} V(D,G) = \mathbb{E}_{x\ p_{data}(x)}[\log D(x)] + \mathbb{E}_{x\ p_z(z)}[\log(1 - D(G(z)))] \tag{1}$$

However, GAN suffers some instability during the training process. Therefore, we use DCGAN architecture (see Fig. 2) which uses a convolution neural network that improves the image feature extraction. *G* uses three hidden layers and one output layer, consisting of transpose convolution, batch normalization, and activation function layer. The architecture of *D* is the same as the normal image classification model.

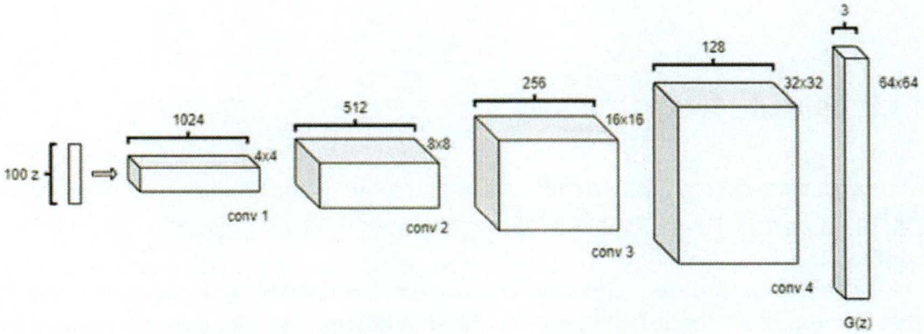

Fig. 2 Generator architecture of DCGAN model

Some of the important features of DCGAN are as follows:

1. Use of CNN strides instead of pooling layers
2. Use of batch normalization in generator and discriminator
3. No use of fully connected hidden layers
4. Use of ReLu activation function in generator layers and Tanh activation for output
5. Use of Leaky ReLu activation function in discriminator layers, except for output.

3.3.2 Convolutional Neural Network (CNN)

Convolution allows the detection of key features in different areas of an image using filters or kernels. CNN, a deep neural network, uses the convolution kernel for layer-by-layer features extraction by learning from the input image. These layers are composed of neurons, which network with other neurons in the adjacent layer. This reduces the complexity of neural networks and enhances calculation efficiency.

3.3.3 Activation Functions

Activation functions are nonlinear to compute complex features and are differentiable for back-propagation. The most common activation functions are ReLu, Leaky ReLu, Sigmoid, and Tanh. ReLu is used in three hidden layers of the generator model, and Tanh is used for the output layer. Leaky ReLu is used for the discriminator model, except for the output layer.

3.3.4 Batch Normalization

Batch normalization is a technique that enables training deep neural networks (DNNs) that normalizes the inputs to a layer for each mini-batch. This stabilizes the learning process and tremendously reduces the number of training epochs required to train DNNs.

4 Proposed Model

We propose a working DCGAN model for anime image generation using PyTorch-Lightning. The working procedure of this proposed model is elaborated stepwise as follows:

STEP 1: The first step is to load and prepare the dataset, and import necessary libraries: torch, PyTorch-Lightning for training model, and matplotlib to plot train loss.

STEP 2: The next step is to represent data as Tensors and prepare batches for the generator.

STEP 3: In this step, the generator class is defined to generate realistic-looking fake images; it consists of four layers, three hidden, and one output layer. It consists of transposed convolution, batch normalization, and activation function layers. We define a noise vector z for the generator by sampling random numbers using PyTorch.

STEP 4: In this step, the discriminator class is defined. It consists of convolution layers, activation layer, and batch normalization. In the DCGAN model, the discriminator uses strides instead of pooling for kernel size reduction. Leaky ReLu is used with a leak slope of 0.2.

STEP 5: In this step, we start the training process; training is done in mini-batch of size 128 and Adam optimizer with a learning rate of 0.0002. Then we define a data loader class. The weights are initialized to a normal distribution with a standard deviation of 0.02 and a mean like 0.

STEP 6: This step is used to conclude by defining the LightningModule to train the model along with defining the generator and discriminator loss for 100 epochs. Using the PyTorch-Lightning module, the train loop is replaced by the trainer as follows:

```
model = GAN(learning_rate = lr, z_dim = z_dim)
trainer = pl.Trainer(max_epochs = 100, gpus = 1)
trainer.fit(model, dataloader)
```

5 Results and Discussion

We trained our DCGAN model with batch size 128, learning rate 0.0002, and epoch 100. The parameter configurations are mentioned in Table 1 in detail.

We build a DCGAN model with a discriminator that maximizes the real image data and minimizes the fake image data. We see that $D(x)$ and $D(G(z))$ yield a value between 0 and 1. The total cost for G and D is expressed by the following equations:

At generator G:

Table 1 Training parameters

Parameter	Meaning	Value
epochs	Number of iterations throughout the dataset while training	100
z_dim	Dimension of the noise vector	100
display_step	How often visualize the images per pass	500
batch_size	Number of images per pass	128
lr	Learning rate	0.0002
beta_1	The momentum terms	0.5
beta_2	Device type	0.999
device		cuda

$$\frac{1}{m}\sum_{i=1}^{m}\log\left(1-D\left(G\left(z^{i}\right)\right)\right) \tag{2}$$

At discriminator D:

$$\frac{1}{m}\sum_{i=1}^{m}\log\left(D\left(x^{i}\right)\right)+\log\left(1-D\left(G\left(z^{i}\right)\right)\right) \tag{3}$$

The training loss for our DCGAN trained model is plotted (see Fig. 3).

We see that DCGAN generates images very similar to the original dataset images (see Fig. 4). The generated images are slightly confusing and less clear than real

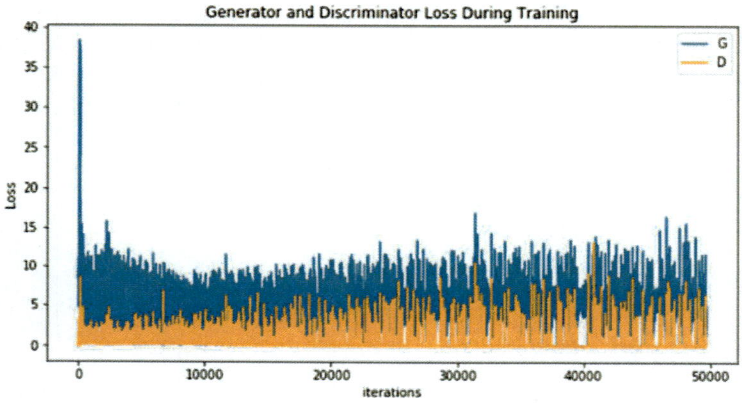

Fig. 3 Training Loss curve for DCGAN

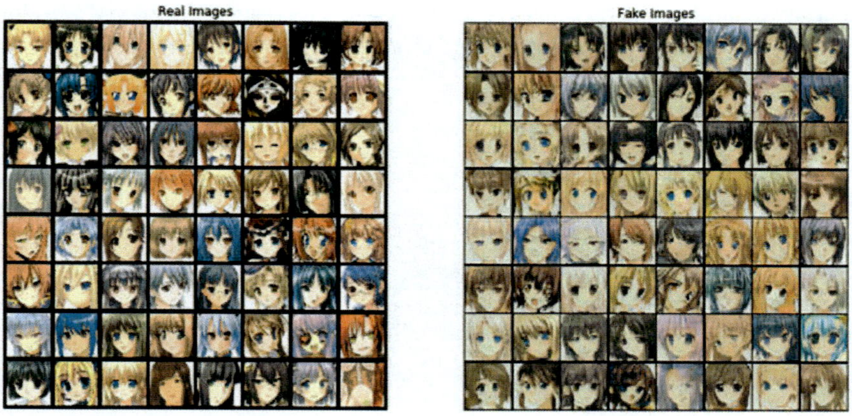

Fig. 4 Visual comparison between real image and fake image

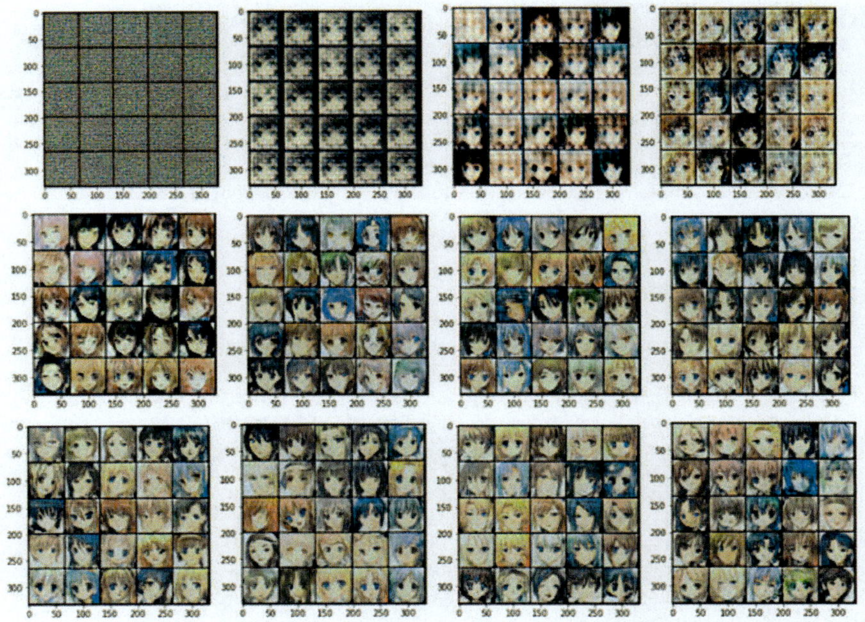

Fig. 5 Image generation at different epochs

images. We can see that generated images improve gradually, eventually reaching the point of equilibrium where the discriminator cannot distinguish between generated and real images, hence the subsequent training has minor effects. Figure 5 shows the new anime faces generated by DCGAN from random noise vector, at different epochs.

6 Conclusions

The rapid improvements in GAN architectures for image generation prove their potential to produce realistic-looking fake images using unlabeled image data. With the help of such deep convolutional GANs, artists and animators can benefit a lot in the entertainment industry. In our experiment with the DCGAN model, we reached the point of the best performance from random noise vectors and achieved satisfactory results using a lightweight framework, called PyTorch-Lightning, to generate new anime face images. In future works, the model can be developed to generate 3D anime images with higher resolutions and pose estimations for better industrial usage.

References

1. Goodfellow, I. J., Pouget-Abadie, J., Mirza, M., Xu, B., Warde-Farley, D., Ozair, S., Courville, A., & Bengio, Y. (10 June 2014). *Generative adversarial networks.* arXiv:1406.2661 [stat.ML].
2. Gui, J., Sun, Z., Wen, Y., Tao, D., & Ye, J., (20 Jan 2020). A review on generative adversarial networks: Algorithms, theory, and applications. arXiv:2001.06937 [cs.LG].
3. Creswell, A., White, T., Dumoulin, V., Arulkumaran, K., Sengupta, B., & Bharath, A. A. (2018). Generative adversarial networks: An overview. *IEEE Signal Processing Magazine., 35*(1), 53–65. https://doi.org/10.1109/MSP.2017.2765202
4. Radford, A., Metz, L., & Chintala, S. (19 Nov 2015). Unsupervised representation learning with deep convolutional generative adversarial networks. arXiv:1511.06434 [cs.LG].
5. Chen, Y., Lai, Y. K., & Liu, Y. J. (2018). CartoonGAN: Generative adversarial networks for photo cartoonization. *IEEE conference on computer vision and pattern recognition (CVPR),* (pp. 9465–9474).
6. Yeh, R. A., Chen, C., Lim, T. Y., Schwing, A. G., Hasegawa-Johnson, M., & Do, M. N. (2017). Semantic image in painting with deep generative models. *IEEE conference on computer vision and pattern recognition (CVPR),* (pp. 5485–5493).
7. Iizuka, S., Simo-Serra, E., & Ishikawa, H. (2017). Globally and locally consistent image completion. *ACM Transactions on Graphics, 36*(4), 107.
8. Wang, J., Yu, L., Zhang, W., Gong, Y., Xu, Y., Wang, B., Zhang, P., & Zhang, D. (August 2017). IRGAN: A minimax game for unifying generative and discriminative information retrieval models. *International ACM SIGIR conference on research and development in information retrieval,* (pp. 515–524).
9. Qiao, T., Zhang, J., Xu, D., & Tao, D. (2019). *IEEE conference on computer vision and pattern recognition (CVPR),* (pp. 1505–1514).
10. Biswal, A., Nanda, S., Panigrahi, C. R., Cowlessur, S. K., Pati, B. (2021). Human activity recognition using machine learning: A review. In: C. R. Panigrahi, B. Pati, B. K. Pattanayak, S. Amic, & K.C. Li (Eds.), *Progress in advanced computing and intelligent engineering. Advances in intelligent systems and computing,* (vol 1299). Springer. https://doi.org/10.1007/978-981-33-4299-6_27
11. Killoran, N., Lee, L. J., Delong, A., Duvenaud, D., & Frey, B. J. (2017). Generating and designing DNA with deep generative models. arXiv:1712.06148
12. Benhenda, M. (2017). ChemGAN challenge for drug discovery: Can AI reproduce natural chemical diversity? arXiv:1708.08227.
13. Choi, E., Biswas, S., Malin, B., Duke, J., Stewart, W. F., & Sun, J. (2017). *Generating multi-label discrete patient records using generative adversarial networks.* arXiv:1703.06490
14. Dai, W., Doyle, J., Liang, X., Zhang, H., Dong, N., Li, Y., & Xing, E. P. (2017). *SCAN: Structure correcting adversarial network for organ segmentation in chest X-rays,* (vol. 1). arXiv:1703.08770
15. Zhang, L., Ji, Y., Lin, X., & Liu, C. (2017). Style transfer for anime sketches with enhanced residual U-net and auxiliary classifier GAN. *2017 4th IAPR Asian conference on pattern recognition (ACPR),* (pp. 506–511).
16. Nayak, S., Panigrahi, C. R., Pati, B., Nanda, S., Hsieh, M. Y. (2021). *Comparative analysis of HAR datasets using classification algorithms, computer science and information systems,* (p. 43).
17. Li, B., Zhu, Y., Wang, Y., Lin, C.-W., Ghanem, B., Shen, L. (2021). *AniGAN: Style-guided generative adversarial networks for unsupervised anime face generation.* arXiv:2102.12593 [cs.CV].
18. Wang, Z. (2021). *Generating anime sketches with C-GAN,* (vol. 1827). Journal of physics: Conference series, 6th international conference on electronic technology and information science (ICETIS). Harbin, China.
19. Chollet, F., & others. (2015). Keras. Available at: https://github.com/fchollet/keras
20. Abadi, M., & others. (2015). *TensorFlow: Large-scale machine learning on heterogeneous systems.* Available at https://www.tensorflow.org

21. Paszke, A., & others. (2019). PyTorch: An imperative style, high-performance deep learning library. *Advances in neural information processing systems 32*, (pp. 8024–8035). Curran Associates, Inc. Available at: http://papers.neurips.cc/paper/9015-pytorch-an-imperative-style-high-performance-deep-learning-library.pdf
22. Falcon, W., & others. (2019). PyTorch lightning. Available at: https://github.com/PyTorchLightning/pytorch-lightning
23. Churchill, S. *Anime face dataset*. Available at: https://www.kaggle.com/splcher/animefacedataset.

An Efficient Service Recommendation with Spatial–Temporal Aware QoS Prediction Mechanism in Cloud Computing Environment

Aktham Youssef, Abhilash Pati, and Manoranjan Parhi

Abstract One of the drawbacks of using predictive quality of service (QoS) in cloud service suggestions is that the values vary rapidly over time, which may result in end-users receiving inadequate services. As a result, the cloud-based recommendation system's performance suffers. In this paper, an efficient service recommendation with a spatial–temporal aware QoS prediction mechanism in a cloud computing environment is proposed. The main contribution of this article is to use the geographical location of the services to help us choose the closest neighbor to show time QoS values sparingly, reducing the range of searches while increasing precision, and then using the Bayesian ridge regression technique to model QoS variations by making a zero-mean Laplace prior distribution assumption on the residuals of the QoS prediction, which corresponds to a Bayesian regression problem. The findings of the experiment show that the proposed approach may enhance the accuracy of time-aware cloud service recommendation by 10% over the previous approaches of temporal QoS prediction.

Keywords Cloud service recommendation · Cloud computing · QoS prediction · Time-aware cloud service · Spatial–temporal QoS prediction

1 Introduction

With the fast advancement of cloud computing technology over the previous decade, cloud services have dominated numerous application sectors. Whether commercial services like Apple's App Store or Tencent's App Store provide a variety of cloud services, many of them do the same or overlapping functions. According to recent researches, it can greatly lower IT costs and raise operational efficiency; for either individual users or SMEs, the adoption of adequate cloud services has resulted in

A. Youssef · A. Pati · M. Parhi (✉)
Department of Computer Science and Engineering, Siksha 'O' Anusandhan (Deemed to Be University), Bhubaneswar, Odisha, India
e-mail: manoranjanparhi@soa.ac.in

B. Pati et al. (eds.), *Proceedings of the 6th International Conference on Advance Computing and Intelligent Engineering*, Lecture Notes in Networks and Systems 428,
https://doi.org/10.1007/978-981-19-2225-1_12

the suggestion and selection of cloud services as one of the most essential responsibilities in the cloud environment [1]. Because a wide range of services with the same or comparable functionality have problems picking appropriate services, cloud customers are increasingly dependent on cloud service suppliers' suggestions.

Cloud service providers may collect non-functional information, such as known quality of service, in addition to the functional information of clouds, to better describe the service (QoS). The service recommendation systems based on QoS, e.g., neighborhood models and clustering algorithms can achieve high recommendations than others only using the functionality information. There are many factors which heavily influence QoS value like the Internet for connectivity and various geographical location for end-users. Therefore, users at the different locations have different QoS values even on the same cloud service which cannot be used directly by others. This makes QoS prediction a hot research. To address this issue, collaborative filtering is becoming an important approach for personalized cloud service recommendation by predicting QoS values of these services. QoS values in a highly dynamic Internet environment usually change with time and status of services like network conditions and the number of clients. Hence, the optimal QoS values of service highly fluctuate during the time. The ARIMA model is one of the temporal models we can use to analyze the behavior of a sequence of QoS values, but due to its stationary stochastic features, it is still unable to predict sudden changes in the sequence values of QoS, resulting in poor recommendations and sabotaging service-oriented application results [2].

In order to cope with an environment whose behavior is defined as dynamic, the prediction of QoS values from other users should be continued in order to gather all current values and exploit them. These approaches which are based on factorization for using newly accumulated QoS information require rebuilding the model which incurs a high cost of computational. For these challenges that are facing the recommendation system in services that are aware of time, a model that is described as spatial and temporal for QoS prediction by dealing with the problem as a Bayesian ridge regression case is proposed in this paper. We choose the closest neighbor efficiently to create the sparse representation by utilizing the geo-location of users and services to make search space small and get better accuracy. The result of this method compared to others getting a 10% improvement on the accuracy.

The remaining paper is arranged accordingly. The relevant studies in this domain are described in Sect. 2. The proposed approach is explained in Sect. 3. Section 4 provides an examination of the results, and Sect. 5 concludes the paper with future scope.

2 Related Works

We may deal with gathered QoS values as a time series in these techniques, so we can forecast values that are temporal of QoS. We can use a variety of methodologies to illustrate the dynamic behaviors of QoS characteristics. Hu et al. [3] presented

QoS prediction's new time-aware technique that assimilates the time information with similarity measures mutually. Song et al. [4] suggested calculating the similarities rather than Euclidean distances using a unique approach for personalized QoS prediction, extracting a functionality point of the QoS sequence and the dynamic time warping distance. Zhang et al. [5] proposed a WSPred model to deliver values of QoS prediction of service for various users of these services.

Singh et al. [6] studied the work of recommendations systems in time-conscious, using a time–frequency allowance for the LSTM and econometrics. Zhang et al. [7] captured complicated user-service-dependency patterns by using a deep learning-based RTF model for the time-aware recommendation of service. Singh et al. [8] experimented using input time series were carried out on many models of the neural network to identify the optimum model to forecast customized QoS-based online services. However, many historical data of QoS values and assumptions are requested for these approaches which result in their experiments showing good accuracy and low computational cost.

3 Proposed Work

This section goes through the suggested work, which is an efficient service recommendation using a spatial–temporal aware QoS prediction mechanism in a cloud computing context.

3.1 Using Bayesian Ridge Regression in QoS Prediction

Bayesian regression methods can be employed in an estimating approach to incorporate regularization parameters. The regularization parameter is not fixed in a hard sense but adjusted to the data. We can accomplish that by offering uninformative priors over the model's hyperparameters. Instead of setting lambda manually in L_2 regularization is utilized in generic regression which is the same as discovering a maximum a posteriori estimation under a Gaussian prior over the coefficients w with precision λ^{-1} We're going to manage it as a random data evaluation variable. The output y is supposed to be Gaussian X_ω to achieve a totally probabilistic model.

$$P(y|X, \omega, \alpha) = N(y|X\omega, \alpha) \tag{1}$$

The random variable which we talked about it before will be expressed as α. A spherical Gaussian gives the prior to the coefficient ω:

$$P(\omega|\lambda) = N\big(\omega|0, \lambda^{-1} I_p\big) \tag{2}$$

The priors over α and λ are chosen to be gamma distributions, the conjugate prior for the precision of the Gaussian. The resulting model is called Bayesian ridge regression and is similar to the classical ridge. The parameters ω, α, and λ are estimated jointly during the fit of the model, the regularization parameters α and λ being estimated by maximizing the log marginal likelihood. The initial value of the maximization procedure can be set with the hyperparameters alpha_init and lambda_init. There are four more hyperparameters, $\alpha_{1,}\alpha_{2,}\lambda_{1,}$ and λ_2 of the gamma prior distributions over α and λ. These are usually chosen to be non-informative. By default, $\alpha_1 = \alpha_2 = \lambda_1 = \lambda_2 = 10^{-6}$. In the next part, we will describe how in the time-aware example cloud service suggestion we may choose the most comparable sequences.

3.2 Space–Time QoS Forecasting

Normalized cross-relationship will be used between every model x in the accumulated QoS data with y to get the dynamic attributes of the given sequence y, the function of similarity is calculated as:

$$S(x, y) = \left[\left(x - x^{-}\right)^{T}\left(y - y^{-}\right)\right] / \left\|x - x^{-}\right\| \left\|y - y^{-}\right\| \tag{3}$$

After calculating the similarity, the top sequences which have a big amount of cross-correlation value will be selected. The huge size of the gathered QoS data does not allow the utilization of a linear scan in all the data. Also, by utilizing the pre-gathered data set which allows mapping each geographical point with its IP address, both users' and services' geo-location can be known. Because user-service pairs which are geographically closed have a big chance to share the same IT infrastructure, like network loads and routers, the QoS values of them may have the same, especially when sudden changes happen during the time. For this point the user-service pairs which have similar spatially it is expected that the values of temporal QoS of them very correlated.

From user-service pairs, we will utilize the spatial data to diminish the looking through range while accomplishing high precision. First, we will prove that likely any temporal QoS sequences of user-service pairs which are spatially close will be correlated for this assumption, from the QoS repository [5] we randomly select a set of test cases, for each one of them we get the QoS sequence y_a which is temporal and also corresponding to the pair of user-service p_a then looking for the most correlated sequence of QoS x_b corresponding to the pair of user-service p_b. We use spatial simS to compute two pairs' space distance for this purpose.

Let $p_a = (u_a, v_a)$ and $p_b = (u_b, v_b)$, to calculate the spatial similarity between them we will employ dist()function which represents the geodesic distance, so we get:

$$\text{simS}(p_a, p_b) = (1/2)(\text{dist}(u_a, v_a) + \text{dist}(u_b, v_b)) \tag{4}$$

Table 1 Discovered correlation coefficients based on the spatial similarity

Spatial similarity (km)	Mean of the discovered maximal correlation coefficient in RT sequence
(0, 500]	0.837
(500, 1000]	0.818
(1000, 1500]	0.784
(1500, 2000]	743
(2000, 2500]	0.598

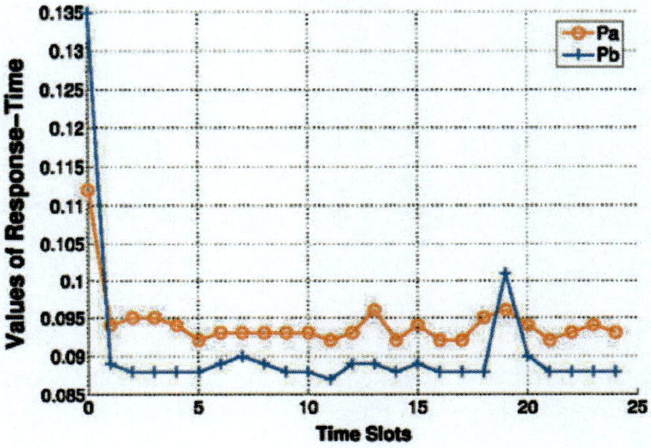

Fig. 1 Temporal response time sequence: y_a (corresponding to p_a) and x_b (Corresponding to p_b), S (y_a; x_b) = 0:955, simS (p_a; p_b) = 83:121 km

It can be noticed that when the pairs spatially close for p_a sequences of them, will be most correlated of y_a, which is depicted in Table 1. Thus, we will employ spatial information for pairs to explore the most correlated sequences which are temporal. Figure 1 denotes the discovered sequence x_b of pair p_b is much correlated to a temporal sequence y_a of pair p_a.

According to the previous analysis, we can reduce the search space to get the sequence which is correlated to y by exploring the sequences which pairs of them spatially closed to pair of y and we can express them as z_i ($1 \leq I \leq L$, where L the amount of sequences those are both temporal and most correlated).

If we have temporal sequence y of pair $P(u, v)$ and want to find the sequences which are most correlated of it we need first to retrieve the set of pairs GS(P) that are spatially closed to p, so for this step, we will depend on the geographical map to great grid representation by dividing the map into many buckets as we can see in Fig. 2, the dimension of each bucket represent the longitude and latitude. We will set the length of each bucket which represents the latitude to 0.1156 km and its width which represents the latitude to 0.1491 km as in Wang et al. [9]. We should notice that the user/service does not exist in the same bucket, then we map each user service of the sequence, we can say the two pairs are spatially closed when u.bucket equal ui.bucket or v.bucket equals vi.bucket, in this way we can retrieve top K' of pairs p_i which are spatially similar to p. After that, we can calculate the correlation coefficient

Fig. 2 Finding GS (P) for P

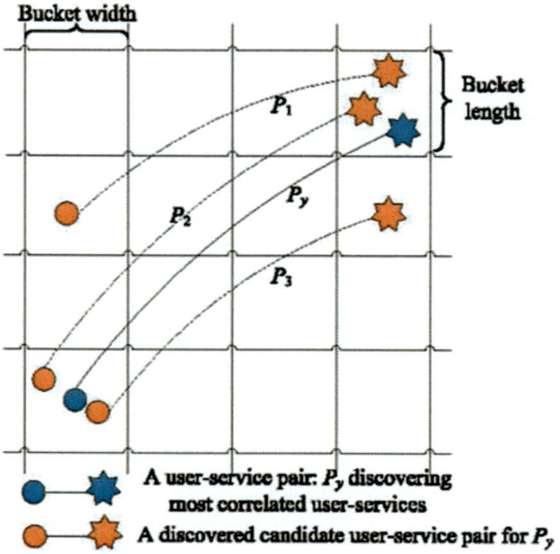

between temporal sequences of p_i and temporal sequence of p and select top K of them with notice that the correlation coefficient must be more than zero.

4 Results and Discussion

This section covers the experimental setup for the proposed work as well as the setting of K and the examination of QoS prediction performance. This section also includes a table with a comparison of outcomes as well as graphical representations of several models.

4.1 Experimental Setup

We employ a QoS performance repository [3] for the evaluation of the suggested strategy, which has a big amount of temporal response time sequences selected from 57 countries containing 142 distributed computers to 4532 distributed services. Each sequence has at most 64 values of QoS which are temporal and gathered from a client for a service, these values have been gathered during time slots each one of them continues 15 min, there are periods between two neighboring time slots also each period continues 15 min. Thus, we have a matrix with three-dimension user-service-time $142 \times 4532 \times 64$, respectively. This matrix includes values that represent the response time value of QoS invocation. We should notice that some values might be not valid so it is assigned to zero and also, we set values that are more than the 20 s to 20 s.

The baseline method is used for evaluating the performance of the proposed approach is the ARIMA method which can satisfy assumptions of normality, stationary, serial dependency, and inevitability [10], for this reason, we use the KPSS test [11] for its normality, the ADF test for checking the stationary, the QLB [10] test to check serial dependency of y. We utilize the average of the last three noticed values in the sequence of QoS to consider them unknown values to be anticipated. We also used for comparison the Lasso regression method [12] with consideration that we set λ to 0.1 and set K to 20. Also, for comparison and achieving good results of prediction we used the UPCC, IPCC, and WSRec which mix IPCC with UPCC that are CF methods also. We use the average method (AVG) for temporal QoS prediction which predicts QoS by taking three recent values which are valid of the temporal sequence for y then calculating the average of them. Due to the CF techniques are not doing well in the dynamic environment we combine their results with the results of the average method to get UPCC* instead of UPCC and also both IPCC* and WSRec* instead of IPCC and WSRec, respectively. We should notice that we set W to 0.5 in all-out experiments. We employ both mean absolute error (MAE) and root mean square error (RMSE) as the evaluation metrics, to evaluate the performance of the proposed approach. The metric MAE is calculated as:

$$\mathrm{MAE} = \sum_i \left|\hat{y}_p - y_p\right| / N \tag{5}$$

And the definition of RMSE is:

$$\mathrm{RMSE} = \sqrt{\left[\sum_i \left(\hat{y}_p - y_p\right)^2 / N\right]} \tag{6}$$

where yn indicates the anticipated value of QoS at a current time tn, which belongs to a sequence of y, $\hat{y}n$ show the real value of y at the same time. The number of anticipated values of QoS is represented by N.

4.2 Setting of K

We have 10,000 test cases that are randomly selected to apply the Bayesian ridge regression technique with different K values to conduct QoS prediction in order to assign the optimal value to K in our suggested model. We also utilize MAE and RMSE to improve prediction accuracy when altering the density of response values for each K value. This allows us to determine which K value is best for our method by comparing different densities.

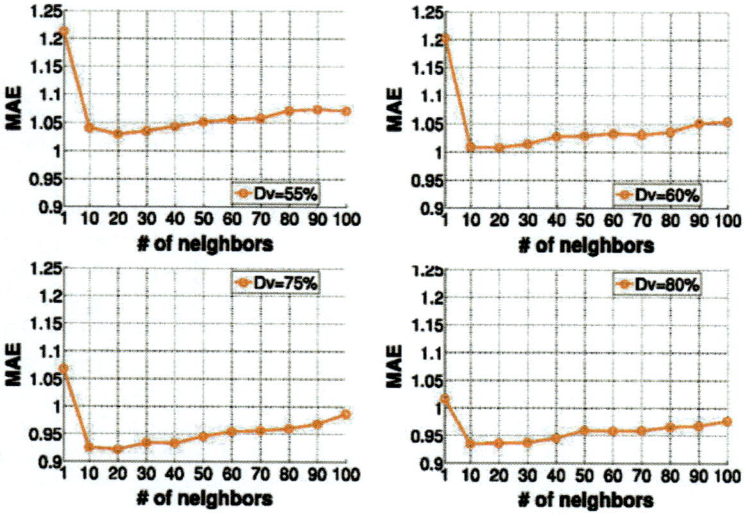

Fig. 3 Best K during changing the densities of RS values

Figure 3 shows that when the density is changed, the suggested Bayesian ridge regression technique achieves the best MAE when the value of k is between 10 and 20.

4.3 Analysis of QoS Prediction Performance

We will use two metrics to evaluate the prediction accuracy of the employed comparing algorithm for each test case. As for grid representation which is shown in Fig. 2, we are configured to 0.1156 and 0.1491 correspondingly for each bucket length and breadth, K' is set to 400, and for K is set to20. Both Fig. 4 and Table 2 present findings of the MAE and RMSE of several reaction time prediction techniques, where the reaction time density varies between 55 and 80%. The following points may be seen from the results:

- Some of the historical values of QoS will not be considered in the sequence which makes the IPCC, UPCC, and WSRec did not achieve better prediction accuracy than other approaches.
- We can notice that when the density is high, the prediction accuracy of AVG is a little bit worse than ARIMA in addition, the prediction accuracy of both AVG and ARIMA is better than UPCC, IPCC, and WSRec methods.
- The Lasso ($K = 20$) may increase by 15% compared to the ARIMA model if the density of QoS values high 80%.

Fig. 4 Comparison between Bayesian ridge regression and QoS prediction algorithms on response time value

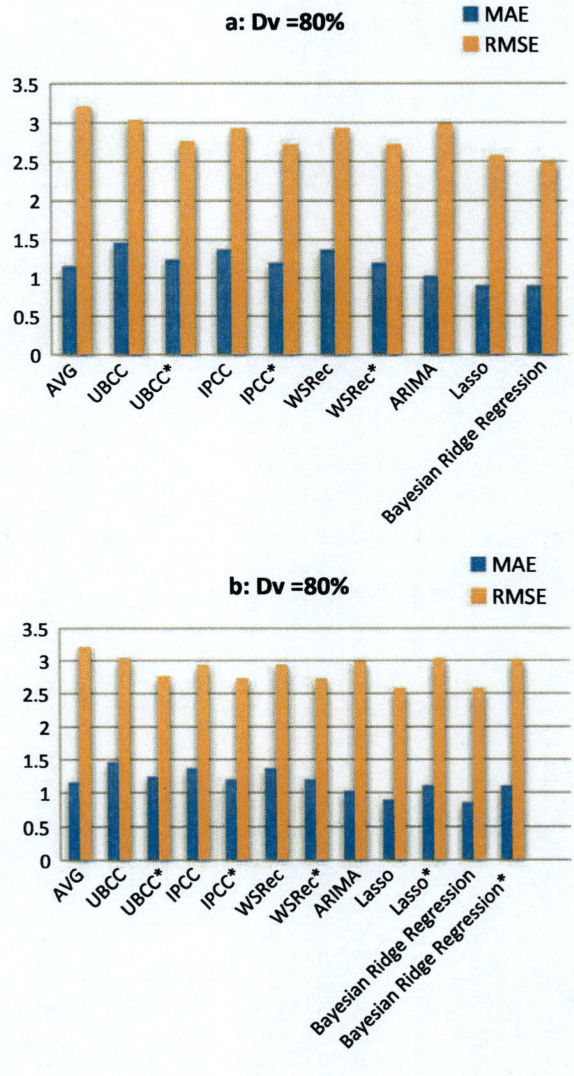

- We can see that our suggested Bayesian ridge regression is slightly better than the Lasso method and its benefits will reach around 16 percent compared with the ARIMA model.
- The importance of using the geolocation technique is when we compared the Lasso and Bayesian Ridge Regression with the two other designed temporal QoS prediction approaches which are Lasso* and Bayesian Ridge Regression* by not using the geolocation technique with them, it is found that the results of prediction accuracy of Lasso and Bayesian Ridge Regression are much better than that of Lasso* and Bayesian Ridge Regression*. This proves that it is very important

Table 2 Comparing prediction approaches on RS with consideration the different densities of QoS value, the increase of performance compared to the ARMI model is represented by Δ

Approaches	$D_v = 55\%$		$D_v = 60\%$		$D_v = 65\%$		$D_v = 70\%$		$D_v = 75\%$		$D_v = 80\%$	
	MAE	RMSE	MAE	RMSE	MAE	RMSE	MAE	RMSE	MAE	RMSE	MAE	RMSE
AVG	1.172	3.226	1.167	3.219	1.164	3.214	1.162	3.211	1.160	3.209	1.159	3.206
UPCC	1.470	3.034	1.467	3.027	1.464	3.019	1.466	3.027	1.464	3.026	1.467	3.032
UPCC*	1.252	2.775	1.251	2.773	1.244	2.759	1.244	2.759	1.242	2.763	1.242	2.753
IPCC	1.396	2.951	1.388	2.937	1.384	2.926	1.378	2.928	1.374	2.923	1.372	2.925
IPCC*	1.221	2.742	1.216	2.736	1.210	2.724	1.206	2.720	1.202	2.717	1.200	2.714
WSRec	1.391	2.951	1.384	2.937	1.381	2.926	1.376	2.928	1.372	2.923	1.372	2.925
WSRec*	1.220	2.747	1.215	2.740	1.208	2.728	1.206	2.724	1.202	2.721	1.200	2.716
ARIMA	1.201	3.332	1.160	3.285	1.089	3.175	1.087	3.042	1.051	3.030	1.028	2.986
LASSO(K = 20)	1.000	2.822	0.984	2.781	0.954	2.723	0.922	2.637	0.909	2.607	0.893	2.572
LASSO(K = 20)*	1.240	3.338	1.208	3.250	1.176	3.148	1.214	3.225	1.160	3.111	1.112	3.026
Bayesian ridge regression	0.995	2.742	0.980	2.775	0.950	2.698	0.911	2.630	0.887	2.600	0.853	2.568
Bayesian ridge regression*	1.125	3.300	1.190	3.112	1.169	3.130	1.200	3.100	1.150	3.099	1.102	3.013

*The improved versions of the saidtechniques by combining their prediction results with the temporal average (AVG) to improve the accuracy

to include the spatial information of users and services for boosting prediction performance. Here, in determining the most correlated sequences to sequence for prediction y, we formed other approaches that are similar to what exists for prediction QoS called Lasso* and Bayesian ridge regression*, except in process of exploring the spatial similar GS(P), so K' is selected randomly from the data set. As shown in Fig. 4b, the accuracy of prediction results shows that both Bayesian ridge regression and Lasso are better than Bayesian ridge regression*, and Lasso*, respectively, which can prove the importance of using the geolocation technique for getting the best results.

5 Conclusion and Future Scope

Through this study, we learned that the general regression problem is related to our task, which is temporal QoS prediction. As a result, we rely on Bayesian Linear regression to make the temporal QoS sequences appear sparse, as well as to find the most comparable QoS sequences. End-users and service geo-locations are used. We discovered that our technique outperforms previous temporal QoS prediction approaches for time-aware cloud service recommendations after deploying it and obtaining results from the comprehensive experimental investigation.

Even though the performance was outstanding, a few limitations should be addressed in the future, such as getting more QoS values during the current time slot and using an online algorithm that anticipates future QoS values based on obtained data.

References

1. Zhang, Y., & Lyu, M. R. (2017). Time-aware model-based QoS prediction. In: *QoS prediction in cloud and service computing*, (pp. 35–53). SpringerBriefs in Computer Science. Springer. https://doi.org/10.1007/978-981-10-5278-1_3
2. Calheiros, R. N., Masoumi, E., Ranjan, R., & Buyya, R. (1 October–December 2014). Workload prediction using ARIMA model and its impact on cloud applications. *QoS, IEEE Transactions on Cloud Computing*, *3*(4), 449–458. https://doi.org/10.1109/TCC.2014.2350475
3. Hu, Y., Peng, Q., & Hu, X. (2014). A time-aware and data sparsity tolerant approach for web service recommendation. In: *2014 IEEE international conference on web services*, (pp. 33–40). IEEE.
4. Song, Y., Hu, L., & Yu, M. (2018) A novel QoS-aware prediction approach for dynamic web services. *Plos one*, *13*(8).
5. Zhang, Y., Zheng, Z., & Lyu, M. R. (2011). WSPred: A time-aware personalized QoS prediction framework for Web services. In: *2011 IEEE 22nd international symposium on software reliability engineering*, (pp. 210–219). IEEE.
6. Singh, V. P., Pandey, M. K., Singh, P. S., Karthikeyan, S. (2019). An empirical mode decomposition (EMD) enabled long sort term memory (LSTM) based time series forecasting framework for web services recommendation. In: *Fuzzy systems and data mining V* (pp. 715–723). IOS Press.

7. Nanda, S., Panigrahi, C., & Pati, B. (2020). Emergency management systems using mobile cloud computing: A survey. *International Journal of Communication Systems*, e4619.
8. Singh, V. P., Pandey, M. K., Singh, P. S., & Karthikeyan, S. (2020). Neural net time series forecasting framework for time-aware web services recommendation. *Procedia Computer Science, 171*, 1313–1322.
9. Wang, X., Zhu, J., & Shen, Y. (2014). Network-aware QoS prediction for service composition using geolocation. *IEEE Transactions on Services Computing, 8*(4), 630–643.
10. Box, G. E., Jenkins, G. M., Reinsel, G. C., & Ljung, G. M. (2015). Time series analysis: Forecasting and control. Wiley.
11. Parida, S., Pati, B., Nayak, S. C., Panigrahi, C. R. (2021). JOB-DCA: A cost minimizing jaya optimization-based data center allocation policy for IaaS cloud model. In: C. R. Panigrahi, B. Pati, B. K. Pattanayak, S. Amic, & K. C. Li (Eds.), *Progress in advanced computing and intelligent engineering. Advances in intelligent systems and computing*, (vol. 1299). Springer. https://doi.org/10.1007/978-981-33-4299-6_5112
12. Wang, X., Zhu, J., Zheng, Z., Song, W., Shen, Y., & Lyu, M. R. (2016). A spatial-temporal QoS prediction approach for time-aware web service recommendation. *ACM Transactions on the Web (TWEB), 10*(1), 1–25.

PUASIoT: Password-Based User Authentication Scheme for IoT Services

Bhawna Narwal, Khushi Gandhi, Revika Anand, and Riya Ghalyan

Abstract In today's scenario where there is an exponential increase in the need for IoT services, user authentication is an essential feature for device security. IoT services allow many devices to be accessed and connected over the Internet anytime and anywhere and thus pose a risk over its privacy and security. A single-factor authentication scheme can be used to provide robust security, privacy, and secure access over devices. To deal with a range of IoT devices with varying storage, we have propounded a lightweight novel password-based user authentication scheme with low computation costs. This scheme uses lightweight XOR and hash operations. AVISPA tool has been used to prove security against various attacks.

Keywords Authentication · IoT · Password · AVISPA · Security

1 Introduction

IoT is a network of physical objects which includes RFID tags, sensors, actuators, etc., which can sense and gather data [1]. This data utilization can provide intelligent services such as remote control, medical aid, optimal indoor environment, surveillance, security . Analyzing the scenario where IoT services can interact with smart devices, combining these IoT networks with devices enables the provision of IoT services to users. Today, most IoT services can be accessed and controlled by smartphones, for example, Google Assistant Smart Home, which allows connected devices to be controlled by the Google Home app and Google Assistant surfaces [2]. Likewise, IoT services can be managed and accessed by remote devices bringing

B. Narwal (✉)
IT Department, Indira Gandhi Delhi Technical University for Women, Delhi, India
e-mail: bhawnanarwal@igdtuw.ac.in

K. Gandhi
ECE Department, Indira Gandhi Delhi Technical University for Women, Delhi, India

R. Anand · R. Ghalyan
CSE Department, Indira Gandhi Delhi Technical University for Women, Delhi, India

B. Pati et al. (eds.), *Proceedings of the 6th International Conference on Advance Computing and Intelligent Engineering*, Lecture Notes in Networks and Systems 428,
https://doi.org/10.1007/978-981-19-2225-1_13

Table 1 Notations

Notations	Description
SIDj/UIDi	Unique identity of IoT sensor node/user identity
Rj	Random number generated by IoT sensor node
K_{GS}	Shared secret key of gateway and IoT sensor node
Sj	IoT node
Ui	User
GW	Gateway
h(.)	hash function
Ri, rj	Random number
T1, T2, T3, T4	Timestamp
⊕ , (l, m)	XOR, data concatenation
Skey	Session key

the risk of information leakage access by unauthorized sources and privacy breaches [3–5]. Therefore, it becomes much easier for attackers to extract the stored information about them. The IoT streaming applications can cause serious privacy concerns especially when the user forgets to discontinue them, leading to privacy and security concerns like confidentiality, authenticity, and integrity [7, 8]. The factors such as limited memory, low computation speed, and compact memory cause variations in the security need of IoT devices (Table 1).

Nowadays using cloud computation, it is possible to hack secure protocols where hackers can cause modification in the original messages to be communicated. With the growing devices connected to IoT services, the vulnerability of privacy breaches and data leakage increases. Taking into consideration, all possible types of platforms over which IoT is built, the performance, storage, and security requirements differ from one application to other. Thus, a good authentication scheme must be devised. The existing related schemes prove to be incapable of promising the security of IoT devices at ground level, and the schemes that can provide security cannot be implemented on small resource-restricted connected devices over IoT networks [9, 10].

Several pieces of research have been presented in the field of IoT security. With the increase in demand to switch from manual to automated, IoT has paved its path to the front by providing promising solutions that grant access and control from remote devices such as a click on a smartphone. The authors Santoso et al. [11] have presented an approach to integrating high security while employing IoT to build smart homes and promising benefits to users while system operation. To ensure high security, authors have used an asymmetric ECC and a Wi-Fi gateway for authentication and authorization of IoT nodes. For ensuring security in IoT entitled smart buildings, architecture related to many security and privacy workings has been propounded [12]. The authors Lee et al. [13] have proposed an encryption scheme for IoT environment based on XOR manipulation to enable anticounterfeiting and to ensure security. Hardware design mechanisms and enhanced security protocols have been demonstrated. An Braeken proposed a PUF-based auth scheme for IoT environment by eradicating the flaws present in [14]. The author has proposed an alternative scheme that promises privacy by providing an efficient key agreement mechanism [14]. Ashok Kumar Das et al. have presented an anonymous signature-based authenticated scheme that uses a key exchange mechanism for an IoT-built smart grid environment [15]. Based on the issues in the existing related schemes, we have proposed a novel authentication scheme for IoT services using password-based mechanism.

The outline is: Sect. 2 mentions PUASIoT scheme. Section 3 mentions security and conclusion in Sect. 4.

2 The Propounded Scheme: PUASIoT

2.1 User Registration

(a) **Registration between Ui and GW**

Ui carries out the registration process with GW, using following steps:

1: Ui generates its identity UIDi, password PWDi, and ri.

2: Ui calculates the shadow password, SPWDi = H(ri || PWDi), and shadow identity, SUIDi = H(ri||UIDi).

3: Ui sends < SUIDi, SPWDi > to GW securely.

4: GW calculates ai = H(SUIDi||T1|| KGW), bi = H(SPWDi||T1|| KGU), and ci = ai ⊕ bi.

5: GW sends < ci, ai, KGU, T1 > to Ui.

6: Ui receives the parameters and checks |T1-T|< ΔT. If yes, then stores ci, ai, KGU, and T1 in smart device

(b) **Registration between Sj and GW**

Sj carries out the registration process with GW, using following steps:

1: Sj calculates cj = H(K_{GS}|| SIDj||T2||rj), dj = K_{GS} ⊕ rj, and ej = cj ⊕ dj.

2: Sj sends < SIDj, ej, dj, T2 > to GW.

3: GW upon receipt of mesg checks |T2 − T| < ΔT? If no, then abort. If yes, then calculates rj* = K_{GS} ⊕ dj and cj = ej ⊕ dj. Then, computes cj* = H (K_{GS}|| SIDj||T2||rj*). Then checks whether cj* = cj? If no, then abort, else GW computes fj = H (SIDj||T3|| K_{GW}), gj = H(cj|| K_{GS}), hj = fj ⊕ gj.

4: GW sends < hj, fj, T3 > to Sj.

5: Sj checks |T3 − T| < ΔT? If yes, the Sj stores hj, fj, and T3 in smart device

2.2 Login

Ui sends login request to Sj to carry out authentication (based on Dolev-Yao threat Model [16, 17]) as per following:

1: Ui submits UIDi* and PWDi*. Further, calculates SPWDi = H (ri || PWDi*).

2: Sj calculates bi* = H (SPWDi||T1|| K_{GU}).

3: Next, Ui computes original bi value, i.e., bi = ci ⊕ ai.

4: Check bi = bi*? Is not the same, abort. Otherwise, computes U1i = H (bi|| K_{GU} ||T1) and generates a random nonce q. Next, U2i = ai ⊕ q. Next, Ui chooses the respective Sj for providing the services and sends the message <SUIDi, ci, U1i, U2i, T1> via an insecure channel to Sj.

2.3 Mutual Authentication

1: User sends authentication message <SUIDi, ci, U1i, U2i, T1> to Sj. Then, verifies |T1 − T|< ΔT? If no, terminate. Otherwise, move to next step.

2: Using stored values of hj and fj, the Sj calculates gj = hj ⊕ fj.

3: Next, Sj computes Vj = H (K_{GS}||T3||T2) ⊕ gj. Sj sends <SUIDi, ci, U1i, T3, T2, SIDj, fj, Vj> to GW.

4: Then, GW verifies |T3 − T|< ΔT? If true, GW calculates fj* = H (SIDj||T3|| K_{GW}) and gj* = fj* ⊕ hj. Now, GW will calculate its own gj = H (K_{GS}||T3||T2) ⊕ Vj. Now, GW checks whether gj = gj*? If same, GW authenticates Sj as a valid registered node.

5: After successful authentication of Sj by GW, GW computes ai* = H (SUIDi||T1|| K_{GW}) and bi* = ci ⊕ ai*. Then, GW calculates Xi = H (bi*|| K_{GU}|| T1). Xi will be used by GW to authenticate Ui.

6: GW checks whether received U1i = Xi? If same, the GW successfully authenticates Ui.

7: GW computes Oij = ai* ⊕ H (fj*|| K_{GS}) to be used by Sj, to derive the value of nonce q. Then, calculate Qj = H (fj*|| K_{GS}||T2||T3||T4). Also, GW computes Wi = H (Xi||T2||T3||T4) to be sent to Ui. The parameter Wi will be used for authentication between GW and Sj. GW then sends auth parameters to Sj and Ui <Oij, Qj, Wi, T2, T3, T4>.

8: Upon receipt of mesg from Sj, Ui verifies |T4 − T|< ΔT? If no, abort. Else, Ui calculates Wi* = H (H(bi*|| K_{GU}|| T1) ||T2||T3||T4) and validate Wi* = Wi? If found same, Ui corroborates authenticity of GW and Sj. Session key will be produced in the next step. If no, then terminate.

9: Upon receipt of mesg <Oij, Qj, Wi, T2, T3, T4> from GW, Sj will calculate Qj* = H (fj|| K_{GS}||T2||T3||T4) and checks whether Qj = Qj*? If same, Sj will calculate ai* = Oij ⊕ H (fj|| K_{GS}) and q = U2i ⊕ ai*.

10: Sj will generate a nonce n and computes Pij = H (ai*||SIDj||T2||T3||T4) ⊕ n and Skey = H (q ⊕ n).

11: Sj will send the authentication message <Pij, SIDj, T2, T3, T4> to Ui.

12: On successful authentication in Step 11, Ui computes n = Pij ⊕ *H* (ai||SIDj||T2||T3||T4). Finally, Ui can compute Skey = *H* (q ⊕ n).

3 Security Analysis of PUASIoT

The HLPSL specification (in AVISPA [17]) for GW (gateway), Ui (user), Sj (sensor node), session, and environment is provided in Figs. 1, 2, 3, 4, and 5, respectively. And, result in (Fig. 6).

4 Conclusion

With the increase in the need for IoT services and simultaneous dropping costs of sensors, connectivity has become an essential feature thereby increasing the risk of privacy and security breaches. Since IoT involves connected things which share data, this large network of connected things becomes a vulnerable focal point of privacy and security attacks. Thus, it becomes pivotal that protocols can provide essential security against attacks related to security and data leakages. Thus, we developed PUASIoT which uses simple one-way hash and XOR operations, making scheme suitable for small devices with low storage and resources. For security analysis, AVISPA has been used and results suggest that our scheme resists security attacks.

```
role gateway(
        Ui,Sj,GW          : agent,
        SKun,SKng,SKug : symmetric_key,
        H                 : hash_func,
        Snd,Rcv           : channel(dy))
played_by GW def=
local
        State                           : nat,
        UIDi,PWDi,Ri,Rj,SPWDi,SUIDi,Kgu,Kgw,T1,T2,T3,T4,Kgs,Ai,Bi,Ci,U1i,U2i,Q,Oij,Qj,Wi,Pij,SIDj,N,Skey,Fj,Cj, Gj,Hj,Ej,Dj,Vj,Xi :text
        const subs1,subs2,subs3,subs4,subs5,alice_bob_rb,bob_alice_ra,alice_gateway_q : protocol_id
    init
        State := 0
    transition
    1. State  = 0 /\ Rcv({SUIDi'.SPWDi'}_SKug) =|>
       % Registration Phase
       State':= 1 /\ secret({Kgu},subs1,{Ui,GW})
                  /\ secret({UIDi,PWDi,Ri},subs2,Ui)
                  /\ secret({Kgw},subs3,{GW})
                  /\ T1':= new()
                  /\ Ai':= H(SUIDi'.T1'.Kgu)
                  /\ Bi':= H(SPWDi'.T1'.Kgu)
                  /\ Ci':= xor(Ai',Bi')
                  /\ Snd({Ai'.Ci'.Kgu.T1'}_SKug)
    2. State  = 1 /\ Rcv({SIDj.Ej'.Dj'.T2'}_SKng) =|>
       State':= 2 /\ secret({Kgs},subs4,{Sj,GW})
                  /\ secret({SIDj.Rj.Kgs},subs5,Sj)
                  /\ Cj':= xor(Ej',Dj')
                  /\ T3':= new()
                  /\ Fj':= H(SIDj.T3'.Kgs)
                  /\ Gj':= H(Cj'.Kgs)
                  /\ Hj':= xor(Fj',Gj')
                  /\ Snd({Hj'.Fj'.T3'}_SKng)
3. State  = 2 /\ Rcv({SUIDi'.Ci'.U1i'.T3'.T2'.SIDj.Fj'.Vj'}_SKng)=|>
       % Authentication Phase
       State':= 3 /\ Fj':= H(SIDj.T3'.Kgw)
                  /\ Gj':= xor(Fj',Hj)
                  /\ Ai':= H(SUIDi'.T1.Kgw)
                  /\ Bi':= xor(Ci',Ai')
                  /\ Xi':= H(Bi'.Kgu.T1)
                  /\ Oij':= xor(Ai',H(Fj'.Kgs))
                  /\ T4':= new()
                  /\ Qj':= H(Fj'.Kgs.T2'.T3'.T4')
                  /\ Wi':= H(Xi.T2'.T3'.T4')
                  /\ Snd({Oij'.Qj'.Wi'.T2'.T3'.T4'}_SKng)
                  /\ Snd({Oij'.Qj'.Wi'.T2'.T3'.T4'}_SKug)
end role
```

Fig. 1 Role gateway

```
role alice(Ui,Sj,GW         : agent,
       SKun,SKng,SKug : symmetric_key,
       H              : hash_func,
       Snd,Rcv        : channel(dy))
played_by Ui
def=
    local
       State                           : nat,
       UIDi,PWDi,SPWDi,SUIDi,T1,T2,T3,T4,Kgu,Ai,Bi,Ci,U1i,U2i,Q,Oij,Qj,Wi,Pij,SIDj,N,Skey,Ri :text
       const subs1,subs2,subs3,subs4,subs5,alice_bob_rb,bob_alice_ra,alice_gateway_q : protocol_id
    init
       State := 0
    transition
    1. State  = 0 /\ Rcv(start) =|>
      % Registration Phase
      State':= 1 /\ SUIDi':= H(Ri.UIDi)
                 /\ SPWDi':= H(Ri.PWDi)
                 /\ Snd({SUIDi'.SPWDi'}_SKug)
                 /\ secret({Kgu},subs1,{Ui,GW})
                 /\ secret({UIDi,PWDi,Ri},subs2,Ui)
    2. State  = 1 /\ Rcv({Ai'.Ci'.Kgu.T1'}_SKug) =|>
      % Login Phase
      State':= 2 /\ Q':= new()
                 /\ SPWDi':= H(Ri.PWDi)
                 /\ SUIDi':= H(Ri.UIDi)
                 /\ Bi':= xor(Ci',Ai')
                 /\ U1i':= H(Bi'.Kgu.T1)
                 /\ U2i':= xor(Ai',Q')
                 /\ Snd({SUIDi'.Ci'.U1i'.U2i'.T1}_SKun)

    3. State  = 4 /\ Rcv({Oij'.Qj'.Wi'.T2'.T3'.T4'}_SKug)=|>
      State':=5   /\ Wi':= H(H(Bi.Kgu.T1).T2'.T3'.T4')
    4. State  = 5 /\ Rcv({Pij'.SIDj.T2'.T3'.T4'}_SKun)=|>
      State':=6   /\ N':= xor(Pij,H(Ai.SIDj.T2'.T3'.T4'))
                 /\ Skey':= H(xor(Q,N'))

end role
```

Fig. 2 Role user

```
role bob(
      Ui,Sj,GW       : agent,
      SKun,SKng,SKug : symmetric_key,
      H              : hash_func,
      Snd,Rcv        : channel(dy))
played_by Sj def=
    local
      State                          : nat,
      UIDi,PWDi,Ri,Rj,SPWDi,SUIDi,Kgu,T1,T2,T3,T4,Kgs,Ai,Bi,Ci,U1i,U2i,Q,Oij,Qj,Wi,Pij,SIDj,N,Skey,Fj,Cj, Gj,Hj,Ej,Dj,Vj :text
      const subs1,subs2,subs3,subs4,subs5,alice_bob_rb,bob_alice_ra, alice_gateway_q: protocol_id
    init
      State := 0
    transition
    1. State  = 0 /\ Rcv(start) =|>
      % Registration Phase
      State':= 1 /\ Rj':= new()
                 /\ T2':= new()
                 /\ Cj':= H(Kgs.SIDj.T2'.Rj')
                 /\ Dj':= xor(Kgs,Rj')
                 /\ Ej':= xor(Cj',Dj')
                 /\ Snd({SIDj.Ej'.Dj'.T2'}_SKng)
                 /\ secret({Kgs},subs4,{Sj,GW})
                 /\ secret({SIDj.Rj'.Kgs},subs5,Sj)
    2. State  = 1 /\ Rcv({Hj'.Fj'.T3'}_SKng)=|>State':=2
      % Login Phase
3. State  = 2 /\ Rcv({SUIDi'.Ci'.U1i'.U2i'.T1'}_SKun) =|>
      % Authentication Phase
      State':= 3 /\ Gj':= xor(Hj,Fj)
                 /\ T3':= new()
                 /\ Vj':= xor(H(Kgs.T3'.T2),Gj')
                 /\ Snd({SUIDi'.Ci'.U1i'.T3'.T2.SIDj.Fj.Vj'}_SKng)

 4. State  = 3 /\ Rcv({Oij'.Qj'.Wi'.T2'.T3'.T4'}_SKng)=|>
      State':= 4 /\ Qj':= H(Fj.Kgs.T2'.T3'.T4')
                 /\ Ai':= xor(Oij,H(Fj.Kgs))
                 /\ Q':= xor(U2i,Ai')
                 /\ N':= new()
                 /\ Pij':= xor(H(Ai'.SIDj.T2'.T3'.T4'),N')
                 /\ Skey':= H(Q'.N')
                 /\ Snd({Pij'.SIDj.T2'.T3'.T4'}_SKun)
    5. State= 4 /\ request(Ui,Sj,alice_gateway_q,Q')=|>
   State':= 5
end role
```

Fig. 3 Role sensor node

```
role session(
        Ui,Sj,GW          : agent,
        SKun,SKng,SKug    : symmetric_key,
        H                 : hash_func)
def=
local Sal,Ral,Sbo,Rbo,Sga,Rga       : channel(dy)
composition
           alice(Ui,GW,Sj,SKun,SKng,SKug,H,Sal,Ral)
        /\ bob(Ui,GW,Sj,SKun,SKng,SKug,H,Sbo,Rbo)
        /\ gateway(Ui,GW,Sj,SKun,SKng,SKug,H,Sga,Rga)
end role
```

Fig. 4 Role session

```
role environment()
def=
        const ui,sj,gw        : agent,
        skun,skng,skug : symmetric_key,
        h              : hash_func,
        alice_bob_rb,bob_alice_ra,subs1,subs2,subs3,subs4,subs5 : protocol_id
        intruder_knowledge= {ui,sj,gw,h}
composition
           session(ui,gw,sj,skun,skng,skug,h)
        /\ session(ui,gw,sj,skun,skng,skug,h)
        /\ session(ui,gw,sj,skun,skng,skug,h)
end role

goal
        secrecy_of subs1
        secrecy_of subs2
        secrecy_of subs3
        secrecy_of subs4
        secrecy_of subs5
        authentication_on alice_gateway_q

end goal
environment()
```

Fig. 5 Role environment

Fig. 6 Simulation result of PUASIoT

```
% OFMC
% Version of 2006/02/13
SUMMARY
  SAFE
DETAILS
  BOUNDED_NUMBER_OF_SESSIONS
PROTOCOL
  /home/span/span/testsuite/results/IOT.if
GOAL
  as_specified
BACKEND
  OFMC
COMMENTS
STATISTICS
  parseTime: 0.00s
  searchTime: 1.78s
  visitedNodes: 64 nodes
  depth: 6 plies
```

References

1. Yan, Z., Zhang, P., & Vasilakos, A. V. (2014). A survey on trust management for internet of things. *Journal of Network and Computer Applications, 42*, 120–134.
2. Dudhe, P. V., Kadam, N. V., Hushangabade, R. M., & Deshmukh, M. S. (2017). Internet of things (IoT): An overview and its applications. In *2017 international conference on energy, communication, data analytics and soft computing (ICECDS)*, (pp. 2650–2653). https://doi.org/10.1109/ICECDS.2017.8389935
3. Atzori, L., Iera, A., & Morabito, G. (2010). The internet of things: A survey. *Computer Networks, 54*(15), 2787–2805. https://doi.org/10.1016/j.comnet.2010.05.010
4. Babar, S., Mahalle, P., Stango, A., Prasad, N., & Prasad, R. (2010). Proposed security model and threat taxonomy for the internet of things (IoT). In: *Recent trends in network security and applications*, (pp. 420–9). Springer. https://doi.org/10.1007/978-3-642-14478-3_42
5. Cassar, G., Barnaghi, P., Wang, W., & Moessner, K. (2012). A hybrid semantic matchmaker for IoT services. In: *Green computing and communications (GreenCom), 2012 IEEE international conference on, November*. IEEE, (pp. 210–16). https://doi.org/10.1109/GreenCom.2012.40
6. Hussain, M., Mehmood, A., Khan, S., Khan, M. A., & Iqbal, Z. (2019). A survey on authentication techniques for wireless body area networks. *Journal of Systems Architecture*, 101655.
7. Mayer, C. P. (2009). Security and privacy challenges in the internet of things. *Electron Communication EASST*, *17*, 1–12. https://doi.org/10.14279/tuj.eceasst.17.208
8. Ham, H. S., Kim, H. H., Kim, M. S., & Choi, M. J. (2014). Linear SVM-based android malware detection for reliable IoT services. *Journal of Application Mathematics, 2014*. https://doi.org/10.1155/2014/594501
9. Kalra, S., & Sood, S. K. (2015). Secure authentication scheme for IoT and cloud servers. *Pervasive and Mobile Computing, 24*, 210–223.
10. Bebortta, S., Senapati, D., Panigrahi, C. R., & Pati, B. An adaptive modeling and performance evaluation framework for edge-enabled green IoT systems. In *IEEE transactions on green communications and networking*. https://doi.org/10.1109/TGCN.2021.3127487
11. Santoso, F. K., & Vun, N. C. H. (2015). Securing IoT for smart home system. *International Symposium on Consumer Electronics (ISCE), 2015*, 1–2. https://doi.org/10.1109/ISCE.2015.7177843
12. Bebortta, S., Senapati, D., Panigrahi, C. R., & Pati, B. An adaptive performance modeling framework for QoS-aware offloading in MEC-based IIoT systems. In *IEEE internet of things journal*. https://doi.org/10.1109/JIOT.2021.3123554.
13. Lee, J., Lin, W., & Huang, Y. (2014). A lightweight authentication protocol for internet of things. *International Symposium on Next-Generation Electronics (ISNE), 2014*, 1–2. https://doi.org/10.1109/ISNE.2014.6839375
14. Braeken, A. (2018). PUF based authentication protocol for IoT. *Symmetry, 10*, 352. https://doi.org/10.3390/sym10080352
15. Srinivas, J., Das, A. K., Li, X., Khan, M. K., & Jo, M. (2021). Designing anonymous signature-based authenticated key exchange scheme for internet of things-enabled smart grid systems. *IEEE Transactions on Industrial Informatics, 17*(7), 4425–4436. https://doi.org/10.1109/TII.2020.3011849
16. Narwal, B., & Mohapatra, A. K. (2020). A survey on security and authentication in wireless body area networks. *Journal of Systems Architecture*, 101883.
17. Narwal, B., & Mohapatra, A. K. (2021). SAMAKA: Secure and anonymous mutual authentication and key agreement scheme for wireless body area networks. *Arabian Journal for Science and Engineering*, 1–23.

Random Connected Graph Generation for Wireless Sensor Networks

Soumyasree Talapatra and Avirup Das

Abstract For analysing networks like social media networks, wireless sensor networks, etc. in many applications, generating random connected graph is very important. As it is time consuming to generate the random connected graph consisting of large nodes it is necessary to generate it in minimum time. Characteristics like dependent edges and non-binomial degree distribution that are absent in many classical random graph models such as the Erdos-Renyi graph model can be captured by random graphs with a given degree range. The problem of random connected graph generation having a prescribed degree range has been addressed here. Random graphs are used to model wireless sensor networks (WSNs) or IoT comprising of sensor nodes with limited power resources. A fast and light-weight algorithm has been proposed in this paper to produce a random connected graph for a real-time multi-hop wireless sensor networks (WSNs). Results show that our method has better performance than other existing methods.

Keywords Graph theory · Random connected graph · Wireless sensor networks · Connectivity · Sensor deployment

1 Introduction

Graph processing is one of the integrable and has significance in research area. Graphs are used as data structure in many applications such as social networking, image processing, data mining and IoT network. In today's world, a rapid growth of complex system is taking place due to the advancement of modern technologies. Systems such as, social networks, Internet are being modelled and analysed by random graphs. To analyse a very large complex system, a massive random network efficiently need to be generated.

S. Talapatra (✉) · A. Das
Advanced Computing and Microelectronics Unit, Indian Statistical Institute, Kolkata, India
e-mail: soumyasree.talapatra@gmail.com

B. Pati et al. (eds.), *Proceedings of the 6th International Conference on Advance Computing and Intelligent Engineering*, Lecture Notes in Networks and Systems 428,
https://doi.org/10.1007/978-981-19-2225-1_14

In 1959, as an extension of the probabilistic method, Paul Erdős and Alfred Renyi introduced the concept of a random graph to discover the existence of certain graph properties [1]. Random graphs have been applied more broadly to solve combinational problems and used for better understanding on graph behaviours. It has been extended to modularise various random-like networks, such as the unpredictable growth of the Internet's web graph, the spread of human population, social relations, and neural networks.

The term "random graph" was first mentioned in a paper by Erdos, where it was used in a remarkable proof of a theorem regarding the existence of certain graphs that demonstrated the power of the probabilistic procedure. For example, it observes that how a structure of random graph can evolve when the number of edges get increased. Since the foundation of the theory of random graphs by Erdos and Rényi, various random graph models have been introduced and studied. A random graph is a graph obtained by adding successive edges with isolated vertices at a random way. The theory of random graphs provides a framework for the understanding of probability distributions over graphs and also useful to understand stochastic processes that happen over a network. A random connected graph is a graph that is connected, i.e. there is a path exists from any vertices to any other vertices in the graph. Random connected graphs are used to capture various aspects, such as communication link and degree distribution. Random graphs are a fundamental property for indicating reliability of multi-hop wireless sensor networks, which are also being analysed and modelled by random graphs.

In various field, nowadays, wireless sensor networks and IoT have gained a great deal of attention. The term wireless has turned into a generic and extensively global term employed to describe communications in which electromagnetic waves (EMWs) are employed in sending signal to several or the entire path of the communication.

A wireless sensor networks (WSNs) can be defined as a self-configured and infrastructure-less wireless networks to monitor physical or surrounding environmental conditions to simultaneously pass their data through the network to a main location or sink, where the data can be observed and analysed which can communicate the information gathered from a monitored field through wireless connection.

In recent years, there has been considerable research interest on algorithms for information exchange, estimation and computation over networks. Such algorithms have a variety of potential applications in sensor networks, peer-to-peer networks, wireless networks and networked control systems.

As collected information needs to be sent to data collection or processing centres, so the connectivity of a WSNs is usually studied by considering a graph associated with that network. This is only possible if a path exists from each node to that collection centre. Due to which connectivity of random graphs is a main subject for theoretical analyses of WSNs because their sensor nodes are usually deployed randomly as shown in Fig. 1, so, they are regarded as wireless ad-hoc networks. Nodes and edges in graphs, respectively, correspond to sensor nodes and communication links in WSNs [2].

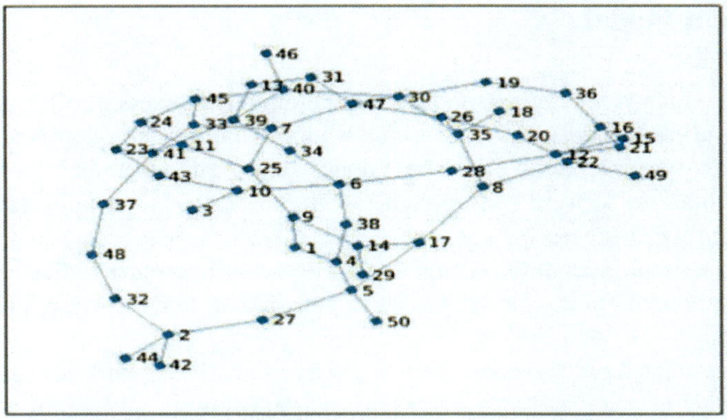

Fig. 1 Representation of a WSN or IoT city using a graph

Due to popularity of random graphs or networks in modelling and simulating, many complex real-world systems such as the Internet [3], social [4] networks, it has gained a significant increase of interest to understand how the systems work through obtaining rigorous mathematical and simulation results. Many random graph models such as the Erdős-Rényi [5], the Chung-Lu [6] models have been proposed to capture various characteristics of real-world systems. Of these systems, one of the important aspect is degree sequence which has a wide and significant application in the areas including structural reliability and communication networks because of the strong ties between the degrees of vertices and the structural properties of dynamics networks [7].

Most previously studied algorithms for this problem either sometimes get stuck or produce loops or multiple edges in the output, requiring frequent restarts. For such algorithms, the probability of a restart needed on a trial rapidly approaches 1 as the degree parameters grow, resulting in an enormous number of trials needed on average to obtain even one graph [8]. A major advantage of our algorithm is that it never gets stuck and a careful selection of order of degree selection is done.

The problem of generating a random connected graph with a given degree sequence becomes considerably easier if self-loops are allowed. Throughout this paper, we considered random connected graphs with no self-loops. By using the Havel-Hakimi method [9], a deterministic graph can be generated following a given degree sequence.

Without any maintenance, sensor nodes are expected to retain required functions of the whole network. In practice, efficient establishment of connectivity is a central issue in design of WSNs [10].

The remainder of this paper is organised as follows. Section 2 graph models are briefly introduced. Section 3 describes the procedure with algorithm. In Sect. 4, we have discussed about results and compared with other existing algorithm. Finally, we have concluded the work in Sect. 5.

2 System Model

A graph in which vertices corresponds to the communication nodes often represents a wireless ad-hoc network or WSNs which compose multiple autonomous, tiny, low cost and low-power sensor nodes. These nodes gather data about their environment and collaborate to forward sensed data to centralised back-end units called base stations for further processing [11].

In such a graph, a connected edge from one node to another indicates that the node corresponding to the former can send data directly to the corresponding node to the latter.

If this associated graph is connected, then we can call the network as connected. A graph G is connected if and only if there exists a path between any pair of its vertices [12]. Any pair of nodes can communicate with each other if a network is connected, possibly taking multiple jumps through relay nodes. Sometimes we can consider stronger forms of connectivity, such as k-connectivity, in which the network remains connected even if $k - 1$ nodes are removed. If a network is k-connected ($k \geq 2$), it has better fault-tolerance than if it is merely 1-connected.

2.1 Graph Model

The Erdős-Gallai theorem is a result in graph theory, a branch of combinatorial mathematics. It provides one of two known approaches to solving the graph realisation problem, i.e. it gives a necessary and sufficient condition for a finite sequence of natural numbers to be the degree sequence of a simple graph. A sequence obeying these conditions is called "graphic." The theorem was published in 1960 by Paul Erdos and Tibor Gallai, after whom it is named.

A graph $G(n, p)$ is a random graph with n nodes where each possible edge has a probability p of existing. The number of edges in a $G(n, p)$ graph is a random variable with expected value np. We will be mostly focussing on the Erdo s-Renyi graph.

- If $np < 1$, a graph in $G(n, p)$ will almost surely have no connected components of size larger than $O(log(n))$
- A graph in $G(n, p)$ will almost surely have a largest component whose size is of order n^2 if $np = 1$.

For a sequence of numbers to be graphic, the conditions of the Erdős-Gallai theorem are necessary. The inequality between the sum of the largest degrees and the sum of the remaining degrees can be established by double counting: the left side gives the numbers of edge-vertex adjacency among the highest-degree vertices, each such adjacency must either be on an edge with one or two high-degree endpoints, the term on the right gives the maximum possible number of edge-vertex adjacency in which both endpoints have high degree, and the remaining term on the right upper bounds the number of edges that have exactly one high-degree endpoint. For any sequence of numbers obeying these conditions, there exists a graph for which it is the degree sequence, and the most difficult part of the proof is shown by Havel-Hakimi theorem.

Havel-Hakimi Theorem: The Havel-Hakimi algorithm is an algorithm in graph theory solving the graph realisation problem. That is, it answers the following question: Given a finite list of non-negative integers, is there a simple graph, such that its degree sequence list. Here, the "degree sequence" is a list of numbers that for each vertex of the graph states how many neighbours it has. For a positive answer, the list of integers is called graphic. The algorithm constructs a special solution if one exists or proves that one cannot find a positive answer. This construction is based on a recursive algorithm. The algorithm was developed by Havel and Hakimi [9].

3 Algorithm

Random connected graph is very necessary for analysing the IoT or WSNs. Here, we have developed a light-weight algorithm for generating random connected graph. First, we are generating a connectivity matrix for the graph G. Then, we can easily plot the graph from the matrix. We considered here that this method has a degree boundation. And highest degree can be $D_{\max}$ and lowest degree can be $D_{\min}$ for each node. Randomly it will be decided the actual degree d of of a node i using this procedure, where, $D_{\min} \leq d \leq D_{\max}$. Number of row is $\frac{N}{D_{\max}}$ and number of column is $D_{\max}$. At first, in each row randomly generated $D_{\max} - 1$ elements in degree d_i of node i is $D_{\min} \leq d_i \leq D_{\max}$. And put 0 in the last position of the row to maintain the size of the row $D_{\max}$. Then, if summation of the row is even then replace the last 0 value with the existing highest even degree (let l) present in that row and increment l elements by 1 in that row. Otherwise, if summation is odd then 0 value is replaced with 1 and any 1 element's value increment by 1. By this procedure, we are getting the connectivity matrix and then we can plot the random connected graph G.

4 Simulation Results

We validated our algorithm using simulation in MATLAB. Also, we have compared our method with [13] and it is clear that our algorithm performs better in terms of execution time. It is very necessary to execute the algorithm in lower time and should be light weight because IoT or sensor device is low computational capability with low storage space.

Figure 2 is showing that average diameter of a random connected graph for different degree range. For example, when degree range is 2–4 then average diameter is about 12. Here, average is taking on 100 runs. Number of nodes is 50 for the the connected graph but different results taken for different degree range.

Algorithm 1: Connectivity matrix generation for random connected graph

Input: Max degree D_{max}, minimum degree D_{min} ($D_{min} \geq 1$), number of nodes N
Output: Connectivity matrix for random connected graph G
Number of row is $\frac{N}{D_{max}}$
Number of column is D_{max}
for *each cluster i for generating connectivity matrix* **do**
    every cluster has D_{max} no. of nodes;
    generate $D_{max} - 1$ random no. in range of D_{max} *to* D_{min};
    add 0 in the last position, to maintain the length of the row D_{max} of connectivity matrix;
    if *sum of the elements of a row is even* **then**
        replace the 0 of last place with highest even degree present in that row;
        if last element is l;
        increment l elements by 1 of that row;
    else
        replace the 0 of last place with 1;
        increment any one element by 1 of that row;
    end
end

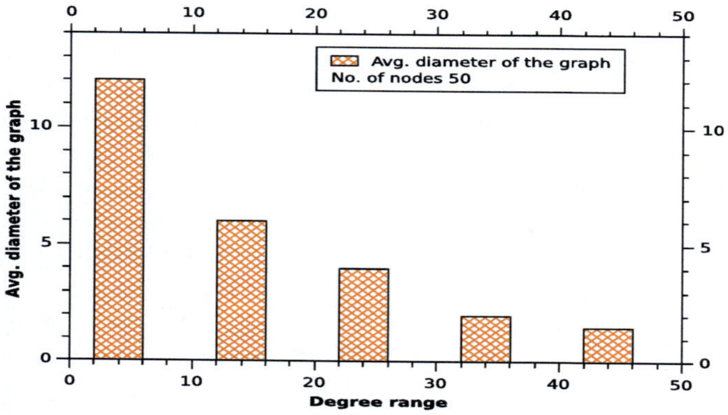

Fig. 2 Avg. diameter of a random connected graph with varying degree range

In Fig. 3, we have shown execution time of the algorithm in sec. and compared with an existing work [13], where our method is taking less time for executing the algorithms or generating the random connected graph. Here, we have taken degree range 2–4 of all the nodes and taken execution time by varying the number of nodes.

In Fig. 4, we have shown execution time of our algorithm with varying the degree range while fixed the number of nodes. Execution time is increasing with the degree range because more degree range means more complex graph, more number of edges in a graph. Our method is taking less time to generate a random connected graph which is feasible to implement in real-time network.

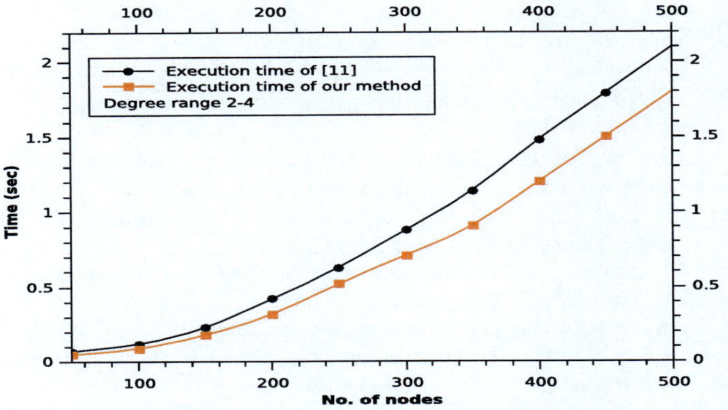

Fig. 3 Execution time to generate a random connected graph with varying no. of nodes

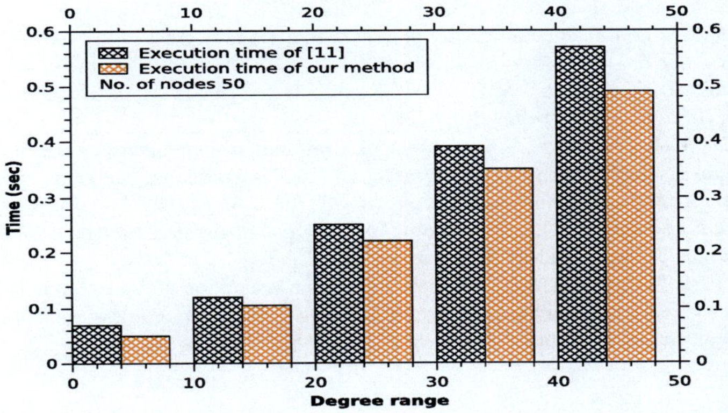

Fig. 4 Execution time to generate a random connected graph with varying degree range

5 Conclusion

We are trying to develop a presented and efficient algorithm for generating random connected graph. It can be used in studying various structural properties and dynamics over a network, sampling graphs uniformly at random from the graphical realisations by estimating the number of possible graphs with a given number of node's. The algorithm never gets stuck, and can generate every possible graph with a positive probability, and executes in less time. Our algorithm to generate a random connected graph is light weight, so it can apply on real-time IoT or sensor devices in a real world. We believe the algorithms will contribute significantly in analysing and mining emerging complex systems and understanding interesting characteristics of such networks.

References

1. Kang, M., & Petrasek, Z. (2015). *Random graphs: theory and applications from nature to society to the brain.*
2. Li, J., Andrew, L. L., Foh, C. H., Zukerman, M., & Chen, H.-H. (2009). Connectivity, coverage and placement in wireless sensor networks. *Sensors, 9*(10), 7664–7693.
3. Siganos, G., Faloutsos, M., Faloutsos, P., & Faloutsos, C. (2003). Power laws and the as-level internet topology. *IEEE/ACM Trans Netw, 11*(4), 514–524.
4. Yang, J., & Leskovec, J. (2015). Defining and evaluating network communities based on ground-truth. *Knowl Inf Syst, 42*(1), 181–213.
5. Bollobás, B. (1998). Random graphs. In *Modern graph theory* (pp. 215–252). Springer.
6. Chung, F., & Lu, L. (2002). The average distances in random graphs with given expected degrees. *Proc Nat Acad Sci, 99*(25), 15879–15882.
7. Boccaletti, S., Latora, V., Moreno, Y., Chavez, M., & Hwang, D.-U. (2006). Complex networks: structure and dynamics. *Phys Rep, 424*(4–5), 175–308.
8. Bhuiyan, M. H., Khan, M., & Marathe, M. (2017). A parallel algorithm for generating a random graph with a prescribed degree sequence. CoRR, abs/1708.07290.
9. Hakimi, S. L. (1962). On realizability of a set of integers as degrees of the vertices of a linear graph i. *J Soc Ind Appl Math, 10*(3), 496–506.
10. Al-Karaki, J. N., & Kamal, A. E. (2004). Routing techniques in wireless sensor networks: a survey. *IEEE Wirel Commun, 11*(6), 6–28.
11. Saha, D., & Das, A. (2015). Coverage area maximization by heterogeneous sensor nodes with minimum displacement in mobile networks. In *IEEE international conference on advanced networks and telecommunications systems (ANTS)* (pp 1–6).
12. Kandel, A., Bunke, H., & Last, M. (2007). *Applied graph theory in computer vision and pattern recognition*. Springer Science & Business Media.
13. Viger, F., & Latapy, M. (2005). Efficient and simple generation of random simple connected graphs with prescribed degree sequence. In L. Wang (Ed.), *Computing and combinatorics* (pp. 440–449). Berlin, Heidelberg: Springer.

A Novel Application of HPSOGWO Trained ANN in Nonlinear Channel Equalization

Pradyumna Kumar Mohapatra, Ravi Narayan Panda, Saroja Kumar Rout, Rojalin Samantaroy, and Pradeep Kumar Jena

Abstract In a communication channel, there is a possibility of distortions such as ISI, CCI, and another source of noise that interfere with useful signals, and the signal becomes corrupted. Therefore, equalizers are needed to counter such types of distortions. In this paper, we presented a nature-inspired hybrid algorithm which is an amalgamation of PSO and GWO. The proposed algorithm is called HPSOGWO. During this work, we pertain to ANN trained with the proposed HPSOGWO in the channel equalization. The foremost initiative is to boost the flexibility of the variants of the proposed algorithm and the utilization of proper weight, topology, and transfer function of ANN in the channel equalization. The performance of the proposed equalizer can be evaluated by estimating MSE and BER by considering popular nonlinear channels and added with nonlinearities. Extensive simulations show the performance of our proposed equalizer, better than existing NN-based equalizers also as neuro-fuzzy equalizers.

Keywords ANN · PSO · HPSOGWO · Nonlinear channel equalization

1 Introduction

Recuperating information from communication channels, adaptive channel equalizers are needed. In [1], restoration of signals from a nonlinear channel, the authors have proposed optimal preprocessing strategies. For perfect reform of discrete data transmission during a channel, Touri et al. [2] have suggested some ideas. Soft computing tools-based adaptive equalizers such as neural networks-based equalizers

P. Kumar Mohapatra (✉) · R. Samantaroy
Vedang Institute of Technology, Bhubaneswar, Odisha, India
e-mail: er_pradyumna@yahoo.co.in

R. Narayan Panda · S. Kumar Rout
Vardhaman College of Engineering, Hyderabad, India

P. Kumar Jena
GITA, Bhubaneswar, Odisha, India

B. Pati et al. (eds.), *Proceedings of the 6th International Conference on Advance Computing and Intelligent Engineering*, Lecture Notes in Networks and Systems 428,
https://doi.org/10.1007/978-981-19-2225-1_15

[3] have been developed in the late 80s. For nonlinear and complex channels, it has been observed that these methods are very suitable. Patra et al. [4] used Chebyshev artificial neural network for the problem of equalization. JPRNN-based equalizer using RTRL developed by Zhao et al. [5] provides some encouraging results. The amalgamation of FIR filter and FLNN with adaptive decision feedback equalizer (DFE) is introduced in [6]. Zhao et al. [7–11] introduced several distinctive types of NN-based equalizers to solve these complex issues. As far as the problem of equalization [5–13] is concerned, ANN has been served as the best tool despite complex problems. Since conventional training algorithms did not succeed in many cases, training ANN with different optimization algorithms involving bio-inspired computation was applied in many applications [14–16]. ANN [17–28] trained successfully with evolutionary algorithms are applied in nonlinear channel equalization. Fascinatingly, in these works, HPSOGWO to neural network training was used to discover ideal weights, transfer functions, and an appropriate topology of a given network's neuron. Thus, with the successful application of HPSOGWO to optimize all network parameters. The author proposed an equalizer whose performance and execution that outperforms contemporary ANN-based [5–7] and neuro-fuzzy [12, 13] equalizers available in the literature which real essence in this paper.

The principal outlines of this article are as follows: Section I deals with introduction after that problem statement in Sect. 2, followed by the proposed system model in Sect. 3. The simulation study is conceded for performance evaluation which is discussed in Sect. 4. Lastly, in Sect. 5, it dealt with a conclusion.

1.1 Contributions

The main contribution can be focused as:

- Development of learning procedure in ANN.
- Make use of HPSOGWO trained with neural networks in channel equalization.
- Nonlinearities used in this works for performance evaluation of different channels are different.

2 Problem Description

The communication system model is described in Fig. 1. The following expression [13] was identified in terms of co-channel and impulse response of the channel.

$$H_i(z) = \sum_{m=0}^{L_i - 1} b_{i,m} z^{-m} \quad 0 \le i \le n \tag{1}$$

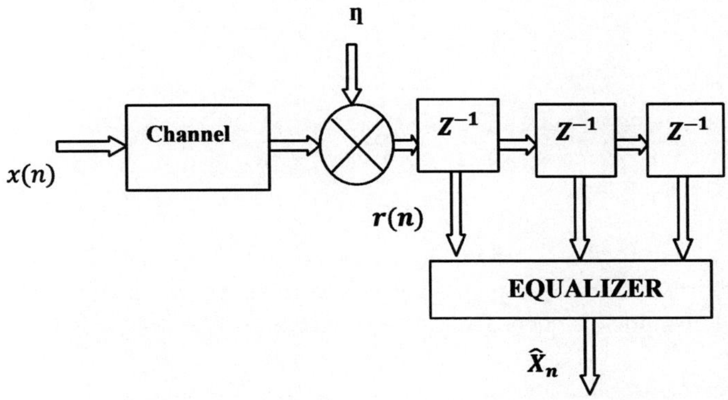

Fig. 1 Model of communication system

where L_i and $b_{i,m}$ represent as length and tap weights of the ith impulse response of the channel. We suppose a parallel model, which creates the study is straightforward, even though the fact that it can be expanded to any communication system in common. The following conditions satisfy for transmitted symbols $x_i(n), 0 \leq i \leq n$ which are set of independent, *i, i, d* dataset consists of $\{\pm 1\}$ which are mutually independent.

$$E[x_i(n)] = 0 \tag{2}$$

$$E[x_i(n_1)x_m(n_2)] = \delta(i - m)\delta(n_1 - n_2) \tag{3}$$

Here, $E[.]$ denotes expectation operator and impulse response can be represented as

$$\delta(n) = \begin{cases} 1n = 0 \\ 0n \neq 0 \end{cases} \tag{4}$$

The output of the channel described as

$$r(n) = d(n) + d_o(n) + \eta(n) \tag{5}$$

where $d(n)$ is received signal of the channel,$d_o(n)$ is nosy signal, and $\eta(n)$ is the element of noise supposed to be Gaussian with variance, i.e.,

$$E[(\eta^2)] = \sigma_n^2$$

The expression of $d(n)$ and $d_o(n)$ can be expressed as

$$d(n) = \sum_{k=0}^{L_0-1} b_{0,m} x_0(n-m) \quad (6)$$

$$d_o(m) = \sum_{j=1}^{n} \sum_{k=0}^{L_0-1} b_{j,m} x_j(n-m) \quad (7)$$

Based on the channel observation vector, $r(n) = [r(n), r(n-1), \ldots, r(n-l+1)]^{\mathrm{T}}$, the objective of the equalizer is to approximate the transmitted sequence $x_0(n-k)$,where equalizer order l is and the delay factor is m.

Error be able to calculated as:

$$e(n) = d(n) - r(n) \quad (8)$$

Because of $e^2(n)$, the instantaneous power of the difference signal can be identified as a cost function that is always positive and is replaced as $e(n)$.We will adopt an algorithm to update the weights iteratively so that $e^2(n)$ must be minimum and reduced to zero.

3 Proposed Model

The proposed equalizer of the system model in this work is an application of the optimized value of HPSOGWO, GWO, and PSO variant, trained with a multilayer artificial neural network where each of its neurons trained with these variants. Specialized points of interest of the paper are proved from its novelty and execution comes about. We experienced the capability of hybrid variants that were run 20 times on cost function. To search for the best solutions, proposed algorithms have to be run at least more than ten to fifteen times. It is once more a common strategy that HPSOGWO is run on a trial issue numerous epoch, and the leading ideal results, in the final creation, are assessed as measurements of execution. The author compares the execution of PSO and GWO with the proposed hybrid algorithm in provisions of best ideal and factual comes about. Based on contrast, HPSOGWO gives results of superior quality with rational iteration and prevents premature convergence.

3.1 ANN Model

Here, the model of ANN like a human brain that is competent of adjusting to varying circumstances and learns rapidly within the redress setting. It is a mechanism for the recreation of the human brain. Neurons are basic elements of ANN. Each neuron of ANN is interconnected in an organized way. To generate output, the neurons get inputs from another neuron by firing through their synapse. The output of neurons can be generated by performing the sum of weighted of inputs of the neuron through a transfer function. Generally, ANN maybe consists of one or more hidden layers including an input layer and output layer. The real computations that ANN performs of the network is done in the hidden layers. The network will be active and will produce desired output when it is given with the proper set of inputs, weights of neurons, and transfer function (it may be different for different neurons but usually the same). To ease the suitable arrangement of a network, training of the ANN is required.

3.1.1 Neural Networks Training

Back propagation is one of the few strategies, as far as training of the ANN is concerned. The information required to train an ANN that takes input so that it will generate possible outputs.

Following procedures are used for training:

1. First take the inputs and possible outputs.
2. Calculate and add the weight of all inputs, then go via transfer functions.
3. Evaluate estimated output using actual output.
4. Based on the comparison, the fitness value can be calculated and updated.
5. Do steps 2 and 3 again unless and until training reached the suitable stage.
6. To optimize fitness, adjust weights in the right direction.
7. Repeat steps 1–6 in anticipation of satisfactory fitness value is set up.

To train a network, the back-propagation method [29] is supposed for modification of weights but it may get a long training time. The training algorithm proposed by the author in this work is an HPSOGWO.

3.1.2 The Transfer Functions

The input of every neuron is connected with a transfer function that can activate it. The weighted sum of the inputs of neurons is calculated. A good choice of the transfer function is a sigmoid function which is applied in this work. Equation (9) shown below expressed as a sigmoid function which is in the range [0, 1]:

$$\text{Sigmoid}(x) = \frac{1}{1 + e^{-ax}} \tag{9}$$

3.2 Variant of PSO

Kennedy and Eberhart [30] and del Valle et al. [31] were initially presented the PSO algorithm. Its primary decision was mainly stimulated by the recreation of the societal behavior of creatures such as fish schooling and bird flocking. When the birds are looking for food, they are either dispersed or have recently settled together in the place or they can discover the nutrition. When the birds are looking for food from one location to another, there is a bird that can smell the food very well on an ongoing basis; that is, the fowl is conscious of where the food can be found, getting the right message of nourishing properties. Since they transmit the information at any point, particularly the valuable information, the birds must eventually run to the place where nourishment can be found when looking for food from one place to the next. This method is erudite from the conduct of animals to measure the functions of worldwide optimization, and each swarm accomplice is called a particle. There are two stages of PSO; velocity and positions of the particles are updated. The global search space of the position of each partner of the swarm is updated by following mathematical expressions.

$$V_i^{n+1} = V_i^n + C_1\mathrm{rand}_1\left(P\mathrm{best}_i^n - X_i^n\right) + C_2\mathrm{rand}_2\left(g\mathrm{best}^n - X_i^n\right) \tag{10}$$

$$X_i^{n+1} = X_i^n + V_i^{n+1} \tag{11}$$

Where V_i^{n+1} is the particle i of velocity at $(n + 1)$ iteration, X_i^{n+1} is the particle's new location value, C_1 and C_2 are factors related to the cognitive weighting and social weighting, respectively, rand_1 and rand_2 are stochastic parameters.

3.3 GWO Algorithm

The algorithm, which is a meta-heuristic algorithm based on population [32] which imitates the hunting strategy of the pack of gray wolves is called the GWO algorithm. They have a place to the first-class family of the nourishment series and hence keep up a social chain of command as takes after. The alpha is considered a pioneer of the pack as a whole. Beta found wolves of the second rank. The most exceedingly bad dim wolves are omega. In case the wolf has no place, the location of the prey to be called delta in any of the over groups. The other oddly social activity of gray wolves in extending to the social level is party hunting. The following points concern the hunting of gray wolves.

- Hunting, tracking, and drawing closer to the prey.
- Search for, surround, and harass the victim until it stops moving.
- Assault upon the victim.

The encompassing approach is represented as follows in the normal GWO:

$$\vec{r} = \overrightarrow{c}.\vec{x}_{p(n)} - \vec{x}(n) \tag{12}$$

$$x(n+1) = \vec{x}_{p(n)} - \vec{a}.\vec{r} \tag{13}$$

From the above equations, n is the number of iterations, random vectors are $\vec{a}$ and $\vec{c}$, the preys location is $\vec{x}_p$, and the wolves position is $\vec{x}$. Alpha is the best solution of the gray wolf technique, the second and third best solutions being beta and delta. The three main arrangements in this algorithm are chronicled, and the majority of the wolves (omegas) upgrade their positions to the best arrangements.

The hunting technique is, for these reasons, modeled as it takes.

$$\vec{r}_\alpha = |\vec{c}_1 \cdot \vec{x}_\alpha - \vec{x}|, \vec{r}_\beta = |\vec{c}_2 \cdot \vec{x}_\beta - \vec{x}|, \vec{r}_\delta = |\vec{c}_3 \cdot \vec{x}_\delta - \vec{x}| \tag{14}$$

$$\vec{x}_1 = \vec{x}_\alpha - \vec{a}_1 \cdot \vec{r}_\alpha, \vec{x}_2 = \vec{x}_\beta - \vec{a}_2 \cdot \vec{r}_\beta, \vec{x}_3 = \vec{x}_\delta - \vec{a}_3 \cdot \vec{r}_\delta \tag{15}$$

$$\vec{x}(n+1) = \frac{\vec{x}_1 + \vec{x}_2 + \vec{x}_3}{3}. \tag{16}$$

Here, the locations of alpha, beta, and delta wolves are identified as $\vec{x}_\alpha$, $\vec{x}_\beta$, and $\vec{x}_\delta$ correspondingly $\vec{a}_1, \vec{a}_2, \vec{a}_3, \vec{c}_1, \vec{c}_2, \vec{c}_3$ are represented as random vectors.

- When random value says $|A| < 1$ the wolves are compelled to attack the prey, and the exploitative ability is to attack the prey otherwise they move away from the victim.
- Discovery capabilities are to hunt for predators.

3.4 *Proposed Hybrid Algorithm*

Talbi's proposed algorithm [33] leveraging low-level coevolution for global mixed hybrid optimization by hybridizing PSO with the algorithm GWO called it HPSOGWO. The main purpose is to develop diversification and to conquer premature convergence. The proposed hybrid algorithm is called as low level since we unite both variants of PSO and GWO. Both algorithms run in parallel. The two variants mixed and developed the concluding results to the problems. Based on this update, we tend to particle swarm optimization to boost the exploitative power with the exploratory power in the gray wolf optimizer to provide the intensity of each model. The author suggested the following equation by updating the location of the three agents in the search space rather than mistreatment normal mathematical equations. We prefer to handle the discovery and exploitation of the gray wolf by inertia constant within the search region. The following equations can be expressed as in modified

Fig. 2 HPSOGWO pseudocode

```
Initialize l, a, w and c
velocity = 0.3 * randn
w = 0.5 + rand()/2
estimate the fitness of agents by taking eqn (5)
do
for every search agent
Modify speed and location by taking eqn (6)
end for
Modify l, a, w and c
estimate suitability of all quest agents
Modify locations of first 3 agents
n = n + 1
while (n < maximum no. of iteration)
Return // Location of first best search agent
```

form. Figure 2 shown the pseudoccode of the proposed HPSOGWO algorithm.

$$\vec{r}_\alpha = |\vec{c}_1 \cdot \vec{x}_\alpha - w \cdot \vec{x}|, \vec{r}_\beta = |\vec{c}_2 \cdot \vec{x}_\beta - w \cdot \vec{x}|, \vec{r}_\delta = |\vec{c}_3 \cdot \vec{x}_\delta - w \cdot \vec{x}| \quad (17)$$

The velocity and new position of the updated equations can be formed by combining PSO with GWO variants can be expressed as follows:

$$V_i^{n+1} = w * [V_i^n + C_1\text{rand}_1(X_1 - X_i^n) + C_2\text{rand}_2(X_2 - X_i^n) + C_3\text{rand}_3(X_3 - X_i^n) \quad (18)$$

$$X_i^{n+1} = X_i^n + V_i^{n+1} \quad (19)$$

4 HPSOGWO, Construction, and ANN Training

Training ANN with HPSOGWO is the main concern of this paper. For this, we must build population-based network. In this research, 30 network populace performs well. Through this population, HPSOGWO communities are created and initialized. This arrangement is termed the topology of the system. The following training procedure is applied for HPSOGWO with ANN:

- Observe the training samples for each network and report the number of network errors.
- Obtain the best problem space network, analyze all the errors.
- Identify the network which has achieved the desired minimum error, exit the program, and record its weights.
- Else, for each network HPSOGWO can change its vectors of position and velocity.

- Do again step 1.

When a particle reaches the desired fitness, which means that a result has been obtained after it transitions from being an employee seeking a resolution to becoming a manager in ANN development. With a several control variables, the equalization of the communication networks is a difficult issue. On this issue, ANN has yet to be an exceptional tool. The reason given is to build and train ANN for calculating channel state. Here, [±1] take as training data. The input to the network which was built used these values. Equation (20) describes network fitness which was to calculate the MSE for the entire training collection.

$$\text{Valu of Fitness} = \sum \text{Recorded Value} - \text{Predicted value of Network} \tag{20}$$

A network is deemed available after it has satisfied any marginal execution needs. The requirement used was the measurable computation identified as the coefficients of correlation which showed in Eq. (21).

$$\mu^2 = 1 - \frac{\text{Value of Fitness}}{\sum \text{Recorded value}--\text{Mean Recorded value}} \tag{21}$$

Here, in this work, HPSOGWO trains the entire network, which is built as a multi-layer artificial neural network whose parameters such as weight, topology, transfer function, are suitably optimized.

5 Training Procedure for HPSOGWO with ANN

The author proposed the training algorithm shown in Fig. 3, ANN defines rules for an organization that acts as a Boss of a company to supply assets(which are nothing but the parameters to be to optimized) to HPSOGWO which behaves as a manager which gives the directions for employee. ANN learning the equalization problem. In this method, ANN acts both as a Boss and as an employee. The flowchart shown below describes pseudocode for the problem. The number of particles and also the number of hidden nodes were identified by P and Q. This training algorithm's flowchart is shown in:

6 Simulations and Discussion

The pseudocodes of all the algorithms are coded in the MATLAB R2015a and implemented on 15.6-inch FHD Intel® HD Graphic with memory 16 GB, 2 × 8 GB, DDR4, 2666 MHz. We have taken 40 numbers as search agents. Number of iterations is 10000 for channel-0 and 500 for channel-1 and channel-2, allowable error are set at 10^{-3}, $c1$

Fig. 3 Training algorithm of the proposed equalizer

```
Initialize ANN-as a BOSS
    For j = 1,2, ......P
        Make HPSOGWO-as manager (j)
        for ANN-as employee k = 1,2, ......P
        create ANN- as employee
        end
    end
 whilst solution is not establish
      evaluate update
       put number of maximum iterations
       for (HPSOGWO-as manager j = 1,2, ......P)
        as (iterations<allocations)
        for (ANN-as employee k = 1,2, ......Q)
         test ANN-as employee(k)
         end
        for (ANN-as employee k = 1,2, ......Q)
        Modify the weights of ANN-as employee (k)
         end
    end
    Return global best
end
Update global best
end
```

$= c2 = 0.5$ and $c3 = 0.7$, $w = 0.7 + \text{rand}()/2$, and $l \in [2, 0]$; such parameter settings are used to measure efficiency of the hybrids and other metaheuristics. Here, the value of P is set at 30 and Q is 5. The performance of the equalizer can be evaluated by using the following three nonlinear channels as shown in Table 1, and the type of nonlinearity introduced as shown in Table 2. The channel input signals are chosen as i.i.d. sequences with a mean of zero. The noise introduced in the channel is additive white Gaussian noise with zero mean and it is not dependent on the channel input.

Table 1 Types of nonlinearity

SL. No	Type of nonlinearity
NLO	$y(n) = \tan h[x(n)]$
NL1	$y(n) = x(n) + 0.2x^2(n) - 0.1x^3(n)$
NL2	$y(n) = x(n) + 0.2x^2(n) - 0.1x^3(n) + 0.5\cos[\pi x(n)]$

Table 2 Types of channel

SL. NO	Channel	Channel type
CH0	$H(Z) = 0.260 + 0.930Z^{-1} + 0.260Z^{-2}$	MIXED
CH1	$H(Z) = 0.303 + 0.9029Z^{-1} + 0.3040Z^{-2}$	MIXED
CH2	$H(Z) = 1 - 0.90Z^{-1} + 0.3850Z^{-2} + 0.7710Z^3$	MIXED

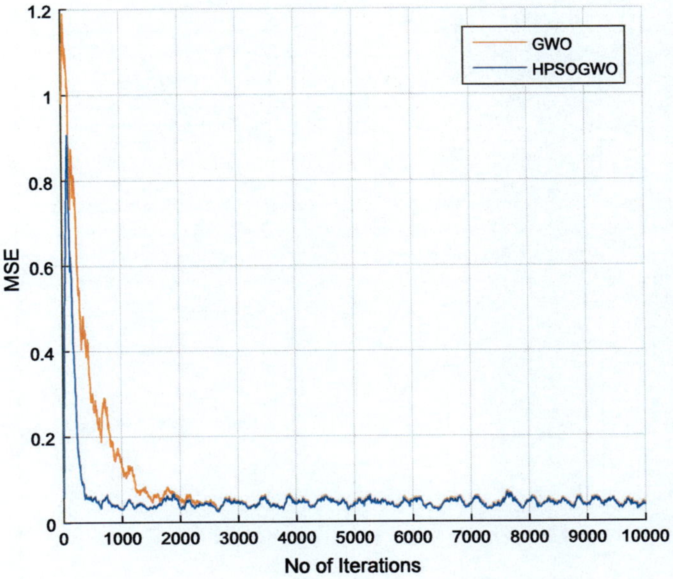

Fig. 4 MSE performance for CH0

The performance evaluation of channel equalizer can be estimated by plotting two parameters, i.e., MSE and BER.

Example 1:

Consider the second-order nonlinear channel (CH0) model having a transfer function as shown in the following Eq. (22).

$$H(Z) = 0.26 + 0.93Z^{-1} + 0.26Z^{-2} \tag{22}$$

The nonlinearity of the above channel model is introduced in the following Eq. (23) to exemplify the consequence on the equalizer performance.

$$y(n) = \tan h[x(n)] \tag{23}$$

We have compared our proposed equalizer with PSO by Das et al. [16], GWO [32] which parameters have already been defined. Mean square error (MSE) was plotted in Fig. 4 with fixed SNR at 15 dB and corresponding bit error rate (BER) plotted in Fig. 5. It is observed from Fig. 5, the performance of our equalizer better than GWO. From Fig. 5 BER performance, it is revealed that up to 5 dB SNR, the performance of all three equalizers is comparable and after that our proposed equalizer outperforms GWO and PSO.

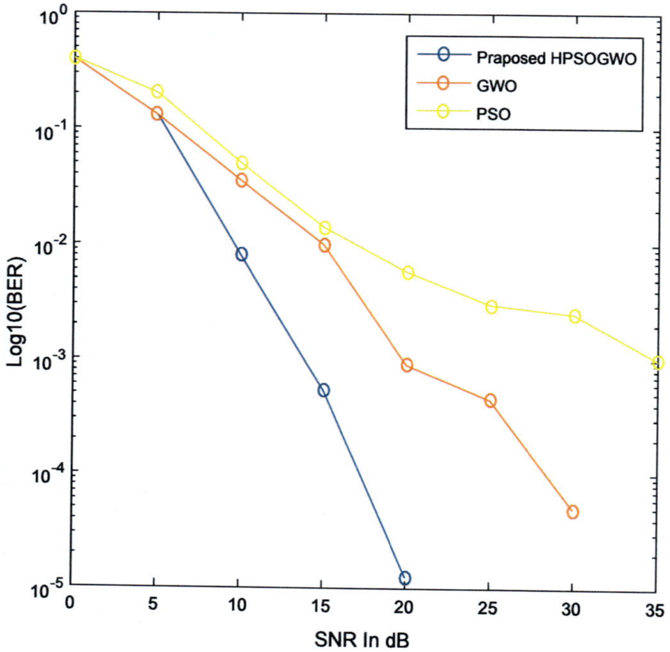

Fig. 5 BER performance for channel CH0

Example 2:

In this example, the author used another popular nonlinear second-order channel model as defined in Table 1 as CH1 and nonlinearity introduced as shown in Table 2 as NL1. The mean square error (MSE) and bit error rate (BER) are plotted in Figs. 6 and 7.

From the simulation result (Figs. 6 and 7) of the above example, it is noted that our planned equalizer is better than BP-ANN [34], PSO, and GWO.

Example 3:

In this example, we have presented widely used third-order nonlinear channel [35] as described in Table 1 as CH2 and nonlinearity introduced as NL2 have shown in Table 2 for simulations.

From Fig. 8, simulations were performed at specific SNR and the plot of MSE of CH2 for both GWO and PSOGWO. Our proposed equalizer compares in this example with the equalizers existing in the literature, neuro-fuzzy equalizers: type-2 TSKFNS [12], GA-NFN [36], PSO, and GWO to compute BER under similar conditions discussed above and result plotted in Fig. 8. From this simulations, it is observed that after 4 dB SNR performance our planned equalizer outperforms GA-NFN, type-2 TSKFNS, PSO, and GWO at all noise conditions.

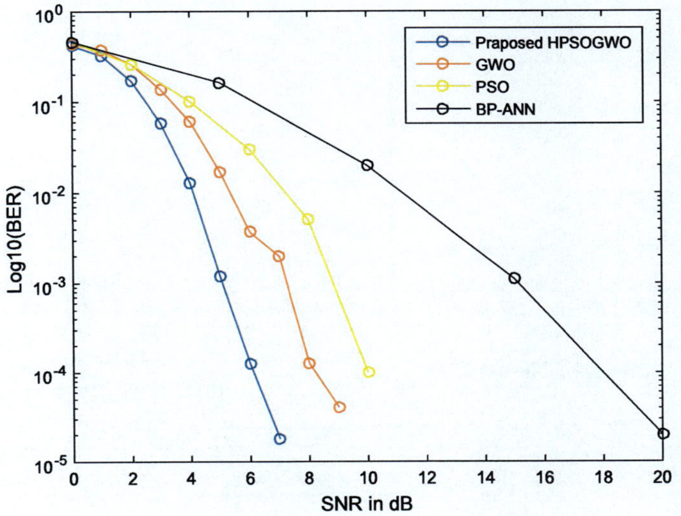

Fig. 6 BER performance for channel CH1

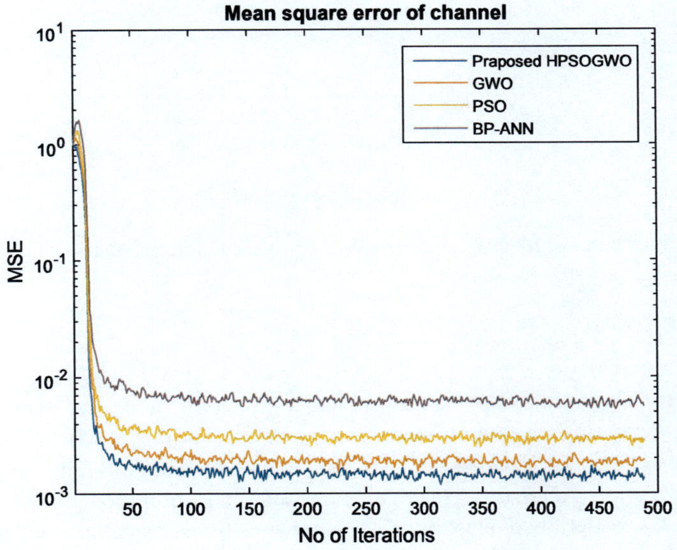

Fig. 7 MSE performance for CH1

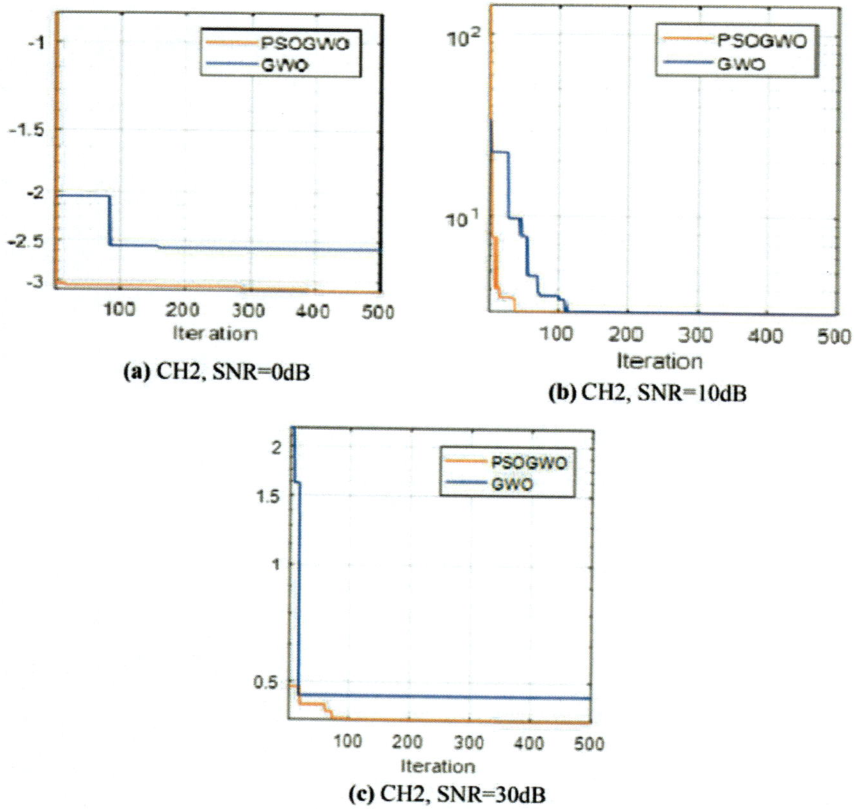

Fig. 8 MSE performance for GWO and PSOGWO equalizer at different SNR

7 Conclusion

The gradient algorithm-based equalizers fail to precisely model the channel characteristics many times in the presence of a burst error. In this article, we proposed a training strategy for proposed HPSOGWO equalizer. The said equalizer trained HPSOGWO with ANN in channel equalization. It is observed that our equalizer executes better than existing neural network-based equalizers in all noise circumstances.

In this works, we explore efficient methods for equalization. This paper makes utilize a hybrid algorithm for the said problem. The use of population-based algorithms, network topology, and transfer function for ANN training is the essence of the contributions made by this paper. Approaches for the proposed equalizer used here in this paper provide thought-provoking results in the literature of equalizers which were found to be better than equalizers based on ANN as well as neuro-fuzzy equalizers. Also, MSE and BER of the proposed equalizers found to perform better in all noise conditions without prior knowledge of SNR.

References

1. Voulgaris, P.G., & Hadjicostics, C. N. (2004). Optimal processing strategies for perfect reconstruction of binary signals under power-constrained transmission. In *Proceedings of IEEE Conference on Decision and Control*, Atlantis, Bahamas, Vol. 4, pp. 4040–4045.
2. Touri, R., Voulgaris, P. G., & Hadjicostis, C. N. (2006) Time varying power limited preprocessing for perfect reconstruction of binary signals. In: *Proceedings of the 2006 American control conference, Minneapdis*, USA, 2006, pp. 5722–5727.
3. Patra, J. C., Pal, R. N., Baliarsingh, R., & Panda, G. (1999). Nonlinear channel equalization for QAM signal constellation using artificial neural network. *IEEE Transactions on Systems, Man, and Cybernetics Part B Cybetnetics, 29*(2).
4. Patra, J. C., Poh, W. B., Chaudhari, N. S., & Das, A. (2005). Nonlinear channel equalization with QAM signal using Chebyshev artificial neural network. In *Proceedings of the International Joint Conference on Neural Networks*, Montreal, Canada, 2005, pp. 3214–3219.
5. Zhao, H., et al. (2011). A novel joint-processing adaptive nonlinear equalizer using a modular recurrent neural network for chaotic communication systems. *Neural Networks, 24*, 12–18.
6. Zhao, H., et al. (2011). An adaptive decision feedback equalizer based on the combination of the FIR and FLNN. *Digittal Signal Processings, 21*, 679–689.
7. Zhao, H., et al. (2011). Pipelined functional link artificial recurrent neural network with the decision feedback structure for nonlinear channel equalization. *Information Sciences, 181*, 3677–3692.
8. Zhao, H., et al. (2009). Adaptively combined FIR and functional link neural network equalizer for nonlinear communication channel. *IEEE Transactions on Neural Networks, 20*(4), 665–674.
9. Zhao, H., et al. (2012). Complex-valued pipelined decision feedback recurrent neural network for nonlinear channel equalization. *IET Communications, 6*(9), 1082–1096.
10. Zhao, H., et al. (2010). Adaptive reduced feedback FLNN nonlinear filter for active control of nonlinear noise processes. *Signal Processings, 90*(3), 834–847.
11. Zhao, H., et al. (2010). Nonlinear adaptive equalizer using a pipelined decision feedback recurrent neural network in communication systems. *IEEE Transactions on Communications, 58*(8), 2193–2198.
12. Abiyev, R. H., et al. (2011). A type-2 neuro-fuzzy system based on clustering and gradient techniques applied to system identification and channel equalization. *Applied Soft Computing, 11*, 1396–1406.
13. Panigrahi, S. P., Nayak, S. K., & Padhy, S. K. (2008). A genetic-based neuro-fuzzy controller for blind equalization of time-varying channels. *Wiley Inter Science International Journal of Adaptive Control and Signal Processing, 22*, 705–716.
14. Yogi, S., Subhashini, K. R., & Satapathy, J. K. (2010). A PSO based functional link artificial neural network training algorithm for equalization of digital communication channels. In *International Conference on Industrial and Information Systems* (pp. 107–112).
15. Chau, K. W. (2006). Particle swarm optimization training algorithm for ANNs in stage prediction of Shing Mun River. *Journal of Hydrology, 329*, 363–367.
16. Das, G., Pattnaik, P. K., & Padhy, S. K. (2014). Artificial neural network trained by particle swarm optimization for non-linear channel equalization. *Expert Systems with Applications, 41*(7), 3491–3496.
17. Lee, C. H., & Lee, Y. C. (2012). Nonlinear systems design by a novel fuzzy neural system via hybridization of electromagnetism-like mechanism and particle swarm optimisation algorithms. *Information Sciences, 186*(1), 59–72.

18. Lin, C.-J., & Liu, Y.-C. (2009). Image backlight compensation using neuro-fuzzy networks with immune particle swarm optimization. *Expert Systems with Applications, 36*(3, Part 1), 5212–5220.
19. Lin, C. J., & Chen, C. H. (2011). Nonlinear system control using self-evolving neural fuzzy inference networks with reinforcement evolutionary learning. *Applied Soft Computing, 11*(8), 5463–5476.
20. Hong, W.-C. (2008). Rainfall forecasting by technological machine learning models. *Applied Mathematics Computation, 200*(1), 41–57.
21. Potter, C., Venayagamoorthy, G. K., & Kosbar, K. (2010). RNN based MIMO channel prediction. *Signal Processing, 90*(2), 440–450.
22. Ingle, K. K., & Jatoth, R. K. (2021). A new training scheme for neural network based non-linear channel equalizers in wireless communication system using Cuckoo Search Algorithm. *AEU International Journal of Electronics and Communications, 138*, 153371.
23. Ingle, K. K., & Jatoth, R. K. (2020). An efficient JAYA algorithm with lévy flight for non-linear channel equalization. *Expert Systems with Applications, 145*, 112970.
24. Panigrahi, S. P., Nayak, S. K., & Padhy, S. K. (2008). Hybrid ANN reducing training time requirements and decision delay for equalization in presence of co-channel interference. *Applied Soft Computing, 8*, 1536–1538.
25. Mohapatra, P., Samantara, T., Panigrahi, S. P., & Nayak, S. K. (2018). Equalization of communication channels using GA-trained RBF networks. In *Progress in Advanced Computing and Intelligent Engineering* (pp. 491–499). Springer, Singapore.
26. Panda, S., Mohapatra, P. K., & Panigrahi, S. P. (2015). A new training scheme for neural networks and application in non-linear channel equalization. *Applied Soft Computing, 27*, 47–52.
27. Mohapatra, P., Panda, S., & Panigrahi, S. P. (2018). Equalizer modeling using FFA trained neural networks. In *Soft computing: Theories and applications* (pp. 569–577). Springer, Singapore.
28. Nanda, S. J., & Jonwal, N. (2017). Robust nonlinear channel equalization using WNN trained by symbiotic organism search algorithm. *Applied Soft Computing, 57*, 197–209.
29. Rumelhart, D. E., Geoffey, E. H., & Ronald, J. W. (1986). Learning representations by back-propagating errors. *Nature, 323*(6088), 533–536.
30. Kennedy, J., & Eberhart, R. (1995). Particle swarm optimization. In *Proceedings of the Sixth International Symposium on Micro Machine and Human Science*, Nagoya, Japan, 1995, (p. 3943).
31. del Valle, Y., Venayagamoorthy, G. K., Mohagheghi, S., Hernandez, J. C., & Harley, R. G. (2008). Particle swarm optimization: Basic concepts, variants and applications in power systems. *IEEE Transactions on Evolutionary Computation, 12*, 171–195.
32. Mirjalili, S., Mirjalili, S. M., & Lewis, A. (2014). Grey wolf optimizer. *Advances in Engineering Software, 69*, 46–61.
33. Talbi, E.-G. (2002). A taxonomy of hybrid metaheuristics. *Journal of Heuristics, 8*(5), 541–564.
34. Zhao, H., Zeng, X., Zhang, J., Li, T., Liu, Y., & Ruan, D. (2011). Pipelined functional link artificial recurrent neural network with the decision feedback structure for nonlinear channel equalization. *Information Sciences, 181*, 3677–3692.
35. Liang, J., & Ding, Z. (2004). FIR channel estimation through generalized cumulant slice weighting. *IEEE Transactions on Signal Processing, 52*(3), 657–667.
36. Panigrahi, S. P., Santanu, K. N., & Sasmita, K. P. (2008a). A genetic-based neuro-fuzzy controller for blind equalization of time-varying channels. *International Journal of Adaptive Control and Signal Processing, 22*, 705–716. WileyInterscience.

IoT-Based Smart Railway Management for Passenger Safety and Comfort

Dethe Tukaram, B. Uma Maheswari, and S. Ullas

Abstract In developing countries such as India, the transportation industry plays a critical role in the economy. Transportation is a vital means of transporting products and people from one location to another. Increased trade and business are facilitated by improved transportation, primarily railways. However, the railway is currently beset by several accident issues. The manual management of such a system is impossible every day. If it is done manually, it will take more time and money. A prototype for automatically examining and identifying cracks, obstacles, and fire in railway tracks is proposed in this paper keeping in mind the passenger safety and comfort. To save electricity, the planned system will turn lights and fans on and off automatically based on the number of people in the coach. The proposed model consists of an IR sensor, ultrasonic sensor, flame sensor, touch sensor, and an emergency switch. If a flame or obstruction is detected, the train will be delayed and eventually stop. The status of all sensors is updated on the Web site to assist railway administrators.

Keywords Internet of Things (IoT) · WSN · Railway infrastructure · Condition-based maintenance (CBM) · Inspection · Smart railway · Power utilization · IR sensor · Ultrasonic sensor · Flame sensor · Touch sensor

D. Tukaram (✉) · B. Uma Maheswari · S. Ullas
Department of Computer Science and Engineering, Amrita School of Engineering, Amrita Vishwa Vidyapeetham, Bengaluru, India
e-mail: tukaramdethe91@gmail.com

B. Uma Maheswari
e-mail: b_uma@blr.amrita.edu

S. Ullas
e-mail: s_ullas@blr.amrita.edu

B. Pati et al. (eds.), *Proceedings of the 6th International Conference on Advance Computing and Intelligent Engineering*, Lecture Notes in Networks and Systems 428,
https://doi.org/10.1007/978-981-19-2225-1_17

1 Introduction

For developing countries, smart transportation systems are critical for transporting goods and moving people from one location to another. Particularly, smart railway system is very much essential to save peoples' lifetime and money. The effective transport system leads to more trade and business along with the safety. There is currently no viable technique for detecting railway track cracks. Some of the currently available technologies, such as infrared thermography, ground-penetrating radar, and conductivity, are useful for detecting railway track cracks; however, they are inaccurate [1]. Sensor networks [2] can be deployed for monitoring sewages [3, 4], health [5, 6], in smart vehicles [7–9], and so on. This study is primarily focused on enabling a smart railway system that can meet public demand for long-distance transport. However, many problems, such as track cracks, obstruction on the track, and train fires, can develop accidentally. Manually detecting these issues on a regular basis necessitates a great amount of manpower and financial planning. Train tracks, carriages, tunnels, bridges, catenary, and electronic devices along the trackside are all part of the railway organization. It is critical for railway management to guarantee that all objects and devices in the railway system are in good condition. Any operating defects are fictitious and strictly prohibited because unanticipated faults could jeopardize the safety of a significant number of passengers. Hence, railway servicing/maintenance has become a major issue for the railway management system. As a result of this issue, the government places a huge pressure on railway officials to fully engage them in the repair process. Consequently, the railway management is expected to have significant difficulties in carrying out the maintenance. According to the survey [8], 42% of bridges and 47% of rail tunnels in South Korea have sustained train loads for more than 30 years. Railway authorities are being compelled to assign extra staff to track maintenance in order to keep track of the railway tracks. In this situation, railway management has started focusing on the concept of condition-based maintenance (CBM) [10–12]. As CBM has certain challenges such as requirement of expertise, this paper addresses this issue using multiple sensors.

Motivation and Contribution

India is advancing in a smart direction these days, supporting smart cities, and smart railway stations. In this regard, this work is motivated and contributed in automatic detection of cracks in the railway track, obstacles, and fire. Furthermore, the system will turn on and off the lights and fans using touch sensors to save energy. It consists of an IR sensor, ultrasonic sensor, flame sensor, and an emergency switch. The ultrasonic sensor detects the obstacle in front of the train. The IR sensor detects the crack in the railway track. This system uses a flame sensor to detect the fire in the train. The train will slow down and stop if it detects impediments or flames. The touch sensor is used to turn lights and fans on and off. If there are passengers in the coach, the lights and fans will automatically turn on; if there are none, the lights and fans will be turned off. If passengers want to turn off the fans and lights, there are also options available. The sensors' status is captured in a database, and status of sensors is updated on a Web page. The crack positions that have been detected will also be sent to the nearest station.

Section 2 describes related work. Section 3 explains the proposed method to detect cracks in the track, obstacles, and fire. Section 4 describes the implementation and snapshots showing the results. Section 5 concludes the work with possible future research direction.

2 Related Work

This section discusses the work carried out by the researchers in railway management. The authors Kovacevic et al. [13] proposes a new approach for the assessment of railways condition. The mainstream of the railway structure in Croatia is more than 100 years old. In the last 30 years, the absence of investments in maintenance and regeneration project has worsen the situation. The authors describe a joint initiative among research and infrastructure managements may boost safety and decrease the cost of the management system solution. To discover parameters influencing track efficiency, researchers used electromagnetic ground-penetrating radar (GPR), drones, seismic refraction, and in-situ geotechnical studies [14]. In this task, sensors with varying energy and processing power are primarily assumed. The existing work in [11] basically focuses on modeling of inspection or replacement policy to observe the performance of stochastically deteriorating systems by using multi-level control boundary rule. In this work, the replacement threshold and examination schedule have been used as decision variables in order to fix the maintenance issues in the system. The train control system (TCS) has been deployed to show the efficiency of LTE in a real-time environment [15]. The work is basically focusing on the quality of service for the railway communication system to assure the reliability over TCS. Many technologies, including infrared thermography, ground-penetrating radar (GPR), and conductivity, have been used to assess the current state of a variety of structures, including highways, bridges, and railway lines [1]. The graphical inspection method was employed in the project [16] to locate and analyze the crack in the railway track. In [17], the authors introduced a Monte Carlo localization technique and ultrasonic sensors for quickly detecting track cracks.

The system described in [17] includes a GPS module and a GSM modem for communication and identification of people crossing railway tracks, as well as crack detection and transmission of crack parameter coordinates to the nearest railway station. Currently, railway track inspection is carried out manually. However, due to a rapid change in the environment, such as rain, night patrolling during the monsoon, hot and cold weather patrolling, manual screening is sometimes not practicable. Ground-penetrating radar (GPR) and seismic methodology are suggested in to evaluate the circumstances of the substructure of the railway track such as ballast, sub-ballast, and subgrade. The work presented in [18] address a few issues that physically challenged people and senior citizens have when traversing the footbridge between the two railway platforms.

Following an examination of existing work, it is understood that several techniques have been used to detect activities occurring in real time on bridges, airport pavement, and railway tracks. Infrared thermography, ground-penetrating radar (GPR), and conductivity are some of the technologies used, but these approaches are insufficient to produce accurate results for useful decision making. This work attempts to develop a smart railway management system that can make reliable decisions in order to avoid accidents by detecting cracks in railways.

3 Proposed Approach

The modules used for railway track crack detection, obstructions and fire detection, and lights and fans switch on and off automatically are described in this section. The modules involved in the overall architecture of the proposed system are depicted in Fig. 1.

Figure 2 illustrates the various sensors connected to Arudino Uno in the proposed system. It is comprised of the components listed below.

- Power Supply
- Arduino Uno
- IR Sensor
- Ultrasonic Sensor
- Flame Sensor
- Touch Sensor
- LCD
- Motor
- IOT Server.

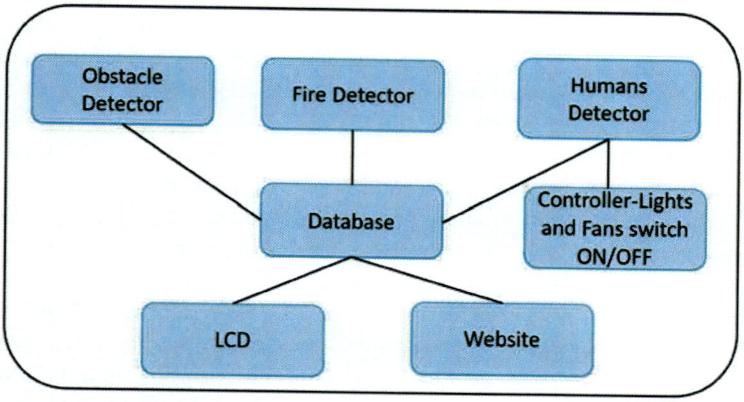

Fig. 1 Modules of the proposed system

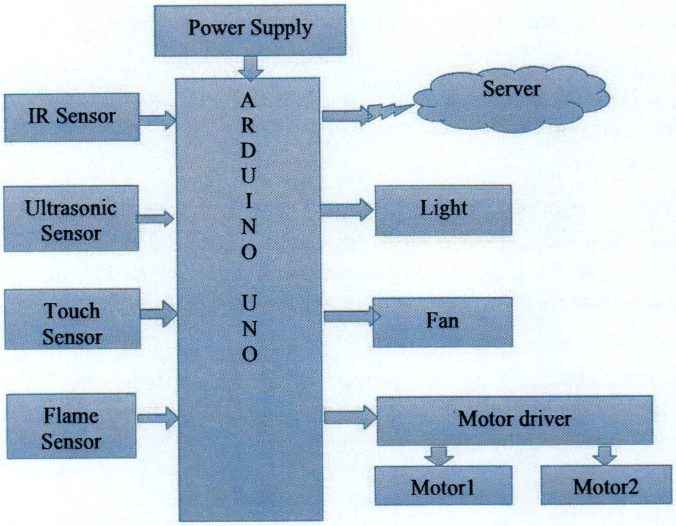

Fig. 2 Illustration of sensors and other components connectivity to the Arduino Uno

Four wireless sensors are linked to an Arduino Uno in Fig. 2: an infrared sensor, an ultrasonic sensor, a flame sensor, and a touch sensor. These sensors will monitor activity within the train and on the railway track, which will then be notified to the user/administrator via Arduino Uno. The LCD also displays the status of these sensors. Furthermore, the information gathered is sent to a database and updated on the Web site at the same time. The following are the descriptions of each module:

A. **Power Supply**

Power supply stores in battery and connected to Arduino Uno. When it is joined to an external circuit, it flows and delivers energy to external devices. When the batteries connect to an external circuit, electrolytes will move as ions within and it will start chemical reaction to be completing at the isolated terminals and thus, it supplies energy to the outside device. Within, the battery ions which permits current to the flow of battery performance work.

B. **Arduino Uno**

Arduino Uno is a microcontroller board, and it is equipped with digital and analog input/output pins. It has totally 14 digital pins and six analog pins. It can be programmed using the Arduino IDE and can communicate with a computer. Multiple sensors such as IR sensors, ultrasonic sensors, flame sensors, and touch sensors are connected to this board.

C. **IR Sensors**

IR sensor monitors and measures infrared radiation in its surroundings. In our proposed system, cracks and the corresponding coordinates of a damaged track are detected using the infrared sensor.

D. **Ultrasonic sensor**

The ultrasonic sensor identifies obstacles on the railway track by producing high-frequency sound waves. If it detects something, the train will slow down and then abruptly stops.

E. **Flame sensor**

The flame sensor is used for flame/fire detection. If a fire is detected anywhere on the train, the train will slow down and come to a complete stop. Flame sensors are useful in preventing catastrophic failure caused by fire.

F. **Touch sensor**

The lights and fans are turned on and off using the touch sensor. If there are passengers inside, the lights and fans will turn on automatically. When passengers do not want the lights or the fans, they can use the manual switch. The sensors' status will be updated on the Web page.

G. **LCD**

On the LCD, all of the results are displayed. A railway track crack, an impediment, or a fire will all be reported if they are discovered. On the LCD board, the status of all sensors will be displayed, as well as the status of the lights and fans, indicating whether they are turned on or off.

H. **IOT server**

Sensor data is saved on an IoT server, and information is relayed to a local railway station to notify the system's current status, including which operations are working and which are not.

I. **Motor**

The robot is propelled by motors. The robot can move forward as well as backward in space. The information obtained from the IoT server is used to operate the robot. Table 1 depicts the robot's movement in forward, backward, right, and left directions.

Table 1 Robot moving direction

Forward	+	–	+	–
Backward	–	+	–	+
Right	+	–	–	–
Left	–	–	+	–

4 Implementation and Results

4.1 Simulation Results

The system is simulated with Proteus, and the hardware prototype is implemented with embedded C. Figure 3 shows connectivity between sensors used and Arduino Uno in Proteus. This simulation process consists of four sensors attached to an Arduino Uno: an infrared sensor, an ultrasonic sensor, a flame sensor, and a touch sensor. The activities/events are sensed using sensors, and the Arduino Uno compiles the data and sends it to the LCD display.

The Arduino Uno is also connected the server. The data captured is also forwarded to a server, where it is updated on the Web site. The robot's movement is controlled by two DC motors. After sensing the activity, the Arduino Uno sends a signal to the DC motor; if it detects any obstacles, cracks, or fire, the motor is instantly halted. On the LCD panel, all of the results are displayed. The LCD in Fig. 4 shows the message **OBST-NO, FLAME-NO, CRACK-DE, L , F-OFF**, when the robot detects a crack. Figure 5 shows the message **OBST-DET, FLAME-NO, CRACK-NO, L,F-OFF**, when the robot detects an obstacles in front of the train.

Figure 6 shows the message **OBST-NO, FLAME-DE, CRACK-NO, L, F-OFF**, when the robot detects fire.

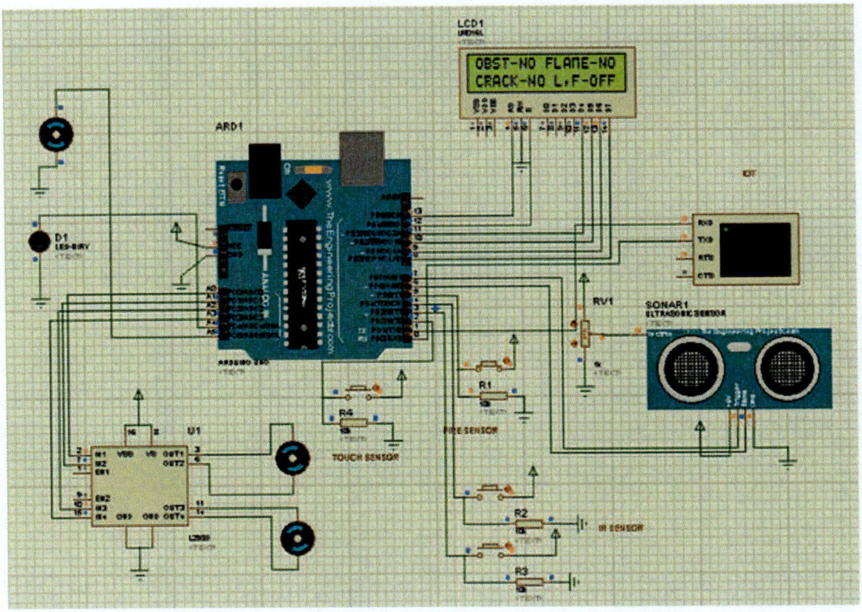

Fig. 3 Illustration of circuit design in Proteus

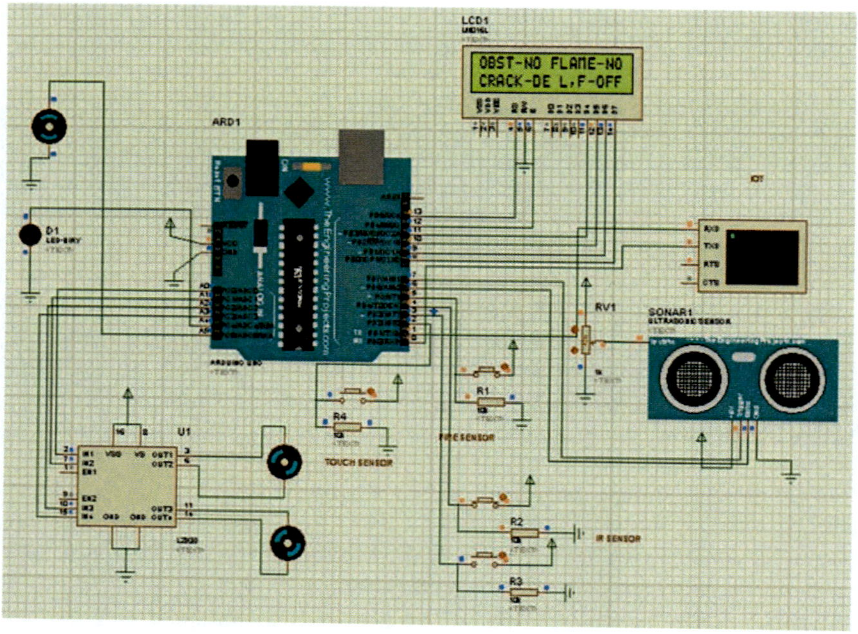

Fig. 4 Railway track crack detection

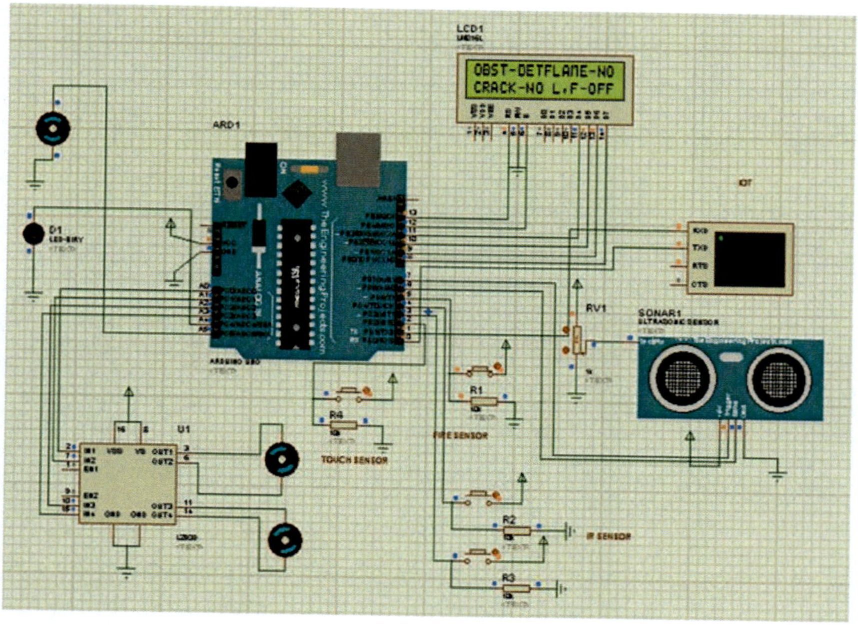

Fig. 5 Obstacles detection in front of the train

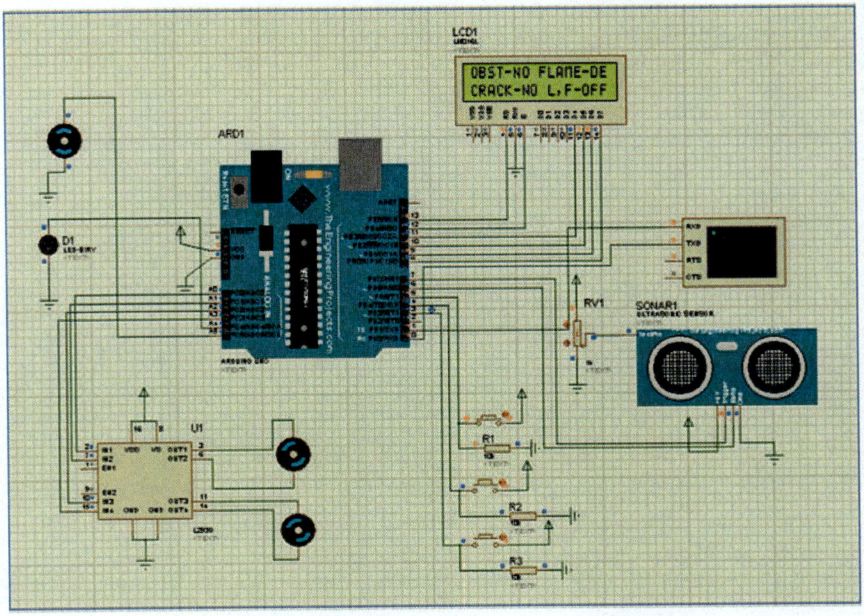

Fig. 6 Fire detection

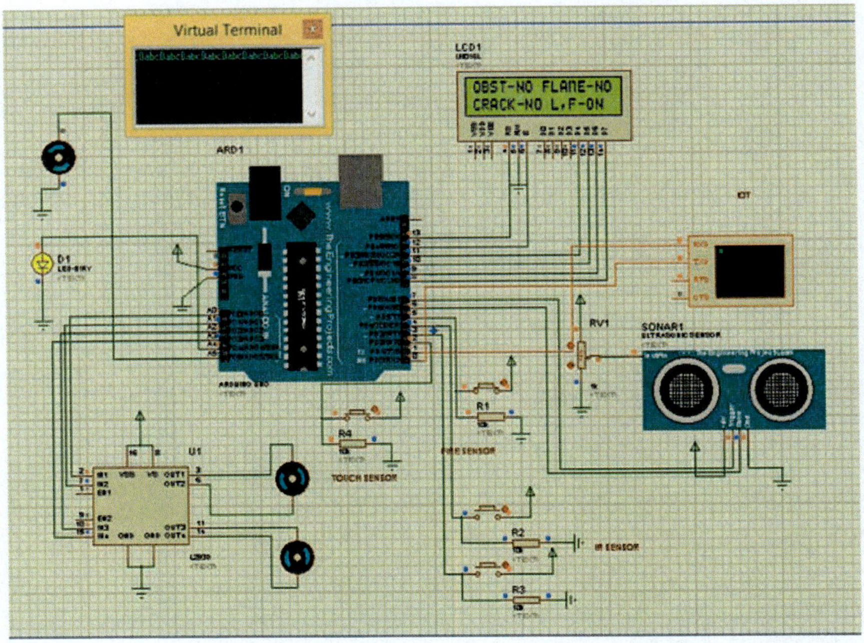

Fig. 7 Light and fan

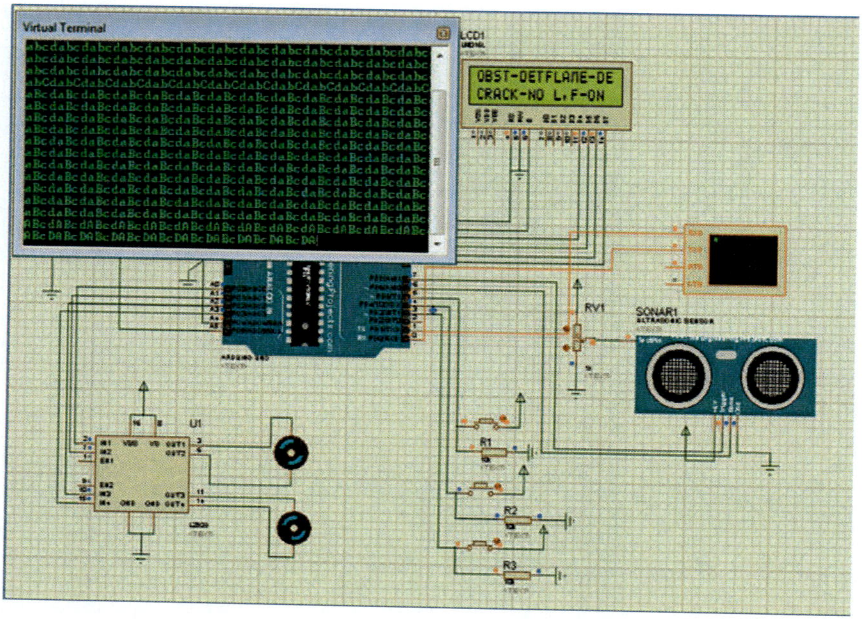

Fig. 8 Result in virtual terminal

Figure 7 shows the message **OBST-NO, FLAME-NO, CRACK-NO, L, F-ON**, when the passengers are inside the train, otherwise light and fan will be OFF.

The result is shown in alphabets on a virtual terminal, as seen in Fig. 8. If the system identifies obstacles, the letter 'A' (capital A) is displayed; otherwise, the letter 'a' (small a) is displayed. If a railway track crack is identified, the letter 'B' (capital B) is displayed; otherwise, the letter 'b' (small b) is displayed. If there is a fire, the letter 'C' (capital C) is displayed; otherwise, the letter 'c' (small c) is displayed. If both the light and the fan are turned on, it displays 'D' (capital D), otherwise 'd' (small d).

4.2 Hardware Prototype

Figure 9 depicts the robot (hardware prototype) designed with various sensors/components and their interconnection. Figures 10, 11, 12, and 13 show the working of all sensors and the corresponding LCD displays.

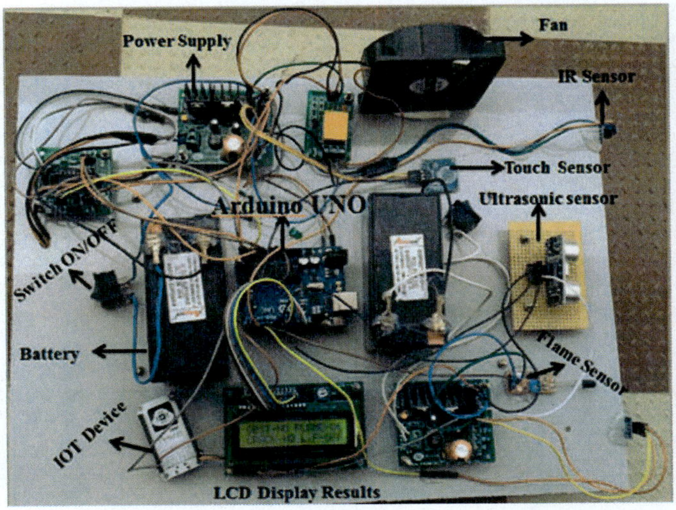

Fig. 9 Illustration of hardware prototype designed

Fig. 10 Demonstration of crack detection

5 Conclusion and Future Scope

In this work, a prototype has been developed to detect railway track cracks, track obstructions, fire detection, and light and fan status. It also introduces IoT and sensor network usages toward achieving smart/better railway system. This paper proposes a smart railway management system that will result in a better railway transportation

Fig. 11 Demonstration obstacle detection

Fig. 12 Demonstration fire detection

system. As a future work, sensors can be used for predictive maintenance, train control, and train and passenger safety and can change the way railways work. This work considers mainly focused on normal speed passenger trains. Smart IoT-enabled railway management system can be deployed in semi-speed and high-speed trains with high accuracy sensors which could detect more precisely the events to prevent sudden train accident and to provide comfortable journey to the people.

Fig. 13 Demonstration of light and fan status

References

1. Clark, M., Gordon, M., & Forde, M. C. (2004). Issues over high-speed non-invasive monitoring of railway trackbed. *NDT & E International, 37*(2), 131–139.
2. Rakesh, N., Shakir, M., Kalamani, P., & Maheswari, B. U. (2017). An energy saving algorithm using heterogeneity aware protocol in wireless sensor networks to sustain lifetime of nodes. In *International Conference on Inventive Systems and Control (ICISC)*, Coimbatore (pp. 1–5).
3. Panigrahi, C. R., Tiwari, M., Pati, B., & Prasath, R. (2014). Malware detection in big data using fast pattern matching: A Hadoop based comparison on GPU. In *Mining intelligence and knowledge exploration* (pp. 407–416). Springer.
4. Rishitha, K., & Ullas, S. (2019). IoT based automation in domestic sewage treatment plant to optimize water quality and power consumption. In *3rd International Conference on Computing Methodologies and Communication (ICCMC)* (pp. 306–310). https://doi.org/10.1109/ICCMC.2019.8819700
5. Nithia, P., & Ullas, S. (2018). An IoT based safety & health advisory system for homemaker with service independency on various sensors. In *Second International Conference on Inventive Communication and Computational Technologies (ICICCT)* (pp. 537–541). https://doi.org/10.1109/ICICCT.2018.8473143
6. Mohanty, J., Mishra, S., Patra, S., Pati, B., & Panigrahi, C. R. (2021). IoT security, challenges, and solutions: A review. In *Progress in advanced computing and intelligent engineering* (pp. 493–504).
7. Singh, D., Pati, B., Panigrahi, C. R., & Swagatika, S. (2020). Security issues in IoT and their countermeasures in smart city applications. *Advanced Computing and Intelligent Engineering, 1089*, 301–313.
8. Sravya, C., Maheswari, B. U., & Sudarshan, T. S. B. (2017). Cluster based fault tolerant architecture for multicasting data in wireless sensor network. In *2017 International Conference*

On Smart Technologies For Smart Nation (SmartTechCon) (pp. 514–517). https://doi.org/10.1109/SmartTechCon.2017.8358426
9. Karanam, V., Maheswari, B. U., & Sudarshan, T. S. B. (2017). Overlay based fault tolerant peer to peer multicasting for emergency data communication in VANETS. In *2017 International Conference on Smart Technologies for Smart Nation (SmartTechCon)* (pp. 510–513). https://doi.org/10.1109/SmartTechCon.2017.8358425
10. Jardine, A. K., Lin, D., & Banjevic, D. (2006). A review on machinery diagnostics and prognostics implementing condition-based maintenance. *Mechanical systems and signal processing, 20*(7), 1483–1510.
11. Bebortta, S., Senapati, D., Panigrahi, C. R., & Pati, B., An adaptive performance modeling framework for QoS-aware offloading in MEC-based IIoT systems. *IEEE Internet of Things Journal*. https://doi.org/10.1109/JIOT.2021.3123554
12. Márquez, F. P. G., Schmid, F., & Collado, J. C. (2003). A reliability centered approach to remote condition monitoring. A railway points case study. *Reliability Engineering & System Safety, 80*(1), 33–40.
13. Bebortta, S., Senapati, D., Panigrahi, C. R., & Pati, B., An adaptive modeling and performance evaluation framework for edge-enabled green IoT systems. *IEEE Transactions on Green Communications and Networking*. https://doi.org/10.1109/TGCN.2021.3127487
14. Hyslip, J. P., et al. (2003). Assessment of railway track substructure condition using ground penetrating radar. AREMA *Chicago, 20*.
15. Kim, J., Choi, S. W., Song, Y., Yoon, Y., & Kim, Y. (2015). Automatic train control over LTE: Design and performance evaluation. *IEEE Communications Magazine, 53*(10), 102–109.
16. Panigrahi, C. R., Sarkar, J. L., Tiwary, M., Pati, B., & Mohapatra, P. (2019). DATALET: An approach to manage big volume of data in cyber foraged environment. *Journal of Parallel and Distributed Computing, 131*, 14–28.
17. Lad, P., & Pawar, M. (2016). Evolution of railway track crack detection system. In *2nd IEEE International Symposium on Robotics and Manufacturing Automation (ROMA)* (pp. 1–6).
18. Bolla, D. R., Rajesh, D. R., Abhiram, M., Kasthuri, B., & BV, M. K. (2018). An innovative smart railway platform assist in domestic railway stations. In *3rd IEEE International Conference on Recent Trends in Electronics, Information & Communication Technology (RTEICT)* (pp. 2608–2612) (Lecture Notes in Computer Science in the ISI SCI Expanded). https://doi.org/10.1109/RTEICT42901.2018.9012272.5

A Novel Cuckoo Search Optimized RBF Trained ANN in a Nonlinear Channel Equalization

Pradyumna Kumar Mohapatra, Ravi Narayan Panda, Saroja Kumar Rout, Rojalin Samantaroy, and Pradeep Kumar Jena

Abstract In this article, a new approach to modeling a nonlinear channel is proposed. In these works, a new cuckoo search algorithm trained with radial basis function neural networks (RBFNN) is applied in the non linear channel for equalization. The efficiency of the proposed algorithm enhanced through the search process by the integration of discovery and exploitation. Therefore, instead of Levy mutated step size operator in CS, the Cauchy mutated operator is used to create the step size which will make a random number which is used to generate a new solution for the global search. The performance of the proposed equalizer can be evaluated by estimating MSE and BER by considering popular nonlinear channels and added with nonlinearities. The consequences of the simulation show the presentation of our projected equalizer better than existing neural networks-based equalizers available in the literature.

Keywords RBFNN · Channel equalization · Cuckoo search algorithm · Cauchy operator

1 Introduction

Signals are transmitted from source to the destination through communication channels. Equalization is a process for countermeasure against distortions (in particular, ISI) introduced by the channel. ANN [1–4] has been commonly used in the equalization over the past few decades. Because of their features, such as the RBFNN equalizers, they work better [5] due to their properties like (a) an easily trainable

P. K. Mohapatra (✉)
Vedang Institute of Technology, Bhubaneswar, Odisha, India
e-mail: er_pradyumna@yahoo.co.in

R. N. Panda · S. Kumar Rout
GIFT, Bhubaneswar, Odisha, India

S. Kumar Rout · R. Samantaroy · P. K. Jena
Vardhaman College of Engineering, Hyderabad, India

B. Pati et al. (eds.), *Proceedings of the 6th International Conference on Advance Computing and Intelligent Engineering*, Lecture Notes in Networks and Systems 428,
https://doi.org/10.1007/978-981-19-2225-1_18

compact structure, thus consuming less computation time [6], (b) better generalization ability, and (c) making precise nonlinear functions approximation. A steady equalization with a higher speed of convergence is also given by RBFNN. This being further substantiated in literatures [3, 7–11]. But, typical RBFNN design approaches use hit and trial training methods and take more time. Using GA and PSO to design RBFNN is illustrated in [12, 13]. Authors of [14] suggest GA and some modified versions in order to boost the local and global search for the RBFNN parameters. The architecture of RBFNNs is proposed in [6] on the basis of neuron activity along with the estimation of the parameters. There are several ANN-based equalizers [15–19] have been successfully applied in channel equalization problem.

The extensive use of this proposed algorithm in numerous research fields has resulted in CSAs enhanced effectiveness in seeking a global optima solution [20, 21]. In this work, we applied new cuckoo search algorithm (NCSA), the modified versions of CS are used for updating the weights of the proposed NCSA-RBF-based adaptive equalizer. Here, NCSA is a modified form of CSA to boost the exploration and exploitation characteristics. This version uses Cauchy operator, rather than Levy flights, to create the step size to efficiently explore the search space. The highlight of this study can be attributed to the use of NCSA-RBF in communication channel equalization. The fundamental distinction among the method proposed and the methods available in the citations is that the method proposed is seen as a classification problem, while the others are based on optimization in the literature. The proposed equalizer outperforms its contemporary ANN [7–9] and swarm [19]-based equalizers as cited in various literatures, which may be viewed as the hallmark of this work.

The principal outlines of this article are as follows: System model in Sect. 2 and in Sect. 3, detail description of proposed algorithm followed by its training procedure in Sect. 4. Simulation study is conceded for performance evaluation which is discussed in Sect. 5. Lastly, in Sect. 6, it dealt with a conclusion.

1.1 Motivation

It has been observed that in the case of burst error, the gradient algorithm-based equalizers have several times failed to correctly model the channel characteristics. Using conventional hit and trial, however, the RBFNN design takes time and is basically sub-optimal. In this article, we proposed a training strategy for RBFNN equalizer. The said equalizer trained with cuckoo search algorithm and its variants in channel equalization.

1.2 Contributions

The main contributions can be focused as

- Development of learning procedure in ANN.
- Make use of NCSA trained with neural networks in channel equalization.

In this paper, we explore efficient methods for equalization. This paper makes utilize of RBFNN for the said problem. Developed equalizers are termed as RBFNN equalizers. Primary objective of this work is training of RBFNN equalizers. Use of population-based algorithms for RBFNN training is the essence of the contributions made by this paper.

2 System Model

Figure 1 portrays a schematic of a digital transmission system.

The FIR model is a traditional linear channel model. The transmitted sequence is binary in this model, defined as $x(k)$ an instance at kth occasion, and the equivalent output at the this moment is $y_1(k)$ can be described as in Eq. (1)

$$y_1(k) = \sum_{i=0}^{N-1} h_i x(k-i) \tag{1}$$

Here, length of the channel and taps of the channel are defined as N and $h_i(i = 0, 1, \ldots, N-1)$, respectively.

Nonlinearity added in the channel which also identified as 'NL'.

The most accepted forms of nonlinear function can be described as

$$y(k) = F(y_1(k)) = y_1(k) + b[y_1(k)]^3 \tag{2}$$

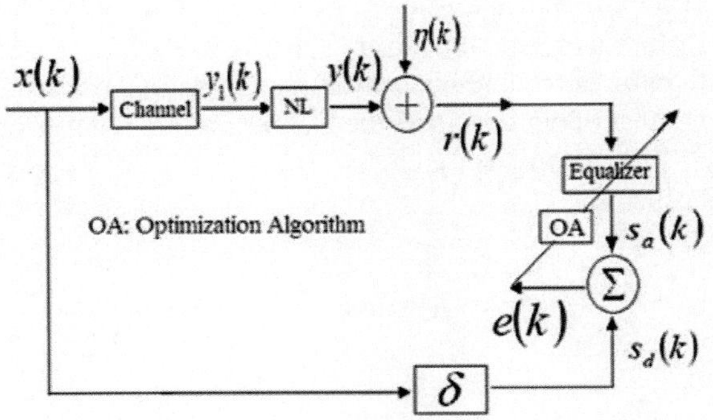

Fig. 1 System model for equalizer

The response of 'NL' can be shown in Eq. (3) where b as a constant.

$$y(k) = \left(\sum_{i=0}^{N-1} h_i x(k-i)\right) + b\left(\sum_{i=0}^{N-1} h_i x(k-i)\right)^3 \tag{3}$$

The response of the receiver $r(k)$ is nothing but sum of the noise included in the channel $\eta(k)$ and $y(k)$ as shown in Eq. (4)

$$r(k) = y(k) + \eta(k) \tag{4}$$

To recover, the transmitted symbol equalizer is used, $x(k-\delta)$, from a priority knowledge of the samples acknowledged, 'δ' identified as delay.

Here, $d(k)$ is the desired signal shown in Eq. (5)

$$d(k) = x(k-\delta) \tag{5}$$

Author considered in this paper, the issue deals with equalization is a classification [6–9] and the input space of the equalizer; $x(k) = [x(k), x(k-1), \ldots, x(k-N+1)]^T$ can be separate in to two distinct regions.

The optimal solution for this is provided by the Bays theory in which the decision function can be described as

$$f_{\text{bay}}(x(k)) = \sum_{j=1}^{n} \beta_j \exp\left(\frac{-\|x(k) - c_j\|}{2\sigma^2}\right) \tag{6}$$

As the transmitted sequence is binary, therefore

$$\beta_j = \begin{cases} +1 & c_j \in C_d^{(+1)} \\ -1 & c_j \in C_d^{(-1)} \end{cases} \tag{7}$$

Equation (7) defines transmitted symbol, and channel states are, respectively, the desired signal, $x(k-\delta) = +1/-1$, and the noise variance represented as σ^2.

In Fig. 1, the 'equalizer' block is ANN. CSA and its variants (NCSA) are used to optimize the number of layers and neurons in each layer. In the input layer, the number of neurons is N, the identical as the number of taps.

The output of radial basis function neural network equalizer which term as RBFNN

$$f_{\text{RBF}}(x(k)) = \sum_{j=1}^{n} w_j \exp\left(\frac{-\|x(k) - t_j\|^2}{\alpha_j}\right) \tag{8}$$

In Eq. (8), terms related as: connecting weights, the spreads, and centers of the hidden layer neurons denoted as w_j, α_j, and t_j, respectively. Equation (8), which is the execution of the decision of the Bays Eq. (6), is assumed, the c_j to be the same as t_j with the connecting weights correspondingly governed.

Hence, the assessment at the output of the equalizer as

$$\hat{x}(k-\delta)=\begin{cases}+1 & f_{ANN}(x(k))\geq 0\\ -1 & \text{elsewhere}\end{cases} \tag{9}$$

The discrepancy amid the output of the equalizer ($s_a(k)=\hat{x}(k-\delta)$) and the expected output ($s_d(k)=x(k-\delta)$) is defined as error and is denoted as $e(k)$, which modifies the weights.

Calculating the mean square error (MSE) measurement which one of the most important parameters for performance evaluation, considering the number of sample 'l':

$$\text{MSE}=\frac{1}{l}E\left[e^2(k)\right] \tag{10}$$

The ratio of error bits and transmitted bits is the bit error rate (BER); In this article, BER and MSE are chosen as the output index.

3 Cuckoo Search Algorithm

One of the recently cited evolutionary algorithms, Yang and Deb [20] introduced the cuckoo search algorithm (CSA). The cuckoos were also observed to follow a 'Lévy flight' mode of motion. CSA introduced in widely applications [21–23]. A particle filter trained with an enhanced CSA has been introduced in [24]. CSA is also used in economic dispatch issues [25].

As shown below are the procedure involved in CSA:

Step (i). Random generation within a given search space of 'N' nests $x_i(i=1,2,\ldots,N)$, where every nest points is a possible solution.

Step (ii). For both of these nests, compute the fitness function f_i.

Step (iii) The new nest population x_i^{new} is collected using Lévy flight as a new nest population.

$$x_i^{\text{new}}=x_i^{\text{old}}+\alpha\oplus\text{Levy}(\chi) \tag{11}$$

Step (iv). Determine the fitness of the new nests f_j^{new} and compared with f_j^{old}.

Step (v). If f_j^{new} is better than f_j^{old}, then put back x_j^{old} with the new nest x_j^{new} otherwise keep the old nest.

Step (vi). Select a fraction of the worst performing nests in the new population collected. Replace these nests within the defined search space with randomly generated ones.

Step (vii). For the nests obtained, determine the fitness function.

Step (viii). Trace the best observed nest, $X_{\text{best}}^{\text{gen}}$ in the recent population set, based on fitness values. If $X_{\text{best}}^{\text{gen}}$ improved than x_{best}, i.e., substitute x_{best} with $X_{\text{best}}^{\text{gen}}$. For the first generation, x_{best} and $X_{\text{best}}^{\text{gen}}$ will essentially be the similar.

Step (ix). If the condition for termination is not satisfied, go to step (ii), or return the solution to x_{best}.

3.1 The Proposed New Cuckoo Search Algorithm (NCSA)

For local search, the CS algorithm based on the likelihood of switching becomes very concentrated and comprehensive for global search. Since past decades, along with evolutionary algorithms [26, 27], the complexity of optimization has increased. The search process of CS will be updated in this paper to boost the location of the new cuckoo. The exploration and exploitation are the two factors for finalization of an efficient algorithm. For a good algorithm, exploration decides in avoidance local minima and the convergence properties are decided by exploitation.

3.1.1 Cauchy-Based Mutation Operator

For exploring the search space, the Levy method is good enough; still room is available for better exploration. Therefore, instead of Levy mutated step size operator, Cauchy mutated operator is used to create the step size which will make a random number δ. On this basis, Eq. (11) changes in the global quest process to Eq. (12):

$$x_i^{\text{new}} = x_i^{\text{old}} + \alpha \oplus \text{Cauchy}(\delta)(x_i^{\text{old}} - x_j^{\text{old}}). \tag{12}$$

3.1.2 Variants of Cuckoo Search Algorithm

Variants of cuckoo search algorithm is proposed with two modifications in standard CS, from the pool of search, the mean of first three solutions is taken to enhance the global pollination phase, based on the current best solution, three solutions are generated. In this regard, the equations suggested are given by

$$\begin{aligned} X_1 &= X_i - B_1\left(C_1 X_{\text{best}} - x_i^{\text{old}}\right); \\ X_2 &= X_i - B_2\left(C_2 X_{\text{best}} - x_i^{\text{old}}\right); \\ X_3 &= X_i - B_3\left(C_2 X_{\text{best}} - x_i^{\text{old}}\right) \end{aligned} \tag{13}$$

$$x_{\text{new}} = \frac{X_1 + X_2 + X_3}{3} \tag{14}$$

The definition is extended to x_{new} approach, then equation is added to the solution, thus obtained (12) of the global phase of the search, resulting in

$$x_i^{\text{new}} = x_i^{\text{old}} + \alpha \oplus \text{Cauchy}(\delta)(x_{\text{best}} - x_i^{\text{old}}) \tag{15}$$

Here, B_1, B_2, B_3 and C_1, C_2, C_3 refer to B and C which can be described as

$$B = 2bd_1 - b; C = 2d_2, 0 \leq b \leq 2 \tag{16}$$

Ranges of d_1 and d_2 are of [0, 1] which defined as consistently dispersed random numbers, $b = a$ random number with iterations that decreases linearly. d_1 and d_2 permit the search agents to boost the algorithm's exploratory tendencies.

4 NCSA and ANN Training

The idea of NCSA, which will give some encouraging result that refers to nonlinear channel equalization was suggested by Salgotra et al. [28]. The population is generated by the solution vectors. The algorithm ends with a predefined limit number of iterations (K).

i. In this step, all five algorithm parameters should be initialized. Parameters are size of population, switch probability, maximum cycles, stopping criteria, and the parameter $\vec{a}$.
ii. Generate randomly, within the specified search space, N nests x_i. In the given optimization problem, a possible solution obtained for each nest.
iii. The fitness function f_i is calculated for each of these nests according to the given problem.
iv. x_i^{new} is obtained using Eq. (15) and defines the new population of nests.
v. Measure the current fitness f_{new} relating to the current nests and evaluate the fitness f_{old} of the preceding nests.
vi. If f_{new} better than f_{old}, substitute x_i^{old} with x_i^{new}, or else keep the earlier one.
vii. Generate new solution using Eqs. (1) and (15). Evaluate and find the best.
viii. If solution not obtained satisfactory, then repeat from step (iii) or else return the solution to x_{best}.

Here, the solution vectors form the best performing nest in current population. The fitness value of nests can be computed through the parametric vector t_i which shown in the following Eq. (17)

$$t_i = \begin{pmatrix} w_{11}^i, w_{12}^i, \cdots, w_{IJ}^i, \alpha_1^i, \alpha_2^i, \cdots, \alpha_I^i, c_{11}^i, c_{12}^i, \\ \cdots c_{1m}^i, \cdots c_{I1}^i, c_{I2}^i, \cdots, c_{\text{Im}}^i, \beta_1^i, \cdots \beta_J^i \end{pmatrix} \tag{17}$$

The parameters used here are α, β, ω and c describes as the spread parameter, bias, weights, and center vector, respectively, and $c_i = (c_{i1}, c_{i2}, \ldots, c_{im})$ defines ith neuron of hidden layers. In this work, the training procedure with NCSA, the vector t_i of nest of related ANN optimizes the fitness function.

$$f(t_i) = \frac{1}{1 + \text{MSE}} = \frac{1}{1 + \frac{1}{Q}\sum_{k=1}^{Q} \|d(k) - y(k)\|^2} \tag{18}$$

For training samples, x_i, $d(k)$, and $y(k)$ are the desired and actual output. 'Q' reflects the number of samples used in the training. The methodology for nonlinear channel equalization of RBFNNs models is detailed in this part. The issues of equalization have been outlined as an optimization problem based on NCSA. The location of the nests is related to the probable principles of the vector, $w(n) = [w_1^T(n), w_2^T(n)]$ that must be calculated in order to accomplish the modeling goal. To obtain the mean squared error (MSE), the fitness value is desirable which is shown in Eq. (19).

$$\xi = \frac{1}{J}\sum_{1}^{j-1}[S_a(k) - S_d(k)]^2 \tag{19}$$

where $S_a(k) = \hat{x}(k - \delta)$ and $S_d(k) = x(k - \delta)$.

4.1 NCSA-Trained ANN for Equalization

The training algorithm which shown in Fig. 2, ANN defines rules for an organization which acts as a boss of a company to supply assets (which are nothing but the parameters to be to optimized) to NCSA which behaves as a manager which gives the directions for employees. ANN learning is the equalization problem. In this method, ANN acts both as a boss and an employee. The flowchart shown in Fig. 2 describes the pseudocode for the problem and depicts the detailed process used for training the ANN using NCSA.

4.2 Parameter Settings

For comparison purpose, the parameters set for the said algorithms have been obtained from respective research articles. For concerned CS and the method proposed, only switching probability needed to be fixed. Initially, a randomly selected value of switching probability is considered. Later on, it was concluded that a probability value of 0.5 suits most conditions well when measuring the effectiveness of the probability transition. The values of all these parameters chosen for the various

Fig. 2 Training ANN with NCSA

```
Initialize ANN-as a BOSS
    For j = 1,2, ... ...P
        Make NCSA-as manager (j)
        for ANN-as employee k = 1,2, ... ... P
        create ANN- as employee
        end
    end
whilst solution is not establish
    evaluate update
    put number of maximum iterations
    for (NCSA-as manager j = 1,2, ... ...P)
        as (iterations<allocations)
        for (ANN-as employee k = 1,2, ... ... Q )
            test ANN-as employee(k)
        end
        for (ANN-as employee k = 1,2, ... ... Q )
        Modify the weights of ANN-as employee (k)
        end
    end
    Return global best
end
Update global best
end
```

Table 1 Parameter selection for simulations

PSO		CSA		NCSA	
Parameter	Value	Parameter	Value	Parameter	Value
Max. no. of iterations	2000	Max. no. of iterations	2000	Max. no. of iterations	2000
Population size	50	Population size	50	Population size	50
Coefficient $C1$	0.7	Switch probability	0.5	Switch probability	0.5
Coefficient $C2$	0.7	Stopping criteria	Max. iteration	Stopping criteria	Max. no. of iteration
				$\vec{a}$	Linearly decreased from 2 to 0

methods are shown in Table 1. For PSO, Das et al. [18] also offered sufficient clarification of the different parameter settings.

5 Simulations and Discussion

Two specific parameters, such as MSE and BER, were used as the index for output estimation. The PSO, CSA, and NCSA simulation parameters are illustrated in Table

1. PSO control parameters are selected in the same way as used by Das et al. [18]. The acceleration parameters for the PSO algorithm, $c1 = c2 = 0.7$ and inertial weight μ_p = (linearly varied from) 0.8 to 0.3, were used. The constraints used by [21] for CSA is $\lambda = 2.5$, $\alpha = 1$ and $p_a = 0.25$, respectively. NCSA control parameters are selected in the same manner as in [28] which is used with the exception of a population size of 50 relative to CSA and PSO. Simulations were carried out in a MATLAB 2015 environment.

Equations (20) to (22) represent three nonlinear channels having transfer functions were illustrated, and simulations were carried for binary signals.

$$H(z) = 0.24 + 0.93Z^{-1} + 0.26Z^{-2} \tag{20}$$

$$H(z) = 0.303 + 0.9029Z^{-1} + 0.304Z^{-2} \tag{21}$$

$$H(z) = 0.3410 + 0.8760Z^{-1} + 0.3410Z^{-2} \tag{22}$$

The consequence of nonlinearity on equalizer performance and the nonlinearity is added which shown in Eq. (23).

$$y(n) = \tanh[x(n)] \tag{23}$$

We also replicated the following three equalizers for comparative purposes. The SNR is constant at 10 dB for evaluation of MSE. MSE and BER are the two important parameters to judge the equalizer performance, and it is plotted in Figs. 3, 4, 5, 6, 7, and 8, respectively, for the channels of Eqs. (20)–(22). From these figures, it is observed that the performance our proposed equalizer is better than the equalizer available in the literature. In Figs. 3 and 4, after 50 iterations, our proposed equalizer outperforms traditional CSA. In Fig. 5, PSO does not converge well and between 0

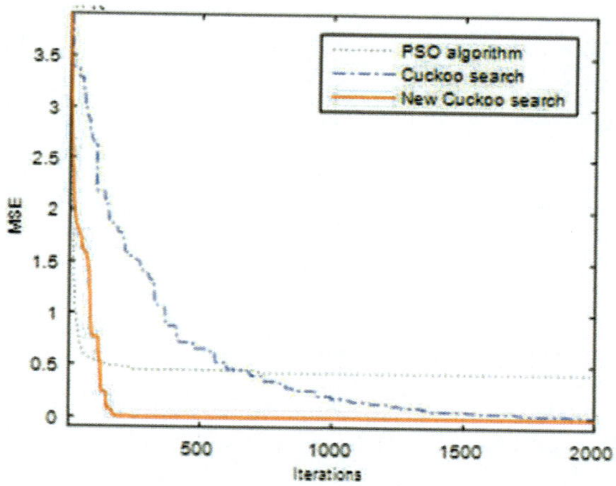

Fig. 3 MSE for channel of Eq. (20)

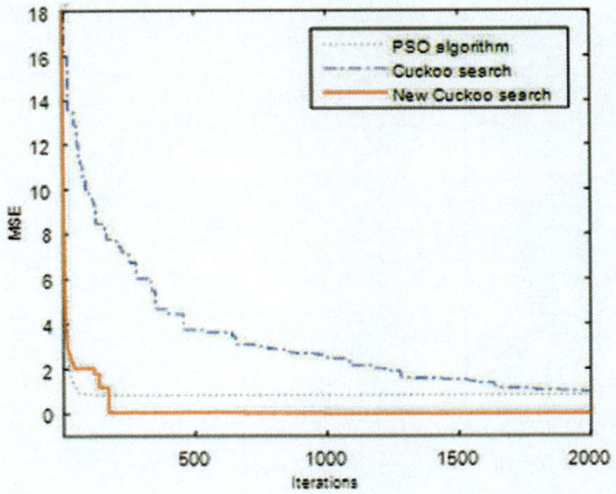

Fig. 4 MSE for channel of Eq. (21)

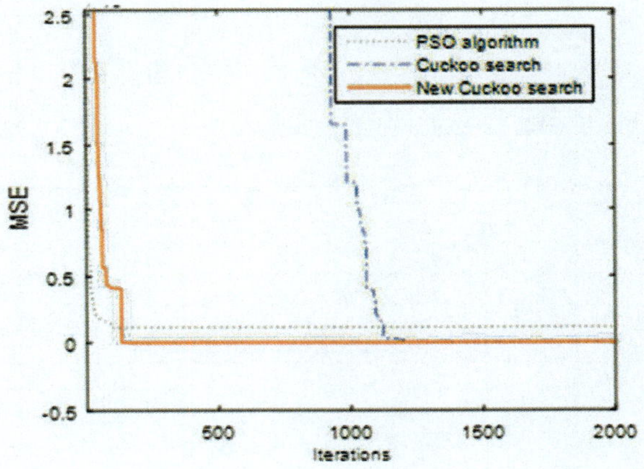

Fig. 5 MSE for channel of Eq. (22)

to 100 iterations, CSA better than NCSA, after that NCSA convergences better than CSA. Figure 6 shows BER simulation for the channel of Eq. (21), and it is observed that after 2 dB SNR, our proposed algorithm better than PSO and CSA. In Figs. 7 and 8, NCSA outperforms PSO and CSA.

6 Conclusion

In this article, we proposed a training strategy for RBFNN equalizer. The said equalizer trained with cuckoo search algorithm and its variants in channel equalization. It is observed that our equalizer executes better than existing neural network-based equalizers in all noise circumstances.

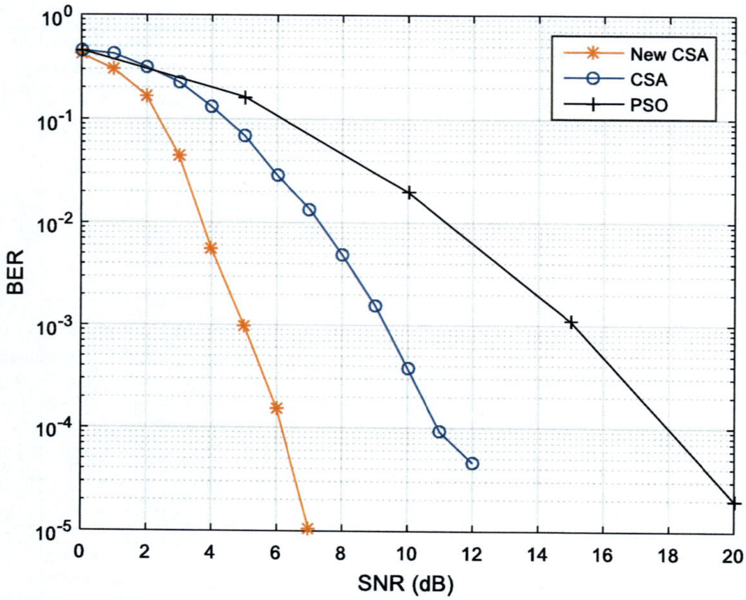

Fig. 6 BER for channel of Eq. (20)

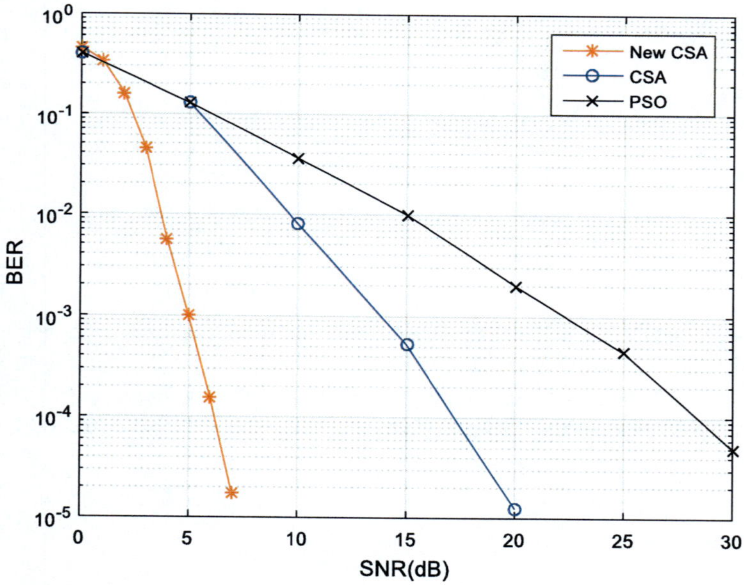

Fig. 7 BER for channel of Eq. (21)

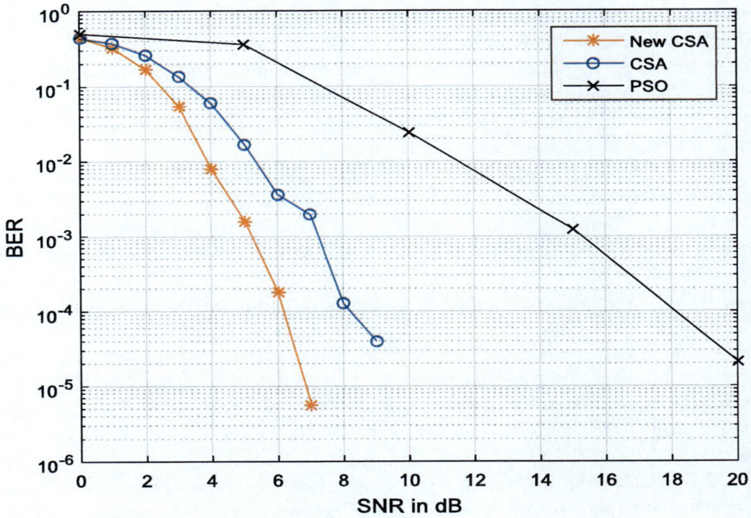

Fig. 8 BER for channel of Eq. (20)

Approaches for RBFNN equalizer training used here in this paper provide thought-provoking results in the literature of equalizers which are found to be better than the contemporary counterparts. In addition, MSE and BER of the proposed equalizers found to perform better in all noise conditions also without prior knowledge of SNR.

References

1. Seyman, M. N., & Taşpınar, N. (2013). Channel estimation based on neural network in space time block coded MIMO–OFDM system. *Digital Signal Processing*, *23*, 275–280.
2. Ruan, X., & Zhang, Y. (2014). Blind sequence estimation of MPSK signals using dynamically driven recurrent neural networks. *Neurocomputing, 129*, 421–427.
3. Rizaner, A. (2013). Radial basis function network assisted single-user channel estimation by using a linear minimum mean square error detector under impulsive noise. *Computers & Electrical Engineering, 39*, 1288–1299.
4. Sahoo, H. K., Dash, P. K., & Rath, N. P. (2013). NARX model based nonlinear dynamic system identification using low complexity neural networks and robust H_∞ filter. *Applied Soft Computing, 13*, 3324–3334.
5. Chen, S., Mulgrew, B., & Grant, M. P. (1993). A clustering technique for digital communication channel equalisation using radial basis function network. *IEEE Transaction on Neural Networks, 4*, 570–579.
6. Qasem, S. N., Shamsuddin, A. M., & Zain, A. M. (2012). Multi-objective hybrid evolutionary algorithms for radial basis function neural network design. *Knowledge-Based Systems, 27*, 475–497.
7. Zeng, X., Zhao, H., Jin, W., He, Z., & Li, T. (2013). Identification of nonlinear dynamic systems using convex combinations of multiple adaptive radius basis function networks. *Measurement, 46*, 628–638.

8. Yee, M. S., Yeap, B. L., & Hanzo, L. (2003). Radial basis function-assisted turbo equalization. *IEEE Transactions on Communications, 51*, 664–675.
9. Xie, N., & Leung, H. (2005). Blind equalization using a predictive radial basis function neural network. *Transactions on Neural Networks, 16*, 709–720.
10. Lee, J., & Sankar, R. (2007). Theoretical derivation of minimum mean square error of RBF based equalizer. *Signal Processing, 87*, 1613–1625.
11. Li, M. N., Huang, G. B., Saratchandran, P., & Sundarrajan, N. (2005). Performance evaluation of GAP-RBF network in channel equalization. *Neural Processing Letters, 22*, 223–233.
12. Barreto, A. M., Barbosa, H. J., & Ebecken, N. F. (2002). Growing compact RBF networks using a genetic algorithm. In *Proceedings of the VII Brazilian Symposium on Neural Networks* (pp. 61–66).
13. Feng, H. M. (2006). Self-generating RBFNs using evolutional PSO learning. *Neurocomputing, 70*, 241–251.
14. Barreto, M. S., Barbosa, H. J. C., & Ebecken, N. F. F. (2006). GOLS—Genetic orthogonal least squares algorithm for training RBF networks. *Neurocomputing, 69*(16), 2041–2064.
15. Mohapatra, P., Samantara, T., Panigrahi, S. P., & Nayak, S. K. (2018). Equalization of communication channels using GA-trained RBF networks. In *Progress in advanced computing and intelligent engineering* (pp. 491–499). Springer.
16. Patra, J. C., Meher, P. K., & Chakraborty, G. (2009). Nonlinear channel equalization for wireless communication systems using Legendre neural networks. *Signal Processing, 89*(11), 2251–2262.
17. Panda, S., Mohapatra, P. K., & Panigrahi, S. P. (2015). A new training scheme for neural networks and application in non-linear channel equalization. *Applied Soft Computing,* 27, 47–52.
18. Das, G., Pattnaik, P. K., & Padhy, S. K. (2014). Artificial neural network trained by particle swarm optimization for non-linear channel equalization. *Expert Systems with Applications, 41*(7), 3491–3496.
19. Mohapatra, P., & Panigrahi, S. P. (2014, December). FFA trained radial basis function neural network in channel equalization. In *International Conference on Swarm, Evolutionary, and Memetic Computing* (pp. 847–855). Springer.
20. Yang, X. -S., & Deb, S. (2009). Cuckoo search via Lévy flights. In *World Congress on Nature Biologically Inspired Computing* (pp. 210–214).
21. Gotmare, A., Patidar, R., & George, N. V. (2015). Nonlinear system identification using a Cuckoo search optimized adaptive Hammerstein model. *Expert Systems with Applications, 42*(5), 2538–2546.
22. Bhandari, A. K., Singh, V. K., Kumar, A., & Singh, G. K. (2014). Cuckoo search algorithm and wind driven optimization based study of satellite image segmentation for multilevel thresholding using Kapur's entropy. *Expert Systems with Applications, 41*(7), 3538–3560.
23. Marichelvam, M., Prabaharan, T., & Yang, X.-S. (2014). Improved Cuckoo search algorithm for hybrid flow shop scheduling problems to minimize makes pan. *Applied Soft Computing, 19*, 93–101.
24. Walia, G. S., & Kapoor, R. (2014). Intelligent video target tracking using an evolutionary particle filter based upon improved cuckoo search. *Expert Systems with Applications, 41*(14), 6315–6326.
25. Vo, D. N., Schegner, P., & Ongsakul, W. (2013). Cuckoo search algorithm for non convex economic dispatch. *IET Generation, Transmission and Distribution, 7*, 645–654 (9).

26. Wolpert, D. H., & Macready, W. G. (1997). No free lunch theorems for optimization. *IEEE Transactions on Evolutionary Computation, 1*(1), 67–82.
27. Crepinšek, M., Liu, S. H., & Mernik, M. (2013). Exploration and exploitation in evolutionary Algorithms: A survey. *ACM Computing Surveys (CSUR), 45*(3), 35.
28. Salgotra, R., Singh, U., & Saha, S. (2018). New Cuckoo search algorithms with enhanced exploration and exploitation properties. *Expert Systems with Applications, 95*, 384–420.

Sentiment Analysis, Machine Learning, and Cloud Computing

User-Based Cloud Service Recommendation System

Suvojit Barick, Satyapragyan Das, and Manoranjan Parhi

Abstract Recommender system is one of the emerging niches of machine learning. They are being used widely in a variety of fields. Netflix, IMDb, and Amazon have been using it commercially for their interests. With the increase in usage of cloud services, cloud service recommendation has attained significant attention in recent years. Cloud service recommendation system aims at helping the user to find services he might prefer. This paper proposes a user-based collaborative filtering method to enhance it. Here, we take the preferences of an active user first and then try to compute users with similar preferences using a popular similarity measure like cosine similarity which computes the similarity scores by computing the cosine distance between users. Then, we recommend the services the active users did not invoke. Finally, we found the root-mean-square error with the various number of recommendations of our proposed approach thirty times and noted the average. We observed that the cosine similarity measure works better than a few other popular measures like Euclidean distance by 14.99% when the number of top recommendations is five.

Keywords Cloud computing · Cloud service · Collaborative filtering · Recommender system · Cosine similarity · Similarity measure · Machine learning · Root-mean-square error

1 Introduction

Recommendation systems have attracted significant attention for their ability to deal with information overloading. The industry has been widely using such algorithms to recommend movies (Netflix), books (Amazon), CDs, and a lot more. In this paper, we have tried to implement a recommender system over cloud services to attain desirable recommendations as per the user's preference. We have proposed a collaborative filtering approach to deal with such requirements.

S. Barick · S. Das · M. Parhi (✉)
Department of Computer Science and Engineering, Siksha 'O' Anusandhan (Deemed to Be University), Bhubaneswar, Odisha, India
e-mail: manoranjanparhi@soa.ac.in

B. Pati et al. (eds.), *Proceedings of the 6th International Conference on Advance Computing and Intelligent Engineering*, Lecture Notes in Networks and Systems 428,
https://doi.org/10.1007/978-981-19-2225-1_19

1.1 Motivation

Cloud services are defined as platforms that are hosted by third-party providers and made available to users through the internet. Cloud computing has emerged as one of the technologies of modern computer science. With such an exponential rise in demand for cloud services, it has been difficult for users to select a cloud service as per their preference. Despite tremendous growth in cloud computing, users have struggled to find services that fit their requirements. To solve this issue, cloud service recommendation system is the need of the hour.

Collaborative filtering is a popular recommendation algorithm that can be broadly divided into (i) memory-based CF and (ii) model-based CF. The memory-based CF further has two approaches, namely (a) user-based CF and (b) item-based CF. The user-based CF takes an active user, finds similar users by considering the similarity of QoS values, and recommends services that these similar users preferred. The item-based CF will consider a service, search users who invoked that service, and find other services that the user liked.

1.2 Contribution

The paper suggests a user-based collaborative filtering method to help the users find the services as per their preference. The work will take a step further in this growing and trending niche of machine learning, recommender systems. The paper tries to give another dimension to this research field.

In this paper, we propose a user-based collaborative filtering approach via analysing the preference of similar users. It has three phases: finding similar users via cosine similarity, finding the mean of the QoS value of the similar users for the services, the active user is yet to invoke, recommending the services in the desired order.

The rest of the paper is structured in the following ways. Section 2 explains the related work of various other proposed approaches on service recommender systems. Section 3 describes the approach proposed in this paper, and Sect. 4 states the methodology and the data set used in this work. Section 5 outlines the result analysis and finally, Sect. 6 concludes this paper with future extensions.

2 Related Work

In recent years, cloud service recommendation system has gained significant attention for which they have been studied extensively. Collaborative filtering, in particular, has been studied widely hence, memory-based collaborative filtering, model-based collaborative filtering, and hybrid approaches have been proposed.

Commercially, companies like Netflix, Amazon, and IMDb have been using collaborative filtering widely. Collaborative filtering has been proposed for the QoS prediction cloud service recommender system. Zhu [1] has proposed a privacy-preserving QoS prediction framework using methods like P-UIPCC and P-PMF. Other approaches have proposed collaborative filtering with PGraph [2] for web service recommendation. Some papers have also suggested recommendations based on personalised hybrid collaborative filtering which uses the personalized version of similarity [3]. Xiong et al. [4] have proposed a hybrid web service recommendation system using deep learning. Wang has suggested a collaborative filtering approach via exploring the usage history of the user [5]. Some approaches even proposed a correlated QoS ranking method [6].

Xu et al. [7] have pointed out a major hindrance in the web service recommendation. They have suggested that QoS prediction is a major hindrance to such a system and have proposed a context-aware QoS prediction and web service recommendation. In their prediction model, they have taken geographical information as user context then identify similar neighbours for each user based on their context's similarity. On the server side, the paper suggested the use of affiliation information as a service context. They suggested QoS prediction by the QoS prediction of users and their neighbours.

Jiang et al. [8] have suggested quite a different approach for web service recommendation. This approach takes unstructured textual information like service description texts into account for their recommendation. It proposed to cluster cloud services into an optimal number of clusters by considering their description. Then, it proposed a personalised PageRank algorithm using service tags for recommendations. But such approaches do not take the QoS values of cloud services into account.

Jian Liu and Youling Chen have suggested a personalised clustering-based cloud service recommendation. They have used past QoS records of users in their approach. A clustering algorithm identifies a set of similar users by considering task similarity, where task similarity can be computed by incorporating both explicit textual information and rating information as well as implicit context information [9]. Then, such an approach suggests a trust-aware collaborative filtering recommendation.

Su et al. [10] have flagged a major shortcoming in existing approaches that they turn blind eye to the problem of data credibility and are quite vulnerable to dishonest users sharing unreliable QoS data. To check such flaws, they have suggested a trust-aware approach for reliable personalized QoS prediction.

Lifang Ren and Weijian Wang have suggested a support vector machine-based collaborative filtering approach for top-N web service recommendations. Such an approach has proved to be effective for web service recommendations [11].

3 Recommendation Approach

Cloud services are a wide range of services delivered on-demand to companies and individuals over the internet. We can straightforward access the application without any need for internal infrastructure.

In our suggested approach, we first compute the user-item interaction matrix. We used the observed QoS value by various users using different cloud services. These values were used to form the user-item interaction matrix. Hence, we have a user-item interaction matrix $R = \{r_{i,j}\}_{m*n}$, where $r_{i,j}$ is the noted QoS value by a user i while invoking service j. Our approach can briefly be described as follows: consider a particular user as an active user, find how similar are other users with the current user based on the QoS values observed by them, find the services the current user has not yet invoked, find the mean of the QoS value of these services observed by the similar users, arrange the services in the desired order and then recommend the top-N services to the active user. The above has been summarized in Fig. 1.

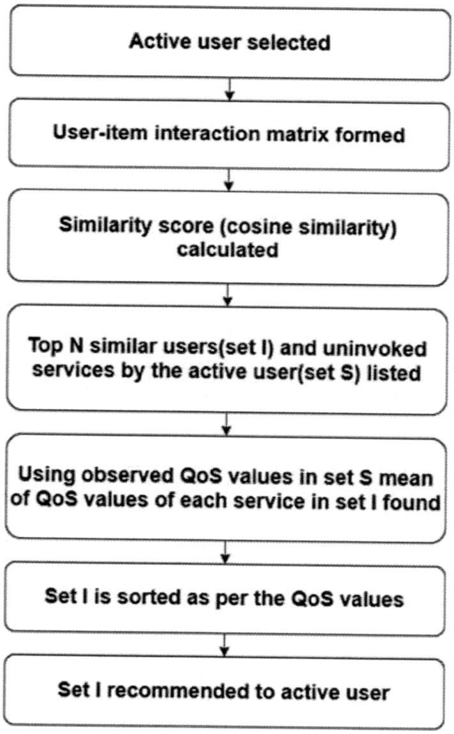

Fig. 1 Recommendation approach

3.1 Similarity Computation

In our suggested step, the most crucial step is the similarity computation. Many related works have used Pearson correlation coefficient and other popular similarity measures for this purpose. In this paper, we will be using the cosine similarity measure for computing similarity. Cosine similarity is one of the popular methods to quantify similarity. This method calculates the cosine distance between two vectors in an n-dimensional plane.

In our suggested approach, each user is considered as a vector and their similarity is quantified by the cosine similarity measure. Mathematically, cosine similarity between two users u_i and u_j is shown in Eq. (1).

$$\text{sim}(u_i, u_j) = \frac{\sum_{k=1}^{n} r_{i,k} . r_{j,k}}{\sqrt{\sum_{k=1}^{n} r_{i,k}^2} . \sqrt{\sum_{k=1}^{n} r_{j,k}^2}} \quad (1)$$

This value lies between the interval [0,1] where higher values signify greater similarity between users.

3.2 Recommendation

Cosine similarity is applied over the user-item interaction matrix. After that we consider the top-*N* similar users from the calculated similarity scores. Let this set be *S*. Then, we deduce the list of services the active users have not yet invoked. Let this set be *I*. Using the observed QoS values in set *S*, we find the mean of QoS values of each service in set *I*. Finally, set *I* is sorted with respect to the calculated QoS values in the desired order and was later recommended to the active user.

4 Methodologies and Datasets

WS-DREAM dataset was used for the suggested approach. Using this dataset user-item interaction matrix was created which was later used for computing similarity scores of users with the help of Python libraries. The recommendation methodology is explained in the following sub-section. Finally, the results were evaluated using the root-mean-squared error technique.

4.1 Dataset Used

We have adopted the WS-DREAM dataset [12] for our proposed approach as shown in Table 1. The dataset contains data of 4500 web services and 142 users. In our experiment, we focus on the response time QoS value of web services for the recommendation system.

We created the above data frame using the pandas library in Python. We created the matrix to calculate similarity scores using the cosine similarity measure.

We form this user-item matrix using pivot_table from pandas library by defining columns as ServiceID, index as UserID and values as mean response time. The formed matrix is shown in Table 2. After the creation of the users-item interaction matrix, we apply the cosine similarity measure over it.

Later, ten similar users were noted and the services the active user (let's say UserID = 3) has not yet invoked were listed. The mean response time of these services is found using the observed response time values of similar users. These services are arranged in ascending order of response time out of which the user is recommended the top-*N* services.

Table 1 First four rows of WS-DREAM dataset

Index	UserID	ServiceID	TimeSliceID	ResponseTime (sec)
0	0	0	0	4.180
1	0	1	0	0.416
2	0	2	0	0.441
3	0	3	0	2.764

Table 2 User-item matrix

UserID	ServiceID					
	0	1	2	....	4498	4499
0	1.955078	0.736922	1.098906	....	20.000000	0.772797
1	1.131172	0.656672	0.350719	....	0.116219	0.650984
2	0.546203	0.236687	0.240500	....	0.115875	0.442734
...	...	...	...	....	...	...
141	1.347844	0.425609	0.387016	....	0.157422	0.353656

4.2 Platform Used: Python

Python3 [13] is a high-level user-friendly language that has been extensively used in machine learning-based projects in recent years. It contains a vast variety of open-source libraries which help in computing high-level mathematical functions effectively. Python3 has been used to implement the proposed approach. We have used Python3 open-source libraries like pandas and math.

Pandas is one of the most efficient libraries of Python. It gives us data structures to deal with numeric data. Here, we use pandas to read the dataset, form the data frame and the matrix (pivot table).

We use cosine_similarity from sklearn.metrics.pairwise to find the similarity scores of different users with respect to an active user. The cosine_similarity finds the similarity scores by considering an n-dimensional space with each user as a vector and finds the cosine distance between these vectors.

mean_squared_error and sqrt from sklearn.metrics and math libraries are used to compute the root-mean-squared error value which is used to find the accuracy of our proposed approach.

4.3 Evaluation Metric

We have the root-mean-square error (RMSE) method to evaluate our approach. RMSE is a highly popular way to measure the accuracy of machine learning models in predicting qualitative data. The RMSE is a frequently used measure to calculate the squared difference between predicted values and actual values of a model.

It is given by Eq. (2)

$$\text{RMSE} = \sqrt{\frac{\sum_{i=1}^{n}(\text{predicted} - \text{actual})^2}{n}} \tag{2}$$

5 Results and Discussion

After computing the list of recommendations and calculating the predicted QoS value (that is the mean observed QoS value for similar users), we find the actual QoS values of the recommended services.

After having a list of predicted and actual QoS values, we evaluate our approach by root-mean-square error. The same was applied over other state of art similarity measures like Euclidean distance and Manhattan distance with different values of N (number of top recommendations) and the performance comparison in Table 3.

Table 3 Comparative study of root-mean-squared error value with different N values

Similarity measures	$N = 5$	$N = 10$	$N = 15$
Cosine similarity	0.10618133428656186	0.10765674883298343	0.11509859261306411
Euclidean distance	0.1220994180170145	0.12450334661810827	0.13691593135711344
Manhattan distance	0.31957585890986434	0.2920865204016519	0.3044826725077073

6 Conclusions and Future Works

In this paper, we proposed cloud service recommendations based on user-based collaborative filtering. Here, we explored the approach to find similar users for active users and then find the mean of the services that the active user has not yet invoked from the observed QoS values of the similar users. We compute similar users by using the cosine similarity measure. The recommendation using the cosine similarity measure is found to be having a lesser root-mean-square error value than the other two compared popular methods.

However, this approach has certain limitations. This approach takes only a single QoS value into account. Future works might also work in finding the weighted mean and include a weighted cosine similarity to improve recommendation. The approach also lacks personalization while doing recommendations. Some other approaches may take privacy into account while doing recommendations.

References

1. Zhu, J., He, P., & Lyu, M. (2015). A privacy-preserving QoS prediction framework. In *IEEE International Conference on Web Services* (pp. 241–248).
2. Zhou, Z., Wang, B., Guo, J., & Pan, J. (2015). QoS-aware web service recommendation using collaborative filtering with PGraph. In *IEEE International Conference on Web Services* (pp. 392–399).
3. Jiang, Y., Liu, J., Tang, M., & Lio, X. (2011). An effective web service recommendation method based on personalized collaborative filtering. In *IEEE International Conference on Web Services* (pp. 211–218).
4. Xiong, R., Wang, J., Zhan, N., Yutao, M. (2018). Deep hybrid collaborative filtering for Web service recommendation. *Expert Systems with Applications,* 191–205.
5. Wang, F., Chen, F., & Li, M. (2019). A collaborative filtering method for cloud service recommendation via exploring usage history. In *Proceeding of the 24th International Conference on Industrial Engineering and Engineering Management 2018* (pp. 716–725).
6. Jayapriya, K., Mary, N., & Rajesh, R. (2015). Cloud service recommendation based on a correlated QoS ranking prediction. *Journal of Network and Systems Management, 24,* 916–943.
7. Xu, Y., Yin, J., Deng, S., Xiong, N.N., Huang, J. (2016). Context-aware QoS prediction for web service recommendation and selection. *Expert Systems with Applications, 53,* 75–86.

8. Jiang, Y., Tao, D., Liu, Y., Sun, J., & Ling, H. (2019). Cloud service recommendation based on unstructured textual information. *Future Generation Computer Systems.*, *97*, 387–396.
9. Liu, J., & Chen, Y. (2019). A personalized clustering-based and reliable trust-aware QoS prediction approach for cloud service recommendation in cloud manufacturing. *Knowledge-Based Systems, 174*, 43–56.
10. Su, K., Xiao, B., Liu, B., Zhang, H., & Zhang, Z. (2017). TAP: A personalized trust-aware QoS prediction approach for web service recommendation. *Knowledge-Based Systems, 115*, 55–65.
11. Ren, L., Wang, W. (2018). An SVM-based collaborative filtering approach for Top-N web services recommendation. *Future Generation Computer Systems, 78*, 531–543.
12. Zheng, Z., Zhang, Y., & Lyu, M. (2014). Investigating QoS of real-world web services. *IEEE Transactions on Services Computing, 7*, 32–39.
13. Python. https://www.python.org/downloads/release/python-370/

Background Subtraction and Singular Value Decomposition Based Movement Epenthesis Detection in Isolated Gestures of Sign Language

Navneet Nayan, Debashis Ghosh, and Pyari Mohan Pradhan

Abstract This paper addresses the issue of movement epenthesis detection in isolated gestures. A novel method based on background subtraction and singular value decomposition to detect movement epenthesis in isolated gestures are proposed in this paper. The singular values of the absolute difference matrix obtained after the background subtraction provide a set of discriminative features to segment the movement epenthesis frames and sign frames. An adaptive threshold value is determined using the statistical properties of singular values for movement epenthesis detection. We tested our method on the ChaLearn LAP IsoGD dataset and the videos containing Indian Sign Language words. Experimental results show that our approach detected movement epenthesis frames with an accuracy of 91.14% and 93.73% on the ChaLearn LAP IsoGD dataset and the Indian Sign Language dataset, respectively.

Keywords Movement epenthesis · Sign language recognition · Singular value decomposition · Background subtraction · Isolated gesture recognition

1 Introduction

In the present context, sign language recognition is an evolving and complex area of research. Its complexity arises due to the diversity in the signs and signers. Various problems like occlusion, variant background, multiple signers in a scene further add complexity to this domain. The hearing and speech impaired community uses sign language as a source of communication for themselves. Sign language is also a way for this community to communicate with the rest of the globe. Broadly, sign language

N. Nayan (✉) · D. Ghosh · P. M. Pradhan
Department of Electronics and Communication Engineering,
Indian Institute of Technology Roorkee, Roorkee, Haridwar 247667, India
e-mail: nnayan@ec.iitr.ac.in

D. Ghosh
e-mail: ghoshfec@iitr.ac.in

P. M. Pradhan
e-mail: pmpradhan@ece.iitr.ac.in

B. Pati et al. (eds.), *Proceedings of the 6th International Conference on Advance Computing and Intelligent Engineering*, Lecture Notes in Networks and Systems 428,
https://doi.org/10.1007/978-981-19-2225-1_20

is divided into two parts, static and dynamic sign language. Further subdivision of dynamic sign language leads to two classes: isolated sign language and continuous sign language. Isolated sign language incorporates the gestures representing the words under its domain.

Sign language recognition is full of several challenges like co-articulation, movement epenthesis (ME), different signs having similar types of gestures and sometimes signer dependency too. Apart from these, there is a lack of universal sign language. Among all these, movement epenthesis is a serious issue that has a negative impact on the sign language recognition rate. Etymologically epenthesis can be seen as *epi* meaning 'in addition', *en* which means 'within', *tithenai* meaning 'to place', a combination of these three parts epi+en+tithenai give *epentithenai* which means 'to insert'. Thus, in simple terms, movement epenthesis is the insertion of movements in between signs and gestures. The movement epenthesis part does not carry any information about gestures and signs. This is the unwanted segment while performing the signs. It may happen that the duration of the movement epenthesis segments may be longer than the sign segments. Also, there are not any properly defined lexicons for movement epenthesis. So, detecting movement epenthesis becomes an important task. The movement epenthesis detection comes handy in enhancing the recognition rate and accuracy.

Here in the present work, we discuss and propose a novel approach for detecting the movement epenthesis in word-level sign language or isolated gestures. In signing a word, hands move from the rest position to perform the sign and again reach the rest position after completing the sign. In this case, movement epenthesis exists between the rest position of hands and the first frame of the sign. Also, it exists between the last frame of sign and rest position of hands. We used background subtraction and singular value decomposition to find the discriminating features to separate the sign frames and movement epenthesis frames. We evaluated our proposed method on the ChaLearn LAP IsoGD dataset [1] and also on the videos containing words of Indian Sign Language(ISL). ChaLearn LAP IsoGD dataset is a user-independent and large-scale dataset. The database of ISL has been made accessible by Indian Sign Language Research and Training Centre (ISLRTC). It contains 3000 words of ISL. Our method gives up to mark performance in finding out the movement epenthesis frames on both of these datasets.

The rest part of this paper is arranged as follows: Sect. 2 discusses the earlier works on movement epenthesis detection and isolated gesture recognition. Section 3 presents the methodology. Section 4 contains the proposed algorithm. Section 5 is an analysis of experimental results and discussion. Section 6 describes the conclusion part.

2 Previous Works

For movement epenthesis detection, researchers used explicit modelling and implicit modelling to model the movement epenthesis. In some of the earlier works in this area, Vogler et al. in [2] used context-dependent hidden Markov models (HMMs) for modelling the movement epenthesis. They applied their method for American Sign Language (ASL). But still, the scalability problem in ASL was a major issue to solve. Authors in [3] handled this problem and used the parallel HMMs to explicitly model movement epenthesis. They showed that the performance of parallel HMMs was better than conventional HMMs. Gao et al. in [4] gave a solution to address the large-scale continuous sign problems. They used transition movement models, temporal clustering and dynamic time warping to cluster the transition movements. In this case, an accuracy of 90.8% was reported on a dataset containing 1500 sentences containing 5113 signs.

One significant issue with explicit modelling of movement epenthesis was the scalability of this approach. For an extensive vocabulary, explicit modelling led to high computational complexity. Explicit modelling also required a large-sized training database. It was another serious issue with this approach. Also, explicit modelling demanded the generalization of sign boundary detection rules for all types of gestures. These limitations motivated the research community to look for some other options, and they headed towards implicit modelling [5].

In implicit modelling, Yang et al. obtained an 83% of recognition rate at the word level in [6]. They segmented the valid signs in continuous sign language sentences and simultaneously separated movement epenthesis part using enhanced level building algorithm. Choudhury et al. in [7] modelled the movement epenthesis in the context of global motion and detected the movement epenthesis using the heights of hand trajectories. Authors in [7] reported 92.8% spotting rate in the continuous gestures. In implicit modelling, authors in [8] used a nested dynamic programming to separate the movement epenthesis frames. Nested dynamic programming was reported to perform better than the classical level building method.

In the last few years, researchers have explored the field of isolated gesture recognition very well and several significant improvements have been reported. In an early work in isolated gesture recognition, authors proposed spatiotemporal feature extraction techniques complemented classification techniques like Bayesian classifiers and K-nearest neighbour (KNN) for Arabic Sign Language [9]. Authors reported 97–100% recognition rate on a dataset of 23 Arabic gestured phrases or words. An issue with their work was the use of less number of signers for preparing the database, and hence, there was a need for large-scale dataset with multiple signers. After some years, ChaLearn LAP IsoGD dataset was developed. Authors in [10] used the pyramidal 3D convolutional network on this dataset. They used this network to obtain multi-scale spatiotemporal features and got an accuracy of 50.93% on the ChaLearn LAP IsoGD dataset. An accuracy of 50.93% was the area that needed an improvement. Authors in [11] proposed effective representation of depth sequences as Dynamic Depth Motion Normal Images (DDMNI), Dynamic Depth Images (DDI) and Dynamic Depth Normal Images (DDNI) and used convolutional neural networks for gesture recognition. They got an accuracy of 55.57%. Hence, there was still a chance of improvement in the accuracy. As further improvement, Lin et al. in [12] obtained state-of-the-art accuracy of 68.42% on ChaLearn LAP IsoGD dataset. They developed a fusion scheme to blend the features via a convolutional layer. However, the authors concluded that though they achieved state-of-the-art accuracy, still a large area is left for improvement in the accuracy.

3 Methodology

There are four basic steps in our proposed methodology: (a) Processing of frames of video clips of isolated gesture, (b) Background subtraction, (c) Singular value decomposition of matrices obtained after getting the absolute value of background subtraction and (d) Deciding threshold value for movement epenthesis detection. While designing our method, we assumed that hands movement is significant compared to other body parts. Also, signs started from rest position of the signer. The object means our gesturing parts like hands and fingers were always in the field of view. The brightness of the scene was constant and didn't change abruptly. The flow diagram of the proposed technique is shown in Fig. 1.

In the proposed approach, we made the first frame of the input video as the reference frame. Then we applied background subtraction on every current frame and calculated the absolute difference after that. Let I_{ref} be the reference frame and I_{curr} be the current frame, then background subtraction (BS) is defined as:

$$\text{BS} = I_{\text{ref}} - I_{\text{curr}}. \tag{1}$$

After calculating BS, we got the absolute values of BS. For this, we obtained the matrices having an absolute value of differences of each pixel of the current frame and reference frame. Further, to quantify the absolute differences of frames, we headed towards the singular value decomposition of these matrices. Let N denotes any matrix of size $p \times q$. Then, singular value decomposition of matrix N can be expressed as:

$$N = U \times S \times V\ . \tag{2}$$

where N is a $p \times q$ matrix to be decomposed, U is an orthogonal matrix of size $p \times p$, V is a $q \times q$ orthogonal matrix and S is a $p \times q$ matrix with the diagonal elements representing the singular values of N. The largest singular value packs most of the energy contained in the image. This largest singular value or energy is also defined as 2-norm value. Further, we stored the largest singular value obtained in each iteration, followed by calculating the threshold value based on the statistical parameters of these stored largest singular values. Let us denote the stored largest singular values as SV. Empirically, we found that threshold value T is described as:

$$T = \text{mean (SV)} + (\text{standard devaition (SV)})^{1/2}. \tag{3}$$

With the help of this threshold value, we separated the movement epenthesis frames and the sign frames with a good accuracy.

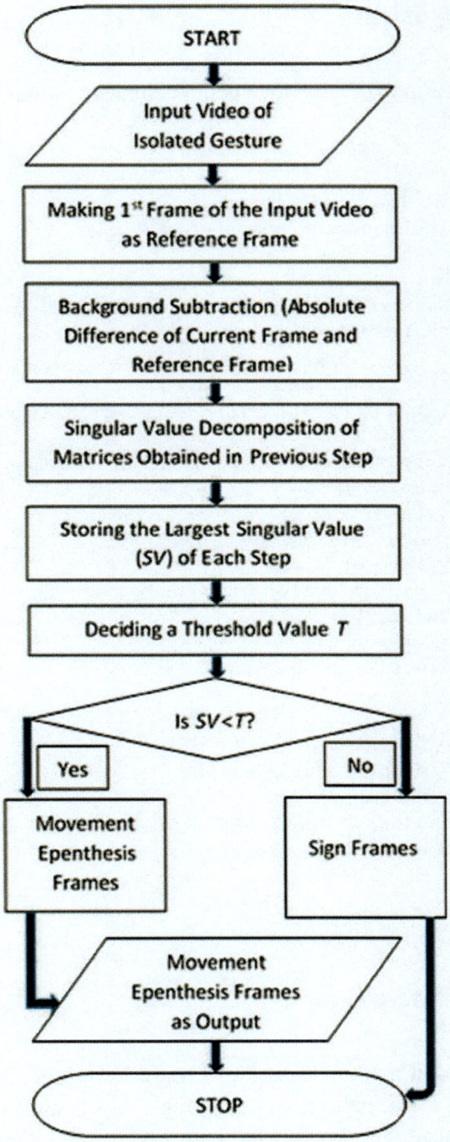

Fig. 1 Flow diagram of methodology of movement epenthesis frame detection

4 Proposed Algorithm

Algorithm 1: Algorithm to detect movement epenthesis frames in isolated gestures.

```
Input: Frames of Isolated Gesture Video
Output: Separated Movement Epenthesis Frames

Initialization:
1. Reading the video and naming the variable as vid
2. Getting total number of frames and naming the variable
    as num
3. Reading frames from video
4. Saving the first frame of video with variable name refframe

Background Subtraction and Singular Value Decomposition):
5.  for n=1:num do
6.  frame=read(vid,n);
7.  grayframe=rgb2gray(frame);
8.  bg_sub=grayframe-refframe;
9.  abs_bg_sub=absolute value of bg_sub;
10. sing_val=svd(abs_bg_sub);
11. max_sing_val=max(max(sing_val));
12. singular_values(k,1)=max_sing_val;
13. end for

Threshold Value Decision:
14. Thresh_val=(mean(singular_values)+
     sqrt(std(singular_values)));
15. movement epenthesis frames=find(singular_values<Thresh_val);
16. return movement epenthesis frames
```

5 Results and Discussion

We tested our proposed method on the ChaLearn LAP IsoGD dataset and videos containing ISL words. ISL dataset videos include words used in everyday life and technical, legal and academic terms. ChaLearn LAP IsoGD dataset contains 47,933 RGB-D gestures in 47,933 RGB-D gesture videos. The gestures in this dataset include sign language for deaf, pantomimes, Italian gestures, underwater sign language, symbolic gestures, body languages and helicopter and traffic signals. Each gesture video represents one gesture. Twenty-one signers were used in preparing this dataset.

5.1 Experimental Results

Experimental Video 1: The video length is 3 s and contains 29 frames. Frame rate is 9.96 frames per second, and each frame is of size 320 × 240. The movement epenthesis frames of this video are shown in Figs. 2 [1] and 3 [1], whereas Fig. 4 [1] presents the sign frame. Figure 5 represents the plot of the largest singular values of absolute difference of reference frames and current frames versus the frame indices. Threshold value obtained in this case is 19.11. The pattern obtained in the plot can be divided into two regions with the help of this threshold, as shown in Fig. 5. One region above the threshold value contains sign frames and the region below the threshold value contains movement epenthesis frames.

Experimental Video 2: Experimental Video 2 is of length 5 s, having 135 frames in it. The frame rate of the video is 25 frames per second. Frames are of size 1920 × 1080. Figures 6 [1] and 7 [1] show movement epenthesis frames of this video, and the sign frame is shown in Fig. 8 [1]. Figure 9 represents the separation of regions containing movement epenthesis frames and sign frames. The threshold value, in this case, is 20.01. Frame numbers 1 to 31 and 97 to 135 are movement epenthesis frames and frame numbers 32 to 96 are sign frames.

Experimental Video 3: The video is of length 4 s. It is made up of 46 frames, and the frame rate is 10 frames per second. Frame size is 320 × 240. Figures 10 [1] and 11 [1] show movement epenthesis frames of this video, and sign frame is shown in Fig. 12 [1]. Figure 13 describes the plot of the largest singular values versus frame indices. The threshold value of 17.49 accurately separates the movement epenthesis

Fig. 2 ME frames

Fig. 3 ME frames

Fig. 4 Sign frames

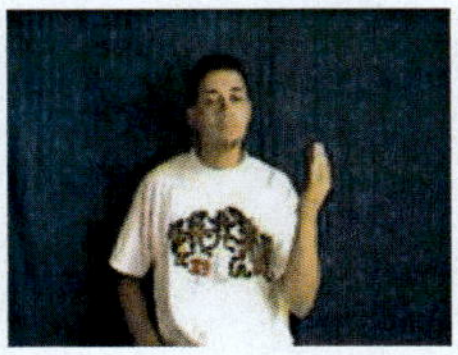

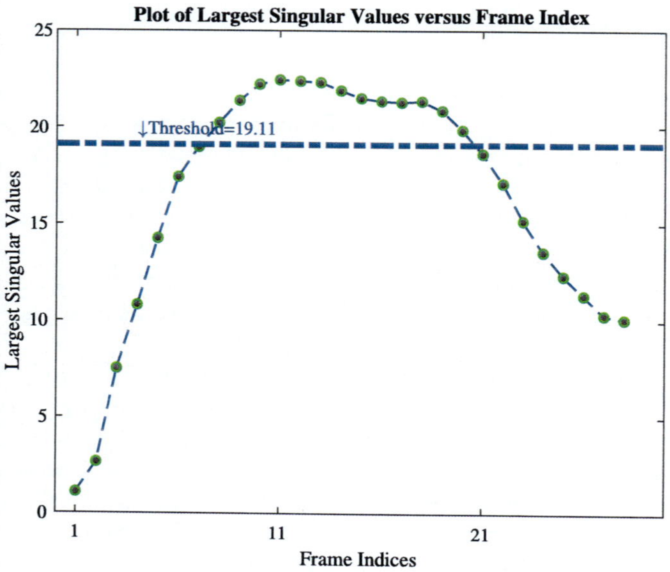

Fig. 5 Plot representing the largest singular values versus frame index. Region over the threshold (19.11 in this case) contains sign frames and regions below this threshold contain movement epenthesis frames

Fig. 6 ME frames

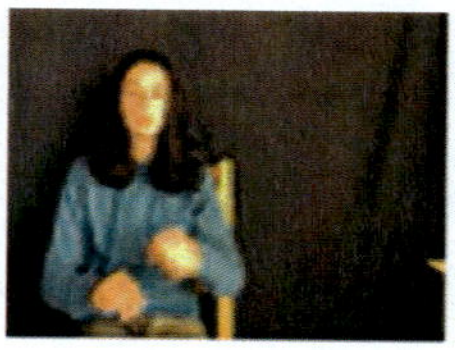

Fig. 7 ME frames

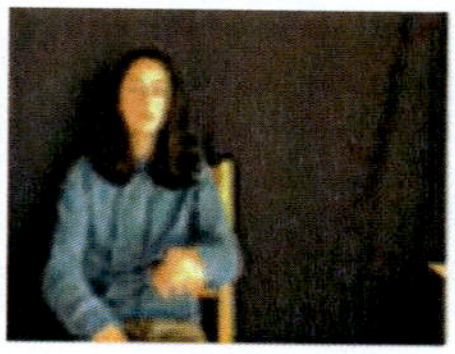

Fig. 8 Sign frames

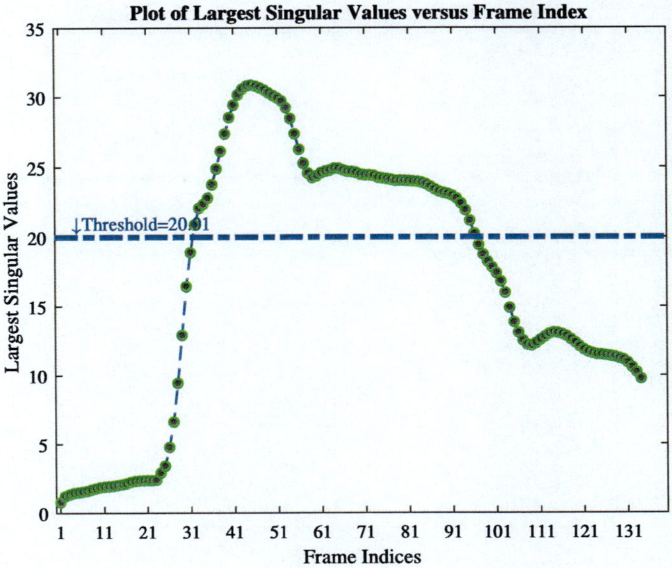

Fig. 9 Plot representing the largest singular values versus frame index. Region over the threshold (20.01 in this case) contains sign frames and regions below this threshold contain movement epenthesis frames

Fig. 10 ME frames

Fig. 11 ME frames

Fig. 12 Sign frames

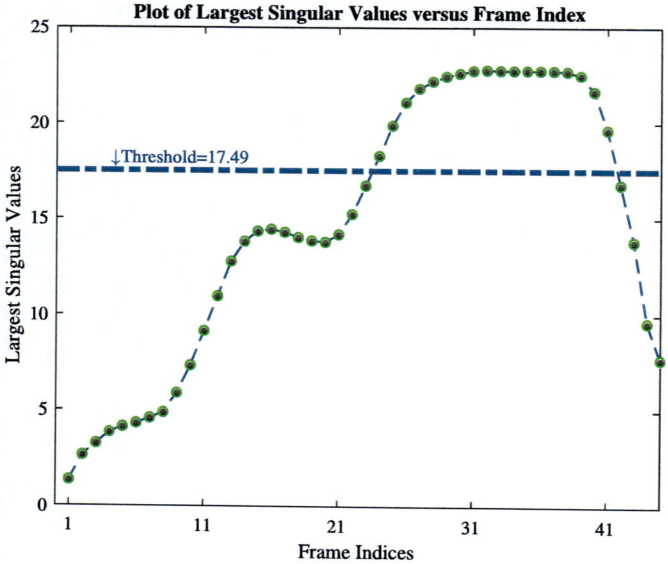

Fig. 13 Plot representing the largest singular values versus frame index. Region over the threshold (17.49 in this case) contains sign frames and regions below this threshold contain movement epenthesis frames

frames from the isolated gestures. The regions below the threshold value contain the frame indices of movement epenthesis frames and the frame indices in the region above the threshold are of sign frames.

6 Conclusion

We proposed a novel method to detect the movement epenthesis frames in isolated gestures or word-level sign language based on the background subtraction and singular value decomposition. The singular values of absolute difference of current frame and reference provide a set of discriminating features to separate the movement epenthesis frames that ultimately led to segmenting the sign frames in the isolated gestures. The calculation of the threshold to separate the frames into two categories was based on the statistical properties of the singular values. The threshold value was adaptive, and it was not dependent on user and dataset. We tested our method on the ChaLearn LAP IsoGD dataset and the videos containing ISL words. On the ChaLearn LAP IsoGD dataset, our method detected movement epenthesis frames with an accuracy of 91.14%. On the ISL dataset, the accuracy of our proposed method was 93.73%. The unwanted shadows, other body parts movement, sudden change in illumination condition or an abrupt change in the scene resulted in the missed detections. In future work, we aim to detect movement epenthesis in the wild.

References

1. Wan, J., Li, S. Z., Zhao, Y., Zhou, S., Guyon, I., & Escalera, S. (2016). ChaLearn looking at people RGB-D isolated and continuous datasets for gesture recognition. In *IEEE Conference on Computer Vision and Pattern Recognition Workshops (CVPRW)* (pp. 761–769).
2. Vogler, C., & Metaxas, D. (1998). ASL recognition based on a coupling between HMMs and 3D Motion analysis. In *ICCV*.
3. Vogler, C., & Metaxas, D. (2001). A framework for recognizing the simultaneous aspects of American sign language. *Journal of Computer Vision and Image Understanding, 81*(3), 358–384.
4. Gao, W., Fang, G., Zhao, D., & Chen, Y. (2004). Transition movement models for large vocabulary continuous sign language recognition. In *6th IEEE International Conference on Automatic Face and Gesture Recognition* (pp. 553–558).
5. Ong, C. W., & Ranganath, S. (2005). Automatic sign language analysis: A survey and the future beyond lexical meaning. *IEEE Transactions on Pattern Analysis and Machine Intelligence, 27*(6), 873–891.
6. Yang, R. D., Sarkar, S., & Loeding, B. (2007). Enhanced level building algorithm for the movement epenthesis problem in sign language recognition. In *IEEE Conference on Computer Vision and Pattern Recognition* (pp. 1–8).
7. Choudhury, A., Talukdar, A. K., Bhuyan, M. K., & Sarma, K. K. (2017). Movement epenthesis detection for continuous sign language recognition. *Journal of Intelligent Systems, 26*(3), 471–481.
8. Yang, R. D., Sarkar, S., & Loeding, B. (2010). Handling movement epenthesis and hand segmentation ambiguities in continuous sign language recognition using nested dynamic programming. *IEEE Transactions on Pattern Analysis and Machine Intelligence, 32*(3), 462–477.
9. Shanableh, T., Assaleh, K., & Al-Rousan, M. (2007). Spatio-temporal feature-extraction techniques for isolated gesture recognition in Arabic sign language. *IEEE Transactions on Systems, Man, and Cybernetics, Part B (Cybernetics), 37*(3), 641–650.
10. Zhu, G., Zhang, L., Mei, L., Shao, J., Song, J., & Shen, P. (2016). Large-scale isolated gesture recognition using pyramidal 3D convolutional networks. In *23rd International Conference on Pattern Recognition (ICPR)* (pp. 19–24).
11. Wang, P., Li, W., Liu, S., Gao, Z., Tang, C., & Ogunbona, P. (2016). Large-scale isolated gesture recognition using convolutional neural networks. In *23rd International Conference on Pattern Recognition (ICPR)* (pp. 7–12).
12. Lin, C., Wan, J., Liang, Y., & Li, S. Z. (2018). Large-scale isolated gesture recognition using a refined fused model based on masked Res-C3D network and Skeleton LSTM. In *13th IEEE International Conference on Automatic Face and Gesture Recognition* (pp. 52–58).

Exploratory and Sentiment Analysis of Classified Drug Groups

Bijayalaxmi Panda, Chhabi Rani Panigrahi, and Bibudhendu Pati

Abstract Health issues in this pandemic situation were most challenging factor for the people where the only path was Internet and the trust worthy Web sites. This research uses a drug review dataset from UCI machine learning repository and filtered a part of dataset based on some psychiatric conditions and identified fourteen drug groups considering suffixes of drugs. By applying exploratory data analysis (EDA) and sentiment analysis (SA) on the drug groups, we have identified the best drug group as well as the less efficient drug group according to its rating and sentiment polarity. From the experimental results on the considered dataset, it was found that barbiturate is the best drug group by customer rating with mean rating of 9.625, and antipsychotics drug group is found to have less rating as per customer suggestion.

Keywords Polarity · Rating · Drug groups · Sentiment analysis

1 Introduction

In this modern era, people are more addicted to social media than their family, friends, and relatives. People are adopting the way of living by following users of social media, and they gather information from different online forums [1]. Similarly, health-related information can be collected from different online medical forums [2]. Patients share information regarding their experience on diseases, drugs, and other activities in forums. A similar kind of information can help other people in decision making. Online medical help became more popular in this pandemic situation. Drug-related information can be gathered from reliable Web sites. Suggestions regarding dosage and other issues can be obtained from blog posts shared by doctors and other health-related consultants [3]. Knowledge of different drug groups is available at most trusted Web site drugs.com [4]. The details of drug groups like antibiotics, analgesics, barbiturate, etc., are explained along with its different drug names, compositions, reasons, side effects, and so on.

B. Panda (✉) · C. R. Panigrahi · B. Pati
Department of Computer Science, Rama Devi Women's University, Bhubaneswar, India
e-mail: bijayalaxmi.panda81@rediffmail.com

B. Pati et al. (eds.), *Proceedings of the 6th International Conference on Advance Computing and Intelligent Engineering*, Lecture Notes in Networks and Systems 428,
https://doi.org/10.1007/978-981-19-2225-1_21

Organization of paper includes Sect. 2 as related work for our research. Methodology for the approach described in Sect. 3 which involves data source, exploratory data analysis, and sentiment analysis. Section 4 describes the experimental set up and result analysis. Conclusion and future work depicted in Sect. 5.

2 Related Work

Analyzing health-related issues on the Internet became a habit of people. So drug categories and their efficiency from a customer point of view are important. Gopalakrishnan et al. [5] made comparison between two different drugs. Classification methods like neural network and SVM are applied on reviews of two drugs to perform sentiment analysis. The neural network proved to be best approach in terms of precision, recall, and F-score. Gurdin et al. [6] applied supervised learning methods on WebMD drug reviews for common, cancer, depression, diabetes, and hypertension drugs and found polarity of drugs. Garg [7] designed a drug recommender system for helping patients suggesting most useful drugs and reducing visit to hospitals by considering reviews of patients as input. LinearSVC using gave best result with 93% accuracy.

Sarchiapone et al. [8] perform preliminary synthesis of findings of tables and quality ratings by searching MEDLINE, Scopus, and Cochrane Library, etc., by following guidelines of PRISMA. Sridevi et al. [9] defined polarity of patient reviews using deep learning models to find out polarity as positive or negative. Logistic regression and LSTM networks are used for sentiment classification of the drug review. Na et al. [10] developed a special approach for sentiment analysis on user-generated content of drug review Web sites.

3 Methodology

This section describes the dataset, and we performed EDA which gives result analysis of classified drug groups and SA which is used to analyze positive and negative polarity of the dataset.

Data Source: UCI machine learning repository contains a large drug review dataset containing around two lakhs fifteen thousand instances [11]. The dataset possesses six attributes such as drug name, condition, review, rating, date, and useful count. In this work, we have extracted a part of this dataset based on certain psychiatric conditions such as insomnia, anxiety, depression, anxiety and stress, ADHD, bipolar disorder, major depressive disorder, and panic disorder. The extracted dataset contains 299 drugs from which we have classified the drugs based on suffixes in the name of drugs and then we have obtained 14 drug groups.

Exploratory Data Analysis (EDA): EDA is the method of evaluating data to solve problems, analyzing the details in graphical and pictorial representations. EDA is applied based on existing or collected dataset. This helps in clearly visualizing the underlying things by implementing EDA on sample dataset.

Sentiment Analysis (SA): SA is the method of analyzing the text into subjective and objective statements where subjective sentence is based on personal opinion and objective sentence is based on facts. The text can be classified as positive and negative sentiment. SA is applied to different areas like movie review, product review, drug review, and so on. In this research, we have focused on classified drug groups related to drugs having psychiatric condition, detecting negative or positive sentiment of patients regarding different drug groups. Classification of positive, negative, and neutral reviews is done based on Python textblob analysis. Polarity is obtained in the range of -1 to $+1$. Then, comparison of several drugs groups are made to determine the best drug group among them.

4 Experimental Set Up and Results

The experimental set up along with the analysis of results obtained is presented in this section.

4.1 Experimental Set Up

In this work, we mainly focused on a drug review dataset from which we have filtered a part of the dataset having certain psychiatric conditions. These conditions cover around 300 medicines. By taking suffixes of these medicines, we have classified drugs into different groups. For example, the suffix "*line*" used in the medicines is coming under *antidepressant* group. Same medicines can be used for treatment of several conditions given in drug review dataset. In this process, we have obtained fourteen groups and then compared them with respect to its rating and polarity (Table 1).

4.2 Results and Analysis

In this section, we have presented the experimental results along with the analysis.

Exploratory Data Analysis (EDA): The filtered dataset contains around 300 medicines from which most important drugs used by patients sharing their reviews coming under fourteen groups. Each group contains more than one medicines used for more than one conditions.

Table 1 Drug name related to drug groups

Drug group	Drug name	Drug group	Drug name
Antidepressants	Sertraline	Antianxiety	Alprazolam
	Amitriptyline		Clonazepam
	Selegiline		Lorazepam
	Nortriptyline		Temazepam
	Maprotiline		Flurazepam
	Protriptyline		Oxazepam
	Pemoline		Triazolam
SSNRIs	Venlafaxine		Estazolam
	Desvenlafaxine		Quazepam
Antifungal (except metronidazole)	Aripiprazole		Diazepam
	Brexpiprazole		
Aminoketone	Bupropion	Oral hypoglycemics	Chlordiazepoxide
Opioid analgesics	Trazodone	Antipsychotics (phenothiazine)	Perphenazine
	Vilazodone		Cariprazine
	Nefazodone		Perphenazine
	Risperidone		Compazine
	Paliperidone		Stelazine
	Lurasidone		Prochlorperazine
	Ziprasidone		Trifluoperazine
	Iloperidone	Alpha blocker	Prazosin
Oral hypoglycemics	Chlordiazepoxide	H2 blockers (anti-ulcers)	Clonidine
Antipsychotics (phenothiazine)	Perphenazine		Amantadine
	Cariprazine	Barbiturate	Secobarbital
	Perphenazine		Pentobarbital
	Compazine		Phenobarbital
	Stelazine		Butabarbital
	Prochlorperazine		
	Trifluoperazine		
SNRIs	Pristiq	Beta blockers	Propranolol
	Nadolol		Atenolol
SSRI	Escitalopram		
	Citalopram		

Sentiment Analysis (SA): Sentiment analysis performs text mining of social media data related to health care, disease diagnosis and gives benefit to same kind of users [12]. Patient experience shared in social media that gives positive impact on other patients [5]. We have obtained fourteen drug groups by performing exploratory data analysis on a part of drug review dataset. Then, sentiment analysis performed using textblob module of Python. The review of drug users of a particular drug group is classified as positive and negative statement. Polarity in the range of –1 to +1 is obtained. We have taken around 14 groups of drugs on the basis of more number of reviews. All fourteen drug groups used for treatment of one or more conditions. The drug groups are compared with respect to their mean rating and mean polarity (Table 2).

The mean rating of fourteen drug groups is shown in Fig. 1 which indicates that barbiturate is the best drug group by customer rating and mean rating is 9.625 which contains group of medicines like secobarbital, pentobarbital, phenobarbital, and butabarbital and mainly used for treatment of insomnia, sedation, hypnotics, antianxiety, and so on. Antipsychotics drug group is found to have less rating that is 6.367 as per customer suggestion which contains group of drugs like perphenazine, cariprazine, perphenazine, Stelazine, prochlorperazine, and trifluoperazine which are used for treatment of anxiety, bipolar disorder, psychosis, etc. So this drug group needs to be improved.

The mean polarity of considered drug groups is shown in Fig. 2 which states that barbiturate is the best drug group according to customer opinion because more positive reviews are found for this group. Mean polarity of this group is found to

Table 2 Mean rating and mean polarity of drug groups

Drug groups	Mean rating	Mean polarity
Antidepressants	7.402	0.074
SSNRIs	6.926	0.066
Antifungal (except metronidazole)	6.383	0.094
Aminoketone	7.331	0.096
Opioid analgesics	6.615	**0.065**
SNRIs	7.175	0.096
SSRIs	7.615	0.08
Antianxiety	8.537	0.123
Oral hypoglycemics	8.593	0.193
Antipsychotics (phenothiazine)	**6.367**	0.093
Alpha blocker	9.286	0.152
H2 blockers (anti-ulcers)	7.899	0.096
Barbiturate	**9.625**	**0.195**
Beta blockers	7.892	0.1

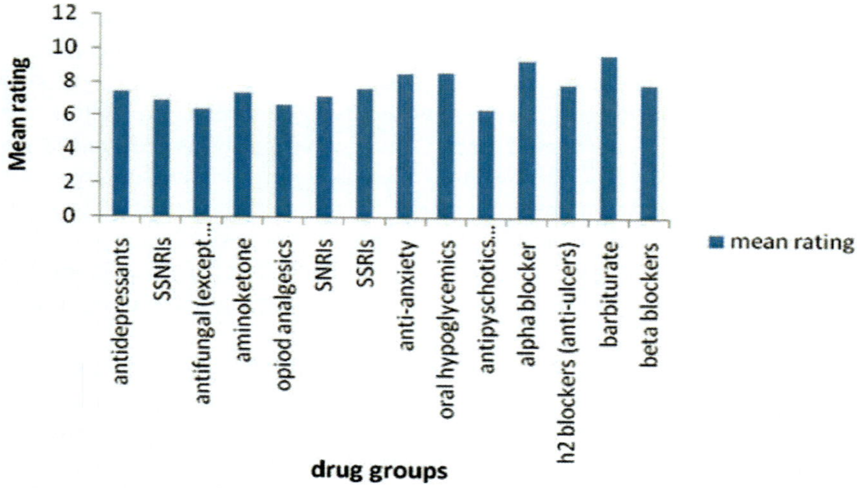

Fig. 1 Mean rating of drug groups

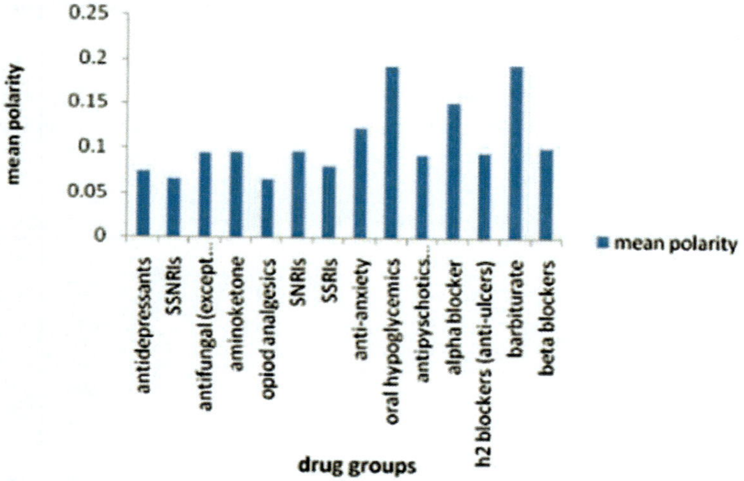

Fig. 2 Mean polarity of considered drug groups

be 0.195. Opioid analgesics group is found to be of less polarity, i.e., 0.065. From this, we can conclude that barbiturate is the best drug group as per both polarity and rating.

The graph as shown in Fig. 3 shows rating verses count of barbiturate drug group and it shows that maximum customers have given ten star rating for this drug group.

Figure 4 shows the data distribution, i.e., sentiment polarity versus density of barbiturate drug group. The frequency distribution is found to be higher in case of

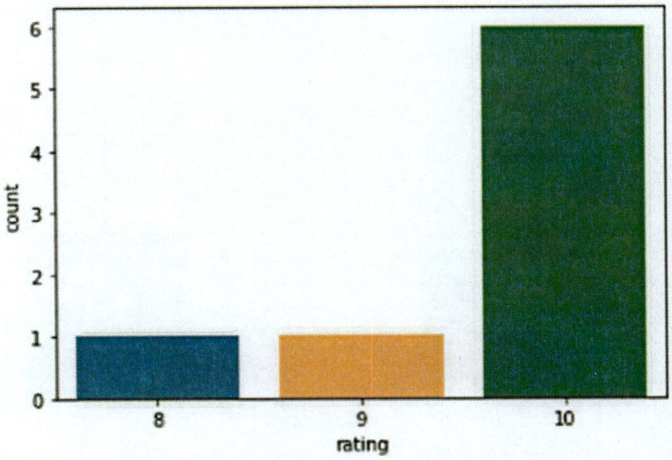

Fig. 3 Rating versus count of barbiturate drug group

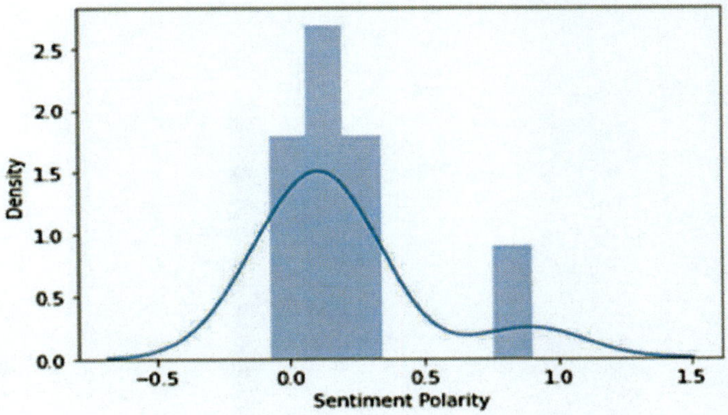

Fig. 4 Sentiment polarity versus density of barbiturate drug group

positive polarity and negligible in case of negative polarity. So, we can assume that users of this medicine have positive sentiment toward it and the highest frequency is found to be between 0 and 0.5.

The graph as shown in Fig. 5 shows the rating vs. polarity of barbiturate drug group. It is found that maximum customers have given rating of ten where sentiment polarity is also highest. Antipsychotics drug group is found to be less effective according to customer rating because most of the customers have given the rating as one and is shown in Fig. 6, and many customers have given negative reviews which is shown as in Fig. 7.

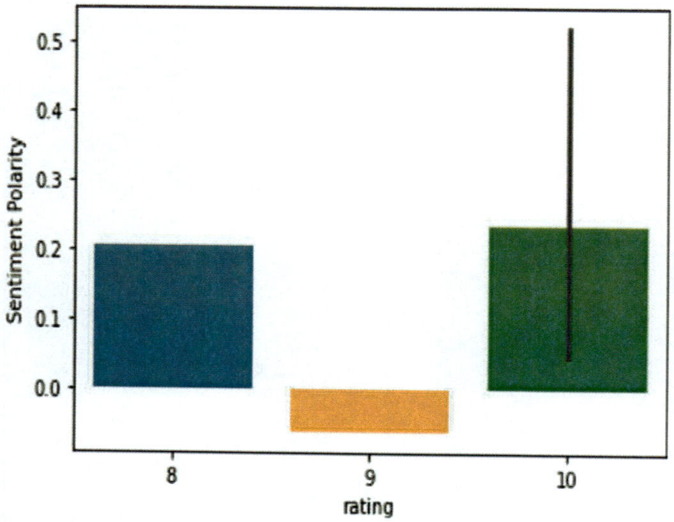

Fig. 5 Rating versus sentiment polarity of barbiturate drug group

Fig. 6 Rating versus count

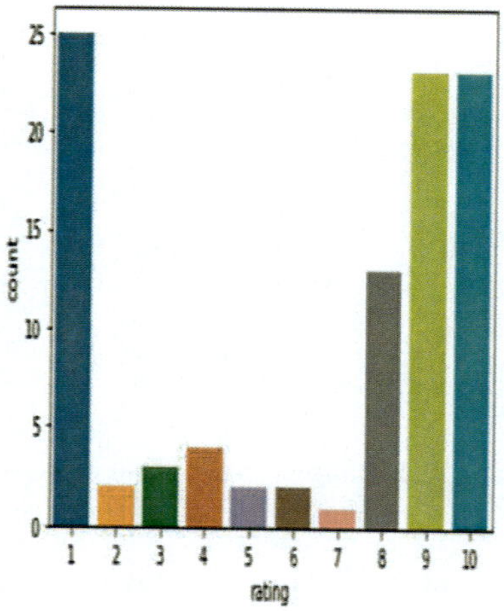

From the experimental results, it was found that opioid analgesics group is less effective according to polarity. Although customers have given ten rating to this group but many negative reviews were obtained in this case (Figs. 8, 9, 10 and 11).

Fig. 7 Polarity versus density

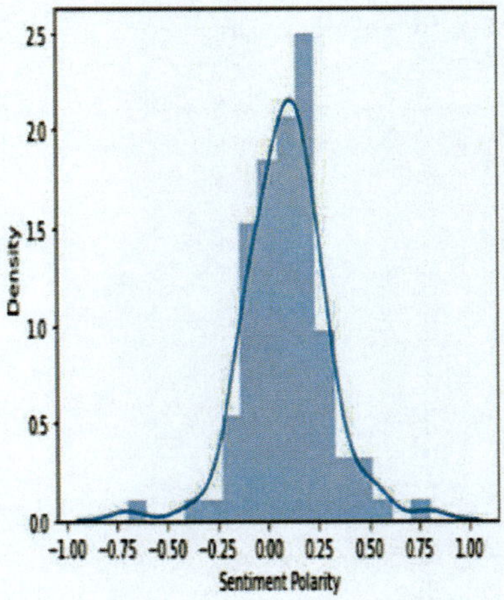

Fig. 8 Rating versus polarity

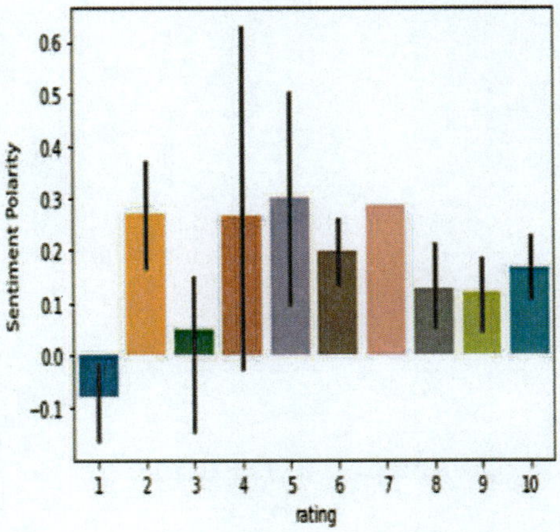

Fig. 9 Rating versus count

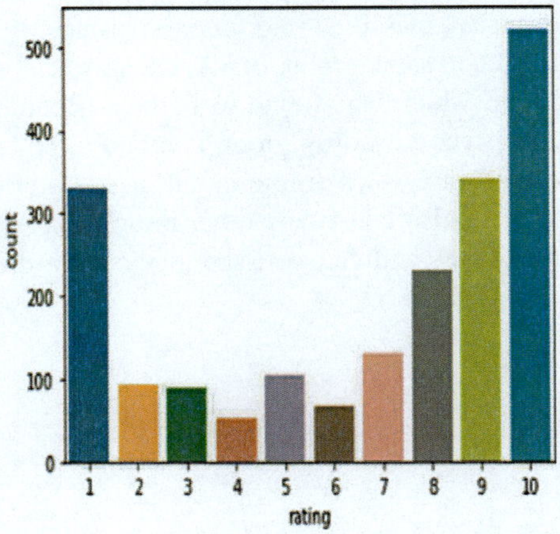

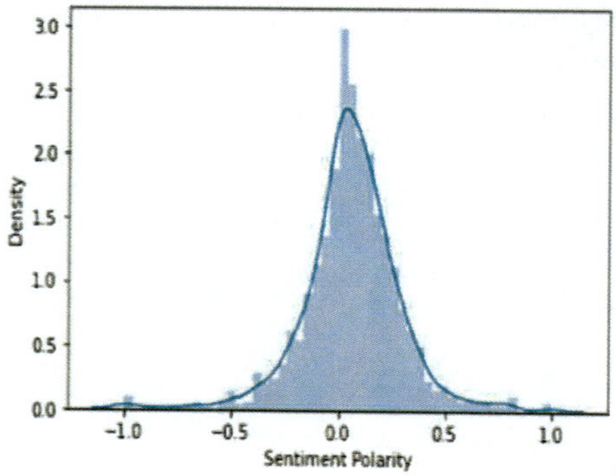

Fig. 10 Polarity versus density

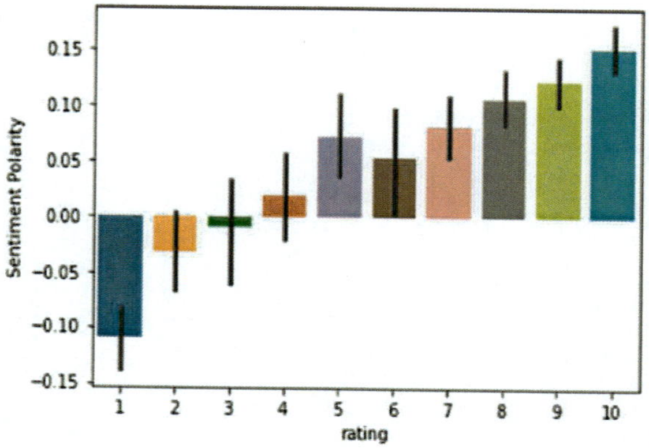

Fig. 11 Rating versus polarity

5 Conclusion and Future Scope

In this work, EDA and SA are applied on a part of original drug dataset where drugs are classified into fourteen groups. In EDA, we have analyzed different drugs related to same group. In SA, we have analyzed the performance of barbiturate drug group which was found to be best according to its rating and sentiment polarity. Antipsychotics drug group is found to be less effective as per customer rating, and opioid analgesics drug group is also less efficient with respect to sentiment polarity. We can also take many other drug groups based on several prefixes and suffixes of medicines and can perform analysis on different drug groups to identify the best among them.

References

1. Drus, Z., & Khalid, H. (2019). Sentiment analysis in social media and its application: Systematic literature review. *Procedia Computer Science, 161*, 707–714.
2. Bounaama, R., & Abderrahim, M. E. A. (2020, February). Sentiment analysis for health monitoring on social media. In *2020 2nd International Conference on Mathematics and Information Technology (ICMIT)* (pp. 149–153). IEEE.
3. https://blog.feedspot.com/doctor_blogs/. Accessed on October 20, 2021.
4. https://www.drugs.com/drug-classes.html. Accessed on December 12, 2021.
5. Gopalakrishnan, V., & Ramaswamy, C. (2017). Patient opinion mining to analyze drugs satisfaction using supervised learning. *Journal of Applied Research and Technology, 15*(4), 311–319.
6. Gurdin, G., Vargas, J. A., Maffey, L. G., Olex, A. L., Lewinski, N. A., & McInnes, B. T. (2020). Analysis of inter-domain and cross-domain drug review polarity classification. *AMIA Summits on Translational Science Proceedings, 2020*, 201.
7. Garg, S. (2021, January). Drug recommendation system based on sentiment analysis of drug reviews using machine learning. In *2021 11th International Conference on Cloud Computing, Data Science & Engineering (Confluence)* (pp. 175–181). IEEE.
8. Sarchiapone, M., Gramaglia, C., Iosue, M., Carli, V., Mandelli, L., Serretti, A., Marangon, D., & Zeppegno, P. (2018). The association between electrodermal activity (EDA), depression and suicidal behaviour: A systematic review and narrative synthesis. *BMC Psychiatry, 18*(1), 1–27.
9. Sridevi, U. K., & Shanthi (2021 February). An ontology-based sentiment analysis model towards classification of drug reviews. In *International Semantic Intelligence Conference* (pp. 164–171).
10. Na, J. C., & Kyaing, W. Y. M. (2015). Sentiment analysis of user-generated content on drug review websites. *Journal of Information Science Theory and Practice, 3*(1), 6–23.
11. https://archive.ics.uci.edu/ml/datasets/Drug+Review+Dataset+%28Drugs.com%29. Accessed on December 20, 2021.
12. Denecke, K. (2015). Sentiment analysis from medical texts. In *Health web science* (pp. 83–98). Springer.
13. https://rehabs.com/blog/classification-of-drugs. Accessed on December 27, 2021.

Survey on Algorithmic Trading Using Sentiment Analysis

Rupali Bagate, Aparna Joshi, Abhilash Trivedi, Anand Pandey, and Deepshikha Tripathi

Abstract In recent years as the computation power and availability of the data has increased exponentially, there has been significant increase in study of human sentiment in various fields. This paper examines the use of sentiment analysis in algorithmic trading. Macroeconomic variables such as GDP, Internet consumption and various other socio-economic factors are also taken into consideration in this paper. The main aim of this paper is to determine all factors and technical indicators that would give us a proper analysis. Human sentiment affects human behaviour adroitly, and thus, market is also not acquitted from its effect. This survey presents current advances in natural language processing (NLP) and prerogative positions of algorithms in market.

Keywords Algorithmic trading · Sentiment analysis · Market indicator · Machine learning

1 Introduction

Sentiment analysis is the study of reverential effect of human emotions in various domains. In sentiment analysis, human emotions are quantified to create colloquial values for evaluation which then are used in concordance with other techniques and algorithm to compute their effect in respective domains [1].

R. Bagate · A. Joshi · A. Trivedi (✉) · A. Pandey · D. Tripathi
Department of Information Technology, Army Institute of Technology, Pune, Maharashtra, India
e-mail: abhilashtrivedi99@outlook.com

R. Bagate
e-mail: rbagate@aitpune.edu.in

A. Joshi
e-mail: ajoshi@aitpune.edu.in

B. Pati et al. (eds.), *Proceedings of the 6th International Conference on Advance Computing and Intelligent Engineering*, Lecture Notes in Networks and Systems 428,
https://doi.org/10.1007/978-981-19-2225-1_22

1.1 Motivations

Algorithmic trading is a peculiar form of trading where computer program follow defined set of instruction to generate profits. The most challenging task for algorithmic trading is to culminate result and process noisy and volatile nature of the stock to gather insights. In recent years, we have witnessed exponential increase in real-time prediction of stock using various algorithms and machine learning technique but because of the capricious nature of the market, they are likely to underperform in many scenarios.

1.2 Contributions

In the statistics community, the growing availability of the data has led to the significant accuracy in performance and explainability of these models. The pious nature of market with statistics has also pushed quantification of various emotions and opinions which further led to the development of model in concept like natural language processing (NLP).

2 Overview

In this survey paper, all vital information is covered regarding algorithmic trading using sentiment analysis. Starting from the existing research done so far in this field or any related field, this paper presents suitable findings in the following sections. A basic introduction about the stock market and the technical indicators are explained. This is followed by a detailed analysis of advantages and disadvantages of algorithmic trading. This paper also includes various algorithms that are known to this day. Lastly, this paper talks about the macroeconomic variables that consist of factors like Net Asset Value per share, price earnings, etc. All this information forms an umbrella under which all aspects of algorithmic trading using sentiment analysis are covered in required depth.

3 State of the Art

Shah et al. [2] have used unigram and dictionary-based approach to analyse the data from pharmaceutical industry. The author described text pre-processing, transformation, comparison and sentiment scores. Kalra and Prasad [3] used support vector machine (SVM) and KNN on stock news data and Yahoo Finance data. The author has described comparison between various machine learning models under similar conditions, described tokenisation, transformation and enumerate sentiments.

Peng [4] has discussed the effect of human sentiment using Pollet and Wilson's theory of volatility decomposition. The author has discussed strategies based on big data collected on Chinese stock markets. The author had used hidden horse model on big data to predict stable that results in volatile exchange. Kaur [5] used Markov decision problem (MDP) approach to represent the trading problem and henceforth solve using Q-Learning and author used augment MDP states to improve the performance. The author also discussed market indicators and their roles as variable in neural network to increase the predictability. Ojo et al. [6] used data from NASDAQ composite (^IXIC) and a LSTM model for predicting behaviour in stock market. The researchers have used autoregressive integrated moving average (ARIMA). Zhnag and Skiena [7] have studied co-relation between Media and Stock Data. The author also proposed relation between Stock Polarity with Shifting Time and Media type. Usage of Diversification, Sentiment Analysis Period, Holding Period and Market Capitalisation for performance evaluation. Birbeck and Cliff [8] proposed a comparative model between multinomial logistic regression, Naive Bayes and support vector machine. For ranking, the author had used Chi-squared and F-value. "A non-statistical method of selection of features is also considered, where a good set of features is explained by duplicate selection of a few and a few features, gradually pruned by those with the lowest contribution to the current set and in addition to the above features". Li et al. [9] proposed an architecture that employs support vector machine (SVM), extreme learning machine (ELM) and back-propagation neural network (BPNN) for both accurate and fast prediction. Ranjan et al. [10] proposed a hybrid model of technical trading and fundamental trading. Bagate et al. [1] did intensive research on NLP used to identify sarcasm.

In order to gather generic sentiment of the market, they used all news instead of using specific news to gather individual sentiment. The author focused on relative comparison between ELM and SVM which resulted in almost similar result but ELM was faster. Ranjit et al. [11] have proposed a model based on artificial neural network (ANN), and the ANN yields the best result measured on the basis of RMSE with six hidden nodes. The author has proposed prepossessing steps of removal of usernames, usage of links, removal of repeated letters, hashtag detection. Usage of accuracy, precision, recall and F-measure as testing metrics for the model. Bell and Gane [12] concluded that the adoption and distribution of algorithmic trading systems have increased, and this is likely to continue, as regulation, competition and innovation drive the development of advanced technology tools. Shirsat et al. [13] had explored the polarity of the article while having quantity of sports and business article almost same similarly quantity of article from technology and entertainment is also same and determined polarity of respective article. Jessica and Raymond [14] proposed a model to work on tweets and predicted the price of Facebook stock. They have used the proposed model on stock exchange and have acquired accuracy of 71.82% with the LibSVM model and 69.01 % with a logistic regression model. A survey done by Bagate and Suguna [15] gives an overall survey for the sarcasm detection, and this can be helpful to analyse various headlines and heading or even the content of online information. Bagate et al. [16] compared various factors like words, punctuation marks, emoticons, the environment and author for understanding speech

Table 1 Summary of different methods of algorithmic trading and sentiment analysis

Author	Domain	Method	Features
Kalra and Prasad [3]	Stock News	KNN, SVM, Naive Bayes	Diverse news on stocks
Shah et al. [2]	Pharmaceutical market	Unigram and dictionary based approach	Dictionary-based sentiment analysis
Peng [4]	Chinese stock market	Hidden horse model	Integrates big data to quantifiable index
Ojo et al. [6]	American stock exchange, NASDAQ Composite (I^XIC)	Multilayer perceptron and LSTM	ipso facto
Zhang and Skiena [7]	Blog and news	Trading strategies to exploit blog and news sentiment	Subjectivity and volumes
Kaur [5]	Yahoo finance	MDP and Q-Learning	Sentiment score
Birbeck and Cliff [8]	Twitter	Bayesian classifier and logistic regression	Trading volume
Li et al. [9]	Finance news article	Back-propagation Neural Network (BPNN), Support Vector Machine (SVM) and Extreme Learning Machine (ELM)	RSI and Bias
Ranjan et al. [10]	Forex	Technical and fundamental trading algorithm	Context consideration from syntactic structure
Ranjit et al. [11]	Foreign currency exchange	Artificial Neural Network (ANN)	Lexical syntactic Feature
Bell and Gane [12]	Market data	Algorithmic Trading	

sarcasm sentiment in the text. A Case Study of the Dhaka Stock Exchange [17] was conducted and the researchers found that for the cement companies some economic factors affected the share price. In a paper by Labidi and Yaakoubi [18], the authors have discussed the various factors affecting the sentiment of all kinds of investors like the big ones and also small investors. A paper by Robert [19] talks about stock market of different nations of the BRIC countries. Researches have studied the Indian Stock Market also as discussed in a paper [20]. This study highlights the aspects specific to the Indian Market. Twitter is a major social media platform, and a study [21] was done to analyse the relationship between the tweet count frequency and telecom companies IPO performance. Foreign Exchange [22] is also an interesting field of study and can prove helpful in determining market trends worldwide. Timing plays a key role in this market and this paper [23] tells how we can work with timing and actually predict the market trends of recent past to far future (Table 1).

4 Introduction to Stock Market

Stock market can be referred to a platform where publicly held company put down the shares of companies for the public. Stock market is the place where general public constantly buy and sell share of these publicly held company. The place where this trading happens is referred to as stock exchange. Stock exchange can be physical or virtual, and there can be multiple stock exchange in countries. Stock markets provide a secure and regulated environment where investor can trade in shares and other financial instruments with low operational risks. People also refer the stock market as a single entity; this single entity is called as "Indexes", and it defines the overall movement of the listed companies stock performance. Investor compare current price with respect to past performance of the market using Index which is a measure of stock market.

The flow of the market can widely be generalised into bear and bull market. When the market flow is in positive direction, i.e. the buying pressure is dominant compared to the selling pressure or when the index is moving up, then the market is referred to as Bullish trend.

If the flow of the market flow is in negative direction, i.e. selling pressure is dominant compared to the buying pressure or when the index is moving down, then the market is referred to as Bullish trend.

In order to maximise the profit, investor from the past have been modelling different patterns and indicators in order to produce buying and selling signals to maximise profit and minimise the loss. Today with no barriers on computational power, it has exponentially increased the use of computer programs for stimulated trading and aided the algorithmic trading systems. Today there are technical indicators defined on patterns followed by the stock market graphs and previous moment of the stock in order to predict the outcome of the stock market and decide accordingly to buy or sell a stock. With decreasing interest rates and easy availability of Internet, people have grown their interest in stock market and hence algorithmic trading.

4.1 *Pre-Trade Analysis*

Nuti [24] in his paper algorithmic trading explains the basic terminology used in stock market, what are some pretrade analysis to look into before making an algorithmic strategy. He explains the main three trade analyses:

- Technical analysis: Aim of technical indicator is to predict the future price or the movement of the stock graph based on the stocks past history and finding out the patterns in graph using the previous data such as volume trade, buying volume, selling volume, opening and closing price. The main objective of technical analysis is to identify and predict the price movement patterns.

- Fundamental analysis: “A fundamental analysis involves a detailed study of related data that could affect price values by determining the fair value of the asset (or potential future price movements)” [25]. Appropriate information may include the general state of the two countries’ economies (such as unemployment figures), interest rates, gross domestic product, or national policies. The notion that the current market price of a commodity is not commensurate with its fair value contradicts the controversial effective market view, suggesting that the current price is an indication of all available information that affects prices [25].

Algorithmic trading is also referred as automated trading, black-box trading, or algo-trading, the structure of algorithmic trading is to use a computer program that follows an algorithm to place a trade. The purpose of algorithmic trading is to generate profits at a speed and frequency that is impossible for a human trader. The main advantage of algorithmic trading is to produce incomparable and more trading in one session compared to human.

4.2 Advantage of Algorithmic Trading

The biggest advantage of algorithmic trading is that it removes the emotional factor of the market and investors. Here all trades are performed according to computerised algorithm. Greed and fear are one of the biggest challenges for investors in case of financial markets. These are the factor that results in loss for the investors. Algorithmic trading also helps in maintaining discipline because it has predefined entry and exit strategy which makes a robust trading in stock market. Another advantage of algorithmic trading and the main reason of growth of algorithmic trading are the speed with which these programs execute orders. Moreover, it can record more data compared to human and execute more accurate trades compared to human brain.

4.3 Disadvantage of Algorithmic Trading

The major disadvantage of algorithmic trading is that it is not accurate and may result in wrong trades which might produce loss. Moreover, it requires a greater study of indicator and technical oscillators to make a model and execute algorithmic trading; hence, it is proven to be dangerous to use if not well-versed with the knowledge of market. The biggest disadvantage of algorithmic trading is during the time of emotional-driven market, with predefined signal and strategies, and there is no scope for emotional indicators which results in low accuracy of algorithmic trading during sentiment-driven market. News and result related to companies instantly changes the flow of market moments of that stock and fails all the indicators and is completely dependent on human emotions, hence resulting in the losses and low accuracy of algorithmic trading.

5 Categorisation of Market Variables

As per the survey, there are various factors which influence any market within specified interval of time. Stock prices are influenced by almost every phenomenon from natural disaster to change in structure of government. Most of these are variables and can easily be accumulated within sentiment analysis but with accumulation comes the problem of effect, some news sentiment change market with much influx as compared to others, some of the factors affect market instantaneously and some show slow and gradual change. Hence, we can say that it's almost impossible to predict the trend in market with absolute certainty.

5.1 Media

During our survey, we have realised different kinds of information and data affects markets differently. The global news like natural disaster and pandemic effect market on global scale adroitly while national news like budget effect the market of that nation and trading nation.

Certain data platforms like Twitter and Facebook show significant predictability for intra-day trading while on the other hand news from news agency concern much larger time frame while this time frame increases significantly when we look forward to blogs.

5.2 Market Capitalisation

Zhnag [7] in his paper proposed how various assets for any company act in favour of and against the company. They argued both large and small firms had better returns as compared to medium-sized firms.

Companies with less stock price are more likely to be affected by trends in market as high buy and sell increase volatility for the stock while small companies are affected as they can be traded in profusion which changes the price for overall stock.

5.3 Indicators

Market indicators are quantitative in nature and seek to interpret stock and financial indicators in attempt to predict the market moves. They have been used in market since it is inception. These technical indicators are used by investor in order to get a better brief about the current market status and future price prediction. There are various technical indicators used together for the formulation of lesser risk and high accuracy technical models in order to maximise the profit.

- Moving Average:

$$\overline{p}_{SM} = \frac{p_M + p_{M-1} + \cdots + p_{M-(n-1)}}{n} = \frac{1}{n}\sum_{i=0}^{n-1} p_{M-i} \tag{1}$$

- Stochastic K:

$$K\%(i) = \frac{cp(i) - L_t}{H_t - L_t} * 100 \tag{2}$$

- Stochastic D:

$$D\%(i) = \frac{K\%(i-2) + K\%(i-1) + K\%(i)}{3} \tag{3}$$

- Relative Strength Index:

$$\text{RSI} = 100 - \frac{100}{1 + \text{RS}} \tag{4}$$

$$\text{where RS} = \frac{\text{Average of T day's up closes}}{\text{Average of T day's down closes}} \tag{5}$$

- Larry William's $R\%$:

$$R\%(i) = \frac{H_t - cp(i)}{H_t - L_t} \tag{6}$$

6 Macroeconomic Variables

Macroeconomic variables have become very popular in determining the dynamics of stock market. These factors are not mandatory when investing in the stock market but are very crucial in determining long-term effects on the stock prices. Some companies, for example cement companies, show a great amount of dependence on the macroeconomic variables for their IPO(s). In a study done by researchers in Bangladesh of Dhaka Stock Exchange, it was found that factors such as Producer Price Index (PPI), IPO first-day return (RIPO), Consumer Price Index (CPI) play an important role.

Table 2 Problems, data sets, algorithms and evaluation measures and methods used [26]

Problem (Object)	Classes	Data set	Algorithms	Evaluation measures	Evaluation methods
NER words	Multi	News	SNER, LingPipe, Illionis	Precision, F-measures and recall	Holdout
Sentiment analysis (reviews)	Binary	Movies and TV	DT, LR, KNN, RF, SVM	Precision, F-measures and recall	Holdout
	Binary	Books	DT, LR, KNN, RF, SVM	Precision, F-measure, Recall	Holdout and Cross-validation (10 folds)
Document classification (documents)	Multi	News	DT, LR, KNN, SVM, Naïve Bayes	Precision, F-measures and recall (micro- and macro-average)	Cross-validation (6 folds)
	Binary	Patents	DT, LR, KNN, SVM, RF	Precision, F-measures and Recall	Holdout and Cross-validation (6 folds)

6.1 *Net Asset Value per Share*

The total value of each asset is determined by dividing the total value of the company's collateral by each residential area each year.

6.2 *Price Earnings*

The price earnings (P/E) in all shares calculated by dividing the market prices by the company quarterly each year as per profit (Table 2).

6.3 *Sentiment*

A sentiment in colloquial sense can be referred as an opinion and when this term is used in computation then this term becomes conducive to quantitative analysis. Sentiment analysis is the term which came along natural language processing and is used to determine to what extent any data is positive or negative.

Current global market is highly influenced by the sentiments as market influence individual's life. Positive news encourages an individual to strive for profit and negative work as a setback.

7 Methods

7.1 *Support Vector Machine*

The support vector machine is a machine learning algorithm used for the purpose of classification of data; the algorithm focuses on dividing the set of data that needs to be classified using a hyperplane by the very colloquial thought of increasing the distance of the hyperplane from the nearest element of both categories.

SVM provides high-dimensional input space like tokenisation of sparse document vector, and it is tolerance to overfitting and underfitting gives advantage over other algorithms.

SVM had better performance in larger data sizes and was the dominant algorithm for rest of the data size experiment [26].

SVM's Prediction accuracy varies between 65.30 and 83.80% [3].

7.2 *Logistic Regression*

Logistic regression is one of the technique which have been inculcated in machine learning from statistics. The algorithm is a method for binary classification of the data. Jessica and Raymond achieved 69.01% accuracy using LR model. [14]

$$\text{Logistic Function} = \frac{1}{1 - e^{-\text{value}}} \tag{7}$$

7.3 *KNN*

KNN has the ability to work as both classification and regression algorithm. KNN considers all the related elements in data that lies in close proximity with each other.

KNN selects k-neighbours based on euclidean distance and then on the basis of similarity group the data points with each other. KNN's prediction accuracy varies between 7 and 91.2% [3].

$$d(p, q) = \sqrt{\sum_{i=0}^{n-1} (p_n - q_n)^2} \tag{8}$$

7.4 *Decision Tree*

Similar to KNN decision tree can also be used for both regression and classification algorithms. Decision tree mimics the reversed tree structure in order to determine the chance of events and decision-making. In decision tree, each node determines a quantitative test to an attribute and determines the outcome of the test.

7.5 *Random Forest*

Random forest is an algorithm that is based on ensemble learning which implies that multiple models will be combined and generated through a strategy to solve a computationally intelligent problem. It contains multiple decision tree on varied data sets whose result is compiled and averaged in order to predict a much more accurate result for the specific problem.

8 Conclusion

Sentiment analysis from social media, news, websites and blogs gives us a clear idea of how human sentiment affects the market. As observed, most of the authors worked on supervised and semi-supervised approaches for sentiment analysis and in terms of statistical approach most of the work was based on mathematical approach of indicators but we realised that very less work has been done to inculcate both domains to produce much more accurate and probable result. For future direction, it will be useful if both these are considered for further development in this field.

9 Future Directions

The efficiency of the model can be increased by improving the prototype. The sentiments that are conveyed by news articles and headlines have a different impact on the market and investors which is usually short-term. By making the model more extensive, sentiment conveyed by blogs and other articles that have long-term effects on the market can be included. Web scraping can be used to automate the entire process and save more time. Hence, this model will become more accurate with these developments.

References

1. Bagate, R., Saini, A., Sethi, K., Tomar, H., & Singh. (2021). Sarcasm detection and explainable AI.
2. Shah, D., Isah, H., & Zulkernine, F. (2018). Predicting the effects of news sentiments on the stock market.
3. Kalra, S., & Prasad, J. S. (2019). Efficacy of news sentiment for stock market prediction.
4. Peng, D. (2019). Analysis of investor sentiment and stock market volatility trend based on big data strategy.
5. Kaur, S. (2018). Algorithmic trading using sentiment analysis and reinforcement learning.
6. Ojo, S. O., Owolawi, P. A., Mphahlele, M., & Adisa, J. A., Stock market behaviour prediction using stacked LSTM networks.
7. Zhang, W., & Skiena, S., Trading strategies to exploit blog and news sentiment.
8. Birbeck, E., & Cliff, D. (2018). Using stock prices as ground truth in sentiment analysis to generate profitable trading signals.
9. Li, X., Xie, H., Wang, R., Cai, Y., Cao, J., Wang, F., Min, H., & Deng, X. (2017). Empirical analysis: Stock market prediction via extreme learning machine.
10. Ranjan, S. K., Trivedi, A., & Rajpoot, D. S. (2015). Sentiment score based algorithmic trading.
11. Ranjit, S., Shrestha, S., & Subedi, S. (2018). Foreign rate exchange prediction using neural network and sentiment analysis.
12. Bell, D., & Gana, L. (2012). Algorithmic trading systems: A multifaceted view of adoption.
13. Shirsat, V. S., Deshmukh, S. N., & Jagdale, R. S. (2017). Document level sentiment analysis from news articles.
14. Jessica, R. S. O. (2019). Sentiment analysis on official news accounts of twitter media in predicting Facebook stock.
15. Bagate, R. A., & Suguna, R. (2019). Different approaches in Sarcasm detection: A survey.
16. Bagate, R., & Suguna, R. (2021). Sarcasm detection of tweets without #sarcasm: Data science approach.
17. Alam, S., Rubel Miah, Md., & Abdul Karim, Md. (2016). Analysis on factors that affect stock prices: A study on listed cement companies at Dhaka stock exchange.
18. Labidi, C., & Yaakoubi, S. (2016). Investor sentiment and aggregate volatility pricing.
19. Gay, R. D., Jr. (2016). Effect of macroeconomic variables on stock market returns for four emerging economies: Brazil, Russia, India, and China.
20. Kumari, J., & Mahakud, J. (2016). Investor sentiment and stock market volatility: Evidence from India.
21. Shelar, A., & Huang, C. (2018). Analyzing relationship: Tweet frequency with the stock prices of telecom companies.
22. Kumar Chandar, S., Sumathi, M., & Sivanandam, S. N. (2015). Forecasting of foreign currency exchange rate using neural network.
23. Merello, S., Ratto, A. P., Ma, Y., Luca, O., & Cambria, E. (2018). Investigating timing and impact of news on the stock market.
24. Nuti, G., Mirghaemi, M., Treleaven, P., & Yingsaeree, C. (2011). UK Centre in financial computing, London.
25. Nuti, G., Mirghaemi, M., Treleaven, P., & Yingsaeree, C. (2011). Algorithmic trading.
26. Baeza-Yates, R., & Liaghat, Z. (2017). Quality-efficiency trade-offs in machine learning for text processing.

Convolutional Neural Networks for Audio Classification: An Ensemble Approach

Bhavi Dave and Kriti Srivastava

Abstract Machine perception has steadily become more accurate with deep learning methodologies. Abundant multimedia data sources have made navigating audio data essential. This work performs environmental sound classification as a step toward integrating artificial intelligence in audio data. Audio files are converted to tensors, resampled and then converted to mel spectrograms to account for human sensitivity to different audio frequencies. Pre-trained and high-performing convolutional neural networks (CNNs) are leveraged to train the ResNet-152 and DenseNet-121 architectures for transfer learning. The custom ensemble model uses these models for inference. The outputs of the models are combined and passed through a dense layer to generate an ensemble capable of inferring correct weightage for each of the models without manual interference. The ensemble model achieves promising results with an accuracy of 91%, and precision and recall of 0.91 and 0.93, respectively. The results demonstrate that a CNN-based ensemble method is adept at extracting and generalizing temporal information from audio signals.

Keywords Artificial intelligence · Environmental sound classification · Deep learning · Ensemble model · Transfer learning · Convolutional neural networks

1 Introduction

This section introduces the motivation for using deep learning in audio classification. It then goes on to outline the contributions made by academia within the domain and lists some of the varied approaches taken. Finally, the existent research gaps in the literature review are listed.

B. Dave (✉) · K. Srivastava
Department of Computer Engineering, Dwarkadas J. Sanghvi College of Engineering, Mumbai, Maharashtra, India
e-mail: bhavidave5@gmail.com

K. Srivastava
e-mail: kriti.srivastava@djsce.ac.in

B. Pati et al. (eds.), *Proceedings of the 6th International Conference on Advance Computing and Intelligent Engineering*, Lecture Notes in Networks and Systems 428,
https://doi.org/10.1007/978-981-19-2225-1_23

Table 1 Table captions should be placed above the tables

References	Dataset	Methodology	Accuracy
[14]	ESC-10, ESC-50	Classification by human listeners	95.7%, 81.3%
[11]	BIRDZ, ESC-50	Ensemble CNN	96.90%, 85.75%
[16]	ESC-50	Compressed and Quantized CNN	83.65%
[15]	ESC-50	Convolutional Restricted Boltzmann Machine + CNN	86.50%
[8]	ESC-50	CNN	83.50%
[19]	ESC-10 and ESC-50	WaveMsNet	95.7%, 81.3%
[16]	ESC-50	CNN + phase-encoded filterbank	84.15%
[7]	ESC-50	Self-teaching + CNN	94.1%
[4]	ESC-50, Speech Commands V2 dataset	CNN + Self Attention	95.6%, 98.1%

1.1 *Organization*

Section 1 outlines the motivation and reviews the literature for audio classification using deep learning. Section 2 then goes on to detail the methodology implemented in this paper—it enumerates a brief data description and then reviews the architecture and training. Section 3 then describes the evaluation metrics and the results obtained. Section 4 then concludes the paper.

1.2 *Motivation*

Audio signals have now become omnipresent in the modern technological landscape. Audio classification can be leveraged in several scenarios like alarm detection for people with an impaired hearing [10], engine monitoring [1] through sound analysis and even detecting the presence of endangered species [9] in an environment. Although significant work has been done in the domains of speech and music, the classification of environmental sounds is relatively scarce. Although audio signals are temporal in nature, advancements in the computer vision domain can be leveraged to use the numerous kinds of pre-trained, high-performing convolutional neural networks for the task.

1.3 Literature Review

The environmental sound classification dataset ESC-50 [14] was published in 2015 and has served as a benchmark for audio classification. Participants from the Crowd-Flower crowdsourcing platform were asked to classify the sounds from the ESC dataset to provide a benchmark for human competence at the task (Table 1).

Nanni et al. [11] used different data augmentation techniques alongside different signal representations that were then transformed into melspectorgrams. Fine-tuned CNNs were combined as an ensemble using the sum rule for inference on the BIRDZ dataset [12] and the ESC-50 dataset. Tak et al. [16] proposed an architecture that can perform classification on edge devices that can perform in an extremely resource-constrained environment. Their reduction mechanism achieved a 97.22% size reduction and 97.28% FLOP reduction.

Sailor et al. [15] proposed the creation of a convolutional restricted Boltzmann machine for learning filterbank from the raw audio signals. The work in [8] explores transductive and inductive transfer learning and proposes a CNN-based framework that archives human accuracy on the task of environmental sound classification and also performs acoustic scene classification. WaveMsNet [19] is a multi-scale convolution operation that improves frequency resolution and learning filters across all frequency areas, leading to better audio representation. Research also posits that most work only processes the magnitude spectrum, ignoring the phase spectrum [16]. Consequently, they propose phase-encoded filterbank energies (PEFBEs) for environmental sound classification. Tak et al. [7] proposed a sequential self-teaching approach that utilizes co-supervision across trained CNN models for effectively classifying sounds. They achieve promising results in a large scale and weakly labeled data points because of effective modeling by the system. In [4], the authors used CNNs alongside a self-attention mechanism to capture long-range global context.

1.4 Research Gaps

- The existing architectures are bulky, requiring deep networks and heavy preprocessing, inevitably increasing the time required for training. Transfer learning can be invaluable for efficiently utilizing resources.
- Several works created complex designs with novel manners of extracting features for improving performance. The trade-off between complexity and performance needs to be explored further as minimal case-specific feature extraction would result in models that can more generally applicable to the domain.

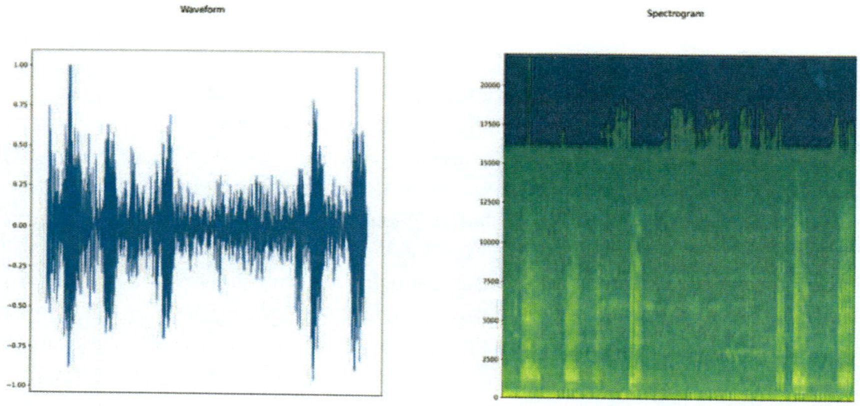

(a) Audio Waveform (b) Mel-Spectogram

Fig. 1 Raw audio waveform and its corresponding mel spectogram

2 Experimental Setup

This section outlines the setup used to perform audio classification. It provides an overview of the dataset and the data processing performed, alongside the motivations for the data transformations undertaken. Post this, the training of the neural network architectures and the design of the custom ensemble model is enumerated. Finally, the methodology of the experiments is also detailed.

2.1 Data Description

The environmental sound classification dataset ESC-50 [14] contains 2000 audio recordings that are commonly found in both a natural and an urban environment. The dataset is balanced with 40 clips for each of the 50 classes present in the dataset. The classes are divided into five major categories: animals, natural noise/soundscapes, human sounds (non-speech), interior/domestic sounds, and exterior/urban noises. The results on this dataset can be an effective testament to the model performance because of a wide variety of classes.

The audio files are read as a tensor and resampled to a fixed sample rate 44,100 Hz. This is because the maximum frequency that can be represented by a sampled signal is at most half the sample rate—called Nyquist frequency. Cochran [2] The time-series signal is then transformed into the image domain by converting it into a spectrogram that represents the frequency content in the audio file as image colors. The work employs a log-scaled mel spectrogram that converts frequencies to the mel scale that takes into account human sensitivity to different ranges of audio frequency. Figure 1 displays an audio file (a) and its corresponding mel spectrogram (b), respectively.

2.2 Architectures

A convolutional neural network (CNN) is a subtype of deep neural network that gains relevant, abstract, and location-invariant features. The rectified linear unit activation function used with CNNs converts negative values to 0, extracting only the prominent features [13]. The max-pooling layer helps reduce complexity while allowing prominent features to pass through to upper layers of the network [17]. CNNs greatly reduce the number of parameters while achieving a similar degree of complexity because of sparse connectivity and weight sharing. The architecture is especially great at learning feature hierarchies from data [18].

ResNet [5] hypothesized that the gradients were not able to flow well through the deep networks that were seen at the time. The scientists found that the information from the input was getting highly morphed as it reached the deeper output layers and consequently proposed a solution of passing the input information repeatedly in stages creating a residual network or the ResNet. In ResNet, after every two layers, the input is given to the first layer along with the output obtained at the second layer. This helped the gradients to flow back better, improving training.

DenseNet [6] connect every layer to every other layer, so there are L(L+1)/2 direct connections for L layers. Every layer uses the feature maps of all the preceding layers as inputs, and consequently, its output feature maps are used as input for subsequent layers. The network is divided into densely connected blocks within which the feature map size remains the same. This facilitates both downsampling and feature concatenation. DenseNets require fewer parameters than a comparable CNN as the need to learn redundant feature maps is eliminated.

Ensemble Model Ensemble deep models are often adept at solving tasks [3]. The models used for transfer learning were pre-trained models trained on the ImageNet dataset. The ensemble model used ResNet-152 and DenseNet-121 individually trained on the ESC-50 dataset. The ensemble model loads the aforementioned models with their weights. The pipeline generates inferences from both the models individually and then combines the output into a one-dimensional tensor signifying the 50 classes. The outputs are then combined and sent through another linear layer that also generates 50 outputs, signifying the probability of the datapoint belonging to each of the 50 classes. The last linear layer in the custom ensemble is meant to learn the context for the combined weights of the ResNet-152 and DenseNet-121 models, allowing for better performance without manually deciding model weights in the ensemble. The ensemble model architecture is displayed in Fig. 2.

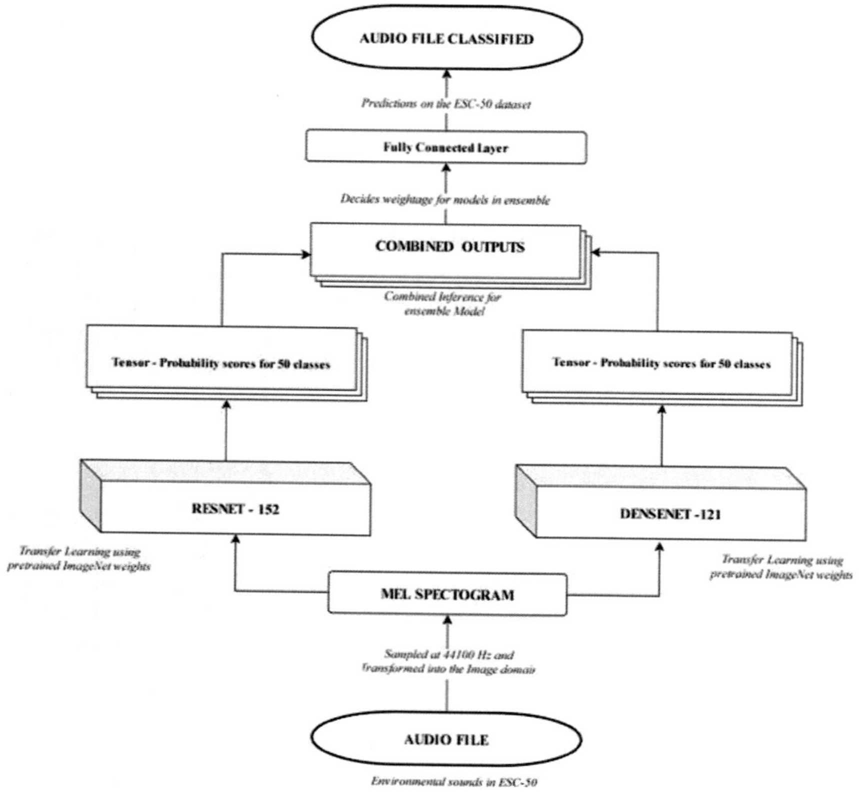

Fig. 2 Ensemble model architecture

2.3 *Methodology*

All three models were trained on the same partition of the dataset to avoid overlap, ensuring that the test data points were unseen by all three models. 15% of the dataset was reserved for testing, and the same other 80% was used to train the three models. All three models used a stochastic gradient descent optimizer with a momentum of 0.9. The criterion used to evaluate loss for backpropagation was cross-entropy loss. All three models also utilized a learning rate scheduler that decayed the learning rate of each parameter group by 0.1 every 3 epochs. Pre-trained ImageNet weights were downloaded for ResNet-152 and DenseNet-121, and both the architectures were modified to take input with a single channel and have 50 out classes. The rest of the architectural design was kept unchanged to preserve the learned weights for transfer learning. ResNet-152, DenseNet-121, and the ensemble model were trained for 15, 8, and 10 epochs, respectively.

3 Results

This section outlines the evaluation measures and the results obtained by the custom model as well as the two convolutional neural networks on the audio classification dataset.

3.1 Evaluation Measures

After training, the model has been evaluated on the unseen test set that comprised 15% of the ESC-50 dataset. The test set was stratified with respect to the class labels.

1. Accuracy: It is the fraction of classifications the model got correct.

$$\text{Accuracy} = \frac{\text{TP+TN}}{\text{TP+TN+FP+FN}} \tag{1}$$

2. Precision: Proportion of the positive identifications which are correct.

$$\text{Precision} = \frac{\text{TP}}{\text{TP+FP}} \tag{2}$$

3. Recall: Proportion of actual positives which were identified correctly.

$$\text{Recall} = \frac{\text{TP}}{\text{TP+FN}} \tag{3}$$

4. F1-score: It is calculated as the harmonic mean of precision and recall.

$$F1 - \text{ score} = \frac{2 \times \text{ precision } \times \text{ recall}}{\text{precision } + \text{ recall}} \tag{4}$$

3.2 Results Obtained

The models were used to classify the test data into the 50 environmental sound classes of the ESC-50 dataset. The results are shown in Table 2, which shows previously described measures for each of the three models trained. All measurements were calculated on test data. The ensemble model is a significant improvement with an accuracy of 91% and the precision and recall of 0.91 and 0.93, respectively. Figure 3 demonstrates classification metrics like precision, recall, and F1-score across all the 50 classes in the EDC-50 dataset.

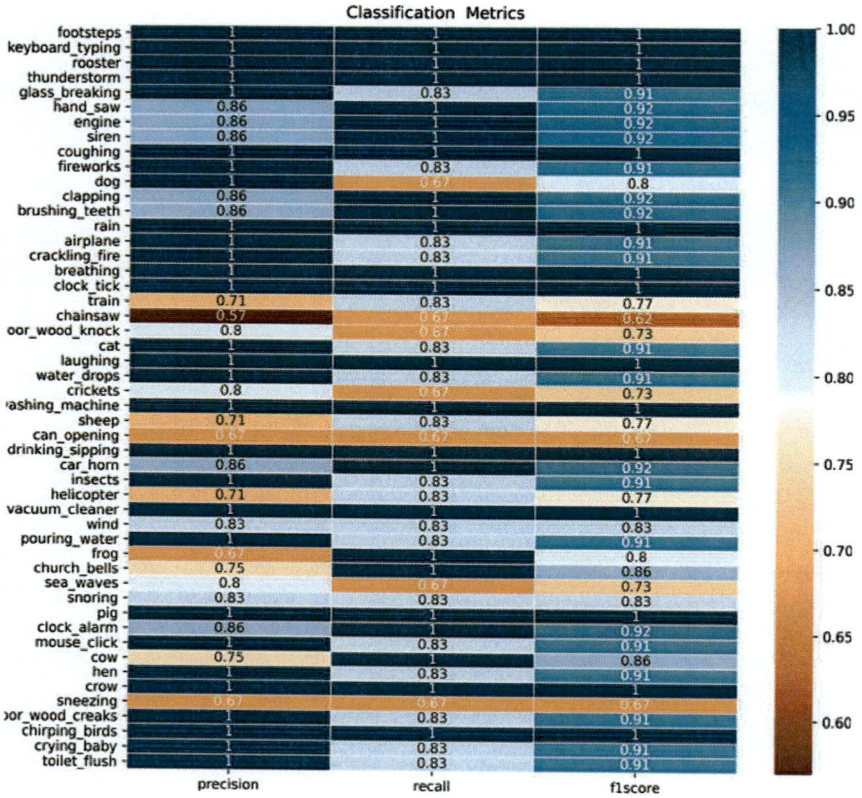

Fig. 3 Classification metrics for ESC-50

Table 2 Result analysis

Metric	ResNet	DenseNet	Ensemble
Accuracy	0.83	0.81	0.91
Precision	0.86	0.85	0.96
Recall	0.83	0.81	0.91
F1- score	0.83	0.80	0.93

4 Conclusion and Future Scope

The work has implemented an environment sound classification pipeline. The paper also provides an exhaustive overview of the existing literature in the domain, allowing researchers to review the possible approaches of solving the problem. When given raw audio signals, the ensemble model classifies the sound into one of 50 classes present in the ESC-50 dataset. To leverage pre-trained, high-performing convolutional neural networks, the audio signals are brought to the frequency domain and then transformed into mel spectrograms to take into account human sensitivity to different ranges of audio frequency. An ensemble model is then built over two popular CNN architectures—ResNet-152 and DenseNet-121. The ensemble model achieves promising results with an accuracy of 91% and precision and recall of 0.91 and 0.93, respectively. This can serve as a step toward tangible progress in the domain of audio classification. This work can be further extended to integrate a wider and more diverse array of audio while maintaining the efficiency and performance of the algorithm. With multimedia continuously becoming an important part of the modern technological landscape, understanding and being able to classify audio will become a crucial skill.

References

1. Chen, J., Kam, A. H., Zhang, J., Liu, N., & Shue, L. (2005). Bathroom activity monitoring based on sound. In *International Conference on Pervasive Computing* (pp. 47–61). Springer.
2. Cochran, W. G. (2007). *Sampling techniques*. Wiley.
3. Ganaie, M., Hu, M., et al. (2021). Ensemble deep learning: A review. arXiv preprint arXiv:2104.02395
4. Gong, Y., Chung, Y. A., & Glass, J. (2021). AST: Audio spectrogram transformer. arXiv preprint arXiv:2104.01778
5. He, K., Zhang, X., Ren, S., & Sun, J. (2016). Deep residual learning for image recognition. In *Proceedings of the IEEE Conference on Computer Vision and Pattern Recognition* (pp. 770–778).
6. Huang, G., Liu, Z., Van Der Maaten, L., & Weinberger, K. Q. (2017). Densely connected convolutional networks. In *Proceedings of the IEEE Conference on Computer Vision and Pattern Recognition* (pp. 4700–4708).
7. Kumar, A., & Ithapu, V. (2020). A sequential self teaching approach for improving generalization in sound event recognition. In *International Conference on Machine Learning* (pp. 5447–5457). PMLR.
8. Kumar, A., Khadkevich, M., & Fügen, C. (2018). Knowledge transfer from weakly labeled audio using convolutional neural network for sound events and scenes. In *2018 IEEE International Conference on Acoustics, Speech and Signal Processing (ICASSP)* (pp. 326–330). IEEE.
9. Kwok, R. (2019). AI empowers conservation biology. *Nature, 567*(7746), 133–135.
10. Lozano, H., Hernáez, I., Picón, A., Camarena, J., & Navas, E. (2010). Audio classification techniques in home environments for elderly/dependant people. In *International Conference on Computers for Handicapped Persons* (pp. 320–323). Springer.
11. Nanni, L., Maguolo, G., Brahnam, S., & Paci, M. (2020). An ensemble of convolutional neural networks for audio classification.
12. Nanni, L., Maguolo, G., & Paci, M. (2020). Data augmentation approaches for improving animal audio classification. *Ecological Informatics, 57*, 101084.
13. O'Shea, K., & Nash, R. (2015). An introduction to convolutional neural networks. arXiv preprint arXiv:1511.08458
14. Piczak, K. J. (2015). ESC: Dataset for environmental sound classification. In *Proceedings of the 23rd ACM International Conference on Multimedia* (pp. 1015–1018).
15. Sailor, H. B., Agrawal, D. M., & Patil, H. A. (2017). Unsupervised filterbank learning using convolutional restricted Boltzmann machine for environmental sound classification. In *Interspeech* (vol. 8, p. 9).

16. Tak, R. N., Agrawal, D. M., & Patil, H. A. (2017). Novel phase encoded mel filterbank energies for environmental sound classification. In *International Conference on Pattern Recognition and Machine Intelligence* (pp. 317–325). Springer.
17. Wang, B., Liu, Y., Xiao, W., Xiong, Z., & Zhang, M. (2013). Positive and negative max pooling for image classification. In *2013 IEEE International Conference on Consumer Electronics (ICCE)* (pp. 278–279). IEEE.
18. Yamashita, R., Nishio, M., Do, R. K. G., & Togashi, K. (2018). Convolutional neural networks: An overview and application in radiology. *Insights Into Imaging, 9*(4), 611–629.
19. Zhu, B., Wang, C., Liu, F., Lei, J., Huang, Z., Peng, Y., & Li, F. (2018). Learning environmental sounds with multi-scale convolutional neural network. In *2018 International Joint Conference on Neural Networks (IJCNN)* (pp. 1–8). IEEE.

A Hybrid Product Recommendation System Based on Weather Analysis

Sangeeth Sajan Baby and S. Siji Rani

Abstract Product recommendation becomes to be one most revenue-generating technologies that every e-commerce website is using. For enhancing the purchase rate and user engagement, various product recommendations are available in the e-commerce website. Most of the external contexts are also taken into consideration for the product recommendation. This type of external context analysis sometimes will provide better recommendations when compared to the user-generated data. This paper proposes an idea of product recommendation using the weather. Here we predict the weather for the coming days using logistic regression and analyze the weather using big data analytics. Based on the analysis, we will sort the products and recommend a product in the same weather.

Keywords Big data · Product recommendation · Content-based recommendation · Logistic regression · Recommendation system

1 Introduction

Modern e-commerce websites are offering various range of items, giving the users a large number of choices to select. But this large amount items sometimes act as a hindrance to the searching of the item. If the search engine is not that efficient, the most desired item of a user may not be the first item in the result. Due to time constraints and the amount of effort, the users want to induce makes the users away from this searching environment. In these cases, the important product recommendation arises, and the user is provided with a personalized product recommendation. Such types of recommendations are based on the user's recent purchase history of

S. S. Baby (✉) · S. Siji Rani
Department of Computer Science and Engineering, Amrita Vishwa Vidyapeetham, Amritapuri, India
e-mail: sangeethsajan13@gmail.com

S. Siji Rani
e-mail: sijiranis@am.amrita.edu

B. Pati et al. (eds.), *Proceedings of the 6th International Conference on Advance Computing and Intelligent Engineering*, Lecture Notes in Networks and Systems 428,
https://doi.org/10.1007/978-981-19-2225-1_24

the user's wish list of the products. These recommendations not only give the user the desired product but also can increase the sales of the e-commerce website. This optimizes the product discovery process and eliminates the issue of long and effortful searches.

Recommendation optimizes the product finding process and eradicates the issue of long and tedious searches. The ease of use and being given more choice enhance the user experience and interaction and as a result enhance customer engagement and brand loyalty. In addition to that, personalized product recommendations tend to trigger a great number of purchases as well as induce a higher average order value. There are many things we need to care about product recommendation such as it is important to differentiate new and returning visitors. Since we do not know new users, we should use universal product suggestions.

2 Related Techniques for Recommendation System

2.1 *Collaborative Filtering*

In collaborative filtering [1–3], the user's historical preferences of contents are taken into consideration. In the case of user preferences, they are taken into account through ratings. There is an explicit rating such that the user is giving the rating on a scale of some value and the implicit rating such that the users will not give the rating but clicks on the content will take into account. Applications of collaborative filtering are generally involved in the process with large datasets. Other than Recommendation engines, this technique is used in mineral exploration, environmental sensing, etc. The process is done as different steps. Initially, it looks for the user who shares the same pattern with the other users. Then use this value to predict the content that might be liked by the other user. The main problem faced by collaborative filtering was that it cannot necessarily succeed in automatically match the content with one's preference.

2.2 *Content-Based Recommendation*

A content-based recommendation [4–6] system looks for similarities before recommending products. Most of the product recommendation engines use this since this does not require the data about other users and recommendations are very specific to a particular user, it is more convenient to scale to a large set of users. In addition to that, this model can capture the specific interests of a particular user and recommend items that very few other users are interested in. But the major problem with this type of system is that it can only make recommendations only based on the current interest of the user that means this model has a limited ability to expand the user interests. Even if this problem exists, we are using this model in our paper to elaborate the project since we are taking care of only the temperature or the weather condition, we do not need to further expand the user interest in these terms

2.3 Association Rule Mining

Association rule mining [7, 8] is used to find the relationship between some set of products in a large dataset. This model mines the frequently occurring items in the transactions. Market basket analysis is a widely used method in association rule mining to find the relationship of items in the transaction. For an instance, bread and butter is bought together by a large number of people, each of them is termed as transactions, and the market basket analysis will find these items bread and butter as related items. Thus the relationship is established. Many terms are included in this such as

- Support—Frequency of occurrence of the item set
- Confidence—Likeliness of the occurrence of consequent on the cart given that the cart already has the antecedents.
- Lift—The ratio of the confidence of the rule and the expected confidence of the rule.

2.4 Using Historical and Weather Data for Marketing and Category Management in E-Commerce

In this paper [9], the authors propose an idea of merging the data across the consumer journey from different small stores. They have identified three sales categories: sales depending on weather, season, and no clear pattern. As an example, they are predicting the sudden spike in the sales of air-conditioning devices. For this, they collected the daily sales report of the AC and temperature on each day. Then, they presented the correlation between sales and temperature.

2.5 Shop Weatherly—A Weather-Based Smart E-Commerce System Using CNN

Here in this paper [10], the authors are concentrating on the convolutional neural network for the recommendation of the products. The significant parts of the methodology are data collection, model training, and application integration. Here the proposed method is majorly done in the summer and winter seasons, and also they are using web scrapping for the data collection.

3 Tools and Techniques for Weather Prediction and Recommendation

Weather prediction is the most important step in this project proposed by the paper. For that, we need to analyze the previous weather data and should predict the weather for the product recommendation. Technologies used for this are,

3.1 Apache Spark

Apache Spark [11, 12] is an analytical engine that is commonly used for large-scale data processing. The architectural foundation of Spark is the resilient distributed dataset (RDD). RDD is a read-only multiset of data items distributed over different clusters, such that it is maintained for fault tolerance. Other components for the Spark are the Spark Core, Spark SQL, Spark Streaming, MLlib library, and GraphX. The Spark core provides the distributed task dispatching, scheduling, and basic I/O functionalities. Spark SQL is a component that is above the spark core used for the data abstractions. Spark streaming uses the fast computation capability for streaming analytics. It takes data as mini-batches and performs RDD transformation on the mini-batch data. MLlib library is used for machine learning operations since it has many common machine learning and statistical algorithms. GraphX is a graph processing framework built on the top of Apache Spark. It handles the knowledge graph and the analysis of those graphs.

3.2 Content-Based Recommendation

A content-based recommendation system [13, 14] looks for similarities before recommending products. These similarities are computed with different techniques such as Euclidean distance and cosine similarity [15].

- Euclidean distance—This distance is used when we have numeric data.
- Cosine Similarity—This type of metric is used to compute the similarity textual data.

A content-based recommender works with the data extracted from the behavior of the users. Two concepts called term frequency and inverse document frequency are used in the content-based filtering mechanism. Term frequency is the count of a particular word in a document. Inverse document frequency is the inverse of the frequency of the word.

3.3 *Logistic Regression*

Logistic regression [16] is a statistical model the uses logistic functions to model a dependent variable. It is done in binary classification problems that classify between two values 'YES' or 'NO'. It is also called a sigmoid function and was developed by mathematicians for describing certain properties related to the field of statistics and economics. It is an 'S'-shaped curve that takes real value number and map between 0 and 1, and in general computing, it can be termed as 'True' or 'False' or 'Yes' or 'No'. The input values (x) of logistic regression are combined using coefficient values to predict the output value (Y). The difference from linear regression is that its output value will be binary data while the linear regression has a numeric value.

The equation for the logistic regression is

$$y = e^\wedge\left(b0 + b1^*x\right) / \left(1 + e^\wedge\left(b0 + b1^*x\right)\right)$$

4 Proposed Methodology

This paper suggests two methods for the product suggestion, one is based on the current weather by getting the weather information from the user's hardware and the second one is by weather forecasting. In some cases, we need to understand the characteristics of the weather, for an instance, the monsoon weather causes intermittent rain for 2–3 months. This can only be understood by weather analysis. So the proposed method is divided into certain parts (Fig. 1).

4.1 *Weather Analysis*

We will collect the data from various sources, and with the help of Apache Spark and Spark SQL, we will analyze the weather of that particular area.

```
weatherdf=(weatherdf.select("MinTemp","MaxTemp","Month",)
.where(col("Rainfall") = "YES"))
```

In this Spark SQL command, we will get the month-wise result for the rainfall. Here weatherdf is the weather data frame, MinTemp is the minimum temperature, MaxTemp is the maximum temperature, and Month is the month itself.

The inferences from the weather analysis can be used in product recommendations in such a way that from the previous year's analysis we can easily find out the seasons. For instance, we can find the rainy season and the umbrella companies can produce

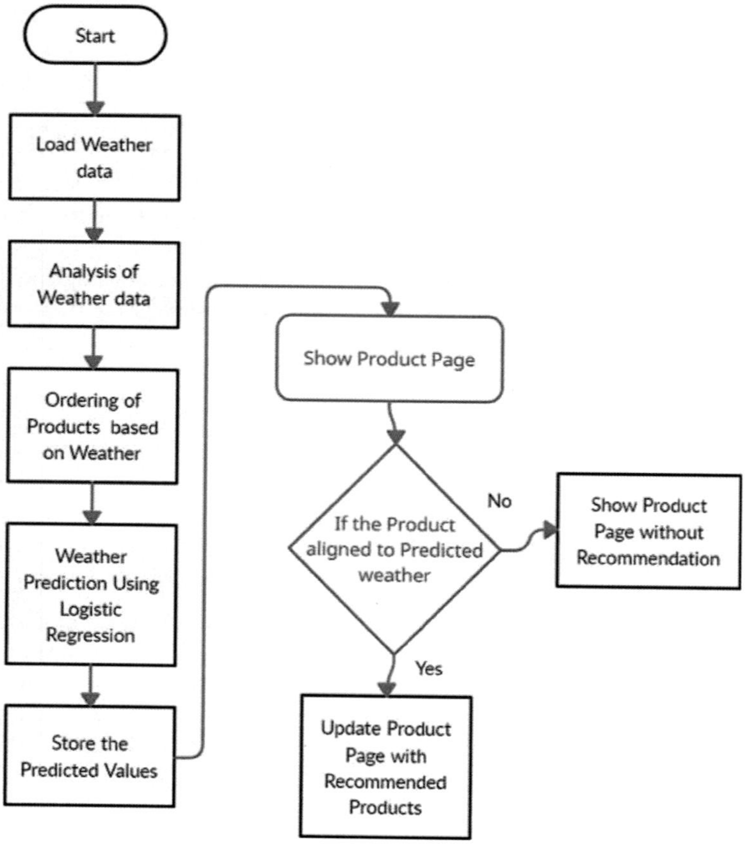

Fig. 1 Flowchart of the methodology

more umbrellas, and e-commerce websites can suggest that particular products before the season arrive.

```
mask = (timedf['Date']
>='2015-01-01')&(timedf['Date']<='2015-12-31')
```

This data frame consists of the minimum temperature and maximum temperature over a region and the code will show it over a year. The figure shows the temperature difference over a region in one year, and with this observation, we can conclude the observation on various aspects. With this observation, various companies can plan to Produce their products, market their products to the users, and more (Fig. 2).

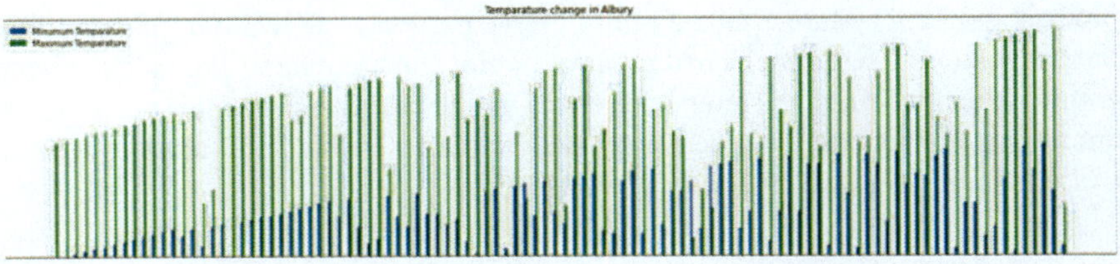

Fig. 2 Temperature change in a region

4.2 *Weather Forecasting*

In the part of weather prediction, we are using logistic regression. In the product recommendation, we will predict and classify whether there will be rain or not or sunny or not in the case of raincoats or umbrellas. We are implementing logistic regression in the Spark platform. Here we will use the vector assembler as the transformer.

```
vecAssembler=VectorAssembler(inputCols=["Rainfall"],
outputCol="features")
vecTrainDF= vecAssembler.transform(trainDF)vecTrainDF.
select("Rainfall","RainToday").show(10)
```

We will be using the columns Rainfall amount, Humidity, and Wind Speed to analyze the Weather

```
lr=LogisticRegression(featuresCol="features",
   labelCol="RainToday")
lrModel = lr.fit(vecTrainDF)
pipeline = Pipeline(stages=[vecAssembler, lr])
pipelineModel = pipeline.fit(trainDF)
```

After the final step, we will get the predicted value of the weather. Whether will there be rain or hailstorm for the next few days. By this, we can recommend the products to the users.

4.3 *Ordering Products*

This is the phase where we add the attributes related to temperature or season or weather for every product that the sale can be affected by the temperature. This phase is done at the product management side where the data of the products are added to the e-commerce website. Along with all other details such as the name of the manufacturer and price we add another column like 'IsRainAffect', 'IsHighTempaffect'. These types of additional attributes are added so that we need to analyze whether a product is a temperature affected or not. Some products sale will be influenced by the temperature like umbrella and raincoat but for some products like mobile phone and washing machines, the temperature will not be influenced. For this purpose, we are adding another column. The newly added columns are boolean values consisting of 1 and 0 representing True and False, respectively.

4.4 *Content-Based Recommendation*

For taking the external details such as temperature and weather into account content-based recommendation is more suitable. In this recommendation engine, the most suitable way of finding the similarity is to use the Euclidean distance. Here for the weather-based recommendation, we are taking care of only the fields in the product details which are weather-related like 'IsRainAffect', 'IsHighTempAffect'. We calculate Euclidean distance between the related fields if the result of the distance comes out to be '0', then the products have the similarity and any other product with similarity other than '0' will not be taken into consideration.

```
sim = np.linalg.norm(Item1 - Item2)
```

The above code will be used to find the Euclidean distance (similarity) between the two products. Using this, we can create a list of products that have similarities in the weather, and with the help of the list, we will recommend the product.

4.5 *Connection to the Web Application*

A set of processes is done to connect the output that we got from Spark should display on the Web application. The steps include

- Connect the Spark and the Web application with Python API
- Request the data from the Web application
- Do the Weather Analysis and Forecasting with the Logistic Regression
- After Ordering do the Content-based recommendation step
- Respond the data to the Web application

5 Result Analysis

The proposed methodology is implemented with Spark and ASP.NET MVC. The whole analysis, logistic regression, and collaborative filtering part were done with Spark, and the Web application implementation is done with the ASP.NET.

Fig. 3 Predicted values after logistic regression

Rainfall	features	prediction
0	[0.0]	0.0
0	[0.0]	0.0
0	[0.0]	0.0
3	[3.0]	1.0
0	[0.0]	0.0
0	[0.0]	0.0
0	[0.0]	0.0
0	[0.0]	0.0
0	[0.0]	0.0
0	[0.0]	0.0
0	[0.0]	0.0
0	[0.0]	0.0
0	[0.0]	0.0
0	[0.0]	0.0
0	[0.0]	0.0
0	[0.0]	0.0
0	[0.0]	0.0
0	[0.0]	0.0

Figure 3 depicts the weather prediction done by the logistic regression function in the PySpark.

For testing purposes, we created a dataset with the context column (Fig. 4). In the final output, the web application shows the correct recommendation. Here in the image, while testing we created the context as rainy season and it recommends the products related to rainy season. In Fig. 5, we can clearly see the product recommendation based on the weather. Some products such as vinyl footwear, raincoat, power bank, towels, tissues are being recommended which are useful for heavy rainy season. In this manner by adding the weather context of purchase to the product table, we are recommend the product by weather forecasting.

6 Conclusion and Future Works

One of the major problems of e-commerce websites is the growing amount of items on their websites. Product recommendation is one of the most viable methods for such problems so that users can easily access the products according to their interests. There is a lot of product recommendation system which works on various levels

	ProductId	ProductName	IsRainAffected
1	3	Rain Coat	1
2	4	Vinyl Footwear	1
3	5	Orange	0
4	6	Towels	1
5	7	Water Heater	1
6	8	Shampoo	0
7	9	Tissues	1
8	10	Earphone	0
9	11	ZipLocks	1
10	12	Mic	0
11	13	Power Bank	1
12	14	Laptop	0
13	15	Charger	0

Fig. 4 Product table in the database

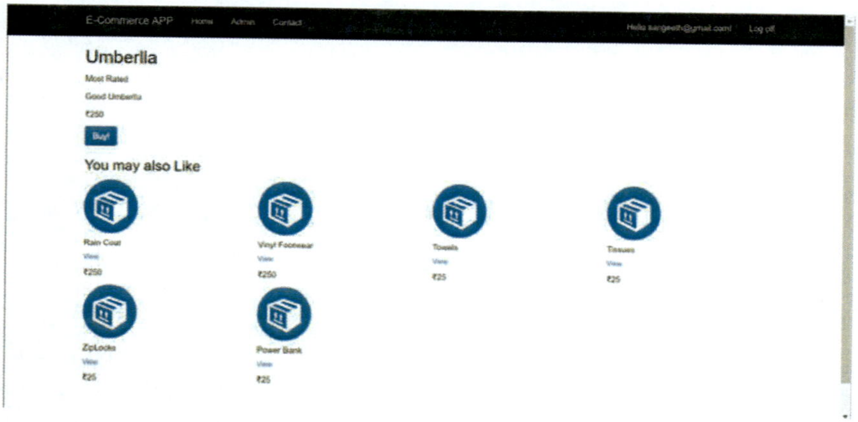

Fig. 5 Product recommendation page of web application

such as some may work of a most viewed item of a person or some others may be based on the items in the wish list of the user, here this paper proposed the recommendation based on the weather which is clubbed with big data analysis and logistic regression. This type of recommendation can provide greater insight to the e-commerce websites to recommend products based on the weather with the help of year-wise weather analysis. In the future, we use more features for the logistic regression so that it can produce more accurate results. On top of that, we can also use association rule mining algorithm such as Apriori algorithm in the initial part for getting the frequently bought items, and this can also produce better product recommendation.

References

1. Sarwar, B., Karypis, G., Konstan, J., & Riedl, J. (2001). Item-based collaborative filtering recommendation algorithms. In *Proceedings of the 10th International Conference on World Wide Web* (pp. 285–295).
2. Herlocker, J. L., Konstan, J. A., & Riedl, J. (2000). Explaining collaborative filtering recommendations. In *Proceedings of the 2000 ACM Conference on Computer Supported Cooperative Work* (pp. 241–250).
3. Zhao, Z.-D., & Shang, M.-S. (2010). User-based collaborative-filtering recommendation algorithms on hadoop. In *2010 Third International Conference on Knowledge Discovery and Data Mining* (pp. 478-481). IEEE.
4. Basu, C., Hirsh, H., & Cohen, W. (1998). Recommendation as classification: Using social and content-based information in recommendation. In *AAAI/IAAI* (pp. 714–720).
5. Devika, P., Jisha, R. C., & Sajeev, G. P. (2016). A novel approach for book recommendation systems. In *2016 IEEE International Conference on Computational Intelligence and Computing Research (ICCIC)* (vol. 1, No. 1, pp. 338–343).
6. Baby, S. S., & Reddy, S. L. (2021). End to end product recommendation system with improvements in Apriori algorithm. In *2021 Third International Conference on Inventive Research in Computing Applications (ICIRCA)* (pp. 1357–1361). IEEE.
7. Rani, S. S., Sreejith, K., & Sanker, A. (2017). A hybrid approach for automatic document summarization. In *2017 International Conference on Advances in Computing, Communications and Informatics (ICACCI)* (pp. 663–669). https://doi.org/10.1109/ICACCI.2017.8125917
8. Ramakrishnan, N., Nair, J., Jayaprakash, D., Ananthakrishnan, H., & Rani, S. (2019). Hypergraph based clustering for document similarity using FP growth algorithm. In *2019 International Conference on Intelligent Computing and Control Systems (ICCS)* (pp. 332–336). https://doi.org/10.1109/ICCS45141.2019.9065630
9. Čreslovnik, D., Košmerlj, A., & MCiavotta, A. (2019). Using historical and weather data for marketing and category management in eCommerce.
10. Sallar, J., Khan, S., Ahmed, S., Kumar, P., Faridy, H., & Rathi, M. (2021). Shop weatherly-A weather based smart E-commerce system using CNN. *Revista Geintec-Gestao Inovacao e Tecnologias, 11*(4), 2785–2800.
11. Spark, Apache. "Apache spark." Retrieved January 17 (2018): 2018.
12. Zaharia, M., Xin, R. S., Wendell, P., Das, T., Armbrust, M., Dave, A., et al. (2016). Apache spark: A unified engine for big data processing. *Communications of the ACM, 59*(11), 56–65.
13. How to build a content-based movie recommendation system in python. https://analyticsindiamag.com/how-to-build-a-content-based-movie-recommendation-system-in-python/
14. What are product recommendation engines? And the various versions of them? https://towardsdatascience.com/what-are-product-recommendation-engines-and-the-various-versions-of-them-9dcab4ee26d5
15. Lal, N. M., Krishnanunni, S., Vijayakumar, V., Vaishnavi, N., Siji Rani, S., & Deepa Raj, K. (2021). A novel approach to text summarisation using topic modelling and noun phrase extraction. In *Advances in Computing and Network Communications* (pp. 285–298). Springer.
16. Sonu, S. B., & Suyampulingam, A. (2021). Linear regression based air quality data analysis and prediction using python. In *2021 IEEE Madras Section Conference (MASCON)* (pp. 1–7). IEEE.

Cryptocurrency Price Prediction Using Machine Learning

Harsh Parikh, Nisarg Panchal, and Ankit Sharma

Abstract Globally, the use of cryptocurrencies to purchase goods and services has been rising. They rely on a secure distributed ledger data structure; mining is an integral part of such systems. The rise of cryptocurrencies' value on the market and the growing popularity around the world open several challenges and concerns for business and industrial economics. Cryptocurrencies have been triggered by the substantial changes in their prices, claims that the market for cryptocurrencies is a bubble without any fundamental value and also concerns about evasion of regulatory and legal oversight. Machine learning is part of artificial intelligence that can make future forecastings based on previous experience. In this paper, methods have been proposed to construct machine learning algorithm-based models such as linear regression, K-nearest neighbour(KNN), and also statistical models like Auto-ARIMA and Facebook's Prophet (Fbprophet). This paper presents a comparative performance of machine learning and statistical modelling algorithms for cryptocurrency forecasting.

Keywords Cryptocurrency · Machine learning · KNN · Linear regression · Auto-ARIMA · Fbprophet · Prediction · Moving average · Time sequence analysis

1 Introduction

In this paper, we have implemented various machine learning models for the price prediction of two cryptocurrencies namely Bitcoin and Ethereum. These cryptocurrencies are decentralized and are available for trade to anyone. Because they are

H. Parikh (✉) · N. Panchal · A. Sharma
Institute of Technology, Nirma University, Ahmedabad, Gujarat 382481, India
e-mail: 19bic036@nirmauni.ac.in

N. Panchal
e-mail: 19bee070@nirmauni.ac.in

A. Sharma
e-mail: ankit.sharma@nirmauni.ac.in

B. Pati et al. (eds.), *Proceedings of the 6th International Conference on Advance Computing and Intelligent Engineering*, Lecture Notes in Networks and Systems 428,
https://doi.org/10.1007/978-981-19-2225-1_25

decentralized, their price fluctuates based on their demand and supply. Moreover the more they are mined, the more difficult it gets to further mine them. These uncertainties affect their value, machine learning methods can unearth patterns in such fluctuations and with a proper dataset, such patterns can be efficiently learned and used for future predictions. We have used machine learning and deep learning algorithms namely KNN, Auto-ARIMA, linear regression, Fbprophet, and moving average to predict their prices. We have also calculated root mean square error (RMSE) for each models' predictions to know how accurate it is in prediction. This study will give us valuable insights into how machine learning performs on real-life problems. However, it is prone to errors as the prices are affected by some events like announcements from an influential person about a specific currency; these events cannot be predicted as they are random. However, generalized long-term patterns can still be predicted up to a certain accuracy.

2 Literature Overview

Vatsal [1] has implemented support vector machine (SVM), linear regression, prediction using decision stumps, expert weighting and online learning along with benefits and drawbacks of each method. Alkhatib, Khalid, et al. have done work in predicting stock prices of six companies using KNN and nonlinear regression approach in [2]. Bini et al. [3] have implemented clustering and regression techniques of data mining for stock prediction to help people identify more profitable companies. In this paper by Zhang et al. [4], they have proposed a new methodology that combines multidimensional KNN and ensemble empirical mode decomposition (EEMD) for prediction of the closing price and high price of stocks. Izzah et al. [5] have implemented improved multiple linear regression (IMLR) in a mobile app for prediction of stock price. Jain, Garima et al. have prepared a study on time series analyses using ARIMA and exponential smoothing (ETS) for weather forecasting in [6]. Yermal et al. [7] have prepared a stock forecasting model using Automatic ARIMA and Eviews 9.5 on a minute by minute dataset for 50 stocks. Brunello, Andrea, et al. have prepared a study on time series handling using decision trees in [8].

Hitam, N. Azizah, et al. implemented various machine learning algorithms for the prediction of cryptocurrency prices and have prepared a comparative study in [9]. Chikkakrishna [10] have implemented Sarima and Fbprophet for short-term prediction of traffic. In this study by Joshi et al. [11], airline prices are predicted using a decision tree regressor and with an accuracy of 82%. Banu and A. Bazilia have prepared a time series analysis for the prediction of Covid-19 infections for the year 2021 using the FbPROPHET model in [12]. Akyildirim et al. [13] have predicted prices for 12 cryptocurrencies at a minute level and daily level using various machine learning algorithms.

3 Dataset Description

The dataset in CSV format was downloaded from Yahoo finance. For Bitcoin, our dataset had day-wise data from 17- 09-2014 to 01-01-2021 while for Ethereum it was from 17-08-2015 to 01-01-2021. There were seven variables available for each dataset namely: 'Date', 'Open', 'High', 'Low', 'Close', 'Adj', 'Close', 'Volume'. Open and close columns represent the starting price and closing price of the currency on particular days. Maximum, minimum, and last price of the currency is described in the 'High', 'Low' and 'Last' columns of the dataset. Also, the market is closed on weekends and bank holidays so the data is not available for them. All the prices were in USD and as we aimed to predict the closing price at a specific date, we have only used the closing price and date from our dataset. The graphs for Ethereum and Bitcoin are shown in Figs. 1 and 2 respectively.

4 Proposed Method

Market analysis is divided into two parts: Fundamental analyses and technical analyses. Fundamental analyses aim at analysing future profitability from the present business environment along with the financial performance of the stock or currency we have to predict. Technical analysis aims at using charts and statistics to predict the trends in the market. In this paper, we have done technical analyses using machine learning and some deep learning algorithms for the prediction of cryptocurrency prices. First, we load the dataset and we will take the closing price as the target variable for prediction. We have split the dataset for training and testing in a ratio of 80:20. For example, Bitcoin has 2399 data points from the start date to the end

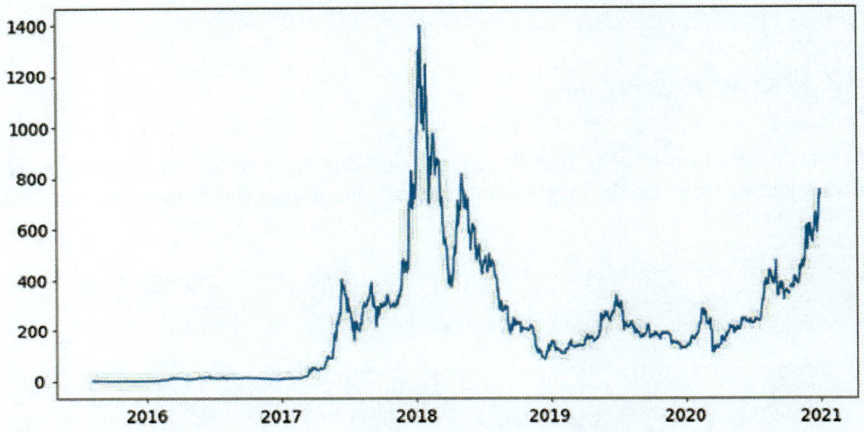

Fig. 1 Closing price of Ethereum from 17-08-2015 to 01-01-2021

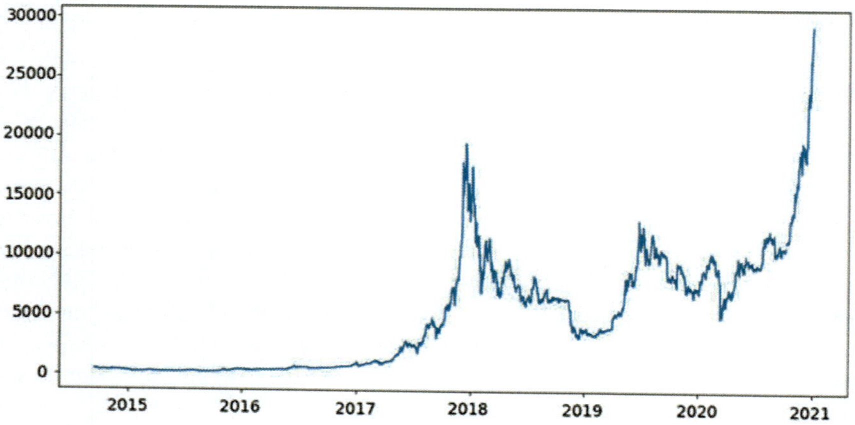

Fig. 2 Closing price of Bitcoin from 17-09-2014 to 01-01-2021

date so the first 1840 data points are taken as training and the remaining are to be predicted. Implementation of each of the algorithms is described below.

4.1 Moving Average

Moving average is slightly different from the normal average. In moving average, the price for the next day is calculated as the average of all the previous values. Once the price of the next day is calculated, its real closing price is then noted and then included in the average to predict the price for the proceeding day. This is explained in Fig. 3. To get an idea about the model performance, RMSE value was also calculated.

4.2 Linear Regression

Linear regression works by forming an equation that has a relationship between the dependent variable and the independent variable. It searches for the equation that best

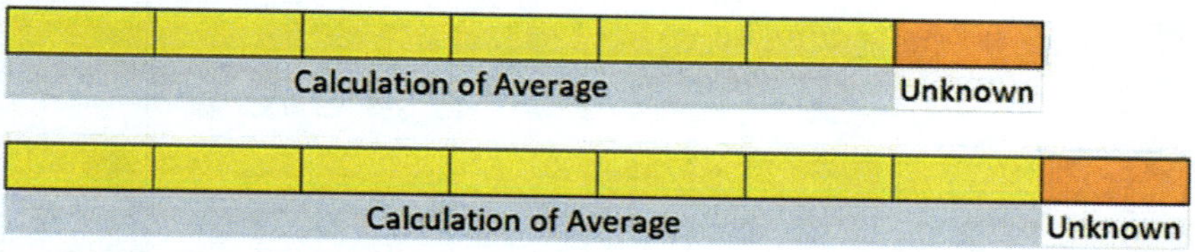

Fig. 3 Moving average algorithm

fits the training dataset and then uses it to predict unknown values. In our problem statement, we have the date as an independent variable. From the date, we extract features like day, month, year and then fit a linear regression model. Moreover, we have also added other features on the idea that the first and the last day of the week would have more impact on the closing price of the stock than any other day. So this feature puts a value as 1 in the column, if the day of the week is 0 or 4; otherwise, it will keep it as 0. Then the dataset was split into train and test, and after training was completed, RMSE value along with a graph that shows predicted and actual values was also plotted.

4.3 K-Nearest Neighbour

KNN works by plotting all the data points and then calculating the value for a new unknown data point as the average of some nearest neighbouring data points. The number of nearest neighbouring data points to look for is determined by a constant K. KNN assumes that a similar class of objects lies together and uses the distance formula to identify the nearest neighbouring element. For KNN also, we have used the same features as Linear Regression.

4.4 Auto-ARIMA

ARIMA is a well-known model for time series analyses; it uses past data and then predicts future values. It uses three parameters p (past values), q (past prediction errors), d (differencing order). Selecting parameters for ARIMA takes a lot of time and so Auto-ARIMA calculates the best parameters for us and speeds up the process. This model can predict whether the trend will go up or down but does not takes into account seasonal trends.

4.5 Fbprophet

Fbprophet was developed by Facebook for time series prediction; it is very easy to implement as it has only two inputs Date and the target variable. In prophet, we have directly given the dates column and the closing price column as an input and it does all the other work on its own. It does not require data preprocessing. Fbprophet also takes into account seasonal trends for better predictions.

5 Results

The RMSE value of moving average is 1840.58 for Bitcoin and 129.20 for Ethereum. RMSE for Bitcoin is observed to be quite high that is because its price is a lot higher compared to that of Ethereum. Only the RMSE value does not give us an idea of a model's performance. If we look at the graph in Figs. 4 and 5, we can see that moving average is doing a good job at predicting prices for both Bitcoin and Ethereum.

The RMSE of linear regression is 3681.86 on bitcoin and 193.14 for Ethereum. Figures 6 and 7 shows performance of linear regression. From the graph, it is observed that linear regression is not able to predict sudden changes in the cryptocurrency market.

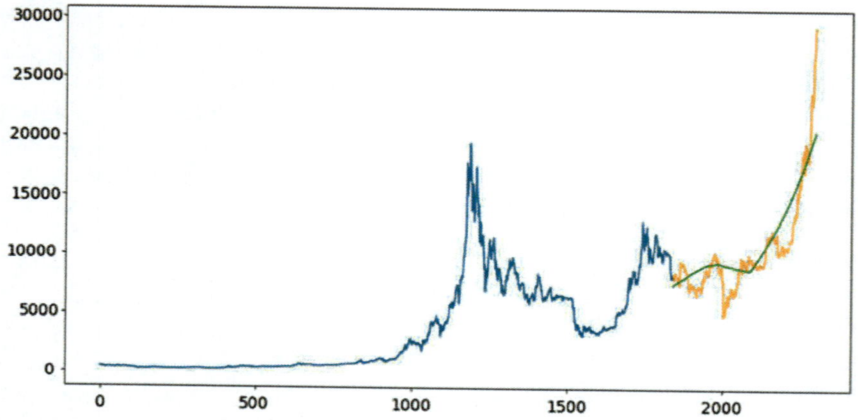

Fig. 4 Prediction of Bitcoin price using moving average

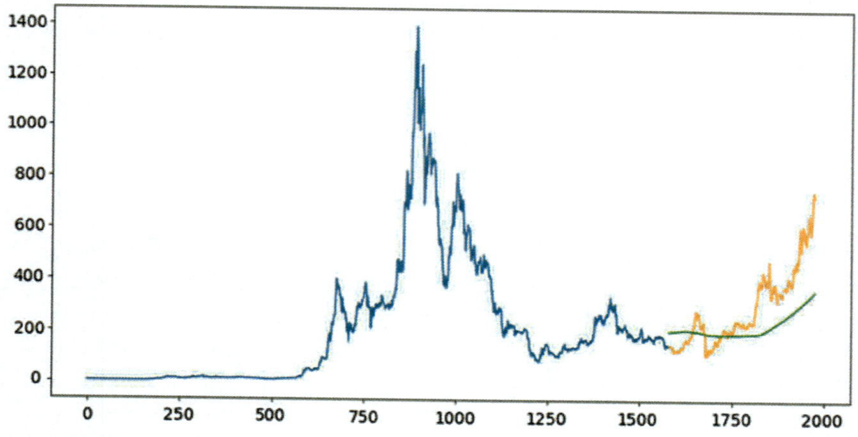

Fig. 5 Prediction of Ethereum price using moving average

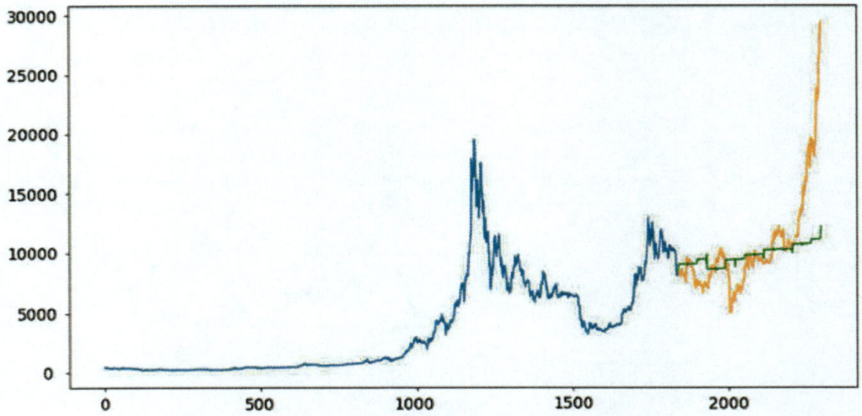

Fig. 6 Prediction of Bitcoin price using linear regression

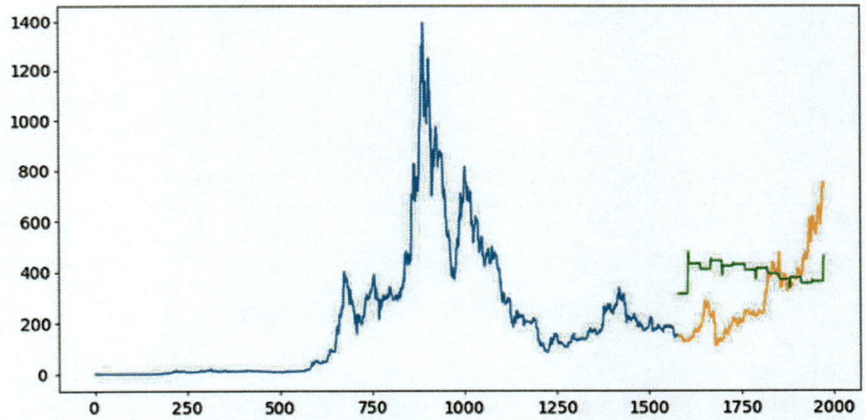

Fig. 7 Prediction of Ethereum price using linear regression

On predicting the prices using KNN, we were getting an RMSE value of 8559.27 for Bitcoin and 131.45 for Ethereum. It is predicting quite good but the graph is very fluctuating and not consistent but the overall trend of predicted values seem to be matching the actual trend. Figures 8 and 9 show their performance in predicting price of Bitcoin and Ethereum, respectively.

Auto-ARIMA model uses past information to comprehend the paradigm in the time series. Utilizing these qualities, it is obvious from the plot that the model has caught a pattern in the series as shown in Figs. 10 and 11; however, it does not zero in on the infrequent part. Prophet attempts to catch the pattern and irregularity from past information. This model normally performs well on time- series datasets but did not perform well for this situation as shown in Figs. 12 and 13.

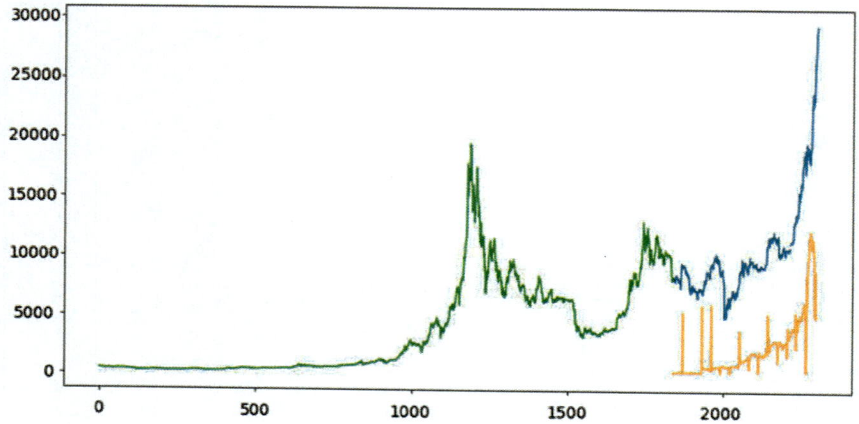

Fig. 8 Prediction of Bitcoin price using KNN

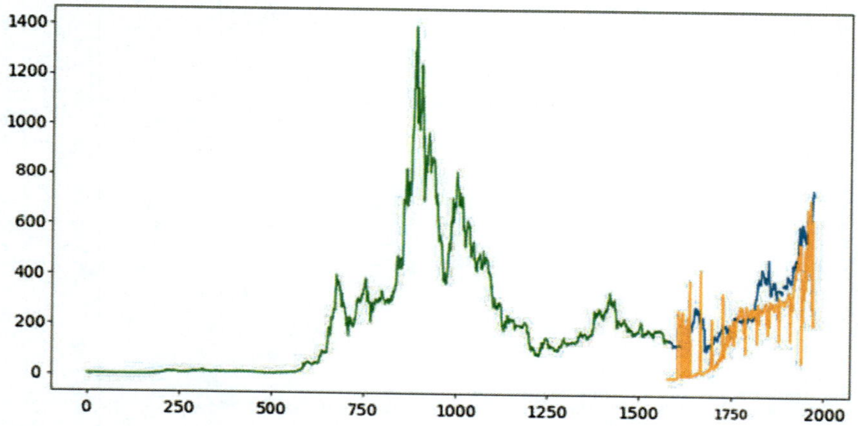

Fig. 9 Prediction of Ethereum price using KNN

The RMSE value of all the employed algorithms for prediction of cyptocurrency prices is shown in Table 1.

6 Conclusion

To foresee Bitcoin and Ethereum prices, a variety of machine learning and statistical models were used. Cryptocurrency and stock prices can be forecasted using this study's findings. Moving average, prophet and KNN has provided us with better

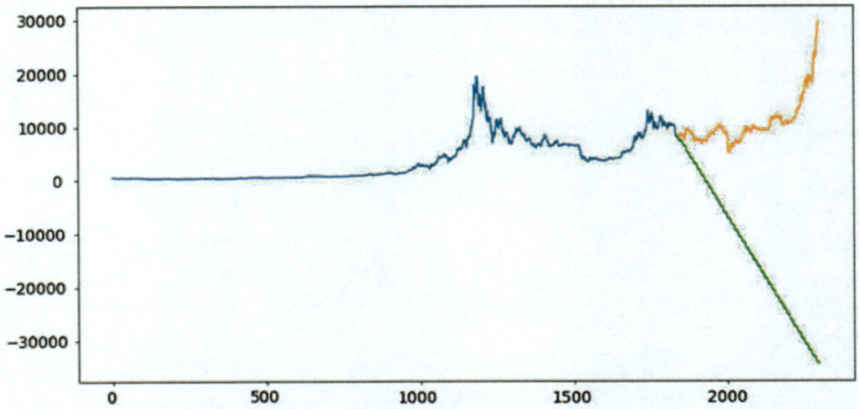

Fig. 10 Prediction of Bitcoin price using Auto-ARIMA

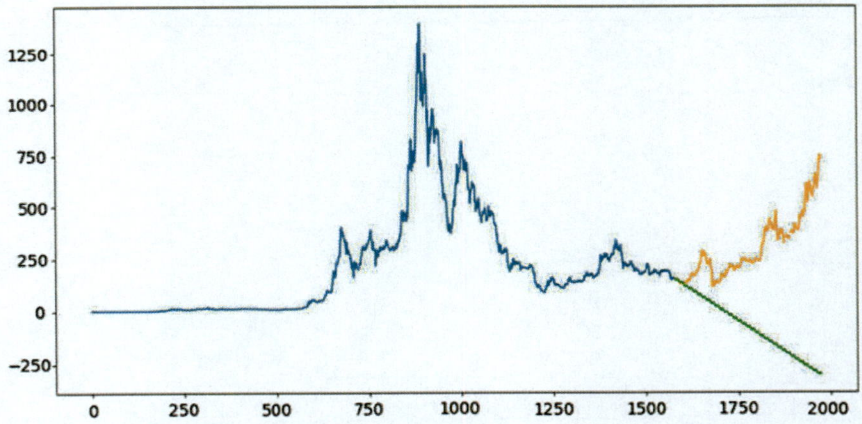

Fig. 11 Prediction of Ethereum price using Auto-Arima

Table 1 RMSE value of all the algorithms

Algorithm	RMSE for Bitcoin	RMSE for Ethereum
Moving average	1840.58	129.20
Linear regression	3681.86	193.14
KNN	8559.27	131.45
Auto-ARIMA	28233.30	459.63
Prophet	3658.74	131.05

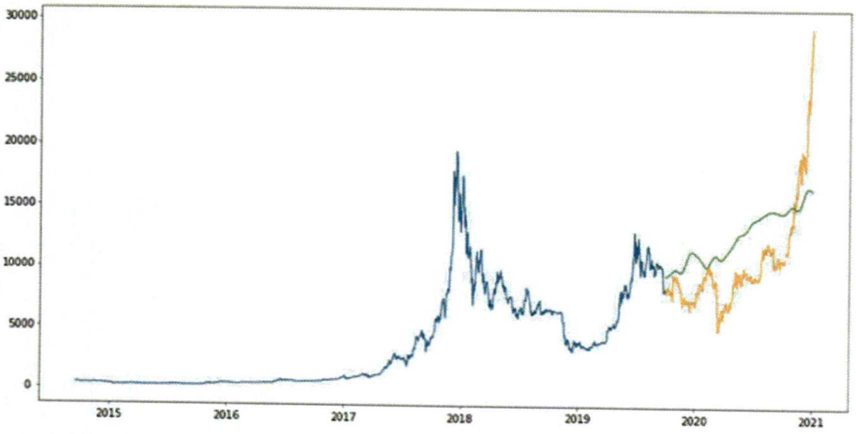

Fig. 12 Prediction of Bitcoin price using Fbprophet

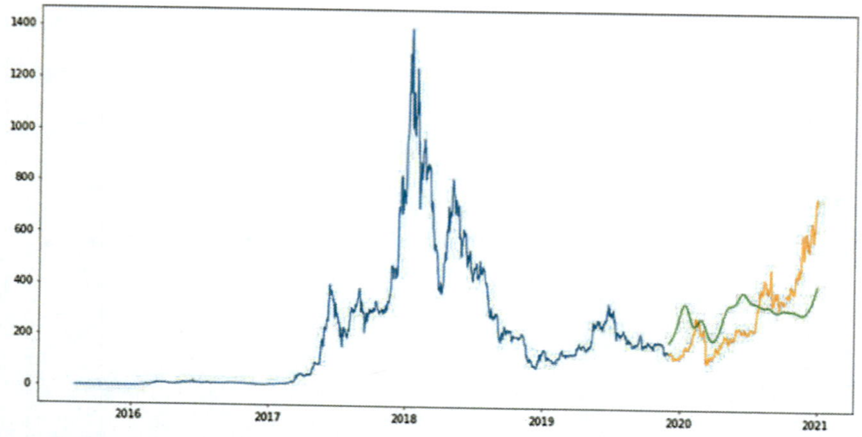

Fig. 13 Prediction of Ethereum price using Fbprophet

results, while Auto-ARIMA and linear regression have shown large deviation from the true prices. This study can also be applied to stock price prediction and other time series forecasting problems.

References

1. Vatsal, H. (2007). Machine learning techniques for stock prediction. *Foundations of Machine Learning, 1*(1), 6–12.
2. Alkhatib, K., et al. (2013). Stock price prediction using K-Nearest Neighbor (KNN) algorithm. *International Journal of Business, Humanities and Technology, 3*(3), 32–44.

3. Bini, B. S., & Mathew, T. (2016). Clustering and regression techniques for stock prediction. *Procedia Technology, 24*, 1248–1255.
4. Zhang, N., Lin, A., & Shang, P. (2017). Multidimensional k-nearest neighbor model based on EEMD for financial time series forecasting. *Physica A: Statistical Mechanics and its Applications, 477*, 161–173.
5. Izzah, A., et al. (2017). Mobile app for stock prediction using improved multiple linear regression. In *2017 International Conference on Sustainable Information Engineering and Technology (SIET)*. IEEE.
6. Jain, G., & Mallick, B. (2017). A study of time series models ARIMA and ETS. Available at SSRN 2898968.
7. Yermal, L., & Balasubramanian, P. (2017). Application of auto ARIMA model for forecasting returns on minute wise amalgamated data in nse. In *2017 IEEE International Conference on Computational Intelligence and Computing Research (ICCIC)*. IEEE.
8. Brunello, A., et al. (2018). A novel decision tree approach for the handling of time series. In *International Conference on Mining Intelligence and Knowledge Exploration*. Springer.
9. Hitam, N. A., & Ismail, A. R. (2018). Comparative performance of machine learning algorithms for cryptocurrency forecasting. *Indonesian Journal of Electrical Engineering and Computer Science, 11*(3), 1121–1128.
10. Chikkakrishna, N. K., et al. (2019). Short-term traffic prediction using ARIMA and Fbprophet. In *2019 IEEE 16th India Council International Conference (INDICON)*. IEEE.
11. Joshi, N., et al. (2019). Airline prices analysis and prediction using decision tree regressor. In *International Conference on Recent Developments in Science, Engineering and Technology*. Springer.
12. Banu, A. B. (2021). Time series analysis for predicting Covid-19 infection using Facebook prophet model. *Turkish Journal of Computer and Mathematics Education (TURCOMAT), 12*(7), 1680–1685.
13. Akyildirim, E., Goncu, A., & Sensoy, A. (2021). Prediction of cryptocurrency returns using machine learning. *Annals of Operations Research, 297*(1), 3–36.

A Deep Learning Framework for Real-Time Indian Sign Language Gesture Recognition and Translation to Text and Audio

Ashwini M. Deshpande, Gayatri Inamdar, Riddhi Kankaria, and Siddhi Katage

Abstract Indian Sign Language (ISL) is used in the deaf community all over India. Development of the ISL recognition system is an active area to aid this community. In ISL, most of the signs are two-handed signs, and thus, it differs from another commonly used American Sign Language (ASL) and seems complex. In this paper, the design and implementation of a system to recognize ISL signs is reported. Building such a system can help specially abled person/people, by providing a medium to communicate with others without human interpreters. The proposed system is built using a deep convolutional neural network (CNN), which performs both feature extraction and classification, preceded by an image preprocessing step. A real-time input (live signs captured from webcam) is given to this system, and the output is delivered in the form of text and audio. Proposed CNN architecture has achieved an accuracy of 98% for a given dataset which comprises 56 items (1–10 digits, A-Z letters, and 20 general words).

Keywords Hand gestures · Indian Sign Language · Contouring · Segmentation · Convolutional neural network

---

A. M. Deshpande (✉) · G. Inamdar · R. Kankaria · S. Katage
Department of Electronics and Telecommunication Engineering, MKSSS's Cummins College of Engineering for Women, Pune, India
e-mail: ashwini.deshpande@cumminscollege.in

G. Inamdar
e-mail: gayatri.inamdar@cumminscollege.in

R. Kankaria
e-mail: riddhi.kankaria@cumminscollege.in

S. Katage
e-mail: siddhi.katage@cumminscollege.in

B. Pati et al. (eds.), *Proceedings of the 6th International Conference on Advance Computing and Intelligent Engineering*, Lecture Notes in Networks and Systems 428,
https://doi.org/10.1007/978-981-19-2225-1_26

1 Introduction

Sign language is a visual-gestural language used by deaf–mute people to communicate with others. They use their hand movements and other body parts like face and arms to convey their messages. Some of the well-known sign languages used across the world are American Sign Language (ASL), British Sign Language (BSL), and Indian Sign Language (ISL). Most of the research works available in the literature focus on ASL rather than ISL or BSL; ISL signs are two-handed, which make it little complex, and the standard datasets available for ISL are also limited. Our aim is to design a computer vision-based system that can understand and interpret the ISL signs in real time and provide a medium for communication between specially abled people with others, without a need of any human interpreter. The proposed system uses a convolution neural network (CNN) to recognize the signs. CNN acts as a feature extractor and classifier both, which makes the system less complex and still achieve acceptable accuracy. Input to this system is live gestures, and output is in the form of text and audio. Image preprocessing steps are applied to the input video frames, namely contouring, thresholding, and segmentation. Followed by which CNN model training on gesture database is carried out.

The main motivation for this work was to develop and implement an end-to-end vision-based system to establish communication between deaf and/or mute person with a normal person without any interpreter. The system accepts signs made by the signer captured through a web camera and interprets the ISL signs in real time.

In this work, a database is created comprising two-handed signs referring to ISL dictionary. The database contains English language alphabets, digits, and words, along with small sentences learned from the signers' signs. The dataset can be extended further to increase the dictionary learning for a greater number of signs corresponding to more words and sentences. Proposed CNN model for sign language recognition is thus mostly useful for educating the learners in their primary phase of sign language learning. System provides output in the form of both text and audio for ease of communication.

2 Literature Review

Sign language recognition using hand gestures is a well-studied area of research. Many researchers have employed different techniques for feature extraction and classification to achieve maximum accuracy for solving hand gesture recognition problem. With the advent of deep learning techniques, use of CNN architecture is prevalent as it does both the functions, i.e., feature extraction and classification during model training.

Singha et al. [1] proposed a computer vision-based system, which has three steps as image preprocessing, feature extraction, and classification. Preprocessing of images involves skin detection and histogram matching. Feature extraction is done using eigenvalues and eigenvectors, and for classification, eigenvalue weighted Euclidean distance is used. With this approach, they have achieved a success rate of 96.25% and are able to recognize 24 alphabets from the live video input. Kòpùklù et al. [2] have published their work, where illustration is given with real-time recognition of dynamic hand gestures from live video. They created two models, one is a detector which is lightweight CNN to detect hand gestures, and another is a classifier which is deep CNN to classify the hand gestures; with this they have achieved an accuracy of 94.04% classification accuracy. Sharvani et al. [3] have presented a static system for sign recognition using speeded-up robust feature (SURF) algorithm followed by image preprocessing steps, such as image segmentation, skin detection, and edge detection. Using the SURF feature extraction algorithm, they achieved an accuracy of 99%.

Based on CNN-based approach, Deshpande and Kalbhor [4] have presented a paper which describes a system, which is able to understand and interpret 25 Marathi sign language alphabets with testing accuracy of 99.29%. Rigorous research has been done on ISL recognition by Kishore et al. [5] to make the system robust in complex backgrounds/environments. Segmentation and tracking of hands in different environments are achieved by using active contour models, and classification of signs is done by an artificial neural network (ANN) using error backpropagation algorithm. With this they have achieved accuracy of 93% for 351 signs. Sharma et al. [6] published their work in ISL recognition systems based on KNN, where feature extraction is done by direct pixel value and hierarchical centroid and classification is carried out by KNN and neural network pattern recognition (NPR tool). This system works well for recognizing numeral ISL static signs with 97.10% accuracy. Different kinds of feature extractors have been used to maximize the efficiency of a recognition system like texture extractor, scale-invariant feature transform (SIFT), or discrete wavelet transform (DWT).

In [7], Kaur et al. presented a SIFT descriptor-based system and artificial bee colony (ABC) and feed forward backpropagation neural networks (FFBPNN) for classification. Their proposed system has given 99.43% accuracy for 26 alphabets and 0–9 digits. ISL recognition work using artificial neural network (ANN) and support vector machine (SVM) classifier with comparison is presented by Rokade et al. in [8]. Skin segmentation is employed to obtain the hand region. Central moments along with HU's moments are used for feature extraction. For classification, with neural networks an accuracy of 94.37%, and with SVM an accuracy of 92.12% was achieved when tested on a dataset of 26 alphabets. Ghotkar and Kharate [9] presented a detailed survey on different vision-based hand gesture recognition methods using ISL. Along with presenting challenges for vision-based system, a summarized overview for hand tracking, segmentation, feature extraction and classification is given in this paper. Itkarkar et al. [10] proposed a vision-based hand gesture for ISL recognition. In their work, database for alphabets and numbers was created using a web camera. With texture feature extraction based on Gabor transform and classification by SVM and K-NN, accuracy reported was 67.69% and 84.90% for SVM and K-NN, respectively.

There are many vision-based approaches employing image preprocessing, feature extraction, and recognition of signs as the primary steps. Deep learning framework with inclusive feature extraction step reduces computational complexity yet gives higher recognition accuracy for large database. Proposed CNN model gives 98% accuracy and proves the usefulness of this framework.

3 Methodology

The proposed system is implemented on a personal computer (PC) with Intel(R) Core (TM) i7 processor, using 1280 × 720P HD web camera and speakers. Python 3.7 and other libraries such as OpenCV, Keras, and TensorFlow are used to develop and test the algorithm for real-time implementation. The webcam captures live gesture inputs followed by video to frame conversion at the rate of 30 frames per seconds (fps). Further, image preprocessing steps are applied, including segmentation, thresholding, and contouring. It is required that the user must adjust hand gestures into the frame to get signs captured correctly. The method used for the proposed system is based on deep learning framework. A convolution neural network (CNN) model is designed to perform ISL gesture recognition task. For creating dataset, 1001 images are captured for each class. Image preprocessing steps are applied on these images.

Figure 1 shows block diagram of the proposed system. A live video input is converted into frames, preprocessed, and then fed to CNN model for training, which is trained on the dataset that we created. Now when these frames are passed to this CNN model, feature extraction is carried out and then classified in respective category by CNN. As the output, predicted results are given in the form of text (sentence) and audio.

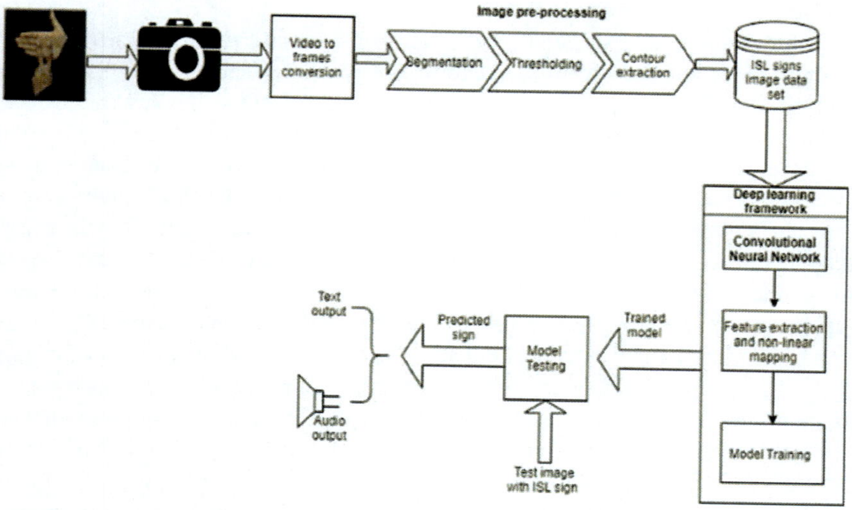

Fig. 1 Block diagram of proposed system

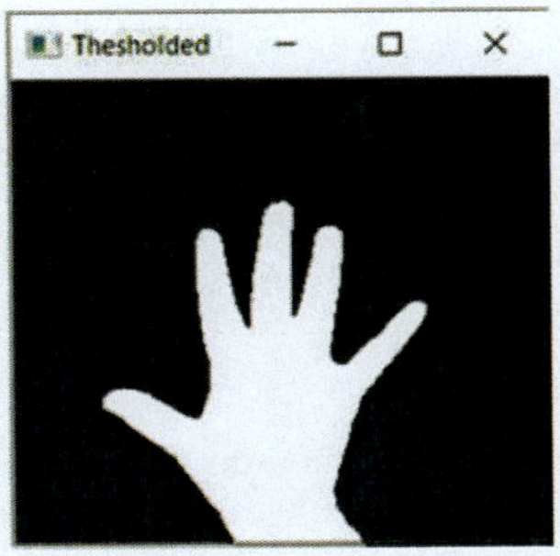

Fig. 2 Thresholding of hand region

3.1 Image Preprocessing

After converting live video input into frames, we perform background subtraction, thresholding, and contour extraction as preprocessing steps to find the hand region in the frames.

3.1.1 Background Subtraction

To separate foreground from background, we use the concept of running averages. We make our system look over a particular background area for 30 frames. During this period, we compute the running average over all the frames. After figuring out the background, we bring in our hand as a foreground object and use the current frame which has a hand in addition to the background. Then, the calculation for absolute difference between the background (updated over time) and the current frame having our hand in it is done to obtain a difference. Thus, the background subtraction takes place.

3.1.2 Thresholding

To detect the hand region from this difference image, we threshold the difference image selecting threshold value as 20, so that only our hand region is visible, and all the other unwanted regions become black as shown in Fig. 2.

3.1.3 Contour Extraction

After thresholding the difference image, we find contours (the outline of an object located in an image) in the image, and the contour with the largest area is assumed to be our hand. Figure 3 shows contour line (shown in red color) around hand region.

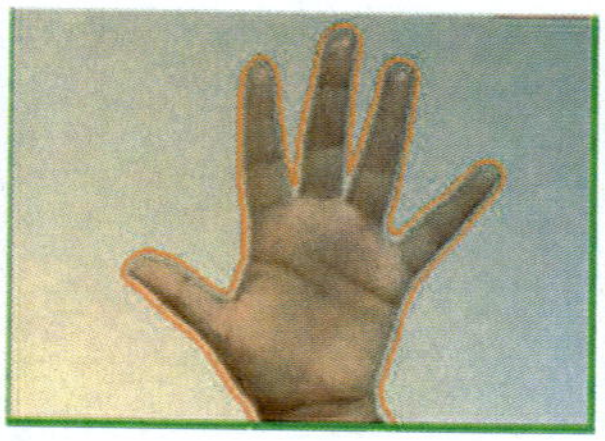

Fig. 3 Contour extraction for hand region

3.2 *CNN Architecture Design Details*

Figure 4 shows the CNN architecture of the proposed system which is used to train the model. CNN model holds combination of convolution + ReLu, pooling, flatten, fully connected, and dense, as five layers.

There is a combination of convolution-ReLu and pooling layer with a filter size from 32, 64, 128, and 256, respectively. We have added two dense layers for better accuracy and reduced trainable parameters, ultimately reducing training time. Also, we used a dropout layer to avoid overfitting of model, added flatten layer to convert the data into a one-dimensional array.

Table 1 shows the CNN model history and how the system has evolved progressively and achieved accuracy of 98%. In the first experiment, three combinations of convolution and max pooling layer were used with one dense layer having activation function sigmoid and a dropout layer with 0.5 value. With these settings, the accuracy come out to be 49%. Next experiment was conducted with only two combinations of convolution and max pooling with one dense layer and no dropout layer. It resulted into 52% accuracy. As there is no dropout layer, there is no loss of neurons; however, time taken for learning is more. In the third experiment, five combinations of convolution and max pooling layers were used with two dense layers having activation function as SoftMax and a dropout layer with 0.25 value. The accuracy obtained was 88%. These experiments and evaluation were performed on the existing ISL dataset [11].

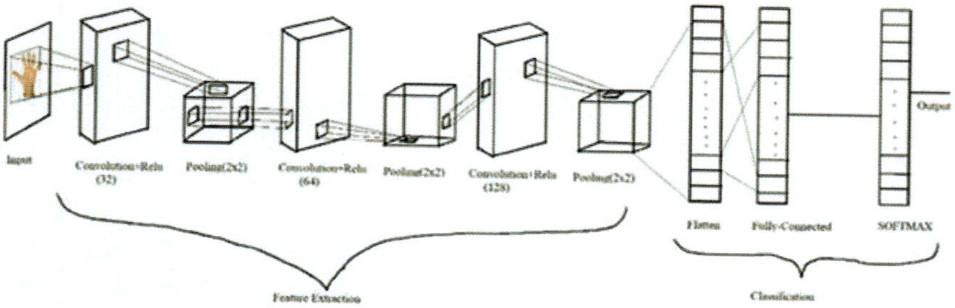

Fig. 4 Proposed CNN architecture

Table 1 CNN model history

Trail no.	Number of convolutional layers	Number of dense layers	Dropout value	Accuracy
1	3, with number of filters as 32, 32, 64	1 dense layer with sigmoid activation	0.5	49
2	2, with number of filters as 32, 32	2 dense layers with Softmax and ReLu activations	–	52
3	5, with number of filters as 32, 64, 64, 128, 256	2 dense layers with Softmax activation	0.25	88
4	5, with number of filters as 32, 64, 64, 128, 256	2 dense layers with Softmax and ReLu activations	0.25	98

Finally, five combinations of convolution and max pooling layers are used for our own created ISL dataset samples with two dense layers having activation functions SoftMax and ReLu, respectively. For this, a dropout layer with 0.25 value is selected. The accuracy achieved here is 98%.

Table 2 represents the detailed architecture of our proposed CNN model. It shows details for each layer, activation function used, stride, and kernel size used in respective layer.

Table 2 Proposed CNN architecture

Layer	Number of filters	Size	Kernel size	Stride	Activation
Input	–	(256, 256)	–	–	–
Convolution	32	(254, 254)	(3, 3)	1	ReLu
Max pooling	32	(127, 127)	(2, 2)	1	–
Convolution	64	(125, 125)	(3, 3)	1	ReLu
Max pooling	64	(62, 62)	(2, 2)	1	–
Convolution	64	(60, 60)	(3, 3)	1	ReLu
Max pooling	64	(30, 30)	(2, 2)	1	–
Convolution	128	(28, 28)	(3, 3)	1	ReLu
Max pooling	128	(14, 14)	(2, 2)	1	–
Convolution	256	(12, 12)	(3, 3)	1	ReLu
Max pooling	256	(6, 6)	(2, 2)	1	–
Flatten	–	9216	–	–	–
Dense	–	150	–	–	ReLu
Dropout	–	150 (Dropout value = 0.25)	–	–	
Dense	–	56	–	–	Softmax

4 Results and Discussion

The dataset is created and then divided into training, testing, and validation. The training, testing and validation set consists of 60%, 20%, and 20% of the aggregate data, respectively. On a given dataset with proposed CNN architecture, we have achieved an accuracy of 98%. This system is built in Python, which provides optimized computer vision tools and libraries.

4.1 Dataset Creation

For this system, the dataset has been created manually by referring to Indian Sign Language Portal [11] and Talking Hands [12]. Figure 5 shows labeled binary images for each ISL sign in the dataset. The images in the dataset are thresholded to reduce trainable parameters and ultimately to reduce the training period.

4.1.1 Dataset Description

Our dataset in total contains 56 static signs of ISL which includes labeled images of 1 to 10 digits, A to Z alphabets, and 20 isolated words such as Add, Good, Density, Learn, Science, and Sphere that are generally useful for training of specially abled school going children to train them.

We separated our dataset into training, testing, and validation dataset with a ratio of 60:20:20. We can customize words as we want and make this system more efficient.

Fig. 5 Sample signs from the created dataset

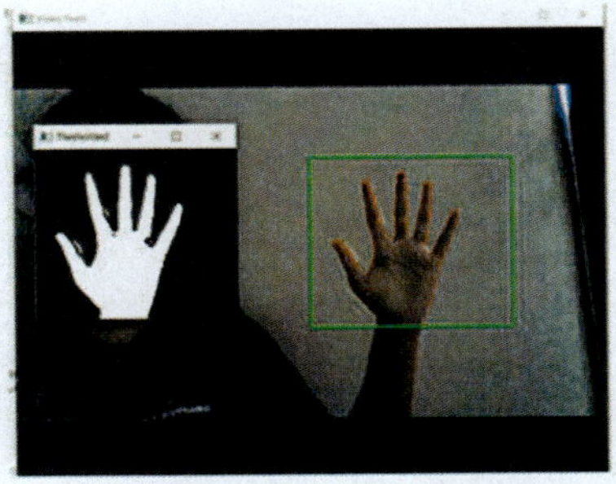

Fig. 6 Preprocessing of live input and ROI

4.1.2 Setup for Creating a Dataset and Data Augmentation

All images in the dataset are captured from three users with similar lighting and backgrounds using the webcam of the computer. We performed the same preprocessing steps (i.e., background subtraction, thresholding, and contouring) before capturing images for each sign. Images in the dataset are resized to (256 × 256) pixels, horizontally flipped, zoomed, shifted, and rotated. This is a data augmentation step which helps to get better accuracy in the predictions, add generalization effect in dataset, and hence is also useful to avoid overfitting.

4.2 Region of Interest (ROI) in Live Input

In preprocessing of live input, thresholding and contour extraction are carried out to get the region of interest. Same steps are followed while creating a dataset. Figure 6 shows results for the same. User will always have to keep his/her hand inside ROI to get perfectly segmented hand images.

4.3 System Output

After getting ROI, during testing phase, frames are passed to CNN model to match with the dataset images and at the end predict the gestures. Figure 7 shows the word "*good*" predicted after user presents ISL sign corresponding to word "*good*" as an input. The system is tested in different environments, although drastically changing lighting conditions and moving background may affect the system performance, as background subtraction method is used.

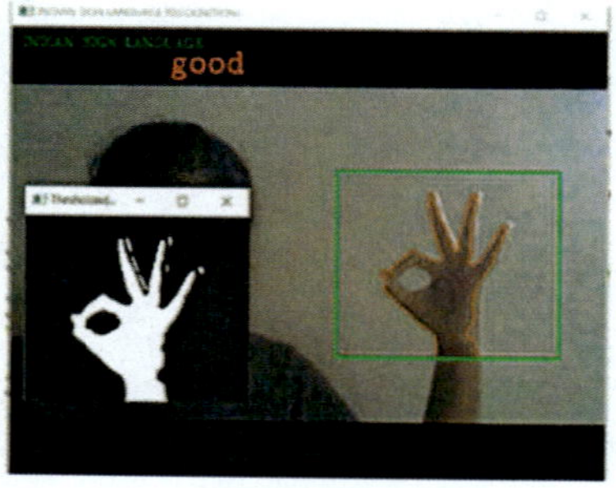

Fig. 7 Real-time gesture prediction—predicted word "good"

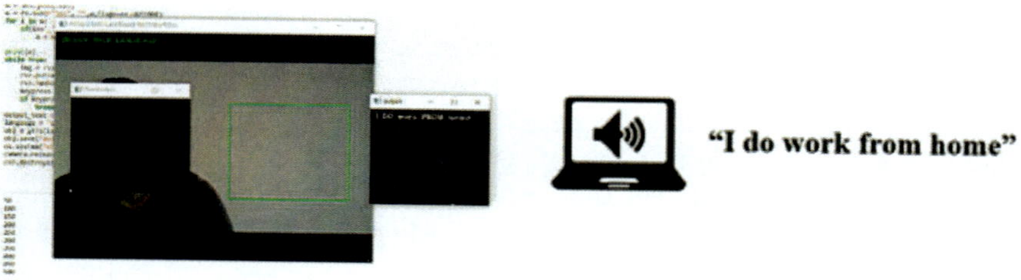

(a) (b)

Fig. 8 **a** Text output—sentence formation, **b** audio of predicted sentence

4.3.1 Sentence Formation and Audio Output from Predicted Words in Live Input

After successful prediction of live input, our aim is to get the output in sentence and audio form. After every 50 frames of live input, the predicted letter or word is appended to the sentence, and at last, the complete sentence is printed and sounded. Figure 8a shows that "**I do work from home**" is formed and displayed on screen, and Fig. 8b shows that the same sentence is sounded immediately in the form of audio through speakers of personal computer.

4.4 Performance Evaluation

Figure 9a shows a plot for loss (training and testing) versus number of epochs. As shown, with the increase in number of epochs, loss decreases, and around 40 epochs, it drops to 2%. Using proposed CNN architecture, we have achieved 98% accuracy, as shown in Fig. 9b where X-axis represents number of epochs, and Y-axis represents accuracy of model. Figure 9c shows a confusion matrix where Y-axis represents true labels, and X-axis represents the predicted labels.

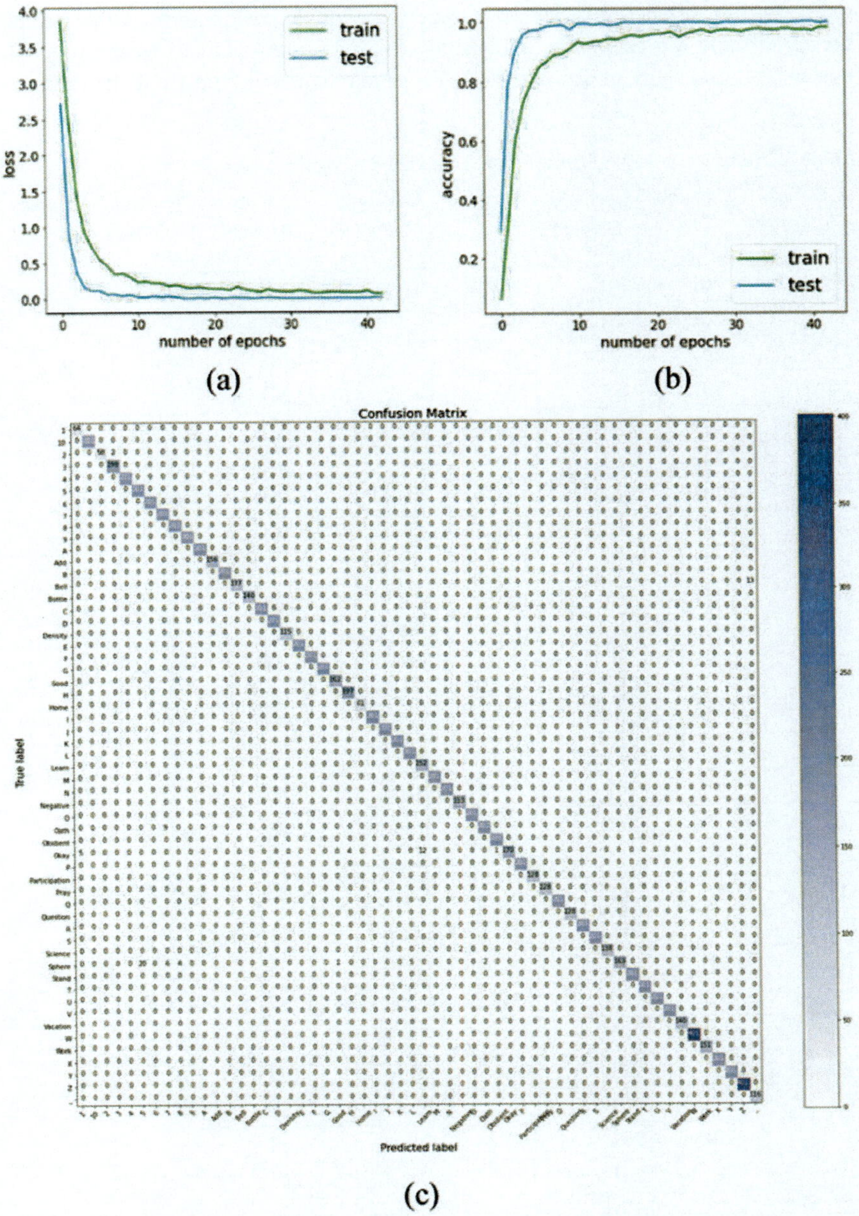

Fig. 9 **a** Loss versus number of epochs, **b** accuracy versus number of epochs, **c** confusion matrix

In the live prediction system, accuracy rate of the recognition is the greatest concern. Table 3 presents the recognition accuracy rate obtained by the proposed system for ISL signs of alphabets (A–Z), numerals (0–9), and English words. As

Table 3 Alphabets (A–Z), numerals (1–10), and word recognition accuracy

ISL signs (Alphabets)	Number of input signs	Correctly predicted signs	Accuracy (%)	ISL signs (Numerals)	Number of input signs	Correctly predicted signs	Accuracy (%)
A	201	201	100	1	201	201	100
B	201	201	100	2	66	58	88
C	201	201	100	3	198	198	100
D	201	201	100	4	201	201	100
E	201	201	100	5	201	201	100
F	201	201	100	6	201	201	100
G	201	201	100	7	201	201	100
H	201	201	100	8	201	201	100
I	201	201	100	9	201	201	100
J	201	201	100	10	201	201	100
K	201	201	100	**ISL Signs (Words)**	Number of input signs	Correctly predicted signs	Accuracy (%)
L	201	201	100	Add	156	156	100
M	201	201	100	Bell	150	137	91
N	201	201	100	Bottle	140	140	100
O	201	201	100	Density	115	115	100
P	201	201	100	Good	162	162	100
Q	201	201	100	Home	201	197	98
R	201	201	100	Learn	201	201	100
S	201	201	100	Negative	152	152	100
T	201	201	100	Oath	153	153	100
U	201	201	100	Obedient	201	201	100
V	201	201	100	Okay	192	170	85.5
W	201	201	100	Participation	128	128	100
X	201	201	100	Pray	128	128	100
Y	201	201	100	Question	128	128	100
Z	201	201	100	Science	140	138	98.5
				Sphere	201	163	81.1
				Stand	201	201	100
				Vacation	140	140	100
				Work	151	151	100

shown in the table, accuracy values for alphabet recognition are highest for predicting signs when compared with actual ones. In case of numerals (digits), average accuracy is 98.8%. System performs fairly accurate for word recognition, although in some cases recognition accuracy drops till 81%.

5 Conclusion

The work presented in the paper intends to predict Indian Sign Language alphanumeric hand gestures in real time and display the recognized sign in terms of text (both words and sentences) and audio outputs. As the system is capturing input directly from webcam, it is very important to train the model to predict the results correctly in real time. A deep CNN framework provides systematic learning approach for the model. Hence, the proposed system is built using CNN, which has achieved accuracy of 98%. The system can predict two-handed sign language gestures from a given dataset pertaining to controlled lighting and plain background conditions. At the output of the system, sentence formation is done with audio of the predicted sentence, and with this it can be used as a medium of communication. Low light conditions can affect the real-time prediction. Also, the system strictly needs a plain background, during live gesture recognition. Proposed system can be extended ahead to communicate in two ways, i.e., to convert spoken words to ISL and vice versa. Also, an Android application can be built to make system portable and increase its usability for communicating with the people having hearing disabilities.

References

1. Singha, J., & Das, K. (2013). Recognition of Indian sign language in live video. *International Journal of Computer Applications, 70*(19), 17–22.
2. Kopuklu, O., Gunduz, A., Köse, N., & Rigoll, G. (2018). Real-time hand gesture detection and classification using convolutional neural networks. In *IEEE International Conference on Automatic Face and Gesture Recognition* (pp. 1–8). arXiv:1901.10323
3. Shravani, K., Sree Lakshmi, A., Sri Geethika, M., Kulkarni, S. B. (2020). Indian sign language character recognition. *IOSR Journal of Computer Engineering, 22*(3), 14–19.
4. Deshpande, A. M., & Kalbhor, S. R. (2020). Video-based marathi sign language recognition and text conversion using convolutional neural network. Lecture Notes in Electrical EngineeringIn S. T. Hitendra, V. Sankar, & R. Shaik (Eds.), *Emerging trends in electrical, communications, and information technologies* (Vol. 569, pp. 761–773). Springer.
5. Ananth Rao, G., Kishore, P. V. V.(2018). Selfie video based continuous Indian sign language recognition system. *Ain Shams Engineering Journal, 9*, 1929–1939.
6. Sahoo, A. K. (2014). Indian sign language recognition using neural networks and KNN classifiers. *Journal of Engineering and Applied Sciences*, 1250–1259.
7. Kaur, J. (2019) An efficient Indian Sign Language recognition system using sift descriptor. *International Journal of Engineering and Advanced Technology*, 1456–1461.
8. Rokade, Y. I., & Jadav, P. M. (2017). Indian sign language recognition system. *International Journal of Engineering and Technology*, 189–196.

9. Archana Ghotkar, A. S., & Kharate, G. K. (2014). Study of vision-based hand gesture recognition using Indian sign language. *International Journal on Smart Sensing and Intelligent Systems*, 1–8.
10. Itkarkar Rajeshri, R., Nandi, A. K. V., & Mungurwadi, V. B. (2019). Indian sign language recognition using gabor feature extraction. *International Journal of Scientific & Technology Research, 8*(12), 3392–3396.
11. https://indiansignlanguage.org/
12. https://www.talkinghands.co.in/

Dense SIFT-Based Facial Expression Recognition Using Machine Learning Techniques

S. Vaijayanthi and J. Arunnehru

Abstract Facial analysis is an active research topic in examining the emotional state of humans over the past few decades. It is still a challenging task in computer vision due to its high intra-class variation, head pose, suitable environment conditions like lighting and illumination factors in behaviour prediction and recommendation systems. This paper proposes a novel facial emotion representation approach based on dense descriptors for recognizing facial dynamics on image sequences. Initially, the face is detected using the Haar cascade classifer to extract the temporal information from the facial frame by applying a scale invariant feature transform by combining a bag of visual words. Later, the extracted high-level features are fed to machine learning algorithms to classify the seven emotions from the MUG dataset. The proposed dense SIFT clustering performance was evaluated on four different machine learning algorithms and achieved a high rate of recognition accuracy in all classes. In the experimental results, K-NN exhibits the proposed architecture's effectiveness with an accuracy rate of 91.8% for the MUG dataset, 89% for SVM, 87.6% for Naive Bayes, and 85.7% decision tree, respectively.

Keywords Facial analysis · Human emotion recognition · Dense SIFT key points · k-means clustering · Bag of words

1 Introduction

Recent artificial intelligence studies related to experimental psychology or subjective experience express that recognizing human emotions is a primary task in decision-making and rational thinking. Over the past few years, progress in emotion recognition has dealt with face tracking [1], voice analysis [2], and body gestures [3]. Facial expression is one of the most fundamental and spontaneous ways to identify a

S. Vaijayanthi (✉) · J. Arunnehru
Department of Computer Science and Engineering, SRM Institute of Science and Technology, Vadapalani, Chennai, Tamil Nadu, India
e-mail: vaijayanthisekar@gmail.com

B. Pati et al. (eds.), *Proceedings of the 6th International Conference on Advance Computing and Intelligent Engineering*, Lecture Notes in Networks and Systems 428,
https://doi.org/10.1007/978-981-19-2225-1_27

person's emotional state over the past decade. The unusual behaviour includes head movement, partially occluded face, temporal actions, and facial micro-expressions. However, human facial expression emotions are challenging to determine the emotional states in illumination conditions, light, pose variations, shadows,etc.

The primary approaches for facial expression recognition (FER) are (i) face component detection [4] using Haar cascade classifier. (ii) Feature selection via spatial-temporal-based feature model [5]. (iii) The extracted dense scale-invariant feature transform (DSIFT) are passed through the machine learning algorithms to recognize the seven emotions namely neutral, anger, fear, surprise, sad, disgust, and happy.

Recently, many machine learning models [6, 7] have been emerging based on visual emotion analysis and gained several advantages over traditional geometric emotion recognition techniques. This paper proposes a dense feature with Haar cascade to classify emotions based on appearance based approaches. Initially, the input RGB frames are converted to grey frames, cropped, and the face is detected using the Haar cascade classifier and passed through dense SIFT for identifying the facial regions using K-means clustering and bag of words; later, the obtained key point feature descriptors are fed to the four different machine learning algorithms like K-nearest neighbour, SVM, decision tree, and Naive Bayes for training and to predict the accuracy of the classifier on the MUG dataset for recognizing the seven facial emotions like anger, fear, disgust, surprise, joy, neutral, and sad.

The paper is organized in the following manner: Sect. 2 provides the previous work done in facial-based emotion recognition. The detailed proposed dense invariant feature workflow is explained in Sect. 3. Section 4 explains the classification approach. Section 5 describes the dataset used, performance metrics, and experimental results on various machine learning algorithms on the MUG dataset, and finally, Sect. 6 recaps the paper with conclusion and future work.

2 Prior Research

This section aims to review the recent aspects of vision-based facial emotion recognition. Typically, facial emotion recognition involves detecting the face in the frame, extracting specific features, and classifying the emotions [8]. To enrich the FER system's performance, preprocessing is done before the feature extraction phase. Preprocessing [9] includes contrast adjustment, image scaling, and supplementary enhancement, which helps boost facial expressions in-depth. Previously, many research articles focus on detecting AU occurrence, AU intensity, and facial action points detection to classify emotions [10]. A brief survey regarding various FER methods involving visual pattern techniques such as geometric-feature-based, appearance-based, texture-based, and hybrid facial image recognition techniques is discussed [11].

Anima et al. [12, 13] developed a self-organizing map describing the 26-dimensional facial point vector using the MMI dataset; especially, the geometric features from eyebrow, eyelid, nose, and lip are considered. The facial image is cropped and scaled with a nose as a midpoint, and other main components of the face are incorporated physically. The paper [14] focused on an active appearance-based framework with fuzzy C-means semi-supervised learning to identify eight different emotional instances using face points and cluster degrees. Tehmina et al. [15] improve the facial expression recognition accuracy by modifying the bag of words with SBoF-SSIFT hybrid feature descriptors in the JAFFE and CK+ datasets. Wang et al. [16] used face descriptors for gender recognition using dense SIFT by combining AdaBoost and Gabor features in three different datasets.

From the above discussion, the recognition of facial emotions, visual information needs greater improvement in the field of emotion recognition. The solution for static facial regions is still preceding and is useful for discriminating between only a few subsets of expressions. Our method mainly focuses on identifying human emotions, in a static sequence by applying Dense SIFT features. These SIFT descriptors encode the magnitude and directional information by mapping the input RGB frame to grey colour space by the property of invariance and rotation. Finally, the key point feature descriptors are passed through four different learning approaches to classify the emotions.

3 Proposed Approach

This paper uses the MUG dataset to recognize facial expressions from static facial action sequences consisting of different emotions such as surprise, anger, fear, joy, neutral, sad, and disgust.

Figure 1 presents the workflow process of entire emotion recognition system using Dense SIFT. Initially, the frame is preprocessed for removing redundant features. The input RGB frames are converted to grey frames, and the face is identified using the Haar cascade, cropped and passed through dense SIFT for recognizing the facial regions through feature descriptors; later, these feature vectors are passed to the machine learning algorithms like SVM, K-nearest neighbour, decision tree and Naive Bayes for training and to predict the accuracy of the classifier.

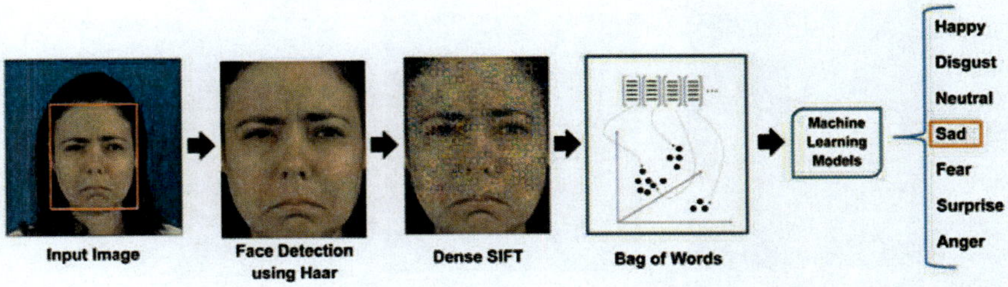

Fig. 1 Proposed approach for emotion recognition

3.1 Face Detection

Haar cascade classifier [17] is a dynamic approach used for object detection with a cascade function. Face detection is achieved easily by training the classifier with positive and negative frames using various weight features and adaboost algorithm. The complete process of the Haar feature extraction for identifying the eyes in facial region is shown in Fig. 2.

3.2 Dense SIFT Descriptors

Scale-invariant feature transform (SIFT) is a keypoint localization algorithm mainly used in computer vision for representing the motion tracking features in a scene and object tracking. The central theme of using SIFT algorithm in descriptive and local features is invariant to brightness conditions, scaling, and rotation transformations with the curse of dimensionality.

SIFT [18] undergoes the following different stages for computing the image features as follows (i) scale-space extrema detection (Gaussian function difference), (ii) keypoint localization (location and scale with suitable stability measures), (iii) orientation assignment (image gradients and transformations), (iv) keypoint descriptor (eliminating distortion and illumination conditions). Finally, the gradient eliminates the edge features in Hessian Matrix, combines the weight and point features, and obtains the pixel position in the computation of the 4×4 matrix sub-regions with eight different bins. The final best SIFT descriptor is $4 \times 4 \times 8 = 128$ bins is generated with the help of a key point generator with maximum gradient scale. Hence, dense SIFT is functional to the full resolution of the face with developing the key points descriptors and robust to distinctive variation in camera and brightness constraint.

As seen in Fig. 3 initially, the dense SIFT extracts the feature descriptors in the facial image; later, these representations are invariant to translation and rotation conditions with illumination factors. The selected features in the face regions are clustered finally with K-means clustering and bag of labelling.

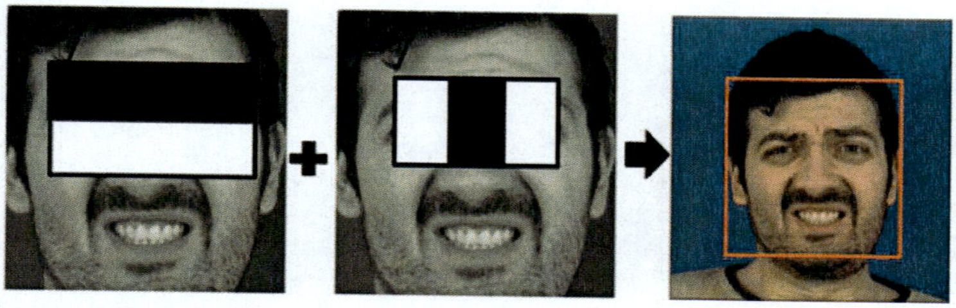

Fig. 2 Face detection using Haar Cascade Classifier

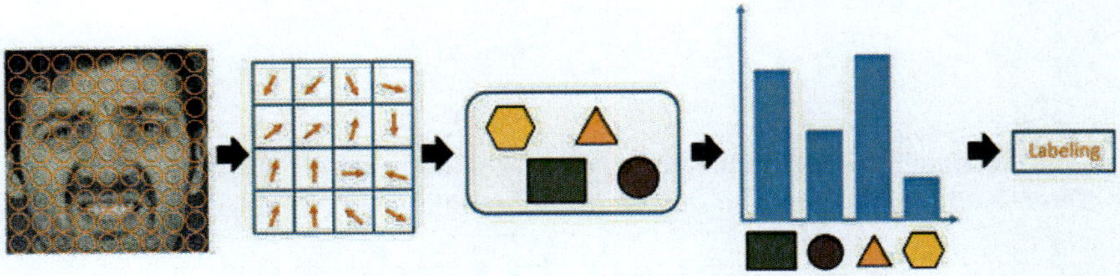

Fig. 3 Bag of words model and labelling

K-means Clustering Features K-means [19] is an unsupervised learning algorithm that reduces the speed and accuracy of the obtained dense SIFT features. Here, k randomly generates data with an exclusive mutual observation of clusters. The centroid k calculates the data between the data points and cluster head. Lastly, it enhances a deterministic way for identifying the cluster head with the minimized objective function representation. In our work, the best fit of k is achieved in a size of 500 in the MUG dataset.

Bag of Words Model (BoW) Bag of words is an image categorization technique mainly used to extract local features, construct visual geometry with labelling, and map the local feature to the visual words in the input facial image as a vector field. Here, the dense SIFT descriptors is considered as an object cluster head in the bag of visual words. The histogram of visual words with a BoW model makes it feasible to represent objects with more minor features.

4 Machine Learning Algorithms

This section briefs the four different classification techniques used in this paper for recognizing seven facial emotion using the MUG dataset. We also compare the accuracy prediction performance of different classification techniques.

K-Nearest Neighbour The K-nearest neighbour is exclusively a statistical analysis and pattern recognition algorithm used in computer vision. This algorithm does not require explicit training of data and performs well both in regression and classification tasks. The training phase saves the data and classifies the data based on the arrival of new data. If the value of k is 1, the assigned nearest neighbours will get a particular data value and vice versa when k is 0. The Euclidean distance (ED) function in Eq. 1 is a continuous distance measurement computed by calculating the approximate distance between any two pairs of points.

$$\mathrm{ED} = \sqrt{\sum_{i=1}^{n} (x_i - y_i)^2} \tag{1}$$

$$D_H = \sum_{j=1}^{n} \left| p_j - q_j \right| \tag{2}$$

The constant fall back on the distance measure obtains the training set containing different scales and the testing set obtains the emotion class instance from the MUG dataset.

Support Vector Machine (SVM) SVM [20] is an efficient supervised learning algorithm mainly explored for multi-class regression and classification tasks for recognizing visual patterns. SVM best suits in recognizing the facial emotions by handling outliers in the MUG dataset. The features obtained from the dense SIFT are clustered as visual words as a point in n dimensional space and optimizing it with maximum margin with a minimum number of support vectors. Here, the training set is not linearly separable; hence, RBF kernel function is used in high dimensional space to classify the suitable hyper-plane in seven classes. The SVM aims to predict a target value of the decision boundary from the testing set consisting of large-margin planes of binary classification. SVM optimizes the trade-off by reducing the error in the pixel location y_i with fine-tuning the score function f, and it is given as

$$f\,(y_i, w, b) = w_{y_i} + b \tag{3}$$

The y_i in Eq. 3 obtains the pixel position of a fattened image in $N \times 1$ dimensional vector, where w represents the weight matrix, and b act as a bias function. Finally, the feature vectors from the MUG dataset are passed through the SVM to recognize the facial emotions.

Decision Tree (DT) Decision tree is a supervised learning algorithm [21] well known for its classification and regression tasks. The vital use of DT is to estimate the value of the facial emotions by learning the bag of feature vectors inferred from the prior training data.

Naive Bayes (NB) Naive Bayes classifier [22], influenced by Bayes theorem, is most suitable for appearance based human facial emotions. The preprocessed features from the MUG dataset containing static image sequence are passed through the probabilistic Naive for classifying the seven emotions with k features. In Eq. 4. (x_i, i) is the frame pixel position, where $i = 1, 2, \ldots, k$ and (y_j, j) implies the final emotional class instance. Naive Bayes estimates the prbability of likelihood $P(y_j/x_i)$ for each potential output class with the highest posterior probability, and it is expressed as

$$Y_{nb} = \arg\max P(y_j) \prod_{i=1}^{n} P(\frac{x_i}{y_j}) \tag{4}$$

The prior probability $P(y_j)$ estimates the symmetry association with the product class y_j in the training data. The conditional probability estimated using Eq. 5 is

$$P\left(\frac{x_i}{y_j}\right) = \frac{1}{\sqrt{2\pi}\,\delta} e^{-(x_i-\mu)^2}/2\delta^2 \tag{5}$$

where μ represents the mean value and δ represents the standard deviation for x_i.

5 Experimental Results and Discussion

The demonstrations are implemented in windows 10 with Intel i7 processor with 3.40 GHz using MATLAB 2019b. As explained in Sect. 3, the dense SIFT with BoW feature vector is extracted from the facial frames and trained with different machine learning approaches to identify the particular emotional states. The proposed dense trajectories performance is tested on K-NN, SVM, Naive Bayes, and decision tree to recognize the accuracy rate of seven different emotional states using the MUG dataset.

5.1 *MUG Dataset*

The multimedia understanding group (MUG) [23] is a facial expression dataset that contains static sequence frames aged between 20 and 35 yrs. Figure 4 shows the seven basic archetypical emotions like anger, disgust, fear, joy, neutral, sad, and surprise expressed by 35 females and 51 males are used for this work. The recorded frames have a pixel size of 896 × 896. In total, the dataset consists of 11,758 image, portraying seven different facial emotional expressions consists of 1587 samples of angry, 1606 disgust, 1638 fear, 1868 happy, 1389 neutral, 1802 sad, and 1868 samples of surprise.

5.2 *Performance Metrics*

The proposed approach uses an Haar cascade with dense SIFT and used a fivefold cross-validation strategy. Those extracted feature vectors are given to the K-NN,

Fig. 4 Sample frames of seven emotions from the MUG Dataset

SVM, Decision tree and Naive Bayes to assess the statistical measures like accuracy (A), precision (P), recall (R), and F-measure performance. Where tp (positive prediction) is true positive, tn (Negative prediction) is true negative, fp (mispredicted as positive) is the false positive, fn (mispredicted as negative) is the false negative.

Accuracy (A) $= \left[\frac{\mathrm{tp}+\mathrm{tn}}{\mathrm{tp}+\mathrm{fp}+\mathrm{tn}+\mathrm{fn}}\right]$ predicts the overall correctness of the facial emotions classified.

Precision (P) $= \left[\frac{\mathrm{tp}}{\mathrm{tp}+\mathrm{fp}}\right]$ gives the measure of perfection.

Recognizing emotions in the exact way defined by Recall (R) $= \left[\frac{\mathrm{tp}}{\mathrm{tp}+\mathrm{fn}}\right]$.

F-measure $= 2\frac{P \times R}{P+R}$ gives the detailed mean of both precision and recall.

Classifier Results The detailed confusion matrix for the K-NN, SVM, Naive Bayes, and decision tree classifier using dense SIFT and BoW on the emotion dataset is shown in Fig. 5. The main diagonal details the instances classified accurately and the rows represent seven different archetypical emotion class, and column represents the facial emotion class predicted by scattering net. Most emotion classes like joy, surprise, and neutral are grouped well with an average recognition rate of 91.8% for the K-NN, 89% for SVM, 87.6% for Naive Bayes and 85.7% for decision tree, respectively. From the results, fear is mispredicted with disgust, which initially appears difficult to separate and require further consideration.

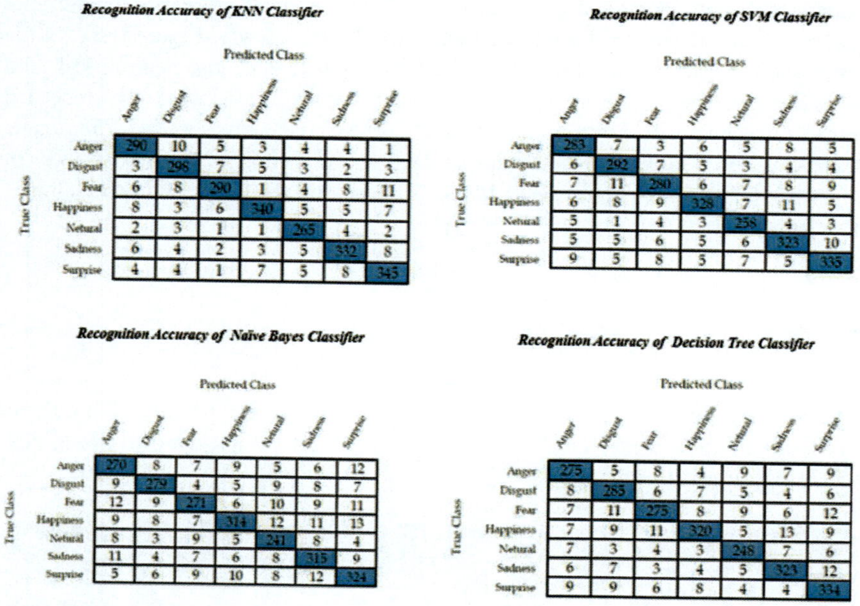

Fig. 5 The detailed confusion matrix obtained for the K-NN, SVM, NB, and DT classifier

Table 1 Average performance measure for dense SIFT-based machine learning algorithms in MUG dataset

Classifier	Precision (%)	Recall (%)	Specificity (%)	F-measure (%)
K-NN	91.8	91.9	98.6	91.8
SVM	89.1	89.3	98.2	89.2
NB	87.5	87.6	97.9	87.5
DT	85.5	85.6	97.6	85.5

Table 1 represents the average emotion performance of precision, specificity, recall, and F-measure values for K-NN, SVM, NB, and DT. From the results, K-NN performs good in identifying the seven emotions when compare to SVM, DT and NB.

6 Conclusion

In this paper, we present a novel method for recognizing emotions based on static facial movements using dense scale-invariant Fourier transform. Demonstrations were carried on the MUG dataset to identify the seven different emotional states. The results show that K-NN gives an overall recognition accuracy of 91.8% for the MUG dataset and 89% for SVM, and 87.6% for Naive Bayes and 85.7% for decision tree, respectively. The features from the dense trajectors help to find the accuracy of quantitative metrics like precision, specificity, recall, and F-measure. From the experiment, it is concluded that the system could not able to differentiate fear and sad with high accuracy. The proposed work requires less processing time comparing to the existing models. Hence, this model is perfect for recognizing visual emotions and increases the system's efficiency and accuracy with a better recognition rate. Our future research extends to recognize the various different emotional states by combining the visual body gestures with facial micro expression recognition using real-time datasets.

References

1. Huang, Y., Chen, F., Lv, S., & Wang, X. (2019). Facial expression recognition: A survey. *Symmetry, 11*(10), 1189.
2. Vaijayanthi, S., & Arunnehru, J. (2021). Synthesis approach for emotion recognition from cepstral and pitch coefficients using machine learning. In *International Conference on Communication, Computing and Electronics Systems* (pp. 515–528). Springer.
3. Arunnehru, J., & Kalaiselvi Geetha, M. (2017). Automatic human emotion recognition in surveillance video. In *Intelligent techniques in signal processing for multimedia security* (pp. 321–342). Springer.

4. Ekman, P., & Rosenberg, E. L. (1997). *What the face reveals: Basic and applied studies of spontaneous expression using the Facial Action Coding System (FACS)*. Oxford University Press, USA.
5. Youssif, A. A. A., & Asker, W. A. A. (2011). Automatic facial expression recognition system based on geometric and appearance features. *Computer and Information Science, 4*(2), 115.
6. Arunnehru, J., Nandhana Davi, A. K., Raghul Sharan, R., & Nambiar, P. G. (2019). Human pose estimation and activity classification using machine learning approach. In *International Conference on Soft Computing and Signal Processing* (pp. 113–123). Springer.
7. Arunnehru, J., Kumar, A., & Verma, J. P. (2019). Early prediction of brain tumor classification using convolution neural networks. In *International Conference on Computational Intelligence, Security and Internet of Things* (pp. 16–25). Springer.
8. Revina, I. M., & Sam Emmanuel, W. R. (2021). A survey on human face expression recognition techniques. *Journal of King Saud University-Computer and Information Sciences, 33*(6), 619–628.
9. Li, S., Deng, W. (2020). Deep facial expression recognition: A survey. *IEEE Transactions on Affective Computing*.
10. Martinez, B., Valstar, M. F., Jiang, B., & Pantic, M. (2017). Automatic analysis of facial actions: A survey. *IEEE transactions on affective computing, 10*(3), 325–347.
11. Patel, K., Mehta, D., Mistry, C., Gupta, R., Tanwar, S., Kumar, N., & Alazab, M. (2020). Facial sentiment analysis using AI techniques: State-of-the-art, taxonomies, and challenges. *IEEE Access, 8*, 90495–90519.
12. Majumder, A., Behera, L., & Subramanian, V. K. (2014). Emotion recognition from geometric facial features using self-organizing map. *Pattern Recognition, 47*(3), 1282–1293.
13. Liu, X., Cheng, X., & Lee, K. (2020). Ga-svm-based facial emotion recognition using facial geometric features. *IEEE Sensors Journal, 21*(10), 11532–11542.
14. Liliana, D. Y., Widyanto, M. R., & Basaruddin, T. (2016). Human emotion recognition based on active appearance model and semi-supervised fuzzy c-means. In *2016 International Conference on Advanced Computer Science and Information Systems (ICACSIS)* (pp. 439–445). IEEE.
15. Kalsum, T., Anwar, S. M., Majid, M., Khan, B., & Ali, S. M. (2018). Emotion recognition from facial expressions using hybrid feature descriptors. *IET Image Processing, 12*(6), 1004–1012.
16. Wang, J. G., Li, J., Lee, C. Y., & Yau, W. Y. (2010). Dense sift and gabor descriptors-based face representation with applications to gender recognition. In *2010 11th International Conference on Control Automation Robotics & Vision* (pp. 1860–1864). IEEE.
17. Li, C., Qi, Z., Jia, N., & Wu, J. (2017). Human face detection algorithm via haar cascade classifier combined with three additional classifiers. In *2017 13th IEEE International Conference on Electronic Measurement & Instruments (ICEMI)* (pp. 483–487). IEEE.
18. Lowe, D. G. (2004). Distinctive image features from scale-invariant keypoints. *International Journal of Computer Vision, 60*(2), 91–110.
19. Kahou, S. E., Bouthillier, X., Lamblin, P., Gulcehre, C., Michalski, V., Konda, K., Jean, S., Froumenty, P., Dauphin, Y., Boulanger-Lewandowski, N., et al. (2016). Emonets: Multimodal deep learning approaches for emotion recognition in video. *Journal on Multimodal User Interfaces, 10*(2), 99–111.
20. Arunnehru, J., Vidhyasagar, B. S., & Basha, H. A. (2020). Plant leaf diseases recognition using convolutional neural network and transfer learning. In *International Conference on Communication, Computing and Electronics Systems* (pp. 221–229). Springer.
21. Arunnehru, J., & Geetha, M. K. (2013). Behavior recognition in surveillance video using temporal features. In *2013 Fourth International Conference on Computing, Communications and Networking Technologies (ICCCNT)* (pp. 1–5). IEEE.
22. Noroozi, F., Corneanu, C. A., Kaminska, D., Sapinski, T., Escalera, S., & Anbarjafari, G. (2018). Survey on emotional body gesture recognition. *IEEE Transactions on Affective Computing*.
23. Aifanti, N., Papachristou, C., & Delopoulos, A. (2010). The mug facial expression database. In *11th International Workshop on Image Analysis for Multimedia Interactive Services WIAMIS 10* (pp. 1–4). IEEE.

A Review on Optical Character Recognition of Gujarati Scripts

Kanal Soni

Abstract Unconstrained handwritten identification is among the toughest situations in recognition and image processing. This appears to be a simple operation for an individual, however, acknowledging handwriting is a time-consuming effort for a system. In the context of a device, the entry should first be digitized from a record, a photograph, or a legitimate device such as a desktop, tablet, or laptop. The digitized text or numeral is then changed into digital form text using the Handwritten Character Identification method. This could be managed and done in two ways: online plus offline. The central target of this survey is offline authentication of Gujarati scripts (characters plus numerals) in paper and electronic materials. Numerous neural and machine learning frameworks with classification techniques were being utilized, however, the bulk of machine methodologies demonstrated efficacy in spotting these scripts in the end.

Keywords Gaussian functionality · Gujarati scripts · Machine learning frameworks · Neural frameworks · OCR—optical character recognition · Textual analysis

1 Introduction

Gujarati's native language is adapted from either the reputed Devanagari ancient scripting. The nomenclature of Gujarati is utilized by well over 45 million unique visitors, primarily Gujarati folks [1]. Gujarat is being addressed in India plus overseas as a Gujarati republic. Individuals are inhabitants of a multitude of locations. Science has progressed rapidly over yesteryears spanning the planet in a fiercely competitive style. The cheap cost point of invention of the Internet has led to expansion to reach any corner of the earth. In almost all of the administration's legacy, the preponderance

K. Soni (✉)
Information Technology Department, Gandhinagar Institute of Technology, Ahmedabad, Gujarat, India
e-mail: kanalsoni015@gmail.com

B. Pati et al. (eds.), *Proceedings of the 6th International Conference on Advance Computing and Intelligent Engineering*, Lecture Notes in Networks and Systems 428,
https://doi.org/10.1007/978-981-19-2225-1_28

of the subject matter is word-based. A substantial number of transcripts and manuals are accessible in two configurations: handwritten verses or rather scanned images.

Gujarati narration might be utilized in a diverse set of orally and in written transcripts. From that a constitutional and legal standpoint, as well as for simultaneous transmission, such details must be secured electronically [2]. Scanning is the most effective tactic for switching paper forms to digital files. Formatting, detecting, and extracting info from digitized textual pictures, on the other finger, is challenging [3]. As a response, obtaining crucial data from a captured page is a critical task. For data retrieval from reports, the following are the two central approaches: recognition-based versus recognition-free. The coherent framework relies on recognition to transform a documentation picture into manuscript (ASCII) information using an Optical Character Recognition setup. The recognition-free scheme utilizes a word actual photo-like question sequence and facilitates the Infrared Spectroscopy procedure by equating the question term image to the source terms representations explicitly [4].

In OCR [5] preparing, an information picture or document is first to peruse in the PC by checking. The filtering pace will be chosen by the nature of the scanning gadgets, quality of that paper, legitimate setting of the OCR framework, and cleanness. In the wake of checking, it reads that record, chooses the threshold or limit esteem, examines for light and dull regions to distinguish each alphabetic letter or numeral. At the point when a letter or numeral is remembered, it is changed over into an ASCII code. The perceiving system is to decipher pictures. The library layouts and setup edge will decide the precision of understanding of the OCR.

A snippet of text recognition software: There seem two sorts of content analysis conflicts: (1) text identification as well as (2) text pairing. Identifying the expression from penned and displayed files is achievable in two levels in the text identification task: (1) Word recognition in the real age, (2) word identification in the era of web information.

Offline identification of a term is associated with terms identified after they have been penned by beings, usually on a pad of paper. Offline content identification is the procedure of comprehending data that has been digitized from a manuscript and retained in electronic formats. The writing is accomplished with any digital device in the online version [6, 7]. Whereas in the matching of a specific term or text, information is received and the need to recognize words is mitigated (Fig. 1).

The remaining paper is coordinated as follows. Section 2 portrays the writing investigation. Section 3 gives short information of reviews in a plain tabular kind. Section 4 depicts the difficulties that actually exist and should be settled soon. Section 5 delineates the end or conclusion and future tasks.

2 Literature Scrutiny

The investigations done by the past analysts are being portrayed underneath.

Desai (2010) [3]: For the classification scheme of Gujarati integers, straight transcend backpropagation neural connections are formulated and solved. Before

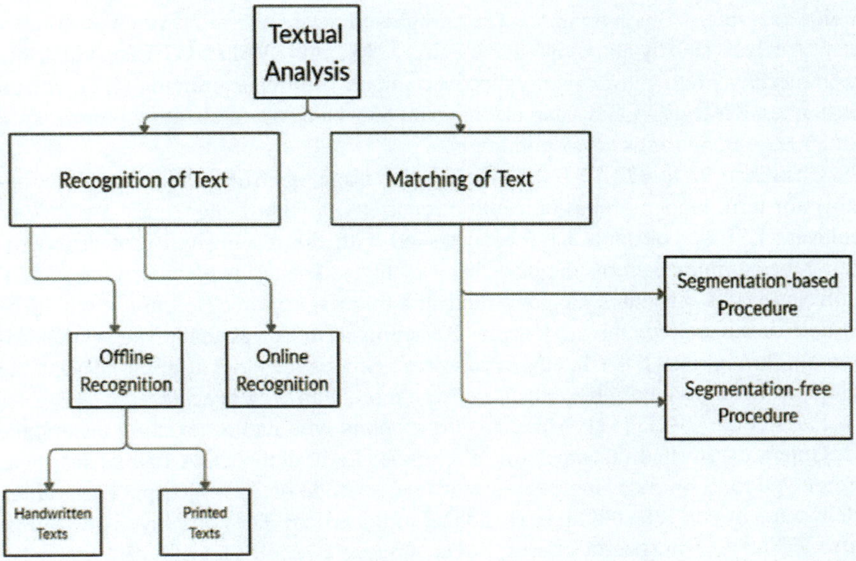

Fig. 1 Flow-chart of textual analysis procedure

deploying number sorting, diverse techniques are employed in the data preparation phase. Although the efficiency of this suggested connection is 81.66%, though, it is not up to par. Any categorization technique's efficiency is significantly governed by attribute abstraction.

Shirke et al. (2021) [8]: The undertaking targets fostering a disconnected (offline) transcribed Gujarati script acknowledgment framework that can be adequately utilized for perceiving manually written Gujarati scripts. This cycle includes the formation of a neural organization to be specific YOLO—You Only Look Once structure that will accept input as a picture and afterward the components will be extricated from it, the neural organization will perceive this text and yield it on the machine. This will help in the acknowledgment of Gujarati contents of various penmanship and will assist with diminishing human mistakes and saving the hour of people.

Vyas et al. (2015) [2]: The distinctive features for Gujarati integrals were recovered applying three distinct approaches in this trial. They consisted of the spatial domain, DCT coefficient determination, and Fourier signifiers. These were utilized on three major categories of classifiers: ANN, KNN, plus SVN, with a backpropagation mechanism. Combining adapted chain code, DFT, with DCT, the greatest prediction accuracy for a massive database of 3000 numbers were 85.6%, 93.6%, and 93.0%, correspondingly.

Baheti et al. (2012) [9]: Gujarati Counts were acknowledged utilizing a data source of manually written native integers. Information was acquired via 16 distinct informants, ranging in age and career. They gleaned 1600 examples in all, with 10 unique samples of each numeral. They employed mechanisms of Principal Component Analysis shortly abbreviated like PCA, KNN, plus SVM and Gaussian distribution mechanism to automatically discover unique specimens utilizing affine variant instances. KNN, PCA Gaussian mechanism, SVM have generalization performance of 84, 92, 90, 87 ratios correspondingly.

Chaudhary et al. (2012) [10]: Through the usage of ESLPP indices as functionality for term recognition is indeed the report's vital contributor. Only a couple of relevant ESLPP constants have been tasked with the responsibility of differentiating the complex patterns the input layer receives. The usage of substantial ESLPP correlations as attributes has amplified this trait even more. Because it would be simplistic but unpredictable, MLP plus Bp neural network was handpicked as the estimator. Throughout all the databases comprises two maybe more resembling printable Gujarati words, a negligible exactness from more than 96% is acquired.

Patel et al. (2013) [11]: Many as 200 scholars who had no existing experience of Gujarati furnished instances for all Gujarati Latin alphabet. A flatbed scan was employed to examine a single bit of evidence at a rate of 200–300 dpi. The authors delineated hybrid selected features in this journal article. They employed tree filters, also; KNN filters at specific phases, and acquired a fewer efficacy of 63% by merging features.

Hassan et al. (2014) [12]: By continuing to learn integration of varying impacts, the thesis promotes a holistic paradigm for large cohort, binary pattern recognition hindrances. From an experimental and computational standpoint, the character categorization obstacle centers primarily on Gujarati plus Bangla terms recognition. The requisite elastic versus non-elastic bending invariant traits are included in a feature descriptor methodology for symbolic images.

Pareek et al. (2020) [13]: For the electronic or automatic acknowledgment of Gujarati scripts, these findings offer a supervised filter strategy relied on MLP plus CNN which had an efficacy of 97 and 65% correspondingly on massive instances of 10,000 photographs.

Prasad et al. (2009) [14]: With plenty of neurons in the framework's hidden nodes, investigators employed a two layers infrastructure that seemed to have feedforward with sigmoid resultant plus hidden neurons that can recognize sequences exceptionally adequately. Conjugate gradient backpropagation was employed to optimize the framework. With this methodology, the mean recognition proportion of 90–95% was acquired.

Goswami et al. (2011) [5]: The assessment of a subset of typed Gujarati words is handled utilizing a Personality Map relied on the KNN learner, and thus the results were conclusive. The strategy is often speedier than other functionality methodologies and can be employed for diverse applications because it does not use component identifying. Nevertheless, there's also a one-time expense dealing with SOM—Self-organizing maps training. Interestingly, it seemed that the tactic may be utilized to classify character customization collections even without fluctuation by just incorporating them in the validation information. On the test source, a mean efficacy of 83% was postulated.

Goswami et al. (2015) [15]: The objective of the project is to construct a reasonably-sized unconstrained handwriting Gujarati notation/numeral data file and identify it by utilizing minimal stroke aspects. Specimens were collected from individuals of varying ages, economic backgrounds, and occupational groups. Endpoints, curvy segments, plus lines are extracted using a novel tactic, and a histogram of low-level stroke features is used to recognize handwritten native integers from famous Indian scripting that are Gujarati plus Devanagari. Based on the KNN system, the investigations were better improvised till 98.46% using the statistical SVM learner framework with RBF—radial basis function kernel.

Goswami et al. (2016) [16]: This investigation's outcome demonstrates how an expedient strategy employed to extract minimal features relied upon strokes like terminals, connector positions, and lines and arc aspects from a thinning captured image could be utilized to characterize a subgroup of printable Gujarati signs. Concerning style, dimension variety, and toner thickness, an experimentation repository contains instances with a broad assortment of variants. The outcomes of the simulation delineate that the aspects are quite resistant to fluctuation in addition to providing a comparable higher efficacy of 96–99%.

Paneri et al. (2017) [17]: Investigators applied Histogram-a graph of Oriented Gradient-hog attributes and predictors like SVN plus KNN to demonstrate a penned Gujarati text assessment methodology in this publication. The measurements were carried out utilizing a repository of typed Gujarati names of each city of a reasonable length. Through employing HoG functionality as well as Classifier, the operation reached a significant precision of 86%.

Solanki et al. (2013) [18]: The Gujarati native words themselves were employed and tested as a sequence by the analysts in this inquiry. Hopfield structure of neural framework was utilized for classifying, and it had a 93.25 percentage detection performance.

Antani et al. (1999) [19]: Under this investigation, a cluster of Gujarati words with similar appearances was carefully selected and categorized utilizing various models. Photographs from sites and scanned documents of hardcopy Gujarati prose were utilized to craft the instance and images for testing for the characters. The Euclidean Shortest Distance and indeed the KNN filter techniques were employed with consistent and persistent aspects to distinguish them. The Hamming Distance descriptor was being utilized to characterize the unique words in the binary feature map. The recognition rates for these systems are demonstrated in the manuscript. A 67% detection value can be obtained.

Macwan et al. (2015) [20]: Investigators went over the HCR procedures for a variety of contexts, ranging from Gujarati, native Devanagari, and lastly Hindi. For the segmentation procedure, a revolutionary scheme is proposed. Unique strategies for Retrieval of attributes have been employed in their segmentation framework, with DFT, DWT, also DCT which are domains of transform surpassing about 90% efficacy. Also, Zernike instant and gradients components yielding 70 and 97% efficacy in the spatial domain. They also advocated incorporating three distinct methodologies, referred to as the freeman network coding, Hu's unchanging moments, as well as point of masses, which yielded a precision of 87.22% for Gujarati text.

Naik et al. (2017) [7]: Developers introduced a methodology for acknowledging electronic printed words for the Gujarati native language in this publication. To categorize strokes employing hybrid features, support vector network with all kernels, the K-nearest neighbor network having ranging figures of k, as well as multi-layer perceptron—MLP are practiced. A batch of 3000 entries was employed to train the network, which was then put to the test by 100 distinct composers. With the SVM-RBF kernel, they attained the maximum efficacy of 91% and the weakest efficacy of 87%. With the SVM Gaussian kernel, a mean functioning time of 0.05 s per stroking and a total peak time consumption of 1.0 s per stroking was reached.

These literary reviews from the past researchers added a superior aid to the field of native Gujarati characters or numerical recognition utilizing varying strategies from past till now.

3 Summarization—Tabular Form

This section delineates the summary of all reviews along with their crucial data in a table format (Table 1).

4 Challenges

Altogether, the impediments seen somewhere along the spans can be delineated as:

- Pixel intensities are compromised due to excessive processing.
- For greater reliability, a large mass is expected.
- A decent set of qualities is still not present
- Character/integer recognition is severely restricted to the boundary zone.
- It is strenuous to distinguish between similar-looking specifications.
- Letter font size is another impediment that can be bypassed by employing OCR since it works with a plethora of fonts.

These all shortcomings should be settled soon to acquire better utilitarian frameworks that can perceive any kind of provincial content with better demonstrating as an aid for humankind.

5 Future Tasks and Conclusion

An evaluation of OCR progress on the Gujarati scripting is introduced in this systematic review. The methodology of analyzing texts is briefly described. To date, alternative approaches for distinguishing Gujarati script letters, figures, and phrases have

Table 1 Reviews of Gujarati scripts detection are sequenced as below

Reference number	Models employed	Group	Efficacy
[3]	ANN	Numerals-Hand written	Accuracy = 81.66%
[8]	NN-YOLO	Letters-Printed	–
[2]	SVM, ANN, KNN	Numerals-Hand written	SVM Accuracy = 93%, ANN Accuracy = 85.6%, KNN Accuracy = 93.6%
[9]	PCA, KNN, Gaussian functionality, SVM	Numerals-Hand written	KNN Accuracy = 84%, PCA Accuracy = 92%, Gaussian mechanism = 90%, SVM = 87%
[10]	MLP with Backpropagation NN	Letters-Print	Accuracy = 96%
[11]	KNN, Tree filters	Letters-Hand written	Accuracy = 63%
[12]	SVM	Letters-Print	Range of accuracy = (97–98) %
[13]	MLP, CNN	Letters-written Hand	MLP Accuracy = 97%, CNN Accuracy = 67%
[14]	Neural Network (NN)	Letters-written Hand	Range of accuracy = (90–95) %
[5]	KNN, Self-organizing mapping	Letters-Print	Accuracy = 83%
[15]	KNN	Numerals-Hand written	Accuracy = 98.46%
[16]	KNN	Letters-Print	Range of accuracy = (96–99) %
[17]	SVN, KNN	Letters-Hand written	Accuracy = 86%
[18]	NN (Hopfield)	Letters-Print	Accuracy = 93.25%
[19]	KNN	Letters-Print	Accuracy = 67%
[20]	SVM	Letters-Hand written	Accuracy = 87.22%
[7]	SVM (RBF Kernels), KNN	Letters-Print	Maximal accuracy = 91%, Minimal accuracy = 87%

indeed been reported. The consequences of comparison among several methodologies illustrate that characters and numeric level identification are exact, while there are fewer shortcomings that should be addressed and rectified. There has been fairly minimal work addressed to Gujarati native language identification, and no piece of writing on Gujarati printed and handwritten word matching has been uncovered.

As a natural outcome, the investigation efforts for Gujarati language textual image retrieval are reasonable.

Growers' future models will surely require expanding and attempting to build a large disorderly collection with higher pixel intensities, which can then be sorted and sorely tested by methodologies that can distinguish letters or digits. Furthermore, other properties, such as transcending the size of the words or numerals, must be implemented for improved service.

References

1. Hassan, E., Chaudhury, S., & Gopal, M. (2014). Feature combination for binary pattern classification. *International Journal on Document Analysis and Recognition (IJDAR), 17*(4), 375–392.
2. Vyas, A. N., & Goswami, M. M. (2015). Classification of handwritten Gujarati numerals. In *2015 International Conference on Advances in Computing, Communications and Informatics (ICACCI)* (pp. 1231–1237). IEEE.
3. Desai, A. A. (2010). Gujarati handwritten numeral optical character reorganization through neural network. *Pattern Recognition, 43*(7), 2582–2589.
4. Kaur, H., & Kumar, M. (2018). A comprehensive survey on word recognition for non-Indic and Indic scripts. *Pattern Analysis and Applications, 21*(4), 897–929.
5. Goswami, M. M., Prajapati, H. B., & Dabhi, V. K. (2011). Classification of printed Gujarati characters using SOM based k-Nearest Neighbor Classifier. In *2011 International Conference on Image Information Processing* (pp. 1–5). IEEE.
6. Dalbir, S. K. S. (2015). Review of online & offline character recognition. *International Journal of Engineering And Computer Science,* 4(5).
7. Naik, V. A., & Desai, A. A. (2017). Online handwritten Gujarati character recognition using SVM, MLP, and K-NN. In *2017 8th International Conference on Computing, Communication and Networking Technologies (ICCCNT)* (pp. 1–6). IEEE.
8. Shirke, A., Gaonkar, N., Pandit, P., & Parab, K. (2021). Handwritten Gujarati script recognition. In *2021 7th International Conference on Advanced Computing and Communication Systems (ICACCS)* (Vol. 1, pp. 1174–1179). IEEE.
9. Baheti, M. J., & Kale, K. V. *Gujarati numeral recognition: Affine invariant moments approach.*
10. Chaudhary, M., Shikkenawis, G., Mitra, S. K., & Goswami, M. (2012). Similar looking Gujarati printed character recognition using locality preserving projection and artificial neural networks. In *2012 Third International Conference on Emerging Applications of Information Technology* (pp. 153–156). IEEE.
11. Patel, C., & Desai, A. (2013). Gujarati handwritten character recognition using hybrid method based on binary tree-classifier and K-nearest neighbour. *International Journal of Engineering Research & Technology (IJERT), 2*(6), 2337–2345.
12. Cybersurf, C. (1994). *Central Intelligence Agency* [Online] on October 4, 1994.
13. Pareek, J., Singhania, D., Kumari, R. R., & Purohit, S. (2020). Gujarati handwritten character recognition from text images. *Procedia Computer Science, 171,* 514–523.
14. Prasad, J. R., Kulkarni, U. V., & Prasad, R. S. (2009). Offline handwritten character recognition of Gujrati script using pattern matching. In *2009 3rd International Conference on Anticounterfeiting, Security, and Identification in Communication* (pp. 611–615). IEEE.
15. Goswami, M. M., & Mitra, S. K. (2015). Offline handwritten Gujarati numeral recognition using low-level strokes. *International Journal of Applied Pattern Recognition,* 2(4), 353379.
16. Goswami, M. M., & Mitra, S. K. (2016). Classification of printed Gujarati characters using low-level stroke features. *ACM Transactions on Asian and Low-Resource Language Information Processing (TALLIP), 15*(4), 1–26.

17. Paneri P. R., Narang, R., & Goswami, M. M. (2017). Offline handwritten gujarati word recognition. In *2017 Fourth International Conference on Image Information Processing (ICIIP)* (pp. 1–5). IEEE.
18. Solanki, P., & Bhatt, M. (2013). Printed Gujarati script OCR using hopfield neural network. *International Journal of Computer Applications, 69*(13).
19. Antani, S., & Agnihotri, L. (1999). Gujarati character recognition. In *Proceedings of the Fifth International Conference on Document Analysis and Recognition. ICDAR'99 (Cat. No. PR00318)* (pp. 418–421). IEEE.
20. Macwan, S. J., & Vyas, A. N. (2015). Classification of offline Gujarati handwritten characters. In *2015 International Conference on Advances in Computing, Communications and Informatics (ICACCI)* (pp. 1535–1541). IEEE.

An Ensemble Sentiment Classification on Multidomain Dataset

S. Madhusudhanan, N. M. Jyothi, A. Vishnukumar, S. Sathya, and N. Arun Vignesh

Abstract With the introduction of web 2.0, individuals now have a perfect platform to communicate their ideas, thoughts, and feelings. Web Opinion Mining / Sentiment Analysis is a text mining job aimed at developing a system that automatically extracts, recognizes, and categorizes user opinions from natural language text, user provided content, or user generated media. We have employed classifiers to analyse data sets from a wide variety of domains. The ensemble algorithm is applied to boost its performance. With an delicacy of 80.93%, the Logical Regression using Ensemble Classifier exceeds the others.

Keywords Text classification · Sentiment classification · Machine learning · Bagging

1 Introduction

Blogs, forums, and Internet community are allowing users to share their opinion on any issue. For example, express their dissatisfaction with a product they purchased, discuss current events, or express their political opinions [5]. This form of user data analysis is required for recommender systems and customization. "Everyone cares

S. Madhusudhanan (✉) · N. M. Jyothi
Department of Computer Science and Engineering, Koneru Lakshmaiah Education Foundation, Vaddeswaram, Guntur (DT), Andhrapradesh, India
e-mail: ssmadhu@kluniversity.in

A. Vishnukumar
Department of Information Technology, Vel Tech High Tech Dr.Rangarajan Dr.Sakunthala Engineering College, Vel Nagar, Avadi,Chennai, Tamilnadu, India

S. Sathya
Department of Electronics & Communication Engineering, Gojan School of Business and Technology, Edapalayam, Redhills, Chennai, Tamilnadu, India

N. Arun Vignesh
Department of Electronics and Communication Engineering, Gokaraju Rangaraju Institute of Engineering and Technology, Hyderabad, India

B. Pati et al. (eds.), *Proceedings of the 6th International Conference on Advance Computing and Intelligent Engineering*, Lecture Notes in Networks and Systems 428,
https://doi.org/10.1007/978-981-19-2225-1_29

about what other people think." A persuasion or a judgement about anything that has a substantial impact on a person's decision-making process is referred to as an opinion. The practise of determining people's feelings or views is known as sentiment analysis or opinion mining. Circumstances affect people's thoughts, feelings, and sentiments. For example, "The film was filled with fun elements" is a positive opinion about the movie, "he made fun with Ram's appearance" is a negative opinion [2, 6].

Sentiment analysis is an understanding of natural language exertion that attempts to categorize the texts predicated on their expressed sentiments on a decisive content [1, 4]. It is a technique for determining a speaker's or writer's attitude towards a topic or the overall impression of a text. Because of its multiple and potential uses, such as automatic product review classification, it has three types of categorization: document, phrase, and feature. At the document classification stage, the authors discuss the entire document and decide whether it is favourable or unfavourable. Sentiment categorization evaluates each sentence separately to determine its polarity at the phrase level. At the feature level, we may categorize sentiment in terms of many different features of things. Aspect level emotion categorization demands a more comprehensive analysis because most features are provided implicitly.

1.1 Motivation

Prior sentiment analysis research resulted in the development of a classification model for a specific domain. So, a multidomain sentiment analysis is offered in order to align the classifier across many domains simultaneously [9, 10]. The main contribution of the proposed work are as follows:

- We employed the TF-IDF feature selection with Chi square (χ^2) method for best feature selection.
- Sentiment classification based on selected feature are implemented with multidomain dataset.
- To increase classification performance, a classifier with bagging is employed.
- Sentiment classification using different regression analysis methods are also employed in this work.
- Finally, based on evaluation measures, the comparisions are made.

Section 2 discussed about the existing work, Sect. 3 implementation of proposed method, Sect. 4 contains result and discussion, and Sect. 5 about conclusion and future ideas.

2 Related Work

The authors [20] attempt to address the issue by developing a sentiment aware dictionary utilizing data from several domains. By using dictionary, they have assorted the target domain's unlabeled reviews. The task was performed in Hindi with a 76% accuracy rate. As an approach for cross-domain sentiment classification, the authors [3] propose a sentiment sensitive distributional thesaurus where sentiment sensitivity in thesaurus is achieved by including sentiment labels in the context vector at the document level. Using a benchmark data set that comprises Amazon user reviews for a variety of product categories, the proposed approach outperforms alternative baselines and achieves results equivalent to previously reported cross-domain sentiment classification algorithms.

The researchers [21] combined sentiment data from four different sources. Sentiment lexicons are a good place to start because they include sentiment polarity for widely used sentiment terms. Sentiment classifiers from diverse source domains are the second source. As the third source is to construct domain-specific sentiment correlations between words.

The target domain's tagged data is the fourth source. They offer a unified architecture for gathering all four forms of sentiment data and training a domain-specific sentiment classifier in the target domain.

The authors [13] suggested a fuzzy technique to describe the polarity learnt from training sets or from a training set. This newly acquired knowledge is combined with additional conceptual knowledge collected from two widely known sentiment analysis resources, SenticNet and the General Inquirer vocabulary. The advised strategy yielded the best results.

Researchers [11] developed a multidomain sentiment classification technique that reduced domain reliance whilst increasing overall performance. The suggested approach employs a way of combining several classifiers. They employ a method that involves training domain classifiers individually with domain-specific data before merging the classifiers to get the desired output.

The authors [8] investigated the capability of four different machine learning classification algorithms utilizing frequently used feature selection methods with three fame datasets are used to evaluate the proposed approaches: IMDb movie reviews, electronics reviews, and kitchen items reviews. The first step is to choose feature subsets from one of three available feature selection methods. After that, set theory concept such as union and intersection are appilied to evoke the top ranking features. The combined approach is achieved the greatest accuracy of 92.31% on the SMO classifier.

The authors [12] investigated the performance of a few feature selection techniques for sentiment analysis. A feature extraction method called as Term Frequency Inverse Document Frequency is utilized to create feature vocabulary. To choose the best set of word vectors, a variety of Feature Selection techniques are employed. Machine learning classifiers are used to teach the required attributes. To improve sentiment analysis performance, classifiers use bagging and random subspace. They proved the effectiveness of feature selection strategies trained with ensemble classifiers outperform neural networks with far less training time and parameters, removing the requirement for hyper-parameter modification.

3 Proposed Approach

Sentiment Classification steps:

Step1: In this work, a multidomain product review database1(Books, DVD, Electronics, Kitchen and Housewares) is collected and used to tackle this problem.

Step2: **Preprocessing** methodology is required to eliminate noisy, inconsistent, and incomplete data by considering tokenization, stop words removal, and stemming approaches.

Step3: **Feature Extraction and Selection**: To begin, we utilize the TF-IDF and Bag of Words (BoW) methods to generate a feature vector in a document. it will receive a score of 1 if it is present, otherwise the score is 0. Following that the CHI feature selection approach is used to pick distinct feature subsets.

Step4: **Classification**: To train the given feature, Support Vector Machines (SVM-RBF), SVM-(Linear Kernel), MNB, Decision Tree (DT), and Linear Regression (LR) are employed.

Step5: **Bagging**: Finally, the ensemble process helps to escalate the classification accurarcy.

3.1 *Methodology*

The proposed architecture is depicted in Fig. 1, with further information on each preliminary function provided in the subsections that follow. The sentiment categorization job in this study is performed using multidomain product review data. In the dataset, there are 1000 positive and 1000 negative tagged reviews in each domain [14, 15] are chosen for implementation. Table 1 summarizes the statistics for this data set.

3.1.1 Preprocessing Task

Tokenization: The tokenizer may separate review into distinct tokens like as words, numbers, special characters, and so forth, making it ready for further processing. **Stop word Removal**: To improve the effectiveness of the feature selection strategy,

Table 1 Multidomain sentiment dataset

Dataset	Total reviews	Positive review	Negative review
Books	2000	1000	1000
DVD	2000	1000	1000
Electronics	2000	1000	1000
Kitchen and housewares	2000	1000	1000

this stage entails deleting commonly used stop words such as prepositions, unnecessary words, special characters, ASCII code, new lines, and excessive white spaces. **Stemming**: it entails transforming each token to its stem, or root, form.

3.1.2 Feature Extraction and Selection

It plays vital role to increase the sentiment categorization process accuracy. TF-IDF and BoW are used to obtain its traits. TF-IDF is known technique to deal the text into feature vocabulary. TF-IDFs are calculated using Eqs. 1, 2 and 3 [7, 16].

$$\text{TF}(t) = \frac{(\text{Number of times term } t \text{ appears in a document})}{(\text{Total number of terms in the document})} \tag{1}$$

$$\text{IDF(t)} = \log_e \frac{\text{Total amount documents}}{\text{Number documents with term } t \text{ in it}} \tag{2}$$

To find the TF-IDF score:

$$\text{TF-IDF}(w) = \text{TF}(w) \times \text{IDF}(w) \tag{3}$$

By calculating the frequency of the entire document, BoW translates text input into numeric numbers. By ignoring word order and focusing on word frequency, it generates feature vocabulary across all pages. Selecting the appropriate feature from the feature lexicon is an important task [13]. A typical statistical test for detecting the relation between a term and the linked class is the Chi square $\left(\chi^2\right)$ statistic. If there is no association between the feature set and class, then it is said to be a null hypothesis. The χ^2 value is calculated by using Eq. (4) [6, 17].

$$\chi^2 = \frac{\sum(\text{observed value} - \text{Expected value})^2}{\text{Expected value}} \tag{4}$$

3.2 *Classification*

3.2.1 Multinomial Naive Bayes (MNB)

It is a popular method for classifying documents based on statistical analysis of their contents. To classify the documents by assessing the probability that a document belongs to the same class as on the same topic. Vectors $\theta_y = \left(\theta_{y1}, \ldots, \theta_{yn}\right)$ are distribution parameters, and θ_{yi} is the probability $P(x_i|y)$ of feature i appearing on the same class y [14]. The parameter θ_y is estimated by Eq. (5),

$$\hat{\theta}_{\text{yi}} = \frac{N_{yi} + \alpha}{N_y + \alpha n} \tag{5}$$

where $N_i = \sum_{\{x \in T\}} xi$ is the number of times feature i appears in a class y in the training set T, and $N_y = \sum_{i=1}^{n} N_{yi}$ is total count of all features for class y and $\alpha \geq 0$.

3.2.2 Support Vector Machine (SVM)

SVM converts the data points into a higher-dimensional space, allowing them to be separated linearly. By determining the optimum hyperplane for dividing the group of data. The main aim is to shorten the distance amongst all data group and the hyperplane. The hyperplanes that should be used are specified by kernel functions. The linear kernel is used if the data can be linearly separated. The Radial Basis Function (RBF) kernel is used for non-linear data. The training data is labelled with (x_i, y_i), $i = 1, 2, 3$. where $x_i \in R^n$ and $y \in \{1, -1\}^l$. The SVM optimization problem can be solved in the following Eq. 6:

$$\min_{w,b,\zeta} \frac{1}{2} W^T W + C \sum_{i}^{i=1} \zeta_i \text{ subject to } y\_i(w^T \phi(x_i) + \text{b} \geq 1 - \zeta_i,$$
$$\zeta_i \geq 0, i = 1, 2, 3, \ldots, n \tag{6}$$

3.2.3 Logistic Regression (LR)

A supervised classifier, the logistic regression model is a rule set deal with multiclass problems. The logistic function determines the relation between two class labels. If the likelihood is greater than 0.5, label "1" is assigned; otherwise, label "0" is assigned [16, 17]. It operates by reducing the loss function by determining the optimal set of weight parameters.

3.2.4 Decision Tree (DT)

The Decision Tree is a binary tree with conditions at the core and class labels at the decedent nodes. Attribute selection method uses information gain or Gini index method to determine the value of each attribute. The highest information gain of attribute is chosen since it gives the most information as a split node [18, 19]. This process is repeated until the last node.

3.3 *Bagging Ensemble Techniques*

It is a bootstrap ensemble that uses re-placement to construct subsets of data from the original data. On each data subset, base classifiers are trained, and the individual predictions are concatenated to get the final prediction [12]. By training many weak classifiers on subsets of the original data, bagging enhances the classifier performance.

4 Experiments

The various classification algorithms is examined in this section. The experiments employed the Multidomain Product Review Dataset [15]. In the Review Dataset, TF-IDF and BoW models are used for feature extraction, and Chi square feature selection models are used to choose top features. Logistic Regression (LR), Support Vector Machines with RBF, Linear Kernels and Grid Search, Decision Tree, Multinomial Naive Bayes (MNB) and Random Forest (RF) are used to train selected features. These classifiers are subsequently trained using bagging techniques.

As performance measures for the classifiers mentioned above, accuracy and F-score are used. Equation 7 depicts the accuracy metric, which is defined as the ratio of correctly predicted numbers to total predicted numbers. The model is perfect if the F-score number is 1. The F-score is calculated using the Eqs. 8, 9 and 10.

$$\text{Accuracy} = \frac{\text{Number of Correctly Predicted}}{\text{Total Number of Predicted}} \tag{7}$$

$$\text{Precision} = \frac{\text{TP}}{\text{TP} + \text{FP}}e \tag{8}$$

where TP is True Positive and FP is False Positive.

$$\text{recall} = \frac{\text{TP}}{\text{TP} + \text{FN}} \tag{9}$$

where TP is True Positive, and FN is False Negative.

$$F1 = 2 * \frac{\text{precision*recall}}{\text{precision} + \text{recall}} \tag{10}$$

5 Result and Discussion

The experiment results depict how each classifier responds to feature selection. The proposed technique's efficacy is shown in Tables 2 and 3. Tables 4 and 5 represent the F1-score of the classifiers.

From the above tables, it shows that LR technique provides better accuracy than other base classifiers. In Bagging, LR classifiers trained on BoW with Chi square feature selection greatly outperformed the other basic classifiers, as shown in Table3. On multidomain datasets, the classifiers SVM (RBF) and DT achieved the same results for TF-IDF feature extraction and Chi square feature selection regardless of any classification methods.

Table 2 Base classifiers accuracy

Feature selection	Classifiers				
	LR	SVM	SVM (RBF)	DT	MNB
TF-IDF	76.5	75.5	51.37	68.25	76.87
TF-IDF + CHI	**80.06**	80.08	51.37	68.25	79.81
BoW	77.93	75.5	51.37	68.93	77.12
BoW + CHI	80.06	77.18	51.43	68.93	78.87

Table 3 Ensemble classifiers accuracy

Feature Selection	Bagging + Classifiers					
	LR	SVM	SVM (RBF)	DT	MNB	KNN
TF-IDF	79	78.31	51.37	75.25	75.93	51.56
TF-IDF + CHI	79	79.03	51.37	75.25	75.93	51.56
BoW	79.17	77.06	51.37	73.37	76.31	55.68
BoW + CHI	**80.93**	80.06	51.43	73.37	78.18	60.12

Table 4 F1-score of base classifier

Feature selection	Classifiers					
	LR	SVM	SVM (RBF)	DT	MNB	KNN
TF-IDF	76.80	75.77	67.28	68.25	76.84	66.77
TF-IDF + CHI	76.80	75.77	67.28	68.25	76.84	66.77
BoW	78.06	75.65	67.28	69.82	77.29	67.68
BoW + CHI	80.17	77.39	67.31	69.82	79.05	67.46

Table 5 F1-score of ensemble classifiers

Feature selection	Bagging + Classifiers					
	LR	SVM	SVM (RBF)	DT	MNB	KNN
TF-IDF	79.02	75.77	67.28	68.25	76.84	67.11
TF-IDF + CHI	79.02	75.77	67.28	68.25	76.84	67.11
BoW	78.06	77.33	67.28	73.89	76.38	65.83
BoW + CHI	80.17	80.12	67.31	73.92	78.62	68.38

6 Conclusion

Sentiment analysis on a multidomain dataset is performed using basic classifiers and ensemble classifiers. According to the experiment results, the LR algorithm outperformed the TF-IDF and BoW models, along with the Chi square feature selection technique, the accuracy of weak classifiers is also improved by using ensemble classifiers. The performance of neural network methods on multidomain datasets will be studied in the future. Other ensemble methods will be investigated in the future, and its performance will also be compared.

References

1. Anjali Devi, S., Sapkota, P., Rohit Kumar, K., Pooja, S., & Sandeep, M. S. (2020). Comparison of classification algorithms on twitter data using sentiment analysis. *International Journal of Advanced Trends in Computer Science and Engineering, 9*(5), 8170–8173.
2. Pang, B., & Lee, L. (2004). A sentimental education: Sentiment analysis using subjectivity summarization based on minimum cuts. In *Proceedings of 42th Annual Meeting on Association for Computational Linguistics (ACL)* (pp. 271–278).
3. Bollegala, D., Weir, D., & Carroll, J. (2013). Cross-domain sentiment classification using a sentiment sensitive thesaurus. *IEEE Transactions on Knowledge and Data Engineering, 25*(8), 1719–1731. ISSN 1041-4347.
4. Pang, b., Lee, L., & Vaithyanathan, S. (2002). Thumbs up? Sentiment classification using machine learning techniques. *Proceedings of EMNLP*.
5. Liu, B., Hu, M., & Cheng, J. (2005). Opinion observer: analyzing and comparing opinions on the web. In *Proceedings of the 14th International Conference on World Wide Web* (pp. 342–351).
6. Bulusu, A., & Sucharita, V. (2019). Research on machine learning techniques for POS tagging in NLP. *International Journal of Recent Technology and Engineering, 8*(1, Special Issue- 4).
7. Manning, C. D., *Introduction to Information Retrieval by Christ Manning*.
8. Ghosh, M., & Sanyal, G. (2018). An ensemble approach to stabilize the features for multi-domain sentiment analysis using supervised machine learning. *Journal of Big Data 5*, 44. https://doi.org/10.1186/s40537-018-0152-5
9. Kotturu, P., Sasank, V. V. S., Supriya, G., Manoj, C. S., & Maheshwarredy, M. V. (2019). Prediction of chronic kidney disease using machine learning techniques. *International Journal of Advanced Science and Technology, 28*(16).
10. Chakraborty, K., Bhattacharyya, S., & Bag, R. (2020). A survey of sentiment analysis from social media data. *IEEE Transactions on Computational Social Systems, 7*(2).

11. Li, S. S., Huang, C. R., & Zong, C. Q. (2011). Multi-domain sentiment classification with classifier combination. *Journal of Computer Science and Technology, 26*, 25–33. https://doi.org/10.1007/s11390-011-9412-y
12. Madasu, A., & Elango, S. (2020). Efficient feature selection techniques for sentiment analysis. *Multimedia Tools Applications, 79*, 6313–6335. https://doi.org/10.1007/s11042-019-08409-z
13. Magoni, M., & Petrucci, G. (2017). A fuzzy-based strategy for multi-domain sentiment analysis. *International Journal of Approximate Reasoning, 93*, 59–73.
14. Nafis, N. S. M., & Awang, S. (2021). Enhanced hybrid feature selection technique using term frequency-ınverse document frequency and support vector machine-recursive feature elimination for sentiment classification. *IEEE Access, 9*, 52171–52192.
15. Ni, J., Li, J., & McAuley, J. (2019). Justifying recommendations using distantly-labeled reviews and fine-grained aspects. In *Proceedings of Conference on Empirical Methods in Natural Language Processing and the 9th International Joint Conference on Natural Language Processing (EMNLP-IJCNLP)* (pp. 188–197).
16. Hou, Q., Han, M., & Cai, Z. (2020). Survey on data analysis in social media: A practical application aspect. *Big Data Mining and Analytics, 3*(4)
17. Sajana, T., & Narasingarao, M. R. (2018). Classification of ımbalanced malaria disease using Naïve Bayesian Algorithm. *International Journal of Engineering & Technology, 7*(2.7), 786–790.
18. Sanjay Bhargav, P., Nagarjuna Reddy, G., Ravi Chand, R. V., Pujitha, K., & Mathur, A. (2019). Sentiment analysis for hotel rating using machine learning algorithms. *International Journal of Innovative Technology and Exploring Engineering, 8*(6), 1225–1228.
19. Srinivas, P. V. V. S., Pavan Sai Sujith, L. V. N., Sarvani, P.M., Kumar, D. S., & Parasa, D. (2019). Prediction of hospital admission using machine learning. *International Journal of Scientific and Technology Research, 8*(12).
20. Jha, V., Savitha, R., Shenoy, P. D., Venugopal, K. R., & Sangaiah, A. K. (2018). A novel sentiment aware dictionary for multi-domain sentiment classification. *Computers & Electrical Engineering, 69*, 585–597.
21. Wu, F., Huang, Y., & Yuan, Z. (2017). Domain-specific sentiment classification via fusing sentiment knowledge from multiple sources. *Information Fusion, 35*, 26–37. ISSN 1566–2535. https://doi.org/10.1016/j.inffus.2016.09.001

Empirical Analysis of Preprocessing Techniques for Imbalanced Dataset Using Logistic Regression

M. Revathi and D. Ramyachitra

Abstract This paper attempts to examine the performance of preprocessing strategies with logistic regression classifier. The goal of this paper is to see if there is a feasible and efficient strategy to enhance the performance of classification techniques on imbalanced datasets for different training dataset percentages. The experiments were conducted on Cleveland dataset—binary class. Several data preprocessing methods like Smote, Borderline-Smote, and ADAYSN were applied to data in order to classify various training dataset percentages. It was necessary to ascertain how the training dataset percentage affected the final classification for preprocessing methods. The experimental results explained that the ratio of 70–30 datasets performed better or better than other ratios when on train and test datasets, respectively. It was found from experimental results that the algorithms gave better accuracy when the training to testing ratio was 70:30 compared to other ratios.

Keywords Classification · Imbalanced datasets · Logistic regression · Smote

1 Introduction

Class imbalance refers [1] to classification issues in which certain classes have many more instances than others. In a two-class situation, the majority class of imbalanced dataset has more instances than the minority class. The dataset imbalance problem[2] is found in a wide range of real-time applications, including telecommunications, text classification, and image retrieval, and is regarded one of the most pressing data mining issues today[3].

Class imbalance typically affects the effectiveness of standard classifiers, especially in terms of the minority class of interest, because most standard classification

M. Revathi (✉) · D. Ramyachitra
Department of Computer Science, Bharathiar University, Coimbatore 641046, India
e-mail: revathiphdcs@gmail.com

D. Ramyachitra
e-mail: ramyachitra@buc.edu.in

B. Pati et al. (eds.), *Proceedings of the 6th International Conference on Advance Computing and Intelligent Engineering*, Lecture Notes in Networks and Systems 428,
https://doi.org/10.1007/978-981-19-2225-1_30

algorithms have a bias towards the dominant class [4]. Many solutions to the problem identification from unbalanced datasets have been presented, which can be grouped into three types [5].

They are as follows:

Datasampling: In this method, training cases are changed to achieve further balanced class distribution, allowing to implement similarly to traditional classification.

Algorithmic modification: This approach aims to adjust basic learning algorithms to be more sensitive to issues of class imbalance.

Cost-sensitive learning: This technique integrated approaches at the algorithmic level, data level, or both levels at the same time, taking into account costs for misclassification of true class instances when compared to false class of datasets, and attempting to minimise the cost errors.

This work focuses mostly on preprocessing approaches on dataset, as preprocessing may be done independently of any classifier. This research evaluates and contrasts many conventional methodology and hybridization of two data preprocessing methods that outperform other methods using ensemble classifiers. In this paper, the results are compared with various percentages of training datasets on Cleveland-0_vs_4.

Oversampling, undersampling, and hybrid solutions are three different types of data preparation strategies that use resampling techniques to remaining the number of cases in majority and minority in imbalance dataset.

- **Undersampling**—a subset of the original dataset is formed, and certain examples from the majority class must be eliminated.
- **Oversampling**—new examples are generated, especially based on minority class.
- **Hybrid**—combination of the two conventional methods [6].

In this paper, various sizes of training dataset classification results for logistic regression classifier [7] are compared. The accuracy of classification is also affected by database preprocessing and model validation approaches. Statistical external validation divides the entire dataset into training and testing sets at random, allowing the classifier's quality to be checked.

To deal with the analysis of various types of data and the building of prediction models, data mining uses methodological and technical methods. Data mining methods and applications have ramifications in a variety of fields, including health informatics, patient care, monitoring systems, and so on. Various data mining algorithms have greatly aided in better understanding industrial, financial, and medical data, as well as decision-making, visualisation, and detection of hidden complicated linkages from various datasets. As a result, even with a tiny amount of data, the data mining process should be able to predict accurately. There are many factors which have an impact on the quality of data. Three such data quality factors are missing values, data noise, and imbalanced classes.

The following sections describe the structure of the paper. Section 2 discusses the associated work. The test findings are presented and examined in Sect. 3, and in the Sect. 4 conclusions and future scope are dealt with.

2 Related Work

For most academic technology enterprises, classifying the imbalance dataset is a difficult undertaking. Logistic regression is a statistical technique that is related to linear regression and also includes a dependent variable. The logistic regression is used to establish a correlation between a categorical definite variable and one or more in definite factors. The logit function is used to calculate the probability. A cumulative logistic distribution is used to compute the probability. Instead of utilising a Gaussian distribution, you can use a non-Gaussian distribution. Logistic regression employs a Bernoulli distribution, followed by linear regression [8].

2.1 Smote

Chawla et al. (2002) proposed synthetic minority over-sampling techniques (SMOTE) [9], in which oversampling strategy employs an iterative exploration and computing approach to generate synthetic samples in minority classes. The minority group's observations will be repeated until the required number has been reached. To create new samples, a user-defined threshold is utilised to select nearest k neighbours.

Then, at random, one of the closest k neighbours is chosen, synthesised, and the process is repeated until all of the required synthetic samples generated. If 30% is used, for example, each examination in the minority group will be awarded 3 knn values. One will be applied in random value, and a new class $x_n + 1$ created based on the characteristics of the random class value x_j and original classes x_i examination is used to generate it.

As shown in Eq. 1, new features created by multiplying attributes in difference of x_i and x_j in a random class value r in betwixt one and zero (1).

$$x_{n+1} = x_{i+r} \cdot (x_j - x_i), \quad 0 \le r \le 1 \tag{1}$$

There are certain disadvantages to this strategy, such as producing noise samples and identifying samples within class overlapping zones. When used on larger datasets, this method has downsides, such as performance degradation. The SMOTE pseudocode is provided below.

Algorithm SMOTE:

Choose N and k, which indicates the number of synthetic notes and the number of nearest neighbours, respectively, then

1. Let x_i, i = 1,...., ns, denote notes that belong to the minority class, and let A denote the set of all xi, such that A belongs to xi.
2. Determine the k-nearest neighbours of xi by computing the Euclidean distance between x_i and all other elements of A.
3. Define S_{ik} as the set of x_i's k-nearest neighbours.

4. Using S_{ik} with replacement, generate N synthetic observations designated x_{ij} and $j = 1, 2 \ldots N$.
5. Let Y represent a number in the range [0,1]. For a specified xij, draw a Y uniformly and then create a synthetic observation by the formula $x_k = x_i + Y$).
6. Execute Step 5 for every x_{ij}.

End of Pseudocode.

2.2 Borderline-Smote

Han et al. 2005 proposed Borderline-Smote [10] method and presented techniques to identify samples that might be misclassified in imbalanced datasets. The proposed system claimed at the time that any given input dataset had three zones: border line, noise, and safe. Given a number of false example g with k nearest neighbours, the region definitions are shown in Table 1.

The Cleveland training dataset considered to X, where $X = (x_i, y_i)$, $x_i \in R_n$, and $y_i \in \{0, 1\}$ is made up of a minority group Y with $y_i = 1$ and a majority group M with $y_I = 0$. The Borderline-SMOTE approach picks just those samples in Y that fall under the border defined in Table 1. The danger group member function, DANGER (Y_i) is picked among those who have more majority group neighbours.

Minority members in the danger group are at risk of being misclassified and are thus used as oversampling candidates to improve the instances closest to the decision boundary. In the first version, only the nearest neighbours discovered in the first stage, the minority group Y, generate new synthetic classes; however, in second stage, the entire applied dataset X generates proposed samples in each number of complicated groups. When total training dataset X is used to produce new classes, also nearest neighbour in M, the generated random classes r defined in Eq. (3) and r value constrained to the $0 \le r \le 0.5$, otherwise r constrained in $0 \le r \le 1$.

The Borderline-Smote algorithm in pseudocode is given below.

1. In the training set, evaluate T's m closest neighbours. For every p in the minority class P, the total number in majority classes in overall m nearest neighbours derived in the m'($0 \le m' \le m$).
2. All of p_i's m nearest neighbours are majority classes, and p_i defined as noise. P_i misclassified and placed a set DANGER if the total values of majority nearest neighbours classes are > total number of minority classes.

Table 1 Region definition

Region	Definition
Safe	$0 \le g < ½\, k$
Borderline	$1/2\, k \le g < k$
Noise	$g = k$

3. In DANGER, the minority class P's borderline statistics are given as illustrations.
   DANGER = $\{p'_1, p'_2, \ldots, p'_{dnum}\}$, DANGER = $0 \leq d_{num} \leq p_{num}$
4. Generate d_{num} samples × synthetic true class values from the dataset in DANGER calculation

$$S_j = P_i + r_{j*}(x_j - x_i), \quad 1 \leq j \leq s$$

End of Pseudocode.

2.3 *ADASYN*

The concept behind ADASYN is to adaptively manufacture minority data samples depending on their distributions: harder to understand minority classes in datasets to generate further synthetic data to make it easier to understand minority classes in calculated datasets. The adaptively reallocated decision boundary in the ADASYN approach turns the focus away from the learning bias induced by the imbalanced data split and toward classes that are the most complicated to analyse [1]. The ADASYN algorithm is depicted in pseudocode below.

ADASYN Algorithm

Choose k and b, which represent the nearest neighbours total counts and the desired level of class balance after creating the synthetic data, respectively. Then,

1. Let n_i represent the number of examinations from the majority values, and ns represent the total counts of examinations from minority values. Determine C = $(n_i - n_s)$*B.
2. Let x_i, i = 1,..., n_s, denote the examination belonging to the minority class, and let ns denoted set of all x_i, such that A belong to x_i. For every x_i,
3. Calculate the distance in Euclidean space between xi and all other elements.
4. Let S_{ik} denote the set of the k-nearest neighbours of x_i.
5. Define Yi as the number of examinations; find the majority samples in datasets in the k-nearest neighbours' region of xi. Determine the ratio Ei, where $E_i = Y_i$ /k, i = 1,..., ns.
6. Normalise E_i according to $\hat{E}_i = E_i / \sum_{nsi} = 1$, so that Ê is a probability ($\sum_i \hat{E} =$ 1).
7. Compute $S_i = \hat{E} \times S$, which is the total counts in synthetic examples for produced for each x_i sample.
8. Randomly sample Si synthetic observations denoted x_{ij}, ($j = 1, \ldots, g_i$) from Sik with replacement.
9. Let Y be a number in the range [0,1]. For a given x_{ij}, generate a synthetic observation according to $x_k = x_i + Y\ (x_i - {}_{xi})$.

This section summarises the three preprocessing techniques, namely ADASYN, B-SMOTE, and SMOTE. Preprocessing is used to segregate the majority and minority classes in the dataset, which is then saved in a database. The stored dataset is taught to analyse the performance of preprocessing procedures using the same learning algorithms.

3 Performance Measures

Generally in confusion matrix, there are four types of outputs. They are as follows:

The TP, TN, FP, and FN stand for true positives (TP), false negatives (FN), true negatives (TN), and false positives (FP), respectively (FN). The majority class is known as 'negative' (FN), and the minority class is known as 'positive' (FP).

These four values provide a more thorough examination and evaluation. These are utilised to analyse the efficacy of all classifiers in classifying the six datasets previously described.

True positive refers to the total count of positive cases positively classified as positives by a classifier (TP).

- The number of false examples accurately categorised as false by a classifier is known as true negative (TN).
- False positive (FP), often known as false, is the number of negative examples that a classifier wrongly classifies as positives.
- False negative (FN), also known as miss, is the number of positive cases that a classifier wrongly classifies as negatives. The following are the performance indicators for the confusion matrix: The term sensitivity, also known as true positive rate or recall, is defined as

$$\text{Sensitivity} = \frac{\text{TP}}{(\text{TP} + \text{FN})} \tag{2}$$

It correctly recognises the positive class. It ranges from 0 to 1, with 1 being the highest possible score.

- The actual negative rate, also known as specificity, is calculated as

$$\text{Specificity} = \frac{\text{TN}}{(\text{TN} + \text{FP})} \tag{3}$$

- It correctly identifies false class as such, the perfect score is 1 and 0 is the worst measure.
- Accuracy is a percentage of true outcomes in the population (both true positives and true negatives).

$$\text{Accuracy} = \frac{(\text{TP} + \text{TN})}{(\text{TP} + \text{FN} + \text{FP} + \text{TN})} \tag{4}$$

4 Empirical Analysis and Comparison

The experimental results of categorising the Cleveland-0 vs 4 dataset with Logistic regression classifiers for various sizes of training datasets are presented in this section. By evaluating varying percentages of the training dataset on the Cleveland-0 vs 4 datasets, the same experiments were repeated. The tests are run on Windows XP with an Intel Core i5 7th Gen CPU T6400 2.00 GHz with Python programming. The Cleveland-0 vs 4 dataset was obtained from the UCI repository.

The Cleveland-0 vs 4 datasets information is presented in Table 2. Maj. and Min. represent the total count of samples in the minority, majority classes, and Att. represents the number of attributes. By using the Cleveland-0 versus 4 dataset, imbalance ratio (IR) is ratio for number of dataset samples in the majority samples to total values of minority samples. The Cleveland-0 vs 4 dataset contains 177 instances and 13 characteristics.

In this section, we looked at how pre-processing approaches performed on an unbalanced dataset. The x-axis defines the parameter value of the training dataset percentage, while the y-axis defines the accuracy. Smote, Borderline, and Smote were compared as preprocessing approaches. Smote, Borderline-Smote and ADASYN were compared as pre-processing approaches for varied training dataset sizes. All accuracy is found to be 70%, which is significantly higher than the other percentages. This suggests that the ideal training and testing dataset ratio for classification is 70:30. The accuracy of the preprocessing techniques is shown in the Fig. 1.

Table 2 Cleveland-0_vs_4 datasets from UCI

Dataset	Att	Features	IR	Maj	Min
Cleveland-0_vs_4	177	13	12.62	164	13

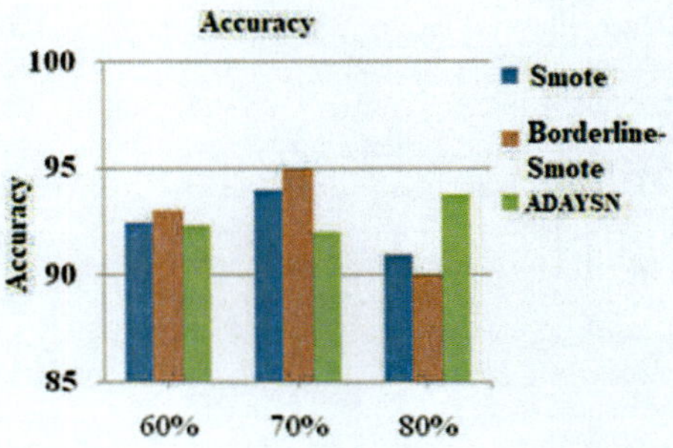

Fig. 1 Illustrate the accuracy for Cleveland-0_vs_4 dataset with the logistic regression classifiers

Fig. 2 Illustrate the for Cleveland-0_vs_4 dataset with the logistic regression classifiers

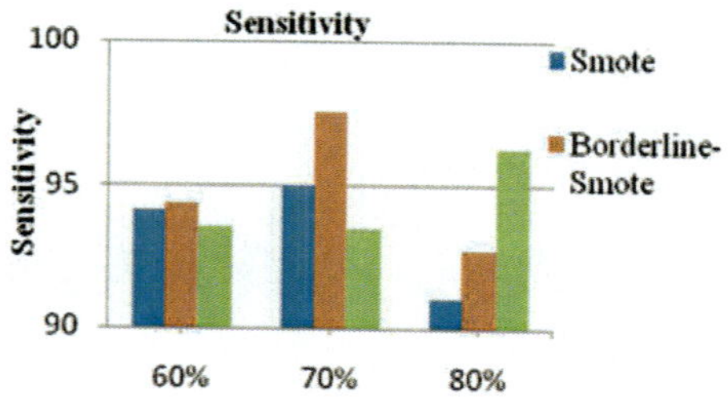

The *x*-axis defines the parameter value of the training dataset as a percentage, while the *y*-axis defines the sensitivity. For various sizes of training datasets, preprocessed approaches Smote, Borderline-Smote, and ADAYSN were compared. All accuracy was found to be 70%, which is significantly higher than the other percentages. Sensitivity for the preprocessing techniques is shown in Fig. 2.

The *x*-axis defines the parameter value of the training dataset %, while the *y*-axis defines the specificity. Smote, Borderline, and Smote were compared as preprocessing approaches. Smote, ADAYSN for varied training dataset sizes. Specificity for the preprocessing techniques is shown in Fig. 3.

When partitioning a dataset into 7:3 for classification, the accuracy, sensitivity, and specificity are shown in Fig. 4. All of the measurements come in at 70%, which is significantly higher than the other percentages.

Fig. 3 Illustrate the specificity for Cleveland-0_vs_4 dataset with the logistic regression classifiers

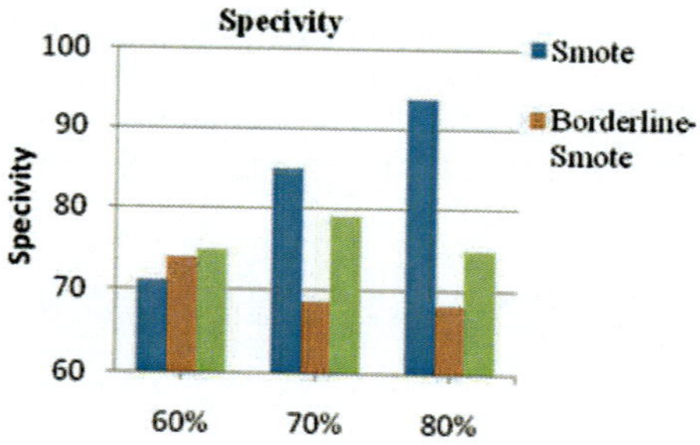

Fig.4 Illustrate the accuracy, sensitivity, and specificity for Cleveland-0_vs_4 dataset for 70% of the training set

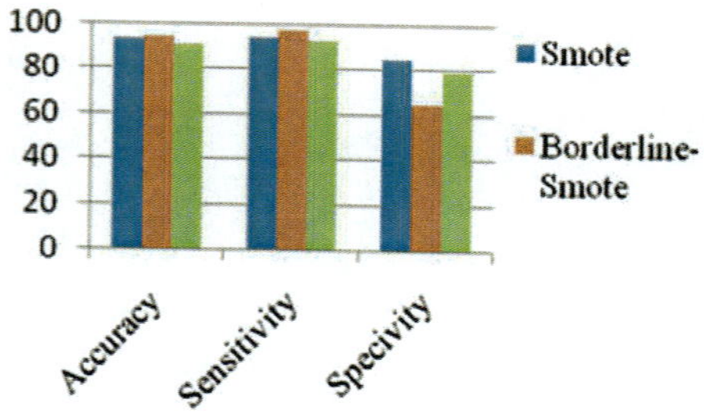

5 Conclusion

In this paper, various experiments are conducted at Cleveland UCI imbalanced datasets by applying logistic regression classifier. The data classification is considerably impacted by training dataset ratio. In reality, not all of the traits have a major impact on categorization accuracy. The trials show that splitting a dataset into a 70:30 ratio beats logistic regression when it comes to correctly categorising the minority cases for the Cleveland datasets (or the majority instances). By altering the amount of data in the training dataset, researchers were able to enhance the accuracy rate of minority events. Other classifiers, which are a common remedy to class imbalance, should be examined in future research. The future work entails using a more detailed statistical data for better data preprocessing on multi-class imbalance.

References

1. He, H., & Garcia, E. (2009). Learning from imbalanced data. *IEEE Transactions on Knowledge and Data Engineering, 21*, 1263–1284.
2. Chawla, N. V., Japkowicz, N., & Kotcz, A. (2004). Editorial: Special issue on learning from imbalanced data sets. *SIGKDD Explorations, 6*(1), 1–6.
3. Yang, Q., & Wu, X. (2006).10 challenging problems in data mining research. *International Journal of Information Technology and Decision Making, 5*(4).
4. Chawla, N., et al. (2004). Special issues on learning from imbalanced data sets. *ACM SigKDD Explorations Newsletter, 6*(1), 1–6. https://doi.org/10.1145/1007730.1007733
5. He, H., Bai, Y., Garcia, E. A., & Li, S.: ADASYN: adaptive synthetic sampling approach for imbalanced learning. In *Proceedings of the 2008 IEEE International Joint Conference Neural Networks (IJCNN'08)* (pp. 1322–1328), Hong Kong.
6. Mao, W., Jiang, M., Wang, J., et al. (2017). Online extreme learning machine with hybrid sampling strategy for sequential imbalanced data. *Cognitive Computation, 9*, 780–800. https://doi.org/10.1007/s12559-017-9504-2
7. Hosmer, D. W., & Lemeshow, S., *Introduction to the logistic regression model.*
8. Fernández, A., García, S., Galar, M., Prati, R. C., Krawczyk, B., & Herrera, F. (2018). *Learning from imbalanced data sets.* https://doi.org/10.1007/978-3-319-98074-4
9. Chawla, N. V., Bowyer, K. W., Hall, L. O., & Philip, W. (2002). SMOTE: Synthetic minority over-sampling technique. *The Journal of Artificial Intelligence Research, 16*, 321–357.
10. Han, H., Wang, W., & Mao, B. (2005). Borderline-SMOTE: a new over-sampling method in imbalanced data sets learning. In *Proceedings of the International Conference on Intelligent Computing* (pp. 878–887).

A Survey on Chatbot in Devanagari Language

Deepak Mane, Mayur Patil, Vrushali Chaudhari, Rutuja Nayakavadi, and Sanjana Pandhe

Abstract Chatbots is the current trending topic in machine learning. It processes the user's queries and gives appropriate response. Chatbots are kind of virtual assistant where we talk with a computer bot not with a real human being. But it feels like we are taking with a real human being. Chatbots are being used heavily in various sectors like in banking, ticket reservations, customer enquiry desks, etc. Chatbots are helping businesses to give quick responses to the user queries, and most importantly, it is saving both time and resources of the businesses and customer. In future, chatbot will play major role for the communication between business and consumers for sure. Most of the chatbots are developed in English language by using various frameworks. This paper explains various techniques and architectures which can be used for developing a real-time chatbot.

Keywords Chatbot · NLP · Deep neural network · Devanagari chatbots · Machine learning

1 Introduction

A chatbot is a computer software that interacts with humans in natural language as if it were a human partner. Chatbots allow humans to communicate with machines via text or speech. Chatbots are now widely used in a variety of sectors, including tourism, health, business, and customer service. Figure depicts the basic architecture of a chatbot. When a user asks a question to the chatbot, the chatbot responds by interacting with the user in human natural language. Chatbots, in general, assess and determine the user's request intent before extracting the relevant entities. Followed by an analysis, the user will receive an appropriate response. English-language chatbots are now becoming increasingly popular. Other languages, on the other hand, have made little progress. However, due to a lack of resources, they are quite difficult to

D. Mane · M. Patil (✉) · V. Chaudhari · R. Nayakavadi · S. Pandhe
JSPM's Rajarshi Shahu College of Engineering, Pune, Maharashtra 411033, India
e-mail: mayurpatil014@gmail.com

B. Pati et al. (eds.), *Proceedings of the 6th International Conference on Advance Computing and Intelligent Engineering*, Lecture Notes in Networks and Systems 428,
https://doi.org/10.1007/978-981-19-2225-1_31

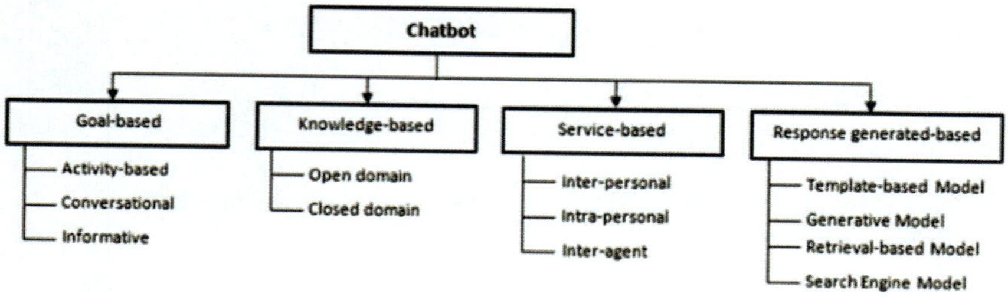

Fig. 1 Different types of chatbot

implement in other Indian languages. Different forms of chatbots are represented in Fig. 1.

An overview of the paper is as follows: Section 2 discusses a few strategies for developing a chatbot, as well as past relevant work and a literature review in the subject. The many strategies that can be utilised to develop a chatbot are described in Sect. 3. Section 4 discusses the difficulties encountered in prior studies. Finally, in Sect. 5, the authors concluded.

2 Related Work

Various strategies are currently being used to create chatbots. The following are a few of them:

1. Rule-based technique
2. AI-based bots
3. Pattern-based bots.

Important rules are hard coded into the code and used to get the right response to each query in the rule-based technique. Consider the following scenario: if the user's query is A, the response is B; similarly, if the question is C, the response is D, and so on. However, in this situation, the programmer must code all of the cases and rules that will be used to process the user's input into the program. If there are 50 different possibilities, the programmer will have to write all of these rules in order to create a bot. Massive volume of data make it practically impossible to draw down rules and/or patterns. AI-based bots, on the other hand, are built using natural language processing techniques and machine learning. When predefined or static rules or patterns do not function, natural language processing (NLP) is utilised.

The existing work regarding chatbots is done in English and Malayalam language. The work on Devanagari languages is extremely less and that they are quite difficult to implement because the resources for that are limited. There are some problems with all the systems designed previously just like the chatbot can give an accurate response to few queries. We are going to address the answer to the problems with the proposed process flow architecture. Related to this yet many researchers have tried to implement chatbots in the English language to supply effective responses. Related to this yet many researchers are tried to implement chatbots in English language to provide the effective responses.

Prasannan has published Hybrid technology which used both AI-based and rule-based chatbot approaches [1]. Chaitrali S. Kulkarni has published a paper that discusses the dataset they compiled from bank Website FAQs, as well as the architecture and technique they utilised to create a conversation bot. A comparison of seven categorisation algorithms is also provided [2]. Bayan AbuShawar has written a study about several types of learning methods in corpora and AI-ML pattern matching approaches on various platforms to test their use [3].

Professor Ela Kumar has published a paper which explains chatbot fundamental design and algorithm, methodology, dialogue act (DA) recognition technique, Bayesian approach to DA model a for developing a chatbot [4]. Professor Ram Manoj Sharma developed a chatbot for college inquiries that was created with AI algorithms. This system analyses the user's query and replies to their messages. This chatbot system includes features such as online noticeboards and more [5]. This document describes a framework for resolving any college-related difficulties. This technology serves as a virtual assistant and a college-focused intelligent machine [6].

This study gives an overview of chatbots that use cloud technology. And describe the code for chatbots, as well as the issues that chatbot programming faces today and, in the future [7]. Professor Raja Rathina publishes a paper proposing a banking chatbot that provides information on the procedures to be followed when obtaining any bank service. In practise, it is extremely difficult for the average person to comprehend the regulations associated with banks. As a result, a user will want assistance to complete the steps required in obtaining bank services [8]. Professor C. B. Lee has written a paper in which he makes numerous recommendations for future research that could aid in understanding chatbots from both a technical and a business perspective. The findings of this study should motivate more research into chatbots in the future [9].

Professor Dinesh Kalla has published a study that describes a chatbot for hospital care that can help in lowering healthcare expenses and boost access to healthcare services. It also aids patients in understanding their medical condition and difficulties [10]. Professor N.G. Kharate has published a paper which explains the method in which we can get the answer to the query given by the student by comparing the query in the database and then giving the output to the user [11]. Sagar Pawar has published a paper that shows how a chatbot for college inquiries works and how it was designed. The chatbot is powered by algorithms that examine user questions and comprehend their messages [12]. Vishal Kisan Borate has done research that explains how we can retrieve the answer to a student's query by comparing the query in the database and then giving the user the output [13].

Professor N. G. Kharate has published a paper which explains the method in which we can get the answer to the query given by the student by comparing the query in the database and then giving the output to the user [14]. In this research, N. Khin and K. Soe propose a chatbot architecture that is both efficient and precise in responding to university-related inquiries in Myanmar. Pandorabots interpreter and Artificial Intelligence Markup Language (AIML) are used [15]. They proposed NLP approaches and a serverless programming model, as well as their incorporation with social media, in this work [16]. This paper offers a chatbot as a possible way to give people with what they are looking for. The suggested chatbot is a cancer chatbot that is solely for cancer patients. People can inquire about any and all aspects of cancer, including symptoms, treatments, and survival [17]. Some researchers have worked in the domain of Devanagari language machine translation and synthesiser systems [18–20].

2.1 *The Summary of the Literature Survey*

Table 1 summarises the findings of the literature survey.

2.2 *Observation on Literature Survey*

We reviewed several research papers on chatbots in this work, and we discovered that the majority of chatbots were developed for English language. There are not many chatbots that are developed in Devanagari languages. The work in the subject of Devanagari chatbots is relatively new, and researchers have faced various challenges in developing one. For example, some researchers have worked on it, but they were unable to obtain the most consistent estimates based on user queries. Those who desire to conduct research can overcome these challenges to create smart chatbot in Devanagari languages.

3 Technique

1. **AI and Rule-based Technique**

   There are four stages involved.

1. User query is taken as input.
2. Tokenisation of sentence.
3. Stop word removal.
4. Root word extraction

Table 1 Literature survey summery

Title	Language	Summary	Challenges/future scope
2020 [1]	Malayalam	***Idea***: A chatbot in Malayalam language is developed to resolve the queries of public during COVID-19 pandemic, and to provide information about orders given by Government of Kerala, and FAQs ***Result***: In this project, the user query is taken as input and the chatbot managed to provide relevant responses ***Technique***: Hybrid technology is used in which both AI-based and rule-based chatbot concept is used	Malayalam is an extremely unique language with its own peculiarities. As a result, modules and platforms for constructing a Malayalam chatbot are few Due to the scarcity of datasets, accuracy was compromised. As a result, accuracy can be increased by expanding the dataset
2017 [2]	English	***Idea***: A chatbot in English language is developed to resolve the customer's queries and to provide the response to the queries ***Result***: In this project, the chatbot provides the quick response to the various user queries by using business logic ***Technique***: The user types in his query, which is tokenised, bigrammed, and sentence similarity scored before fetching and displaying the database's relevant response	It is possible to extend the domain. Images, audio notes, and videos can also be used to respond to the user's request. The banks' gateway can provide account-related details
2015 [3]	English	***Idea***: Trials and testing of ALICE chatbot is done using various techniques to find their outputs ***Result***: In this project, they have described the purpose of various chatbots based on ALICE architecture ***Technique***: In this they have used different types of learning methods in corpora and AI ML pattern matching techniques on various to check its usage	The primary challenge of the ALICE chatbot is to structure sentences using grammatical and sentiment analysis Process repeated inputs and, in general, maintain the structure of all phrases. Enhancing its manual knowledge development

(continued)

Table 1 (continued)

Title	Language	Summary	Challenges/future scope
2018 [4]	English	***Idea***: This paper provides an overview of the methodology and design utilised to construct a voice and text-based college inquiry chatbot system. Both students and teachers can use the initiative to ask any questions they have about college ***Result***: They use the notion of confusion matrix to check the accuracy of text in this project. They also detect accuracy at various input parameters ***Technique***: To design chatbot system, they used AIML, pattern matching Technique, Parse Technique, and Chat Script. They used relational database and SQL to remember previous conversations	When there is a large amount of pattern data, it is difficult to understand user utterances using the pattern matching approach
2020 [5]	English	***Idea***: Professor Ram Manoj Sharma developed a chatbot for college inquiries that was created with AI algorithms. This system analyses the user's query and replies to their messages. This chatbot system includes features such as online noticeboards, and more **Result**: The terms in the user's query will be found and matched using a pattern matching algorithm, and the user will receive responses ***Technique***: Developers set the rules for the chatbot to follow in an XML-based language, and bots can be developed using languages like AIML	It takes a long time to write rules for all potential and unique cases. They state in this paper that their chatbot project can handle simple queries but cannot handle complicated ones

(continued)

Table 1 (continued)

Title	Language	Summary	Challenges/future scope
2017 [6]	English	***Idea***: This project addresses students' and professors' questions about college life. This device communicates with a virtual human who answers the student's questions ***Result***: This system took the keywords from the user's query and used a keyword matching algorithm to match them with terms in the knowledge base, giving the user the best possible outcome ***Technique***: AIML can be used to develop Android applications such as the chatbot system. It can comprehend, analyse, and answer to the user's question	This method is more expensive to implement since it requires recordings of competent humans matching user queries to predefined rules. However, it is suitable for limited domain speech processing
2017 [7]	English	***Idea***: Chatbots are conversational dialogue engines that are written in the Python programming language and have the capacity to respond based on a database of all previous interactions ***Result***: A chatbot can understand what the user is saying and select or provide a response depending on the user's input and interactions. There are two types of chatbot responses: static and dynamic ***Techniques***: To train a model that is evaluated on input and delivers an answer, a chatbot adopts deep learning approaches	1. It is tough when dealing with NLP issues, for example, when we ask the bot "what is the weather?" the bot will respond correctly, but when the user asks "could you check the weather?" the bot will not respond correctly 2. Complex query-handling demands a high level of attention to the singular and plural forms of words, synonyms, hyponyms, and sentiment analysis
2020 [8]	English	***Idea***: A banking chatbot is developed to resolve the customer's queries and to provide the response to the queries ***Result***: In this project, the user query is taken as input, and the relevant information is provided to user ***Techniques***: Customers will access the system through a Web application, and the chatbot will respond to them based on their query	The researcher intends to build on this work by utilising a massive dataset collected from real-time bank interactions. Additionally, it can be implemented on the Website of any bank. By granting each client a private, password-protected login ID, they can have access to services such as transaction history and other sensitive data. As a result, banking is available to people worldwide

(continued)

Table 1 (continued)

Title	Language	Summary	Challenges/future scope
2017 [9]	English	***Idea***: Bibliometric analysis and research on chatbots and conversational agents is done ***Result***: They discovered new conventions in this study, such as how organisations may benefit from the usage of chatbots and conversational agents, how human attitudes will be impacted by chatbots, what role chatbots will play in the future, and other related topics ***Techniques***: In this they have used deep learning and natural processing language to develop chatbots	Businesses can use innovative concepts to build chatbots in order to save money and deliver a better online experience for their customers
2020 [10]	English	***Idea***: A chatbot for hospital care is being built utilising the NLTK library to deliver necessary information about the patient's disease ***Result***: In this project, the user query is taken as input and the chatbot managed to provide relevant responses about patient's illness ***Techniques***: This chatbot makes use of artificial intelligence and the NLTK Library to assist patients in diagnosing their illnesses	A disease training question answering system that is more concerned with locating the textual response to a given query than with speaking with humans. Due to the fact that this is a small application created from scratch, the responses may not always be perfect
2019 [11]	Hindi	***Idea***: A chatbot in Hindi language ***Result***: The user's query in Hindi language is processed and response is given ***Technique***: Using POS, NER tagging and ranking, pattern matching algorithms are used	Currently the project only supports text-based queries which can be extended for the voice-based queries and answers

(continued)

Table 1 (continued)

Title	Language	Summary	Challenges/future scope
2018 [12]	English	***Idea***: A chatbot for college enquiry powered by algorithms that analyse the user queries and comprehend their messages ***Result***: The suggested technology would be used to gather more information about question submitted by users ***Technique***: The user types in his query, which is tokenised, bigrammed, and sentence similarity scored before fetching and displaying the database's relevant response	Complex query-handling chatbots demand a high level of attention to the singular and plural forms of words, synonyms, hyponyms
2020 [13]	English	***Idea***: The college enquiry chatbot using Rasa ***Result***: The suggested system will compare user queries and return answers that are relevant to the user inquiry ***Technique***: The user's query is preprocessed by removing all stop words and then matched it with the knowledge base to provide appropriate response	We can employ speech and face recognition algorithms to automate increasingly complex processes, such as ATM withdrawals. Additionally, these chatbots' efficiency can be improved through the use of various machine learning models
2019 [14]	Marathi	***Idea***: A chatbot in Marathi language ***Result***: The user's query in Marathi language is processed, and response is given ***Technique***: Using POS, NER tagging and ranking, pattern matching algorithms are used	Currently, the project only supports text-based queries which can be extended for the voice-based queries and answers

(continued)

Table 1 (continued)

Title	Language	Summary	Challenges/future scope
2020 [15]	Myanmar	***Idea***: This is a university chatbot that connects the university with its users using Myanmar language. For the University of Computer Studies, they created a chatbot. This is a simple chatbot built on the Pandorabots server using Artificial Intelligence Markup Language ***Result***: In this project, the user query is taken as input and the chatbot managed to provide relevant responses ***Technique***: A rule-based chatbot has a relatively short time, indicating it has a greater influence on user interaction. The AIML language was used to develop this rule-based chatbot	Unexpected bot responses can occur as a result of normalisation mistakes. The system's alternative file can assist with this. Additionally, we will need to gather additional data for the chatbot's knowledge base
2018 [16]	English	***Idea***: The purpose of natural language processing in booking confirmation is to identify numerous patterns characterising ordering requests such as city and date. This study's contribution is to simulate a scenario that occurs when ordering tickets ***Result***: Using the chatbot, tickets can be booked ***Technique***: Natural language processing (NLP), which is applied in the deployment, is a technology for analysing and extracting information from keywords	In the future, the NLP algorithm may be strengthened with the addition of a memory cell that retains information about previous interactions. The memory cell can be used to store and measure user interaction using a hidden Markov model
2019 [17]	English	***Idea***: Cancer bot will allow patients to share their cancer-related difficulties as well as their helplessness. Chatbots are increasingly being utilised to automate activities, freeing humans to concentrate on their health ***Result***: A chatbot named "C Bot" was constructed and taught by feeding it a few questions from the cancer forum and their related responses. It reacts in accordance with the data that has been programmed into it ***Technique***: Natural language processing (NLP), which is applied in the deployment, is a technology for analysing and extracting information from keywords	The bot contained only a few features The functionality can be enhanced by the addition of the following: (1) Features for voice recognition (2) Sentiment analysis can be used to ascertain the user's emotion (3) A bot can be programmed to retrieve data directly from the internet (4) Multiple languages could be taught to the bot

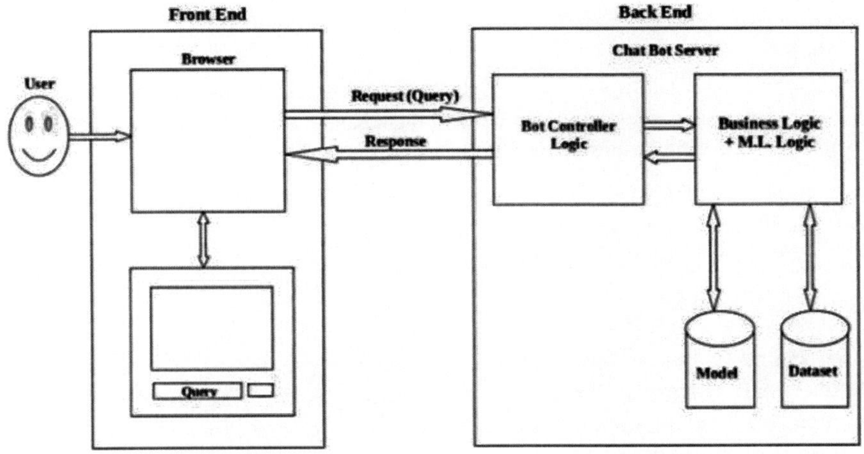

Fig. 2 Chatbot workflow

I. If extracted word is matched, then the chatbot replies.
II. If keyword is not found, then Jaccard similarity measurement is performed to find the most similar word.

2. **Sentence Similarity Score**

Sentence similarity score is a metric that measures how similar two sentences are. The intersection of two sentences is used to calculate it. The first statement (S1) is the user query, and the second sentence (S2) is the pattern from the database that is matched. The following is the formula for calculating the similarity score:

$$\frac{\text{Count}(S1 \wedge S2) \cup \text{Count}(S1 \wedge S2)}{\text{Count}(S1) \cup \text{Count}(S2)}$$

3. **Deep Neural Network**

The DNN model proposed has three hidden layers to process the input. Each of these hidden layers are made of ten neurons. SoftMax is a classifier and function that is being used. With SoftMax, the data can be categorised by direct classifier.

4. **Natural Language Processing**

In NLP, a machine picks up the audio of human speech. After that, the audio to text dialogue is done, and the text is analysed before the data is converted to audio. The technology then uses audio to respond to humans. Natural language processing is used to ensure that the data given to the neural network is of the correct type.

5. **Bibliometric Analysis**

A basic and powerful tool for exploring the patterns and future tendencies of a study topic is bibliometric analysis. This strategy was utilised by many academics to look into research trends in a variety of domains, including the role of IT innovation, project management research, and business intelligence and analytics. They employed three bibliometric analytic tools in this work.

a. Literature databases are utilised to collect publication data and citation reports.
b. CiteSpace is used to evaluate and cluster data
c. Bibliometrix is used to find co-occurrence patterns.

6. **NLTK LIB**

The Natural Language Toolkit is used to find synonyms for any partial response's terms found in a text file. Wordnet is a word database for English and an NLTK collection reader. It can be used to look for word definitions and synonyms. A synonym set, or a collection of synonym words, is referred to as a synset. When employing text analysis in artificial intelligence, it is extremely helpful.

4 Challenges

In this paper, we reviewed several research papers on chatbots and discovered that there is very little work in the field of Devanagari chatbots and that the architecture of the system needs to be reformed to accommodate the languages. In addition, the researchers attempted to construct a chatbot and encountered into a number of issues, including the following:

- Older systems have a lower level of accuracy.
- Writing rules for all possible and unique instances takes a long time in rule-based systems.
- There is a scarcity of study and sources on Devanagari.
- Complex query-handling chatbots necessitate paying close attention to singular and plural forms of words, synonyms, and hyponyms, which is challenging to manage.
- Sentiment analysis can be used to ascertain the user's emotion, which is important in determining how to answer to the user's query.

5 Conclusion

In this study, we summarised and analysed a number of research papers that are relevant to the development of a chatbot in languages such as English, Marathi, Hindi, and others. We have also discussed the many challenges and approaches that can be utilised to create a chatbot. The development of a chatbot in Devanagari languages still poses certain challenges like complex query-handling chatbots demand a high level of attention to the singular and plural forms of words, synonyms, hyponyms, and sentiment analysis. Currently, the above projects only support text-based queries which can be extended for the voice-based queries and answers. We can employ speech and face recognition algorithms to automate increasingly complex processes, such as ATM withdrawals. Sentiment analysis can be used to ascertain the user's emotion. Those who want to do research can circumvent the above limitations by developing a smart chatbot in Devanagari languages.

References

1. Prasannan, P., Joseph, S., & Rajeev, R. R. (2020). A Chatbot in Malayalam using hybrid approach. In *17th International Conference on Natural Language Processing: System Demonstrations.*
2. Kulkarni, C. S., Bhavsar, A. U., Pingale, S. R., & Kumbhar, S. S. (2017). Bank Chat Bot—an intelligent assistant system using NLP and machine learning. *International Research Journal of Engineering and Technology (IRJET), 4.*
3. AbuShawar, B., & Atwell, E. (2015). ALICE Chatbot: Trials and outputs. *Computación y Sistemas, 19.*
4. Das, S., & Kumar, E. (2018). Determining accuracy of chatbot by applying algorithm design and defined process. In *4th International Conference on Computing Communication and Automation (ICCCA).*
5. Gawade, H., Patil, V., Vishe, P., & Kolpe, S. (2020). College enquiry chat-bot system. *International Journal of Engineering Research & Technology (IJERT), 9.*
6. Nitesh, T., Akshay, H., Selote, S., Shinde, A., & Mahakalkar, N. (2017). *Artificially intelligent Chatbot.*
7. AbuShawar, B., & Atwell, E., *Automatic extraction of chatbot training data from natural dialogue corpora.*
8. Raja Rathina, G., Syed Shenaz, S., Vani, R., & Gowri, V. (2020). A banking chat bot—conversational bot for customer care using deep neural networks. *International Research Journal of Engineering and Technology (IRJET), 7.*
9. Io, H. N., & Lee, C. B. (2017). *Chatbots and conversational agents: A bibliometric analysis.*
10. Kalla, D., & Samaah, F. (2020). Chatbot for medical treatment using NLTK lib. *IOSR Journal of Computer Engineering (IOSR-JCE), 22.*
11. Bhagwat, V., Nagarkar, M., Paramane, P., Jindam, S., & Kharate, N. G. (2019). Review of Chatbot system in Hindi language. *International Research Journal of Engineering and Technology (IRJET), 6.*
12. Pawar, S., Rane, O., Wankhade, O., & Mehta, P. (2018). A web based college enquiry Chatbot with results. *International Journal of Innovative Research in Science, Engineering and Technology, 7.*
13. Ingale, N., Jha, T. A., Dixit, R., & Borate, V. K. (2020). College enquiry Chatbot using rasa. *International Journal of Scientific Research in Science and Technology, 5.*
14. Navalakha, D., Pittule, M., Mane, R., Rathod, A., & Kharate, N. G. (2019). Review of Chatbot system in Marathi language. *International Research Journal of Engineering and Technology (IRJET), 6.*
15. Khin, N., & Soe, K., *University Chatbot using artificial intelligence mark-up language.*
16. Handoyo, E., Arfan, M., Somantri, M., & Sofwan, A. (2018). Ticketing Chatbot service using serverless NLP technology. In *Proceedings of 2018 5th International Conference on Information Technology, Computer, and Electrical Engineering (ICITACEE).*

17. Belvin, R. V., Shobana, A. J., Babu, B., & Mathew, A. (2019). A graph based Chatbot for cancer. In *2019 5th International Conference on Advanced Computing & Communication Systems (ICACCS)*.
18. Suryawanshi, D., Itkarkar, R. R., & Mane, D. T. (2014). High quality text to speech synthesizer using phonetic integration. *International Journal of Advanced Research in Electronics and Communication Engineering (IJARECE), 3*(2).
19. Mane, D., & Hirve, A. (2013). Study of various approaches in machine translation for Sanskrit language. *International Journal of Advancements in Research & Technology, 2*(4), 383. ISSN 2278-7763.
20. Mane, D. T., Devale, P. R., & Suryawanshi, S. D. (2010). A design towards English to Sanskrit machine translation and sythesizer system using rule base approach. *International Journal of Multidisciplinary Research and Advances in Engineering (IJMRAE), 2*(2), 405–414.

Image Processing Applications and Pattern Recognition

Detection of Image Forgery Through Novel Hybrid Technique

Tanush Shekharappa Gouda, M. Ravishankar, and H. A. Dinesha

Abstract The rapid development of image editing applications has brought about an immense number of doctored photos coursing in our day by day lives, encouraging an exorbitant interest for automated forgery detection algorithms that can rapidly check the legitimacy of the source image. A robust forgery detection system needs to process the images without the need for prior knowledge about the image or any integral watermark. Forgery detection systems can be developed based on several indicators extracted using the image production approach and by analysing the doctored photos for any abnormal behaviours when compared to the original image. Digital images are considered to portray visual information in the current era and have become quiet predominant in our day-to-day life. Digital images are predominantly used in several sectors for analysing purpose. Few of these sectors include clinical science, criminal assessment, etc., where the legitimacy of an image plays a crucial role. Prediction of forgery in digital images is considered to be the most crucial part of any investigation. In the subject of digital image fraud identification, adequate quality work has depleted the first decade. Detection of forgeries in a digital image and the application standard tools seems to be more optimal with respect to time, space, and ease of work. The methodology proposed over here applies a hybrid combination of deep learning (DL) and machine learning. The proposed hybrid combinations segregate the image as forged and non-forged images. The proposed approach also explores the application and effectiveness of neural network to predict forgery images.

Keywords Image forgery detection system · Crime investigation · Deep learning and machine learning · Neural network

T. Shekharappa Gouda (✉) · M. Ravishankar
VVIET Mysuru, Affiliation to VTU Belagavi, Mysore, India
e-mail: tanush.gouda@gmail.com

M. Ravishankar
e-mail: ravishankarmcn@gmail.com

H. A. Dinesha
NCET Bangalore, Affiliation to VTU Belagavi, Beedaganahalli, India
e-mail: sridini@gmail.com

B. Pati et al. (eds.), *Proceedings of the 6th International Conference on Advance Computing and Intelligent Engineering*, Lecture Notes in Networks and Systems 428,
https://doi.org/10.1007/978-981-19-2225-1_32

1 Introduction

Even for the unskilled user, digital image processing has become a rather simple process, because of the fast advancement of digital image processing technology and the widespread usage of digital recording devices. With modern software, any user can edit digital photos, making it nearly hard to visually differentiate a fake from the genuine data. In the last ten years, due to the advancement in digital image technologies, distorted digital photographs have regularly surfaced, sparking continuous debates about the accuracy of the details supplied.

The deliberate change of details present in an image is referred to as artificial modifications (distortions) or attacks. Copy-move, re-sampling, and splicing are the three most common types of artificial distortion. These are used to obscure or distort information on satellite images. The first sort of distortion (embedding duplicates) is replicating a segment of a satellite image, applying any distortion (affinal transformation, re-quantizing brightness levels, additive noise, etc.), and then embedding the modified piece into another section of the same image (that part of it, which must be hidden). The second most prevalent type of artificial distortion is re-sampling, which is the affinal change of digital image components and embedding them into new images. The third most common type of distortion is splicing. To create a new satellite image or a detailed distortion of an existing one, parts of various satellite images are combined.

2 Literature Review

In this period of digitalization, everyone is encircled by digital items such as digital photographs and digital movies through the usage of electronic gadgets. By encouraging camera-enabled gadgets such as cell phones, laptops, tablets, and other similar devices, recent technical breakthroughs have reduced the cost of photography. Every minute, millions of photographs are submitted by people all around the world. In many cases, these contents make life easier in other way. Technological improvements have also made it feasible to create high-quality image editing software, smartphone apps, and Web interfaces that can readily make unnoticeable changes to images and films. There is presently a lot of image alteration software's available in the market that may be used for image manipulation. Some of the most popular tools include GIMP, Adobe Lightroom, Affinity Photo, Adobe Photoshop, Acorn, and others. Additionally, the Play Store has a number of photo editing apps for Android phones, including (i) Pixlr, (ii) PicsArt, (iii) Instagram, (iv) Cymera, (v) Aviary, (vi) Snapseed, and others. These tools/apps can be used for both good and bad purposes. Spreading rumours, influencing political debates, unethical thinking, and illegal purposes are all examples of negative use. As a result of these nefarious activities, forensic examination of photographs to establish their authenticity is in high demand. After the source has been verified, a digital watermark can be used to

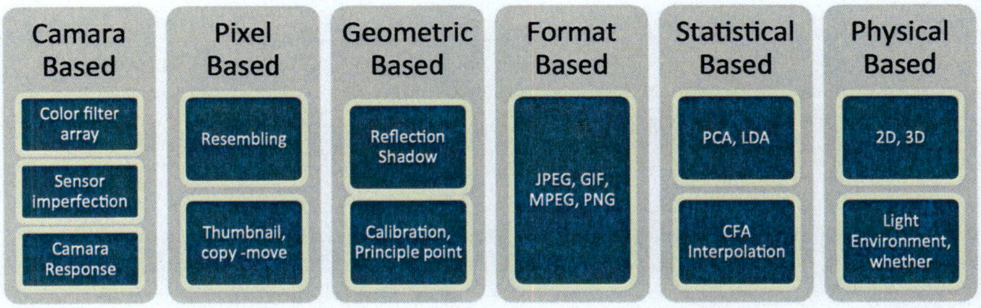

Fig. 1 Various forms of digital forensic tools

verify the image's origin and integrity, which was part of the older stenography technique's active forensic investigation. Photographs' origins are frequently unknown. It is a form of passive/blind forensic investigation in which we have no prior knowledge other than images. In this review study, we focus particularly on passive forensic analysis approaches and their issues because blind forensic analysis is in significant demand right now (Fig. 1).

Let us go over the two methods for making fake images: splicing and copy-move, before we get into the mechanics of blind forensics. Splicing is the method of faking an image by mixing two or more images of different sorts. A copy-move involves manipulating some picture content from the same image to create a fictitious image. Section 3 shows how to spot forgeries and how to spot them. When making realistic-looking fake images, several techniques are used, such as contrast and brightness enhancement, median filtering, imitating, cropping, and so on.

Understanding these procedures from a forensic angle can also help with forgery detection. Furthermore, a number of anti-forensics techniques have been developed to hide counterfeit artefacts. However, literature describes numerous phenomena to comprehend picture counterfeiting, including (i) geometric-based forensics, (ii) statistical-based forensics, (iii) physics-based forensics, (iv) camera-based forensics, and (v) pixel-based forensics. Many forensic procedures have been proposed in the literature, as seen in Table 1. Many digital forensic methodologies, as well as their focus, distinguishing traits, and limitations, are represented.

The next section goes through the copy-move forgery actions and detection procedures.

3 Copy-move Forgery Technique

Images can be manipulated in a number of ways. Several touch-ups and image modification procedures are utilised to improve or enhance the image. Instagram, for example, is a popular app that allows users to modify the appearance of their images by applying "filters" to them. Images are routinely reduced and recompressed to

Table 1 Existing methodologies

Digital forensic technique	Source authors	Addressed problems	Special feature	Constraint
Steganography Dataset: normal image, e.g. Lena, Barbara, Baboon	Sun et al.	Securing information	Good retention of PSNR	Outcomes not benchmarked
JPEG compression Dataset: Corel, NJIT, NRCS, UCID	Luo et al.	Analysis of JPEG error	Effective operation towards error analysis	Computational complexity is not addressed
Preserving prediction direction Dataset: UCID-v2 corpus	Li et al.	Anti-forensic (hiding artefacts of lossy compression)	Higher success rate	– No comparative analysis – Iterative process leading to storage complexity
Detection of Interest point Dataset: Image manipulation Dataset	Zandi et al.	Copy-move attack	– Simple detection Technique – Faster response time	Not applicable for other forms of image attacks
– Segmentation – Image hash Dataset: CASIA	Pun et al.	Tempered region localization	Good precision over coloured image	– Involves computational cost
Visual description, statistical approach, K-Nearest neighbourhood Dataset: DSO-I, DSI-I	Carvalho et al.	Impers orating image regions	– Higher accuracy	– Leads to iterative approach – Computationally complex
– Sliding window	Korus et al.	Detection of tampered	– supports both double	– Only effective for splicing
Linear Filtering JPEG Compression support Vector machine Dataset: UCID	Connoter et al.	Analysis chain of operators Recovery of original tampering operation	96.6% accuracy	Highly dependent upon training operation
Median filter, Gaussian Model Dataset: UCID-v2corpus	Fan et al.	Artefact hiding	Good accuracy, Simple approach	Involves higher computational cost
Expectation–Maximization, Segmentation Dataset: MICC-F600	Li et al.	Copy-move attack	Good accuracy	Involves higher computational cost due to iterative process
Artifacts of histogram. JPEG compression Dataset: BOSS Public dataset, UCID	Cao et al.	Forged image to be detected using contrast	High performance	Does not address complexities associated with dataset

(continued)

Table 1 (continued)

Digital forensic technique	Source authors	Addressed problems	Special feature	Constraint
Key-Point Injection, region classification Dataset: IRIA Holidays, UCID dataset	Costanzo et al.	Identification of SIFT key-point	Solves majority of key-point-based problems	Outcomes not benchmarked

make them more easily exchanged over the Internet, thanks to the advent of cloud-based picture sharing and editing websites like Flickr and Picasa, which are aided by social media apps like WhatsApp, Instagram, and Snapchat. These modifications may not be regarded as image tampering because the goal is to change the information content of an image. On the other hand, image processing tools and toolboxes such as Photoshop and GIMP have made it feasible to change images for both criminal and legitimate purposes.

Knowing the difference between picture improvements and image tampering is crucial. Local manipulations are rarely employed in picture improvement (bilateral filtering for touch-ups being one notable exception), but they are occasionally utilised in tampering. Contrast and brightness adjustments are not normally useful for picture tampering, while sharpening and blurring operations can help to some amount, and copy-move, splicing processes are on the dark side. Furthermore, the intent to harm varies from case to instance.

A bank that accepts customer check photos for photo-based check transactions, for example, would accept changes made during the transfer process, such as quality loss due to compression. Manipulation of the image's digits, on the other hand, would be terrible and must be avoided. A cosmetic advertisement featuring model photographs, on the other hand, is likely to include a variety of changes and touch-ups that a consumer would consider harmful.

It is vital to be able to detect tampering with reasonable accuracy and confidence when modifying photographs and documents for criminal reasons. Image forensics is the process of detecting harmful changes. A signal classification result from an image forensics algorithm can indicate whether or not the image has been tampered with, as well as identify and locate the altered areas of an image. To resist and detect such adjustments, active image tampering detection techniques are commonly used. A watermark is placed in an image when it is taken, and the image is validated by confirming the watermark. These approaches are excessively expensive and need high-end gear.

Manipulation can also be detected using passive methods. A watermark is not required for the legitimacy of passive techniques. Passive methods, on the other hand, attempt to determine the image's integrity without requiring any further data. In terms of techniques, "blindness" implies that picture semantics are ignored while detecting image forgeries. Tampering tactics usually leave indications of the modification in the form of changes in picture statistics, artefacts, and inconsistencies, even if the image was not explicitly edited. These discrepancies are utilised to assess whether

Table 2 Copy-move forgery and its detection techniques

Source	Forgery operations on images	Tamper detection techniques
Digital image tamper detection techniques—A comprehensive study [1]	Retouching spelling, copy paste, cropping, cloning	Edge blurring
Digital image tampering—A threat to security [2, 3]	Copy-move, resize, image splicing, noising blurring	Laplace filter, PCA, DCT, DWT, SVD
Tampering and copy-move forgery detection using SIFT feature [4]	Copy-move, block, feature-based method	PCA, DCT, DWT, SIFT
Efficient copy-move forgery detection for digital image [5, 6]	Copy-move, image splicing	Statistical & block characteristics
Survey of image forgery detection [7]	Copy-move, splicing, resize, cropping cloning	Pixel, format, camera physically, geometric-based
Comparison and analysis of photo image forgery detection techniques [8]	Copy-move, copy create, copy paste	JPEG compression analysis, edge detection, localization

or not a photograph has been tampered with, as well as to pinpoint the area that has been altered. The disadvantage of passive detection algorithms is that they only detect one type of image modification, potentially leaving other image manipulations undetected. These algorithms are also not exact enough. As a result, systems and methods for detecting and locating photo and document forgeries that meet the above requirements are needed (Table 2).

4 Literature Survey

This module contains a collection of ways for identifying cloning in photos that have been discussed by a variety of authors.

Wu et al. [9] Working on block-based matching, in which attacks were immediately discovered using affinal transformation and JPEG compression, and feature extraction was done using a neural network and the point-by-point operation. Gupta and Girdhar [10] developed a precise and robust match algorithm-based key-based matching detection approach. Robust match employs principle component analysis (PCA) coefficients to correctly detect the falsified sector. Yeap et al. [11] The featured approach, two closest neighbour (2NN) by categorised agglomerative clustering, was presented (HAC). On passive forgery detection, altering of images was conceivable with these techniques, and featured methods were utilised to spot geometrical charges. Dhanya and Selvi [12] developed a copy-move image forensic method that utilised techniques like DCT, SVD, DWT, SURF, PHT, and PCA to detect replicating through blurriness, rotation, and scaling in a pre-post operation. Dixit et al.

[13] proposed a method for dividing images into predetermined-size overlapping parts. For each individual block, this technique took into account the frequency domain as well as the image's statistical features, such as mean and variance. The performance was evaluated using the detection accuracy (DA) and false positive rate (FPR) matrices. Shabanian et al. [14] The copy–move zones were discovered using a Gaussian pyramid decomposition method, which proved resilient and sensitive to additive noise. Furthermore, both the block-based and key-point-based approaches used rotation and scale to detect picture duplication, making it straightforward to detect. Elaskily et al. [15] developed an approach for locating image cloning using the DCT technique.

In addition, an algorithm based on image moments was created and used to locate texture and intensity, making duplicity detection straightforward. Khan et al. [1] To minimise the number of dimensions in an image, the discrete wavelet transform (DWT) method was used.

The pictures were then divided into overlapping blocks and put in a grid. Lexicographic sorting and phase correlation were used to sort the data, identifying the delicateness of the blocks and then duplicating them to create a map that can be used to spot frauds. Peng et al. [2] Their findings on cloning fraud were presented utilising a passive and blind technique, with compound statistical features being used.

The pattern noise was decreased using wavelet transforms with sensor pattern noise (SPN) and signal noise ratio (SNR) techniques to generate a feature vector, allowing the images to be segregated into blocks and the forged area of the image to be easily identified.

As indicated by above analysis, multiple researchers have improved the methodologies used to recognise text from natural scene images. Listed above solutions, on the other hand, have alleviated a number of performance difficulties.

5 Methodology

For the detection of clone forgeries, the researchers used the principal component analysis (PCA) algorithm. Using the PCA approach, the copied section will be marked black. The GLCM technique is employed in this study to detect textural components in the input image, which is then used to determine the most recent area for cloning forgery detection (Fig. 2).

(i) **Input Image**: To start, the user chooses an image from the database to detect copy-move cloning.

(ii) **Conversion phase**: It is the process of turning a colour image into a grayscale image. The culled input image will be converted to grayscale if it is a colour image, such as RGB (Red–Green–Blue).

$$I = (0.229R) + (0.587G) + (0.1140B) \tag{1}$$

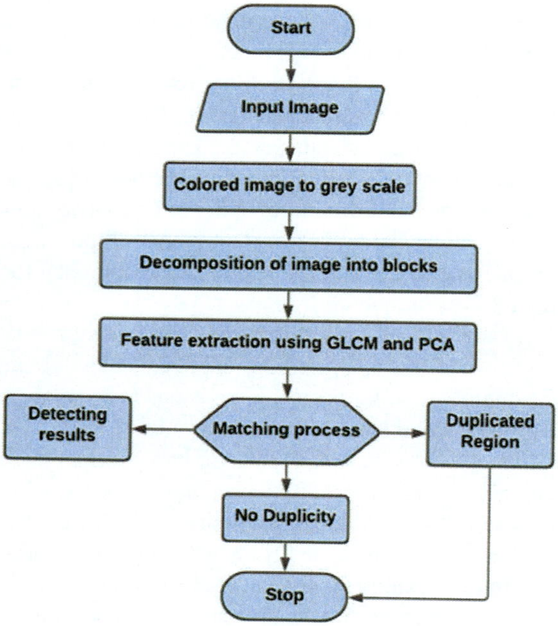

Fig. 2 Flow chart showcasing methodology adopted

(iii) **Fragmenting image into overlapping blocks**: For feature extraction, the image is separated into a number of fine-tuned pieces. When an image has a size of $M \times N$ pixels, it is segmented into blocks of $b \times b$ pixels, yielding N_b blocks, with N_b being the total number of blocks in the image.

$$N_b = (M - b + 1) * (N - b + 1) \tag{2}$$

(iv) **Features extraction**: The process of extracting or discovering features from blocks of an input image in order to accurately characterise an image is known as feature extraction. Principal component analysis (PCA) and grey-level cooccurrence matrix (GLCM) approaches are used to extract the feature. The most efficient method of image analysis has been found to be the extraction of GLCM features. The grey-level covariance matrix is a table that gives statistical measurements for texture analysis.

In a grey-level image, this technique examines the spatial relationship between the brightness of pixels. In this research, the GLCM features were estimated to explore the inconsistencies between the original image and the results. $P(I\,j\,|\,d,)$ is the relative frequency with which two pixels are separated by a distance d, the direction of which is determined by a specified angle (), and the intensity levels of both pixels are defined by I and j.

The steps for a basic GLCM algorithm are as follows:

1. All pairs of pixels with the values *I* and *j* that are used to match the same part of the image
2. The count is entered into the *i*th row and *j*th column of matrix $Pd[i, j]$.
3. If the grey levels of the pairs of pixels $[i, j]$ are not equal, the value of $Pd\ [i, j]$ is not symmetric.
4. $Pd\ [i, j]$ elements can be normalised by dividing each item by the total number of pixel pairings.

$$N[i, j] = P[i, j] / \sum i \sum j\ P[i, j] \tag{3}$$

(v) **Matching Process**: Matching is the process of discovering a match between feature descriptors, and if there is a likeness, it is identified as a sign of duplicated regions. Lexicographic sorting is employed as a similarity metric. The matrix is created by using feature vectors to construct feature vectors in each row, which are then sorted in ascending order. As a result, the most similar or identical features are shown in subsequent rows. The positions of both blocks are saved if the two subsequent rows are identical.

(vi) **Locate and Detect Duplicated Regions**: Finally, blocks with comparable properties are discovered, as well as tampered regions.

Experimental Results

The original image was chosen from Google's database. As a result, the image is tempered first with the Adobe Photoshop application. For future implementation, the original image is no longer needed. The edited image is used as an input to determine the forgery in the image. The block size is 32, 32 because it reduces the number of erroneous matches, resulting in more accurate results. The photos were processed using the R2015b version of the MATLAB toolbox, which was then used to find the results. On an outlying PC with 4GB of RAM, Windows 8.0, and an Intel(R) Core(TM) i5-200M CPU running at 2.50 GHz, the offered method was tested. The proposed work is carried out in MATLAB, and the results are compared in terms of PSNR, accuracy, and precision to existing and proposed approaches. A modified image is precision, and recognising the tampered image is accuracy (Figs. 3, 4, 5, and 6).

Fig. 3 Original image

Fig. 4 Tampered image

Fig. 5 Forensic detection of original image

Fig. 6 Forensic detection of tampered image

The PCA and GLCM with PCA algorithms are employed for clone detection, as illustrated in Fig. 7. The PSNR value of the PCA plus GLCM algorithm is higher than that of the PCA algorithm.

The precise values of the proposed and present methods are compared for the performance analysis, as illustrated in Fig. 8. The accuracy value of the presented method is found to be high when compared to the exact match approach.

The precision values of the existing and suggested methods are compared for the performance analysis, as illustrated in Fig. 9.

The precision value of the researched work is found to be high when compared to existing work.

Fig. 7 PSNR comparison

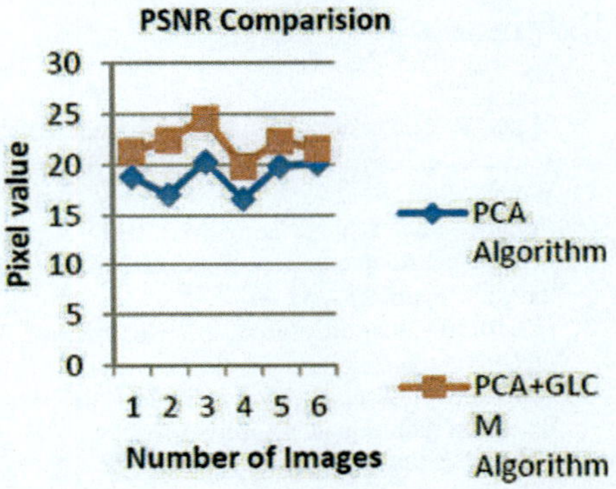

Fig. 8 Accuracy comparison

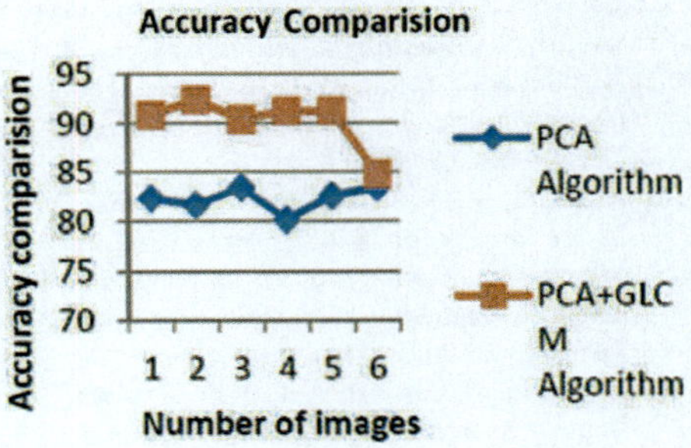

Fig. 9 Precision comparison

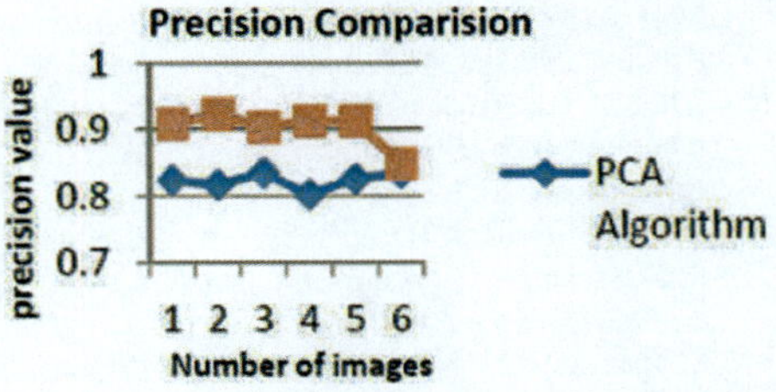

6 Conclusion and Scope for Future Work

In this study, the exact match methodology and the proposed method are compared.

In the exact match approach, the current methodology, the PCA algorithm, was employed to detect cloning. The proposed method employs the GLCM algorithm in conjunction with the PCA algorithm to detect forgeries. The PCA algorithm, along with the GLCM technique, which employs a cooccurrence matrix to detect picture textural aspects, made it easy to pinpoint counterfeit pixels on the image with high accuracy and fewer attacks.

References

1. Khan, S., & Kulkarni, A. (2011). Robust method for detection of copy-move forgery in digital images. In *International Conference on Signaland Image Processing* (pp. 69–73), Chennai, India.
2. Peng, F., Nie, Y. Y., & Long, M. (2011). A complete passive blind image copy-move forensics scheme based on compound statistics features. In *International Forensic Science Journal*, (Vol. 212, no. 1, pp. 21–25).
3. CoMoFoD-Image Database for Copy-Move Forgery Detection [Online]. Accessed on April 2018.
4. Kanwal, N., Girdhar, A., & Kaur, L. (2017). A taxonomy and analysis of digital image forgery detection techniques. *Engineering Science and Management Education, 10*(I), 35–41.
5. Hilal, A., Hamzeh, T., & Chantaf, S. (2017). Copy-move fogery detection using principal component analysis and discreete cosine transform. In *Sensors Networks Smart and Emerging Technologies (SENSET)*, Beirut.
6. Suresh, G., & Rao, C. S. (2016). Copy-move fogery detection using GLCM Based statistical features. *International Journal on Cybernetics and Informatics (IJCI), 5*, 165–171.
7. Kushol, R., Salekin, M. S., Kabir, M. H., & Khan, A. A. (2016). *Copy-move forgery detection using colour space and moment invariants-based features*, (pp. 1–6). https://doi.org/10.1109/DICTA.2016.7797027.
8. Agarwal, V., & Mane, V. (2016). *Reflective SIFT for improving the detection of copy-move image forgery* (pp. 84–88). https://doi.org/10.1109/ICRCICN.2016.7813636.
9. Wu, Y., Abd-Almageed, W., & Natarajan, P. (2018). Image copy-move forgery detection via an end-to-end deep neural network. In *IEEE Winter Conference on Applications of Computer Vision* (pp. 1907–1914), Lake Tahoe, NV, USA.
10. Gupta, G., & Girdhar, A. (2018). A robust passive method for detection of copy-move forgery in images. In *Second IEEE International Conference on Inventive Computation Technologies* (pp. 1–6).
11. Yeap, Y. Y., Sheikh, U. U., & Ab Rahman, A. A. H. (2018). Image forensic for digital image copy-move forgery detection. In *14th International Colloquium on Signal Processing & its Applications* (pp. 239–244), Penang, Malaysia.
12. Dhanya, R., & KalaiSelvi, R. (2018). A state of the art review on copy-move forgery detection techniques. In *IEEE International Conference on Circuits and Systems, Thiruvananthapuram* (pp. 58–65), India.
13. Dixit, R., Naskar, R., & Sahoo, A. (2017).Copy–move forgery detection exploiting statistical image features. In *International Conference on Wireless Communications, Signal Processing and Networking (WISPNET)* (pp. 2277–2281), Chennai.
14. Shabanian, H., & Mashhadi, F. (2017). A new approach for detecting copy-move forgery in digital images. In *IEEE Singapore International Conference on Communication Systems* (pp. 1–6), Guangzhou, China.

15. Elaskily, M. A., Aslan, H. K., Elshakankiry, O. A., Faragallah, O. S., Abd El-Samie, F. E., & Dessouky, M. M. (2018). Comparative study of copy-move forgery detection techniques. In *International Conference on Advanced Control Circuits Systems (ACCS) Systems & International Conference on New Paradigms in Electronics & Information Technology (PEIT)* (Vol. 017, pp. 193–203), Alexandria, Egypt.

Survey on Accent Correction and Region Prediction

Rupali Bagate, Aparna Joshi, Narendra Kumar Yadav, Aman Singh, Gaurav Singh Dhek, and Naincy Rathore

Abstract Background: In recent years, speech recognition technology has become a dominant part of our everyday lives, and as most of the future technology being developed can easily be integrated with the help of speech recognition. To make a digital future, technological advancement of everyone is necessary, and to make this technological advancement not so technical, speech recognition serves its role. Although speech recognition has made significant advances at certain languages, what has been achieved is a drop and what is left is an ocean. This technology has failed miserably in recognizing different accents of a single language or a voice disorder, and this has led to various questions on the authenticity of progress of the process. This paper documents the drawbacks of this technology and the areas where its immediate progress is possible. It talks about the limitations of various existing and popular and under radar ASR technologies with insights of their flaws which need to be considered immediately to avoid various social dilemmas and insecurities.

Keywords Accent · Region prediction · MFCC · Naive Bayes classifier · Stemming and lemmatisation

R. Bagate (✉) · A. Joshi · N. K. Yadav · A. Singh · G. S. Dhek · N. Rathore
Department of Information Technology, Army Institute of Technology, Pune, Maharashtra, India
e-mail: rupali.bagate@gmail.com

A. Joshi
e-mail: ajoshi@aitpune.edu.in

A. Singh
e-mail: amansingh_18376@aitpune.edu.in

G. S. Dhek
e-mail: gauravsingh_18347@aitpune.edu.in

N. Rathore
e-mail: naincyrathore_18422@aitpune.edu.in

B. Pati et al. (eds.), *Proceedings of the 6th International Conference on Advance Computing and Intelligent Engineering*, Lecture Notes in Networks and Systems 428,
https://doi.org/10.1007/978-981-19-2225-1_33

1 Introduction

1.1 What Is Voice-to-Text Conversion?

Voice-to-text conversion is a technique, in which software converts the audio-to-text format. The same process can be referred as speech recognition. These days you can achieve 99 accuracies in these softwares. Voice-to-text conversion is based on AI and ML algorithms. To develop more powerful, cheap and more accurate, software companies are investing in this technology. The toughest work in this conversion is to remove the unwanted noise, so that it can give us more accuracy. In this with the help of the other sounds and accent of the person, we will try to tell the region of that person. As the software will adapt person's voice, so with time it will become more accurate.

1.2 Why Do We Need Voice-to-Text Conversion?

The answer is very simple, we all know that the voice is the easiest and fastest way to communicate with people rather than the text, and this technology is used in many devices and software like Google speech recognition, Alexa, Siri in iPhone and many other applications too. It is also very useful for some of the disabled people who are not able to see, but with the help of this, they can also send their text messages to their loved once in their day-to-day life. There are many useful examples of this voice-to-text conversion, some of which are mentioned below.

1.3 Alexa/Google Assistant

These are the software which will convert your voice to text, and then they try to search those words in browser, so it becomes very easy for a person to surf things, he just has said those words, and the result will in front of him even without typing anything in the search engine. This is the latest technology with the help of which humans can save much more time and can do something more productive by saving this time. These Alexa, Siri can play music for us, these can send messages, they can call, and they can add events with the help of more advanced tools they have become smarter. Even we can talk to these software, and sometimes it feels like we are talking to some person itself.

1.4 Voice Mails for Deaf Students

As deaf students cannot see, so it becomes very difficult for them to send some text messages, but with the help of voice to text, it becomes very easy for them to send emails, messages. In this, they just have to press a key and as they speak their voice gets recorded and converted into the text format and then the computer reads what has been written, and after finalizing their text, they just have to press a button to send their mails.

1.5 Automatic Speech Recognition

Every individual has his own metonymy that depends on many factors, such as their vernacular and accent as well. These differences generally originate modelling problems for wide-reaching speaker-autonomous systems intended to process input from any individual of a given language. This thesis focuses on identifying the region associated with the accent of a speaker given an instance of their speech and illustrates how such a technology can be practised to enhance automatic speech recognition (ASR). In this thesis, we discuss a range of techniques that make use of different libraries in the acoustic signal to extract phonetic features from audio source and identify the regional dialect of a speaker. In particular, we used HMM, phonetic features (MFCC), as well as stemming, lemmatization algorithms and naive Bayes classifiers to produce the best results and end point detection to make the performance of the algorithm much better and reliable (Fig. 1).

1.6 Overview

The previous work that has been done on foreign accent classification with the help of traditional machine learning classifiers, Mozilla Common Voice dataset and applied Liberosa library to extract phonetic features from an audio source to develop MFCC. MFCC calculations are done to lower the data size of speech signal before pattern categorization. Stemming, lemmatization algorithm and naive Bayes classifier are applied on reduced data size to remove meaningless words, classify repeated words, unordered words and to make the recognized sentence clearer.

Fig. 1 ASR diagram

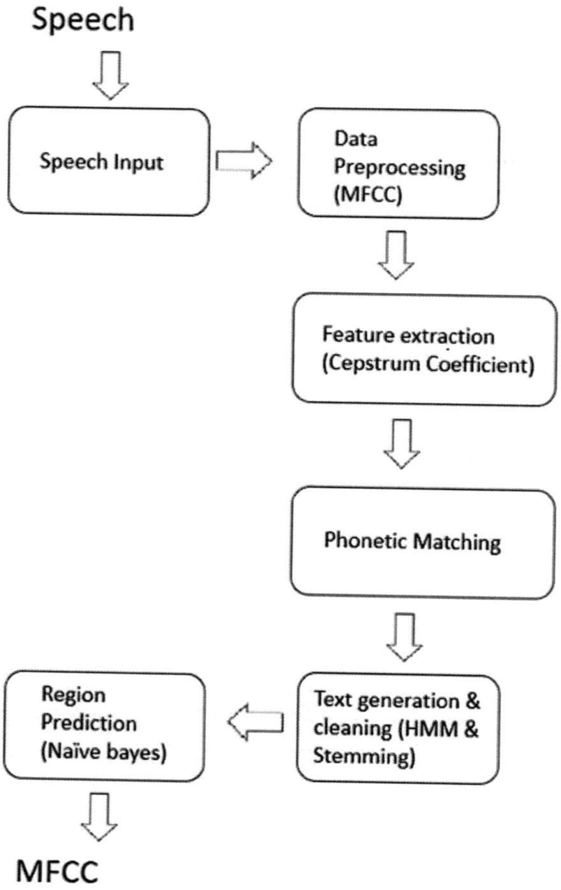

2 Previous Works

Previous work that has been done on foreign accent classification with the help of traditional machine learning classifiers shows that there is a huge scope of research in this field.

Chen et al. [1] in their work used basic sci-kit learn classifiers like SVM, naive Bayes and regression modified with different features to obtain 57.14 test accuracy for Mandarin and German non-native speakers, and they used the CSLU database for their research.

Wang et al. [2] in their work used various layered classifications to generalize that model trained on male data do not work on female data. So, working on various layered classification models he worked on by first classifying gender and then further on accent on word-level occurrences.

Ge et al. [3] in their work used PLP features instead of the more favoured MFCCs focussed on vowel extraction instead of consonants because it is observed that most accents appeared in the pronunciation of vowels instead of consonants.

Upadhyay in his work scrapped audio from online videos of five speakers from China, India, France, Germany and Spain and used a very different approach from already existing research by using deep belief networks to perform classification.

Ma et al. [4] after extensive research with different models and classifiers identified that Gaussian mixture model approach along with modified or direct use of HMMs is the best approach to accent classification.

3 Dataset and Feature Extraction

Mozilla Common Voice dataset is chosen as the dataset which includes demographic meta data like age, sex and accent that can help train the accuracy of speech recognition engines. As we are considering English as a first language to implement this approach on, and the dataset contains 780 h of validated audio recordings of 39,577 different voices. The features that are pre-sets with the datasets like age and sex and ones that we extract from the data when we differentiate different acoustic features of voice from an audio. These features can be generally classified as stops in speeches, affricates, fricatives, nasals, semivowels and vowels. As to the extraction of features from the audio file, we used the Liberosa library to extract MFCCs from each data file. MFCC is extracted because they can be converted to specific feature on the basis of frequencies and is best suited for speaker/speech recognition [5].

3.1 End Point Detection

Categorisation of audio into two voiced and un-voiced audio gives us the benefit for successive processing. In this, we can divide this sound in three broad domains: (1) Voice, (2) Unvoiced, (3) Silence, and with the help of these, we can further process for "end point detection and consonant identification". In turbulent domain audio specimen contains undesired signals, and atmosphere noise can be removed with the help of "end point perception". End point technique is build on "short-term log energy" and "Short-term 0 intersection allowance". Now these two are calculated with the help following equations [6]:

$$E_{\log} = \sum_{n=1}^{n} \log(s(n)^2) \tag{1}$$

$$\text{ZCR} = \frac{1}{2} * \sum_{n=1}^{n} \quad \text{mod } sgn[s * (n+1)] - sgn[s * (n)] \tag{2}$$

$$E_{\log} = \text{Log} - \text{Short Term energy}$$
$$n = \text{speech signal}$$

ZCR = short-term zero crossing.

3.2 Feature Extraction MFCC

The current strategy for feature extraction in mel frequency cepstral coefficient is to extract phonetic features from an audio source as the first step in developing an ASR (MFCC). MFCC has a major benefit over LPCC in that it uses mel frequency warping in addition to extracting phonetic information from an audio input. Human perception of sound frequencies for speech communications does not follow a linear scale, according to study . The frequency "f" is measured in Hz on a distinct scale known as the mel scale, which has a linear frequency spacing below 1000 Hz and a logarithmic frequency spacing above 1000 Hz [7]. To find the mel scale value for a given frequency f in kHz, it can use the following format:

$$\text{mel}(f) = 2595 * \log_{10}(1 + \frac{f}{700}) \tag{3}$$

4 Mel Frequency Cepstral Coefficient

Characteristics production is one of the most important tasks in this whole system. So, the main aim of this characteristics production is to lower the "data size of the speech signal before recognition or pattern classification". The main steps for (MFCC) calculations are: discrete cosine transform (DCT), logarithmic functions, mel frequency functions, discrete Fourier transform (DFT), windowing, framing. So, to convert the input audio into the output text, we mainly use four steps with the help of MATLAB. And those steps are recognition, characteristics production, pre-initialization and speech database (Fig. 2).

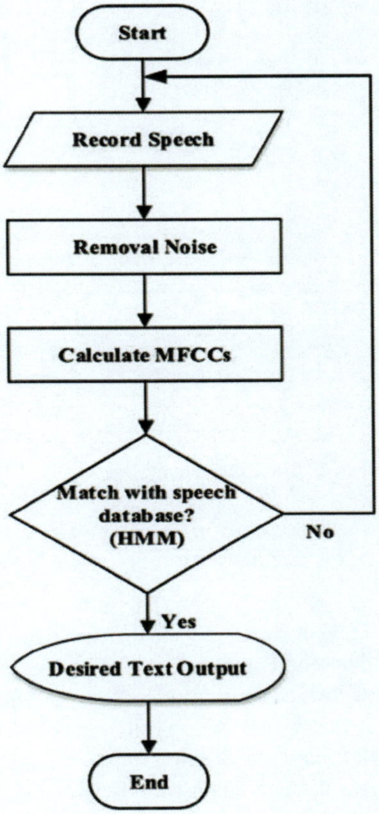

Fig. 2 Flowchart of speech-to-text conversion

5 Stemming and Lemmatisation

While processing text or speech to text, it is important to understand that multiple words, which to the algorithm might seem to be different, can be similar in the sense that their existing user needs them to be grammatically that way. The grammatical change in the verb just to fit into the sentence time space might add upto a number of extra dimensions to the word vector and might result in a more complicated and inefficient solution. In order to reduce these multiple words into their common root word stemming is used [8]. For example, WALK, WALKING, WALKED are three words with different spelling but the same semantic meaning and to us, only the root of the above tree word matters which is WALK.

A lot of time words with same semantic meaning, also called synonyms, which add same context and value to the existing data vectors, tend to reduce the accuracy of the algorithm because they add extra dimensions to the data vectors which alters the parameters and the vectors which should have been similar give different results. [9].

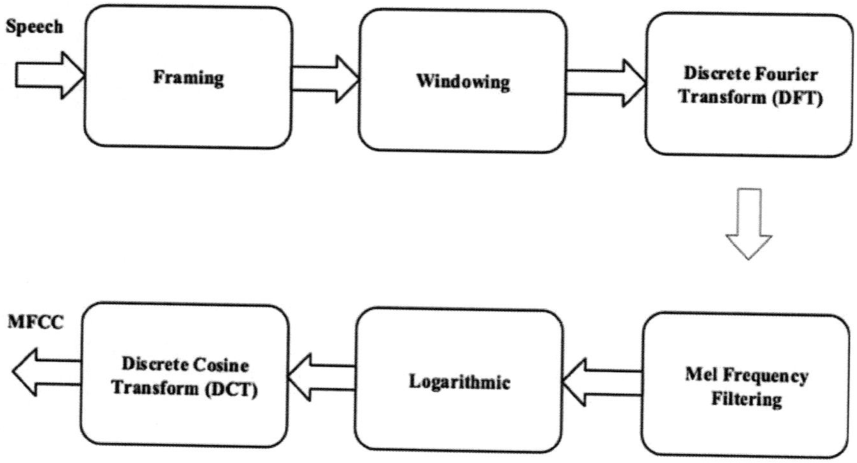

Fig. 3 Block diagram of mfcc

5.1 Stemming

Stem of word refers to the core word that derives other form of tenses of the word, and this might not work on different words with semantic meaning but it is certainly very useful for a common word in different forms of tenses. This idea behind stemming is that words with similar stem can be grouped together for better vectorisation of the sentence. Stemming manipulates a word in ways such as removing the prefixes or suffixes to their root of so-called stem so that the word becomes more analysable [10]. As in the processing of natural language processing, the less the number of dimensions in the vector, the better and efficient will be the processing and prediction (Fig. 3).

Multiple stemming algorithms exist but still there exist problems which can lead to meaningless words. The reason for this could be because even though the words have been reduced to their stems it is not necessary that the stem makes sense. For example, words like FINALLY, FINAL, FINALE are being stemmed are reduced to FINA, which actually does not makes any sense.

5.2 Lemmatisation

Lemmatisation can be considered an upgrade to stemming process as it rectifies the mistakes and errors of stemming. While stemming only focussed upon reducing the word to its stem following a crude heuristic process, meanwhile losing the dictionary meaning of the word a number of times [11]. Lemmatisation focusses more on the vocabulary and meaning of the word and thus reducing the words to their dictionary spelling. Different lemmatisation algorithms are implemented differently and thus use different English or any other language databases like WordNext. Lemmatisation is an extensive process unlike stemming so more processing power is time and is used in order to produce accurate results. For example, lemmatised word for LEAVES and LEAFS is LEAF [12].

6 Naïve Bayes Classifier

Nowadays, voice assistants have become an integral part of our life, and speech recognition is the most important part of a voice assistant. Naive Bayes helps in correct recognition of speech by classifying the accent and words in a proper way [13]. Naive Bayes classifier is basically based on the theory of probability, where all of them share a common principle. It is used with other techniques and algorithms to compute its effect in respective domains. It is used in speech recognition to classify repeated words, unordered words and to make the recognized sentence clearer. It is one of the most effective text classification methods. For example, by remembering how many times a word has occurred, if the word exists in the vocabulary or not, the speaker's recognized word is not what it should be. So, basically the technique used to process and analyse the text spoken by the speaker is to represent a bag of words, which consists of an unordered set of words, i.e. by ignoring their stance and only remembering, the number of times a word has been occurred in the sentence. For example, "I am extremely happy" and "I would be happy to help you", we simply note that the word I has occurred 2 times in the entire section, the word happy has occurred 2 times, and the words would, help, and you have appeared once, or more [14]. As we know, naive Bayes is basically based on the theory of probability, that means for a doc a, out of all classes c which belongs to C the classifier returns the class which has the max probability in the document. Basically, in Eq. 1 we have used the hat symbol ^ to symbolize "our evaluation of the true class".

$$C = \text{argmax}(P(\frac{C}{a}))\ldots c\epsilon C \tag{4}$$

Bayes' rule represented in Eq. 2; we can break this into three more probabilities:

$$P(\frac{x}{y}) = \frac{p(\frac{y}{x}) * p(x)}{P(y)} \tag{5}$$

In straight English, using Bayesian probability, the above equation can be depicted as:

$$\text{Posterior} = \frac{\text{prior} * \text{likelihood}}{\text{evidence}} \tag{6}$$

In practice, only numerator of the fraction is considered as the denominator is independent of C. So, this is the basic conditional probability model on which the naïve Bayes classifier works. The words which have occurred in the sentence but not in the original vocabulary are neglected. Remove those words from the test document and do not include any likelihood for them at all. In the speech recognition experiment, the naive Bayes model for classification was applied to text categorization tasks. The probabilities were calculated exactly as described in the previous section. Naive Bayes is an innovative model that makes the bag of words presumption (stance does not matter) and the conditional independence assumption (words are independent of each other given the class) and completely neglects the unknown word. Naive Bayes model can be very useful in speech recognition technology. As the probability approximations of the naive Bayes rule are such that they get smaller and smaller for longer and longer segments.

7 Conclusion

Mentioned speech-to-text convertor program has been implemented using the mel frequency cepstral coefficient for extracting features and HMM for recognizing. For the database, more than five audio clips are being used and analysed to make feature vectors. The features are first modelled in HMM, and then the input query is addressed by the next algorithm of HMM. From the results of the simulation, it can be observed that recognition rate of input data is greater than 87% by the number of states ($n = 5$) than any other number of states. State ($n = 5$) is very specific to this result because having any less will give a very tightly packed model while having any more states will need much more data to predict accurate results or a better model, and thus, a perfect balance needed to be struck so that the model overall does not degrade. In this model, the accuracy and performance of the algorithm are much better and reliable using "end point detection" algorithm for pre-processing data.

References

1. Chen, P., Lee, J., & Neidert, J. (2011). *Foreign accent classification* cs 229, fall 2011.
2. Sood, S. K., Sarje, A. K., & Singh, K. (2010). *An improvement of wang et al.'s authentication scheme using smart cards.*
3. Ge, Z., Tan, Y., & Ganapathiraju, A. (2015). *Accent classification with phonetic vowel representation.*
4. Jahangir, R., Teh, Y. W., Nweke, H. F., Mujtaba, G., Al-Garadi, M. A., & Ali, I. (2021). *Speaker identification through artificial intelligence techniques: A comprehensive review and research challenges.*
5. Johnson, F., Garza, S., & Gutierrez, K. (2016). Research on the use of voice to text applications for professional writing. In *2016 IEEE International Professional Communication Conference (IPCC)* (pp. 1–6).

6. Garza, S. (2015). Extended abstract: Write like you talk? research on the effects of voice to text applications when used as part of the writing process. In *2015 IEEE International Professional Communication Conference (IPCC)* (pp. 1–3).
7. Shirodkar, N. S. (2016). *Speech to text recognition using Hiddenmarkov model toolkit*. Ph.D. thesis.
8. Wieting, J., Gimpel, K., Neubig, G., & Berg-Kirkpatrick, T. (2021). *Paraphrastic representations at scale*
9. Liu, H., Christiansen, T., Baumgartner, W. A. Jr., & Verspoor, K. (2012). A lemmatization toolfor morphological processing of biomedical text. *Journal of Biomedical Semantics, 3*, 1–29.
10. Saito, K., Nakamura, T., Yatabe, K., Koizumi, Y., & Saruwatari, H. (2021). *Sampling-frequency-independent audio source separation using convolution layer based on impulse invariant method.*
11. Chen, X., Hsieh, C.-J., & Gong, B. (2021). *When vision transformers outperform resnets without pretraining or strong data augmentations.*
12. Badkobeh, G., Charalampopoulos, P., Kosolobov, D., & Pissis, S. (2021). *Internal shortest absent word queries in constant time and linear space.*
13. Huang, Y., & Li, L. (2011). Naive bayes classification algorithm based on small sample set. In *2011 IEEE International Conference on Cloud Computing and Intelligence Systems* (pp. 34–39).
14. Yang, F.-J. (2018). An implementation of Naive Bayes classifier. In *2018 International Conference on Computational Science and Computational Intelligence (CSCI)* (pp. 301–306).

A Two-Stage Convolutional Neural Network for Hand Gesture Recognition

Garg Mallika, Debashis Ghosh, and Pyari Mohan Pradhan

Abstract Gesture recognition task is a challenging research topic due to different hand sizes and poses. A hand gesture recognition convolutional residual network that learns to partition the hand region from the background is proposed. The second stage employs two convolutional neural networks which takes appearance information and the shape information as input to perform the gesture recognition. The appearance information is given by RGB images, and shape information is given by segmented gesture region obtained from first stage. The classification is computed based on accuracy, precision, and F1-score. The results manifest a high accuracy of 98.75%, F-score of 0.94, recall of 0.94, and a precision of 0.95 which is better than other methods on OUHANDS dataset.

Keywords ConvNet · Gesture recognition · Gesture segmentation · Convolutional neural network · Isolated gestures · Hand gesture

1 Introduction

For the disabled people, gestures are an important way of communication. Gestures can also be used to facilitate human–computer interactions. These can be interactive gestures, controlling gestures, or communicative gestures. Hand gestures that are used for communication convey some form of information. Hand gestures are

G. Mallika (✉) · D. Ghosh · P. M. Pradhan
Department of Electronics and Communication Engineering, Indian Institute of Technology, Roorkee, Uttarakhand 247667, India
e-mail: mallika@ec.iitr.ac.in

D. Ghosh
e-mail: debashis.ghosh@ece.iitr.ac.in

P. M. Pradhan
e-mail: pmpradhan@ece.iitr.ac.in

B. Pati et al. (eds.), *Proceedings of the 6th International Conference on Advance Computing and Intelligent Engineering*, Lecture Notes in Networks and Systems 428,
https://doi.org/10.1007/978-981-19-2225-1_34

also used in autonomous vehicles. Traditionally, handcrafted features were used for gesture recognition but with the advancements in vision-based pattern recognition technology, self-learned features are used by the researchers to recognize gestures. The most common preprocessing technique in gesture recognition system is the segmentation of hand regions, which produces an attention to where the hand is located in the image that results in better performance of the system.

Recently, many deep learning algorithms have been developed that facilitates the gesture recognition task [15, 21, 22]. Hand region is first segmented from the background, and then the classifier classifies the gesture performed [3]. Attention-based mechanisms [8] also add to the performance of the system. Features like joint angles [1], color-coded joint velocity feature map [14], joint angle displacement map [13], spatial 3D relational geometric features [12], quad joint relational feature [10] can be extracted for gesture recognition system.

The initial stage consists of segmentation that leans to segment the background and gesture portion. The second stage employees two convolutional neural networks which take appearance information and the shape information as input to perform the gesture recognition. The appearance information is conveyed by RGB images, and shape information is conveyed by the segmented hand region obtained from the first stage.

1.1 Contribution

We perform gesture recognition task which uses segmentation as the preprocessing stage. The main contributions are:

(i) The proposed gesture recognition system focuses on increasing the performance of the recognition system by using appearance and shape information of isolated hand gestures.
(ii) Experiments are performed on OUHANDS datasets, and results are computed based on accuracy, precision, and F1-score which shows our method performs better for gesture recognition.

1.2 Organization

The paper consists of literature review in Sect. 2. In Sect. 3, the method proposed in this paper is described. Experiments and discussions of the results are presented in Sects. 4 and 5. Lastly, conclusion is drawn in Sect. 6.

2 Literature Review

Numerous methods have so far been proposed for static as well as dynamic gesture recognition. Jiang et al. [7] proposed a Sign Language Graph Convolution Network (SL-GCN) and a Separable Spatial–Temporal Convolution Network (SSTCN). Both local and global features are used to take the advantage of multimodality. Tur and Keles [24] present an HMM-based classifier with pretrained ResNet50 features and a CNN-based dimentionality reduction model with an accuracy of 90.15% in Montalbano dataset which is comparable to other deep learning models.

In [25], Xiao et al. addressed the two-way communication between hearing impaired and normal people using a skeleton-based gesture recognition system and RNN-based generation framework. Recognition accuracy of 82.55 and 79.12% has been obtained, respectively, on the real and generated Chinese Sign Language. In [2], Chen et al. used temporal hierarchical dictionary to guide the hidden Markov model (HMM) for online gesture recognition. HMM hidden state has been estimated through Bi-LSTM network. Since large redundancy is seen in the online gesture, relative entropy is introduced to observe the redundancy.

In [11], a two-level hierarchical structure with a detector and a classifier uses sliding window to enable offline convolutional neural network (CNN) architectures to work efficiently online. The author used Levenshtein distance to measure misclassifications, multiple detection, and missing detections at the same time. Jie Huang et al. described an attention-based 3D-convolutional neural networks (3D-CNNs) for sign language recognition in [5]. Maximum cosine similarity and fast nearest neighbor (NN) techniques [20] also offer high-recognition accuracy and computational advantages.

Pavllo et al. [18] presented a framework to overcome the mutual hand occlusion. First, occlusion is resolved and then joint transformations are tracked. Kirishima et al. [9] examined eighteen hand gestures using swarm-based integration and found that the multiview skeleton recognition strategy is more robust and better than individual fixed-view recognition. A probabilistic appearance-based framework [23] has been proposed on IMGH dataset with a detection rate of 76.42% for multiview gestures. Support vector machine-based solution to tackle occlusion faced by bimanual (two hands) gestures are proposed in [19].

So, in the literature, it can be found that high accuracy is still to be achieved while gesture recognition is considered in case of static gestures. So, first segmentation of image is done, and then both the original image and segmented image are used for recognition. This is how the proposed algorithm is used to perform recognition with high accuracy.

3 Proposed Method

A deep neural network consists of hierarchical arrangement of layers that learns to extract the features from the image. This paper proposes a deep model trained on training and augmented data, and evaluated on test data. The network consists of two

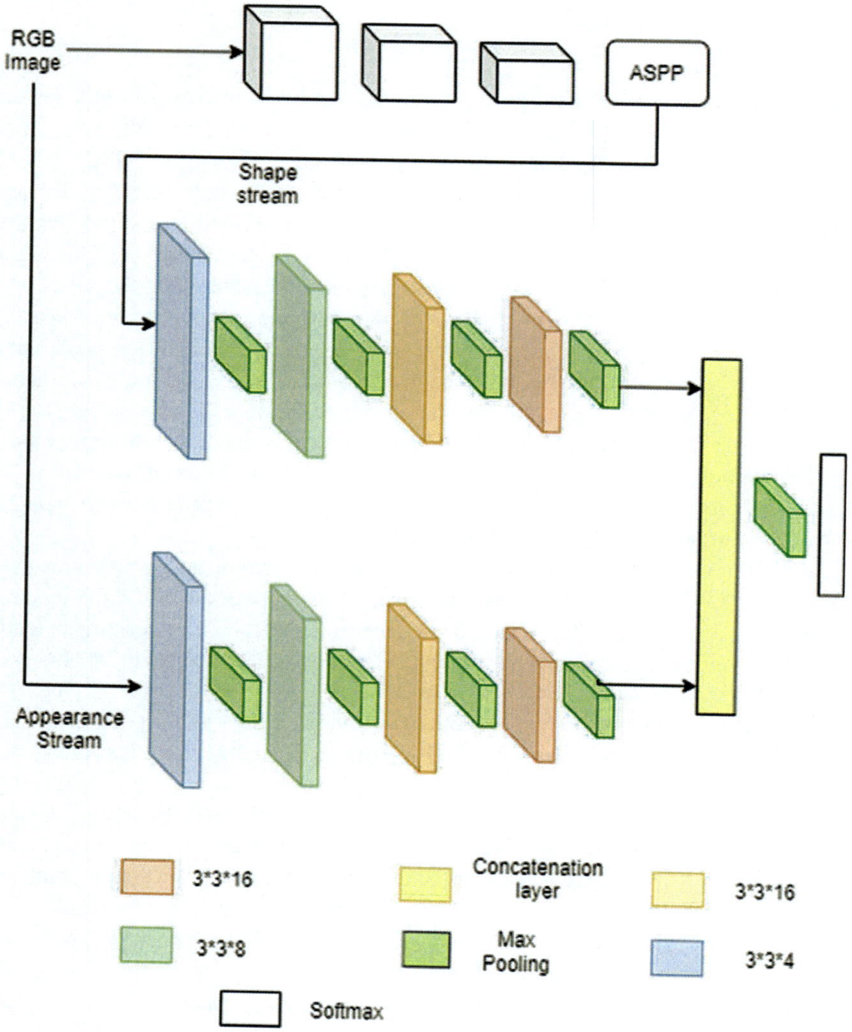

Fig. 1 Architecture of proposed two-stage hand gesture recognition system

stages. Hand region is first segmented from the image, and the segmented image is given as input to the second stage. The second stage takes the RGB image and the segmentation image from the first stage as input. The second stage is the recognition stage which outputs the classification result using softmax classifier. So, the whole network first segments the hand region from the image and then classifies the gesture. The proposed network is shown in Fig. 1.

3.1 Hand Segmentation Network

Hand segmentation network consists of a fully convolutional residual network and an Atrous Spatial Pyramid Pooling (ASPP) module [3]. Neural networks sometimes face the problem of overfitting and vanishing gradients while training. So, residual networks [4] were proposed to help gradient to surpass layers and avoid the problem of overfitting and vanishing gradients. Residual networks use identity mappings as skip connections. So, for hand segmentation purpose, we have used the fully connected residual layers to construct a deep convolutional neural network (CNN) with 28 convolution layers. The network consists of three residual blocks, an input, and a convolutional layer. Each block consists of three convolutional layers, batch normalization layers, and ReLU layer. Each residual unit can be expressed as

$$y_k = h(x_k) + F(x_k, W_k), \tag{1}$$

where x_k and y_k are the input and output of the kth residual unit, $h(x_k)$ is an identity mapping, and F is the residual function defined by weights W_k.

The contextual information plays a very important role for semantic segmentation tasks [16]. The atrous convolution learns multi-scale contextual information without losing resolution. The dilated convolutions perform exponential expansion of the receptive field which are able to capture multi-scale information. The basic ASPP module has five convolutional layers with different dilation factors. The dilations are 1, 3, 6, 12, and 18. The output of each layer is concatenated and fed to a 1×1 convolution. The output of the convolutional layer is then upsampled 4 times to give the multi-scale contextual information.

3.2 Recognition Network

The proposed network predicts the class of the gesture performed in the image. This network takes two input streams. One takes the appearance-based information and the other takes the shape-based information as input. The appearance information is conveyed by the RGB images, and the shape information is conveyed by the segmented images obtained from the first stage. Each stream branch consists of four convolution layers followed by a fully connected layer. Each convolution layer uses ReLU for nonlinear activation, same padding, and max-pooling layer with a kernel size of three. The feature maps from each branch are then concatenated and dropped with a ratio of 0.2 followed by the fully connected layer. The output is finally fed to the softmax layer for classification. The architecture detail is given in Table 1.

The convolution layers convert the hand gesture image into its feature representation while max-pooling layer downsamples the feature map. N_l distinct kernels have N_l feature maps of size M_l, where M_l is the product of height and width of the

Table 1 Architecture details of the second stage

Layer type	Output feature
Input	320 × 320 × 3
Convolution1	320 × 320 × 16
Max-pooling1	106 × 106 × 16
Convolution2	106 × 106 × 32
Max-pooling2	35 × 35 × 32
Convolution3	35 × 35 × 64
Max-pooling3	11 × 11 × 64
Convolution4	11 × 11 × 128
Max-pooling4	3 × 3 × 128
Concatenate	3 × 3 × 256
Max-pooling5	1 × 1 × 128
Flatten	256
Dropout	256
Dense	10

Here only one stream is mentioned before concatenation for convenience

feature map. Let, there are K number of convolution and max-pooling layers each, and then the feature map of lth convolution layer is given as

$$I^l(n_1, n_2) = \sum_{k_1=0}^{W-1} \sum_{k_2=0}^{H-1} f(k_1, k_2) I(n_1 - k_1, n_2 - k_2), \tag{2}$$

where $0 \leq n_1, n_2 < M_l$, $f(k_1, k_2)$ are the filter coefficients. The detailed architecture is given in Table 1.

Let, the outputs of the shape stream, $f^S \in R^d$ and appearance stream, $f^A \in R^d$ are fused using fusion function, respectively. The total output f^{sum} then becomes

$$f_i^{\text{sum}} = f_i^S + f_i^A, \tag{3}$$

where $1 < i < d$ and $f^{\text{sum}} \in R^d$, and d is number of the neurons. The feature vector f^{sum} is then fed into a softmax classifier for gesture recognition.

4 Experimental Setup

The experiments are performed on Nvidia Tesla K80 GPU machine with 24GB RAM, CUDA 10.1 with cuDNN 8.1.1. The deep learning libraries utilized in this implementation are Tensorflow-gpu = 1.14 and Keras. The batch size is set to two images per batch which are selected by performing different experiments on different

Table 2 Details of transformation applied for augmentation

Rotation (%)	Shearing (%)	Zooming (%)	Translation (%)
40	25	15	15

batches, and 200 epochs are used to train the model on the complete training datasets. The network is trained using Adam optimize $\beta 1 = 0.9$ and $\beta 2 = 0.999$. Dynamic learning rate is used while training which is 0.01 for epochs less than 25 and is reduced to 0.001 for epochs upto 50. However, it is further slowed down to 0.0001 and 0.00001 upto 75 and 100 epochs, respectively.

4.1 Data Augmentation

Since the images in the dataset are less, our model faces the problem of overfitting and data biasing. To deal with this problem, we add the training set fourfold by applying rotation, shearing, zooming, and vertical/horizontal translations. The details are given in Table 2. This helps the model to converge and avoid overfitting.

4.2 Datasets

The OUHANDS dataset [17] is used in this paper for gesture recognition. This dataset consists of static hand pose and non-hand images. This dataset consists of ten different hand pose captured by 23 signers. The objective of providing hand images is that it can be used for hand pose classification and hand detection task. Along with this, non-hand images are also included in the dataset for hand detection purpose. The hand images contain segmentation masks, bounding boxes and orientation normalization while non-hand images do not.

The dataset contains 3150 hand images (2150 training and 1000 test images) and 5288 non-hand samples (3412 training and 1876 test images). Each image has resolution 640×480. Data is acquired under various conditions such as various angles, various hand shapes and sizes. Along with the RGB images, the depth images were also captured. These were used to produce the binary hand segmentation masks.

5 Result Analysis

Table 3 presents the comparison results obtained from different experiments. The matrices used for comparison include precision, F1-score, recall, and the accuracy of the model. The model when fed with shape and appearance information gives

Table 3 Hand gesture recognition comparison results for the proposed model

Model	Precision	F1-score	Recall
Shape (through stage one) + appearance	0.95	0.94	0.94
Only shape (through stage one)	0.86	0.85	0.83
Only appearance	0.74	0.77	0.72

Table 4 Hand gesture recognition accuracy comparison for the proposed model

Model	Accuracy (%)
Shape (through stage one) + appearance	98.75
Only shape (through stage one)	93.24
Only appearance	90.10

Table 5 Comparison of the proposed model with other model

Method	F-score
Jianchun et al. [6]	0.8621
HGR-Net [3]	0.9375
Our	**0.94**

precision = 0.95, recall = 0.94, and F1-score = 0.94 which is better than when the model is fed with only shape or appearance information.

Similarly, comparison of the model is shown based on the accuracy of the model in Table 4. Our proposed model has the accuracy of 98.75%, 93.24%, and 90.10 for both appearance and shape information, only shape information and only appearance information, when given as input to the network, respectively.

In Table 5, we have compared our results with HGR-Net [3, 6]. The segmentation network for both HGR-Net and the proposed model is same. The difference lies in the recognition network. In [6], an attention mechanism is proposed for complex backgrounds. Results show that the proposed model performs better with F-score of 0.94, whereas the HGR-Net has F-score 0.9375 and [6] has F-score of 0.8621.

6 Conclusion and Future Scope

In this paper, the RGB image is first segmented and then the segmented image and RGB image are used to classify the gesture performed in the image. Results show that proposed approach is better than other approaches. The proposed approach works better when both streams are used as input with an accuracy of 98.75%. In the proposed architecture, the main components are convolutional neural network and atrous spatial pyramid pooling. In the future, deeper and advanced networks may help to increase the accuracy. Various other well-known architectures can also be used to increase the accuracy of the system.

References

1. Avola, D., Bernardi, M., Cinque, L., Foresti, G. L., & Massaroni, C. (2018). Exploiting recurrent neural networks and leap motion controller for the recognition of sign language and semaphoric hand gestures. *IEEE Transactions on Multimedia, 21*(1), 234–245.
2. Chen, H., Liu, X., Shi, J., & Zhao, G. Temporal hierarchical dictionary guided decoding for online gesture segmentation and recognition. *IEEE Transactions on Image Processing: A Publication of the IEEE Signal Processing Society*
3. Dadashzadeh, A., Targhi, A. T., Tahmasbi, M., & Mirmehdi, M. (2019). Hgr-net: A fusion network for hand gesture segmentation and recognition. *IET Computer Vision, 13*(8), 700–707.
4. He, K., Zhang, X., Ren, S., & Sun, J. (2016). Deep residual learning for image recognition. In *Proceedings of the IEEE Conference on Computer Vision and Pattern Recognition* (pp. 770–778).
5. Huang, J., Zhou, W., Li, H., & Li, W. (2018). Attention-based 3d-cnns for large-vocabulary sign language recognition. *IEEE Transactions on Circuits and Systems for Video Technology, 29*(9), 2822–2832.
6. Jianchun, G., Jiannuan, G., & Lili, W. (2021). Gesture recognition method based on attention mechanism for complex background. *Journal of Physics: Conference Series, 1873*, 012009. IOP Publishing.
7. Jiang, S., Sun, B., Wang, L., Bai, Y., Li, K., & Fu, Y. (2021). *Skeleton aware multi-modal sign language recognition*. arXiv preprint arXiv:2103.08833
8. Khodabandelou, G., Jung, P. G., Amirat, Y., & Mohammed, S. (2020). Attention-based gated recurrent unit for gesture recognition. *IEEE Transactions on Automation Science and Engineering*.
9. Kirishima, T., Manabe, Y., Sato, K., & Chihara, K. (2009). Real-time multiview recognition of human gestures by distributed image processing. *EURASIP Journal on Image and Video Processing, 2010*, 1–13.
10. Kishore, P., Perera, D. G., Kumar, M. T. K., Kumar, D. A., & Kumar, E. K. (2020). A quad joint relational feature for 3d skeletal action recognition with circular cnns. In *2020 IEEE International Symposium on Circuits and Systems (ISCAS)* (pp. 1–5). IEEE.
11. Köpüklü, O., Gunduz, A., Kose, N., & Rigoll, G. (2020). Online dynamic hand gesture recognition including efficiency analysis. *IEEE Transactions on Biometrics, Behavior, and Identity Science, 2*(2), 85–97.
12. Kumar, D. A., Sastry, A., Kishore, P., Kumar, E. K., & Kumar, M. T. K. (2018). S3drgf: Spatial 3-d relational geometric features for 3-d sign language representation and recognition. *IEEE Signal Processing Letters, 26*(1), 169–173.
13. Kumar, E. K., Kishore, P., Sastry, A., Kumar, M. T. K., & Kumar, D. A. (2018). Training cnns for 3-d sign language recognition with color texture coded joint angular displacement maps. *IEEE Signal Processing Letters, 25*(5), 645–649.
14. Kumar, E. K., Kishore, P., Kumar, M. T. K., Kumar, D. A., & Sastry, A. (2018). Three-dimensional sign language recognition with angular velocity maps and connived feature resnet. *IEEE Signal Processing Letters, 25*(12), 1860–1864.
15. Lai, K., Yanushkevich, S. N. (2018). Cnn+ rnn depth and skeleton based dynamic hand gesture recognition. In *2018 24th International Conference on Pattern Recognition (ICPR)* (pp. 3451–3456). IEEE.

16. Liu, Z., Li, X., Luo, P., Loy, C. C., & Tang, X. (2015). Semantic image segmentation via deep parsing network. In *Proceedings of the IEEE International Conference on Computer Vision* (pp. 1377–1385).
17. Matilainen, M., Sangi, P., Holappa, J., & Silven, O. (2016). Ouhands database for hand detection and pose recognition. In *2016 Sixth International Conference on Image Processing Theory, Tools and Applications (IPTA)* (pp. 1–5). IEEE.
18. Pavllo, D., Delahaye, M., Porssut, T., Herbelin, B., & Boulic, R. (2019). Real-time neural network prediction for handling two-hands mutual occlusions. *Computers & Graphics: X, 2*, 100011.
19. Poon, G., Kwan, K. C., & Pang, W. M. (2018). Real-time multi-view bimanual gesture recognition. In *2018 IEEE 3rd International Conference on Signal and Image Processing (ICSIP)* (pp. 19–23). IEEE.
20. Poularakis, S., & Katsavounidis, I. (2015). Low-complexity hand gesture recognition system for continuous streams of digits and letters. *IEEE Transactions on Cybernetics, 46*(9), 2094–2108.
21. Santhalingam, P. S., Pathak, P., Košecká, J., Rangwala, H., et al. (2019). Sign language recognition analysis using multimodal data. In *2019 IEEE International Conference on Data Science and Advanced Analytics (DSAA)* (pp. 203–210). IEEE.
22. Sarma, D., Kavyasree, V., & Bhuyan, M. (2020). *Two-stream fusion model for dynamic hand gesture recognition using 3d-cnn and 2d-cnn optical flow guided motion template*. arXiv preprint arXiv:2007.08847
23. Shukla, D., Erkent, Ö., & Piater, J. (2016). A multi-view hand gesture rgb-d dataset for human-robot interaction scenarios. In *2016 25th IEEE International Symposium on Robot and Human Interactive Communication (RO-MAN)* (pp. 1084–1091). IEEE (2016).
24. Tur, A. O., & Keles, H. Y. (2021). Evaluation of hidden markov models using deep cnn features in isolated sign recognition. In *Multimedia Tools and Applications* (pp. 1–19).
25. Xiao, Q., Qin, M., & Yin, Y. (2020). Skeleton-based Chinese sign language recognition and generation for bidirectional communication between deaf and hearing people. *Neural Networks, 125*, 41–55.

K-Means Clustering in Image Compression

Kanika Khatri, Vidushi Singhal, and Shefali Singhal

Abstract Two techniques of machine learning are supervised learning and unsupervised learning. In Supervised learning, data is labelled whereas in a non-supervised learning algorithm input data is unlabelled. Supervised learning is further divided into classification and regression and unsupervised into clustering. This paper discusses the idea of clustering and its classification into hierarchical and partitional clustering, further discussing the types of partitional clustering, mainly K-means, and its difference over partition around medoids (PAM) and clustering large applications (CLARA) is also explained. This paper talks about the application of K-means clustering in image compression, and a practical case of compressing an image is also discussed.

Keywords Partitional clustering · K-means · Centroids · Image compression

1 Introduction

Data mining is the procedure of collecting valuable information and recognizing patterns from a huge aggregate of data. It extracts value out of raw data. With the elevation in the amount of data being encountered, the demand for classifying and analysing that data is also increasing.

This classification of data can either be supervised or unsupervised.

In supervised clustering, the input and preferred results are included within the data. Neural network, multilayer perceptron, and decision trees are some of the supervised models (Chitre and Dr. D. Maheswar, A Comparative Study of Various Clustering Algorithms in Data Mining).

K. Khatri (✉) · V. Singhal · S. Singhal
Manav Rachna International Institute of Research and Studies, Faridabad, India
e-mail: kanikakhatri2001@gmail.com

S. Singhal
e-mail: shefali.fet@mriu.edu.in

B. Pati et al. (eds.), *Proceedings of the 6th International Conference on Advance Computing and Intelligent Engineering*, Lecture Notes in Networks and Systems 428,
https://doi.org/10.1007/978-981-19-2225-1_35

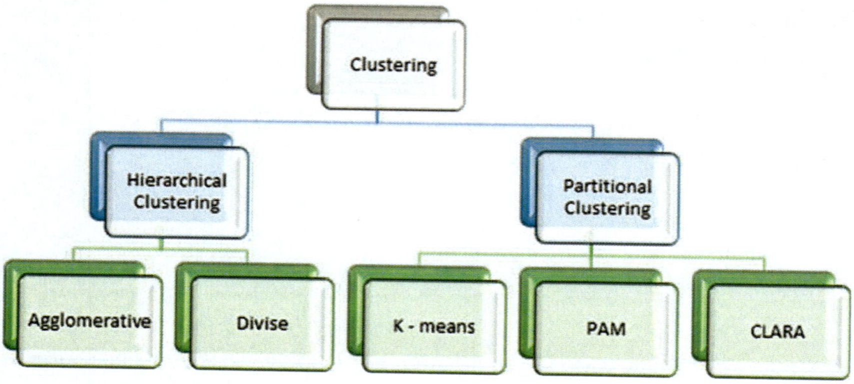

Fig. 1 Types of clustering

In unsupervised classification, data with similar observation is clubbed together rather than predicting a target. A requisite means for handling the data is to classify or group it into clusters [1].

The main idea of clustering is to increase proximity in data points belonging to the same cluster and increase dissimilarity among various clusters [2]. Clustering is an unsupervised technique of grouping similar data in the form of clusters, such that each object has high similarity within the cluster but differs in comparison to objects in another cluster. Clustering has its applications in a lot of areas, with data mining, machine learning, biology, and statistics [3] To have a clear understanding of clustering take an example of a grocery store, where similar vegetables are grouped in the same section and fruits are grouped in a section different from that of vegetable so that it becomes easy to find things.

Clustering can be categorized into three main approaches, namely hierarchical, partitional, and density-based approaches (Fig. 1).

Hierarchical clustering is done in two ways either by merging two similar clusters into one known as agglomerative or by breaking one cluster known as divisive. For categorical data, hierarchical clustering is considered to be more suitable, as long as the similarity measures are defined. Typical methods of hierarchical clustering are Diana, Agnes, balanced iterative reducing and clustering using hierarchies (BIRCH), robust clustering algorithm for categorical attributes (ROCK), and CHAMELEON. In partitional clustering, iterative relocation algorithms are used.

According to this algorithm, a given clustering criterion is minimized by iteratively relocating data points within a cluster to achieve an optimal partition. Partitional clustering is considered to be a better and faster approach for clustering as compared to hierarchical, since it looks for stronger assumptions like the number of k-clusters and their initial centres, and has comparatively better running time, resultant clusters, and input parameters. Some of the most important partitional clustering algorithms are K-means, partition around medoids (K-medoid) and clustering large applications (CLARA) [4].

In this paper, we have discussed the K-Means clustering algorithm, and why it is more preferable to PAM and CLARA, and mainly its application in the field of image compression [5].

Algorithm	Scalability and efficiency	Noise	Input data	Shape of the cluster
K-means	Scalable and efficient in processing a large amount of data set	Sensitive towards outliers and noise	Numerical data	Cannot handle arbitrary structures and focuses on convex shaped clusters
PAM (Chitre & Dr. D. Maheswar, a comparative study of various clustering algorithms in data mining)	Scalable only on comparatively small data sets	Not so sensitive towards noise and outliers	Data of all attributes	Convex shape is not must
CLARA	Scalable in processing large data sets and efficiency depends on sample size	Not so sensitive towards noise and outliers	Data of all attributes	Convex shape is not must

1.1 K-Means Clustering

One of the most commonly used partitional clustering algorithms is the K-means, in which a given set of data is divided into k number of groups also called k-clusters [6]. This grouping of data is done keeping the intracluster similarity high, thus making data in one cluster highly dissimilar to that of another cluster.

The following steps are involved in the K-means algorithm:

(1) Indicating the total number of clusters that are going to be generated.

1. The algorithm initiates by randomly choosing the k number of objects from a data set and initializing them as the centres of their particular cluster, also known as the centroids or cluster means.
2. After defining the centroids, the next step is to categorize the other remaining objects on the basis of closeness to the centroids. Objects are assigned their closest centroid for a cluster as per the Euclidean distance between that object and its cluster mean.

Where p refers to the centroid point and q refers to the object point.

1. After assigning the objects to the clusters, the mean value is recomputed for each cluster. This step is termed cluster centroid update. Since every cluster now has an updated centroid or cluster mean, each object distance is re-evaluated and checked if any change in the cluster is required.
2. These two steps for assigning clusters and updating centroid are iteratively repeated until there comes a point when no further changes in the cluster are being observed (or until convergence is reached).

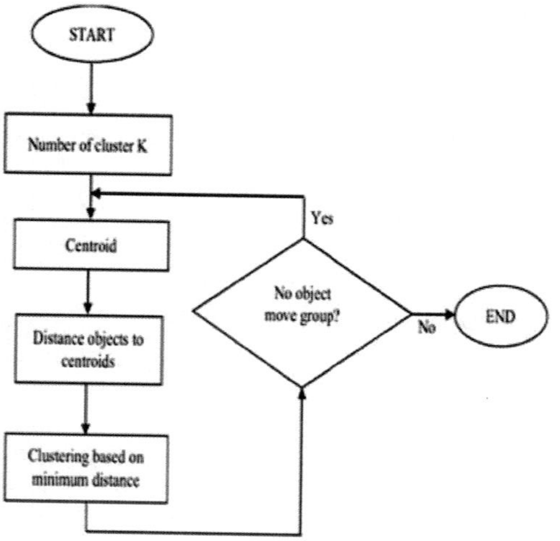

K-means algorithms do not guarantee convergence to the global superlative. The result of the algorithm completely varies with the selection of the cluster centroids; thus, it is preferred to run the algorithm multiple times by taking different initial centroids. The time complexity of K-Means is 0 (nkt), where n refers to the total number of objects in the data set, k is for the number of clusters and t for the total number of iterations. In general, the value of n is much larger than that of k and t. Thus, this method is considered to be relatively more scalable and efficient in processing a large number of data sets.

1.2 *Image Compression*

Image compression is the type of data compression done by reducing the number of bits required to represent an image to reduce their cost for transmission and storage. Image compression can be lossy and lossless. Lossless compression techniques can be used when raw images contain vital information and information might be destroyed by compression. With the knowledge of K-means clustering lossy image compression can be done. As the compression done is lossy the original image cannot be recovered back [7].

With the increase of compression ratio, the difference between the compressed image and the original image also increases. During image compression, only minor changes are done keeping in mind that dimension remains the same and just size changes [8].

1.3 Implementing K-means Clustering for Image Compression: Analysing the Properties of an Image

For accessing an image, it is necessary to set a standard view size and dimensions, thus attributes like size, rows, columns, and channels in an image are identified first. The image size, here indicates the total number of pixels and the image shape indicates the row, column, and channels of the image. Let us assume that an image contains three channels, considering it is a coloured image, but in the case of a black and white image the channel number would have been one [8].

Calculations

But practically the human eye can visualize very few colours so the K-means algorithm is used for image compression for clustering colours having different intensities but are indistinguishable by the human eye.

With the help of K-means clustering, these 16 million combinations are reduced to only 160 colours and the colour space is visualized again.

Here, reducing to 160 colour combinations means that 160 centroids are identified, making 160 clusters and examining them in our application of image.

Originally in a coloured image, each pixel is made up of primary colours RGB, i.e. red, green, and blue. Pixel would have a size of 3 bytes, i.e. 1 byte for each colour.

Since, 1 byte = 8 bit

This means, 28 combinations = 255 combinations of one colour

Since RGB consists of 3 bytes, the total number of combinations will be

= 255 * 255 * 255 combinations

=> 16,581,375 combinations

compression, the number of multiple unique colours is clustered and reduced to 160 keeping the same overall appearance. To all 160 clusters, a different RGB value will be assigned.

The principle of K-means clustering followed in image compression is

- Selected K-clusters < the number of image pixels N.
- Each pixel of the image is considered as a data point, clustering is done with the K-means algorithm to obtain the centroid $\mu\square$.
- Storing the centroid and the index of the centroid of each pixel, so it does not need to keep all the original data [9].

2 Conclusion

Due to increasing technology, data is also increasing and becoming difficult for humans to manage. Data mining can be considered as a solution to this problem; thus, it is becoming an area of interest for many researchers. It provides many algorithms and methods. Clustering is one of the most useful and upcoming data mining algorithms.

K-Means clustering is considered to be the most convenient approach for solving clustering tasks especially spherical and for getting an idea about the data set by iteratively grouping similar data points together around a centroid. K-Means clustering does a great job in image manipulation, i.e. image compression and image segmentation. K-means algorithms perform quantization of colours present in an image, thus compressing it.

3 Future Scope

A large portion of this increasing data is filled with images. Millions of pictures are being uploaded on the web on a daily basis through social media sites and cloud storage platforms. Thus, it becomes necessary to group and analyse these images. K-Means clustering algorithm is becoming extremely popular due to its application in image compression by grouping similar images, compressing them and thus reducing the storage.

Acknowledgements This work is supported by Manav Rachna International Institute of Research and Studies. This paper could not be completed without the constant guidance of our teacher Ms. Shefali Singhal. And we would also like to thank all the researchers mentioned in the reference for providing us with their excellent work in this field.

References

1. Sinaga, K. P., & Yang, M. (2020). Unsupervised K-means clustering algorithm. *IEEE Access, 8*, 80716–80727. https://doi.org/10.1109/ACCESS.2020.2988796
2. Bindra, K., & Mishra, A. (2017). A detailed study of clustering algorithms. In *2017 6th International Conference on Reliability, Infocom Technologies and Optimization (Trends and Future Directions) (ICRITO)*, Noida, India, pp. 371–376. https://doi.org/10.1109/ICRITO.2017.8342454
3. Patel, D., Modi, R., & Sarvakar, K. (2014). *A comparative study of clustering data mining: Techniques and research challenges* (vol. iii, pp. 67–70).
4. Kutbay, U. (2018). *Partitional clustering*. https://doi.org/10.5772/intechopen.75836
5. A Comparative Study of Various Clustering Algorithms in Data Mining K. Chitra1 , Dr. D.Maheswari2 Research Scholar, School of Computer Studies, Rathnavel Subramaniam College of Arts and Science, Sulur, Coimbatore1 Head, Research Coordinator, School

of Computer Studies- PG, Rathnavel Subramaniam College of Arts and Science, Sulur, Coimbatore

6. Anitha Elavarasi, S., & Akilandeswari, J. (2011). *A survey on partition clustering algorithms, international journal of enterprise computing and business systems.*
7. Wang, C.-W., & Jeng, J.-H. (2012). Image compression using PCA with clustering. In *2012 International Symposium on Intelligent Signal Processing and Communications Systems.* https://doi.org/10.1109/ispacs.2012.6473533
8. Zheng, X., Lei, Q., Yao, R., et al. (2018). Image segmentation based on adaptive K-means algorithm. *Journal of Image Video Processing, 2018*, 68.
9. Na, S., Xumin, L., & Yong, G. (2010). Research on k-means clustering algorithm: An improved k-means clustering algorithm. In *2010 Third international symposium on intelligent information (Chitre, A Comparative Study of Various Clustering Algorithms in Data Mining) Technology and Security Informatics*, Jian, China, 2010, pp. 63-67. https://doi.org/10.1109/IITSI.2010.74
10. Banerjee, A., Pujari, A. K., Panigrahi, C. R., Pati, B., Nayak, S. C., Weng, T. H. (2021). *A new method for weighted ensemble clustering and coupled ensemble selection, connection science* (pp. 1–22, 2021). Taylor & Francis.

Application of Watershed Algorithm in Digital Image Processing

Sumant Sekhar Mohanty and Sushreeta Tripathy

Abstract Segmentation of image is the method of separating objects from its background. It is helpful finding and deciding about which pixel belong to which objects. It is the process of making pixels of every region has similar visual characteristics. The watershed algorithm provides a complementary approach to the segmentation of an object. It is essential for segmenting objects where they touch their boundaries. Watershed is a dividing ridge between drainage areas. In digital image processing, the banks are the watershed lines and the drainage areas are catchment basins. Watershed is the representation of the grayscale image as topographic relief. Watersheds with adjacent catchment basins are being built during the repeated flooding of the gray value relief. This flooding method is performed on the gradient image. The magnitude of gradient value is intensely sensitive to image noise. So noise value plays a vital role in watershed processing. In computer-world watershed is a classical algorithm used for image segmentation. In this paper, we will witness the approaches of the watershed transformation in an image with a proper analysis of its advantages over other methods.

Keywords Image segmentation · Watershed · Catchment basins · Digital image

1 Introduction

Image segmentation is an essential process for most image analytics task. The objects in the study of an image may be: blood cells, stars, micro array elements, or quantum dots. It is the method of dividing images into regions according to their properties. Different colors and objects are the parts of the stuff of an image. This process is

S. Sekhar Mohanty
Department of CSE, Siksha 'O' Anusandhan (DU), Bhubaneswar, Odisha, India

S. Tripathy (✉)
Department of CA, Siksha 'O' Anusandhan (DU), Bhubaneswar, Odisha, India
e-mail: sushreetatripathy@soa.ac.in

B. Pati et al. (eds.), *Proceedings of the 6th International Conference on Advance Computing and Intelligent Engineering*, Lecture Notes in Networks and Systems 428,
https://doi.org/10.1007/978-981-19-2225-1_36

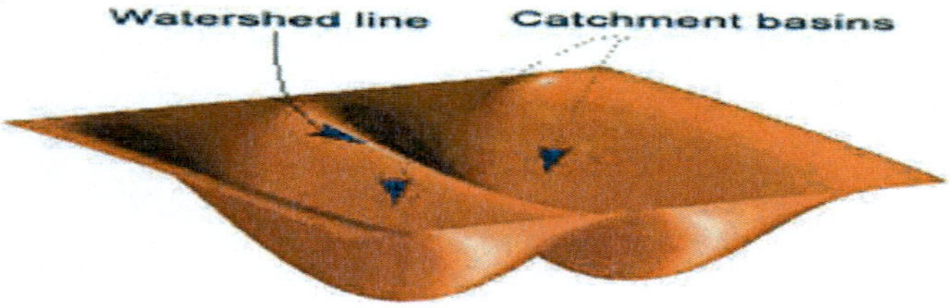

Fig. 1 Catchment basins and watershed line of an image having two objects [https://in.mathworks.com]

beneficial for the identification of homogeneous areas of the picture [1–3]. Homogeneous is pixel areas that are belonging to the same thing on that image. Techniques of image segmentation have been the subject of considerable research activity over the last three decades [4, 5].

Watershed segmentation is a morphological-based method of image segmentation. The magnitude of the steepest slope of the gradient value of an image is the topographic surface (surface shape and roughness) for the watershed transformation. Flooding of gray value relief starts from the lowest topographic point as shows in Fig. 1. If you imagine that shining areas are "high" and the dark regions are "low", then a 2D image can represent as a 3D surface, taking height as grayscale brightness magnitudes and the ridge line as watershed line and darker regions as catchment basins [6].

The initial step of the watershed algorithm is to extracting background and foreground, and then it runs and detects the exact boundaries. This algorithm generally helps in detecting touching and overlapping objects in an image.

2 Related Literature

The watershed transformation follows two types of water filling methods, continues and discrete. Broad application of digital image comes under the discrete technique as it has no unique definition of the path a drop of water would follow. Numerous algorithms have evolved to evaluate watershed transformation in discrete cases [7]. These algorithms classified into two types, immersion and topographical distance techniques. Immersion technique introduced by vincent soile where topographical distance technique proposed by Meyer. There are a few applications of continues function in the digital images. In the other hand, the discrete methods accept widely in day to day life application of the digital image. A difficulty which arises for digital images is the appearance of plateaus. Plateaus are regions of constant gray value which extend over large pixels of the picture. This unpleasant occurrence of gray value leads to a problem for parallel computation of watershed [8, 9]. In 2020, S. Tripathy and T. Swarnkar proposed median filter is a reasonable methodology while contrasted with different techniques, since picture quality of median filter is better.

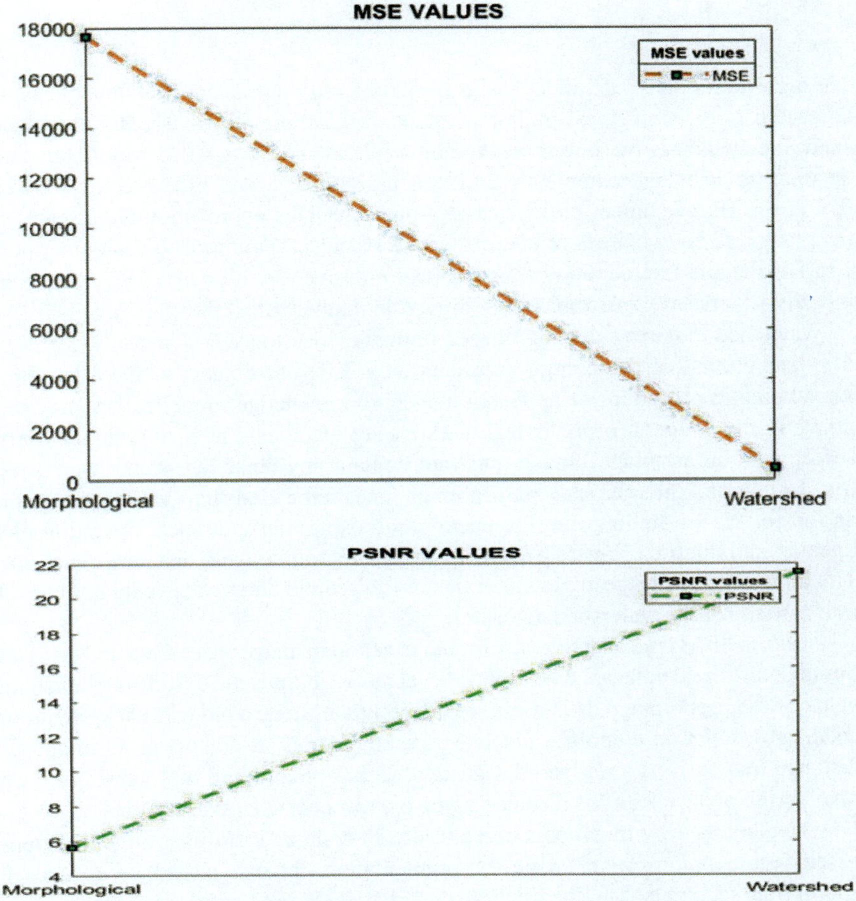

Fig. 2 **a** Watershed shows lower MSE value. **b** Watershed shows higher PSNR value

A near examination is complete by execution of filters dependent simulated output parameters PSNR and MSE [10]. In 2020, S. Tripathy and T. Swarnkar proposed a DWAMF filter inadequate to clear the excessive amount of impulse noise. It performs well overall anyway it fumbles when the likelihood of non-impulse noise and impulse noise occurrence turns out to be high [11].

3 Methodology

The algorithm floods catchment basins from the local minima until catchment basins attributed to different local minima of another. Catchment basins meet on watershed lines which is the line of separation of them. The watershed transforms and decompose an image completely and thus assign each pixel either to a region of the object, IE, catchment basin, or watershed line. This approach aims to segment the picture when two areas of interest (pixels of objects) are near to each other [12, 13]. Digabe and Lantuejoul (1977) and later improved by Li et al. (2003) proposed initially watershed transformation in grayscale mathematical morphology [14].

Watershed requires selection of seed points interior to each object of the image. The background of the picture considers as a different object for the algorithm. An automatic procedure of application-specific knowledge model marks the seed points in the image. Morphological methods are taken care after drawing of every seed points on an image. Image must not contain any noise before the watershed transformation. The watershed algorithm includes three vital steps which operate on an image. We are finding the sure background using morphological operation like opening and dilation. We are finding the sure foreground using distance transform. The unknown area is the area neither lies in foreground and background nor used it as a marker for the watershed algorithm.

Otsu's method is an adaptive way for binarization in image processing and takes all possible threshold values, i.e., 0–255 for evaluation. It finds the best fitting threshold value of the input image. It is useful to differentiate foreground and background of an images. Dilation combines pixels to the boundaries of objects in an image. A distance transform is a derived design of a digital picture and map labels of each pixel of the picture with the distance to the nearest obstacle pixel [15, 16].

Watershed is very much conducive to use in contour identification and region-based segmentation on an image. Its segmentation depends on ridges to achieve appropriate segmentation. The ridges in contour detection of an image are the boundaries of the object. However, the edges of the object consider as ridges in region-based segmentation. There are various watershed methods which use its distance measure techniques. However, they all agree to fulfill the first common goal to avoid over-segmentation is that the lines of watershed seems as the points of equidistant between two adjacent seeds or minima [17].

3.1 Pseudo Code for Watershed Transformation

Steps

1. Img = imread(I)
2. Rows = len(Img).Height Columns =len(Img[0]).Width
3. R=numpy.zeros([Rows,Columns])
   G=numpy.zeros([Rows,Columns])
   B=numpy.zeros([Rows,Columns])
4. B,G,R = cv2.split(Img)
5. Img = cv2.merge((B,G,R))
6. Gray = cv2.cvtColor(Img, cv2.COLOR BGR2GRAY)
7. Apply of Otsu's binary threshold in the grayscale image
   Optimal threshold=0
   I = 0
   Finding variance of each pixel's gray scale value by the function

$$\sigma^2 = \frac{\sum_{n=0}^{N}(x_i - \mu)^2}{N} \quad (1)$$

   REPEAT UNTIL I (iteration count) = N (total pixels)
   IF Variance < *Optimal threshold*
   Then *optimal threshold* = Variance
   ELSE I = I+1
   Otsu's binary threshold = Variance

8. Filling little holes inside the foreground objects and tiny black spots object
   Closing=cv2.morphologyEx(Img, cv2.MORPH CLOSE, kernel)

9. Apply morphology function
   (1) W F = weight of foreground
   V F = variance foreground
   W B = weight of foreground
   V B = variance background
   (2) K = Min([(W F * V F) + (W B * V B)])

10. Sure background = cv2.dilate(closing, kernel, iterations= 3)
11. Distance transform = cv2.distanceTransform(K, cv2, DIST L2,3)
   New_threshold=0.1×(maximum() of distance matrix(Distance transform))
   Image pixelrowsno = rows no. of input image pixel
   Image pixelcolumnsno. = columns no. of input image pixel
   N = Image pixelrowsno × Image pixelcolumnsno.
   Max trheshold value = 255

Iteration = 0
REPEAT UNTIL Iteration = N
IF Pixel value > threshold value then Pixel value = 1
ELSE Pixel value = 0 and Iteration = +Iterations

12. Unknown regions = Image – (Foreground region + Background region)
13. Background regions = 1
    Unknown regions = 0
14. Resultant image= Watershed flooding (Image)

4 Experimental and Theoretical Application

The watershed algorithm mainly focuses on the field where the specific object and its precise boundary detection are essential. Some of its vast areas of application include medical, material science, pattern recognition, visualization, and image compression.

With the increasing use of medical imaging approaches such as MRI, pathology research, and surgical guidance, picture segmentation has become a important tools for medical image analysis. However, due to the ambiguity of the medical image itself and the complexity of the human anatomy, how to effectively remove the corresponding human tissue is still a challenging task. Watershed is a common technique which is used in image segmentation, and it is also a well-known application in the field of preprocessing of digital pictures [18, 19].

5 Result Description

The proposed algorithm works on an image in various chronological steps. Some of them do as a preprocessing step for an accurate watershed transformation later. Detection of the threshold value is used as a value to differentiate foreground, background, and object lines. Morphological operations entirely take some of the processing time for the shaping of the objects in an image. Then, the dilation process takes place for adding extra pixels on the boundaries of the items in the picture. Distance transform is what takes place to map every pixel of the resulting output from previous steps and counts the distance to the nearest obstacle pixel. Foreground and background are designed separately with the contrasting colors for easy identification. Finally, the watershed transformation takes place by flooding basins and ridge lines. The outcome of the watershed technique is broad segmentation, border finish, and larger accuracy. It can achieve one-pixel wide, connected, closed, and exact position of the outline. Providing a closed shape of the objects, requiring low computational time, finding fast, simple and intuitive methods, and being able to produce separated division in separated regions are some of the advantages of watershed transformation. However, there is couple of disadvantages of the watershed as over-segmentation and under segmentation.

Table 1 Evaluation of PSNR and MSE on dissimilar images

Images	PSNR	MSE
Morphological image	5.6672	17,634.3735
Watershed image	21.4677	463.7740

In this paper, we witness the above process implemented by Python. After this process, PSNR values of source image, morphological image, and resulted watershed image counts as implemented by MATLAB (Table 1).

As we discussed watershed image is high with accuracy, as its PSNR value is high compared with other images. Figure 3a shows the original satellite image of Hong Kong International Airport which is taken as the source image for implementing watershed. Figure 3b displays the output of watershed transformation on the source image. Figure 4a shows the output image of Otsu's binary threshold implementation on the grayscale image described on seventh step of algorithm. Figure 4b shows resultant image of morphology implementation on the Otsu's threshold image described on ninth step of algorithm. Figure 5a is the distance matrix image on the dilation implemented image as described on 11th step of algorithm. Figure 5b shows foreground and background detections of the image from distance transformed image. Figure 6a shows implementation of the dilation on the image on the morphology applied image as described on tenth step of algorithm. Figure 6b is application of

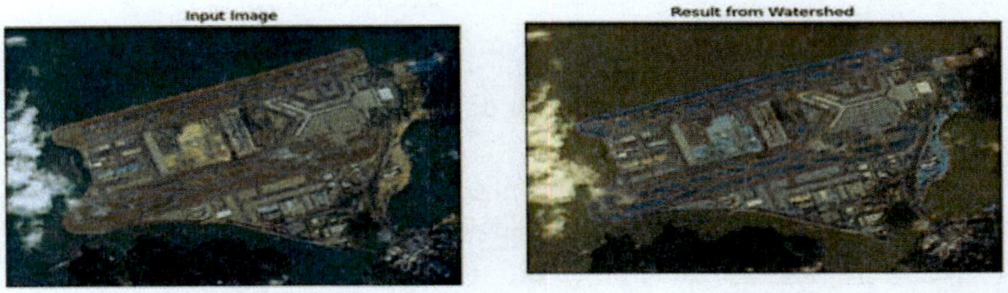

Fig. 3 **a** Input image—satellite image of Hong Kong. **b** Final result from the watershed transformation international airport image

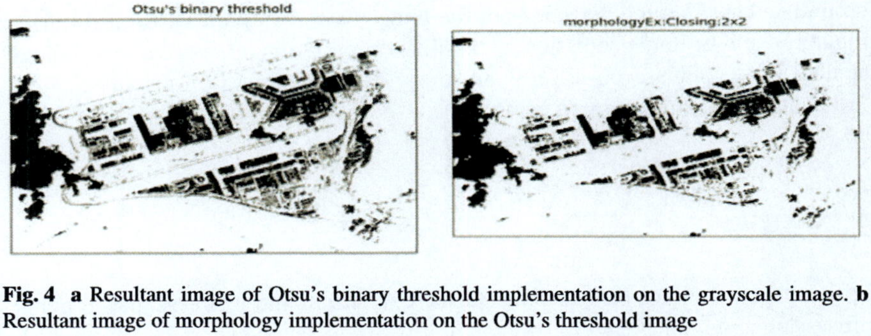

Fig. 4 **a** Resultant image of Otsu's binary threshold implementation on the grayscale image. **b** Resultant image of morphology implementation on the Otsu's threshold image

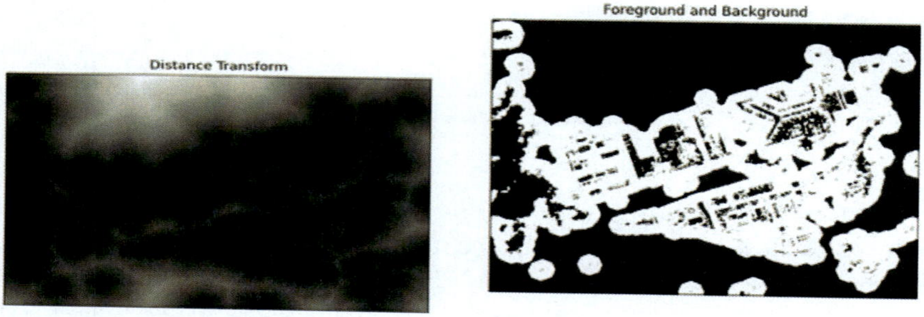

Fig. 5 **a** Distance matrix transformation image. **b** Foreground and background detections of distance matrix transformation image

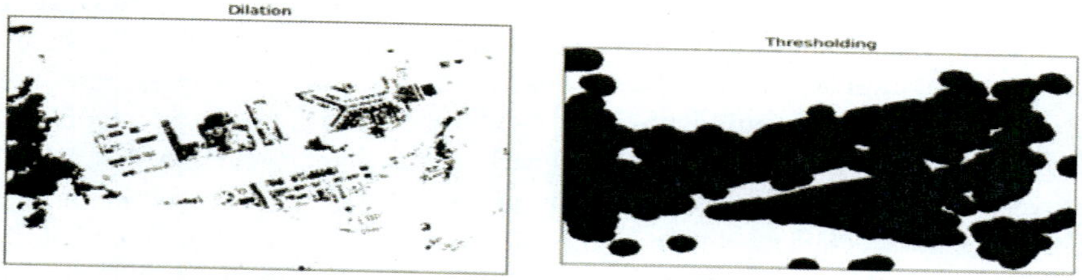

Fig. 6 **a** Implementation of the dilation on the image of the morphological image. **b** Application of thresholding on the image from resultant distance transformed image

thresholding on the image from resultant distance transformed image as described on 12th step of algorithm.

Figure 3a shows the original satellite image of Hong Kong International Airport which is taken as the source image for implementing watershed where Fig. 3b is the output of watershed transformation on the source image. Figure 4a shows the resultant image after applying of Otsu's binary threshold implementation on the grayscale image described earlier. Figure 4b is resultant image of morphology implementation on the Otsu's threshold image. Figure 5a, in the page seven, shows resultant image of the distance matrix after implemented on the dilated image. Figure 5b shows foreground and background detections of the image from distance transformed image. Figure 6a shows implementation of the dilation on the image on the morphology applied image. Figure 6b is resultant image of thresholding on the image from resultant distance transformed image.

6 Conclusion

In this paper, we have beheld the watershed algorithm and it is essential in image processing and image learning. Although it is beneficial and conducive for various fields of image processing, it has its disadvantages of over and under segmentation. We have over viewed all these above advantages and disadvantages shortly. We have implemented the watershed algorithm on the satellite image of Hongkong airport and analyzed its effectiveness. The same algorithm can be applied to other sensitive image datasets as well.

References

1. Kornilov, A. S., & Safonov, I. V. (2018). An overview of watershed algorithm im- plementations in open source libraries. *Journal of Imaging* , *4*(10), 123.
2. Tripathy, S., & Swarnkar, T. (2020). *Application of big data problem solving framework in healthcare sector- recent advancement. Intelligent and cloud computing*. Springer.
3. Tripathy, S., & Swarnkar, T. (2019). Imaging and machine learning techniques used for early identification of cancer in breast mammogram. *International Journal of Recent Technology and Engineering*, 7376–7383.
4. Tripathy, S., & Swarnkar, T. (2020). *Investigation of the FFANN model for mammogram classification using an improved gray level co-occurances matrix* (pp. 4214–4226)
5. Tripathy, S. (2019). Performance evaluation of several machine learning techniques used in the diagnosis of mammograms. *International Journal of Innovative Technology and Exploring Engineering (IJITEE)*, 228–232.
6. Lu, Y., et al. (2019). An improved watershed segmentation algorithm of medical tumor image. In *IOP Conference Series : Materials Science and Engineering*, Vol. 677. No. 4. IOP Publishing.
7. Tai, X.-C., et al. (2007). Level set methods for watershed image segmentation. In *International Conference on Scale Space and Variational methods in computer Vision*. Springer.
8. Ng, H. P., et al. (2006). Medical image segmentation using k-means clustering and improved watershed algorithm. In *IEEE Southwest Symposium on Image Analysis and interpretation*. IEEE.
9. Tripathy, S. (2014). Solving muitiple travelling salesman problem using modified ant colony optimization algorithm: A theoretical approach. *International Journal of Engineering Technologies in Computation and Applied Sciences*, pp. 146–151.
10. Tripathy, S., & Swarnkar, T. (2020). A comparative analysis on filtering techniques used in preprocessing of mammogram image. In *Advanced Computing and Intelligent Engineering*, Springer, pp. 455–464.
11. Tripathy, S., & Swarnkar, T. (2020). Performance observation of mammograms using an improved dynamic window based adaptive median filter. *Journal of Discrete Mathematical Sciences and Cryptography, 23*(1), 167–175.
12. .Tripathy, S., Hota, S., & Satapathy, P. (2013). MTACO-miner: Modified threshold ant colony optimization miner for classification rule mining. *Emerging Reserch in Computing, Information, Communication and Application*, 1–6.
13. Tripathy, S., & Hota, S. (2012). A survey on partitioning and parallel partitioning clustering algorithms. In *International Conference on Computing and Control Engineering*, Vol. 40.
14. Li, H., Elmoataz, A., Fadili, J., & Ruan. (2003). *An improved image segmentation approach based on level set and mathematical morphology*.
15. Tripathy, S., & Singh, R. (2021). Convolutional neural network: An overview and application in image classification. In *Advances in intelligent systems and computing*, Springer.
16. Tripathy, S., & Swarnkar, T. (2020). Unified preprocessing and enhancement technique for mammogram images. *Procedia Computer Science, 167*, 285–292.
17. Zhang, H., Tang, Z., Xie, Y., Gao, X., & Chen, Q. (2019). A watershed segmentation algorithm based on an optimal marker for bubble size measurement. *Measurement, 138*, 182–193.

18. Tripathy, S. (2021). Detection of cotton leaf disease using image processing techniques. In *International Conference on Computational intelligence & Energy Advancements (ICCIEA)*
19. Mohanty, S., & Tripathy, S. (2021). Application of different filtering techniques in digital image processing. In *International Conference on Computational intelligence & Energy Advancements (ICCIEA)*.

Facial Expression Recognition Using Hyper-Complex Wavelet Scattering and Machine Learning Techniques

S. Vaijayanthi and J. Arunnehru

Abstract Human emotion recognition is an active research topic in analysing the emotional state of humans over the past few decades. It is still a challenging task in artificial intelligence and human–computer interaction due to its high intra-class variation. Facial emotion analysis achieved more appreciation in academic and commercial potential challenges mainly in the field of behaviour prediction and recommendation systems. This paper proposes a novel scattering approach for recognizing facial dynamics using image sequences. Initially, we extract the temporal information from the facial frame by applying a saliency map and hyper-complex Fourier transform (HFT). Later the extracted high-level features are fed to the scattering transform method and machine learning algorithms to classify the seven emotions from the MUG dataset. The performance of proposed wavelet scattering network was evaluated on four different machine learning algorithms and achieved a high rate of recognition accuracy in all classes. In the experimental results, K-NN exhibits the proposed architecture's effectiveness with an accuracy rate of 97% for the MUG dataset, 95.7% for SVM, 93.7% for decision tree and 91.2% naive Bayes, respectively.

Keywords Facial emotion analysis · Hyper-complex Fourier transform · Feature descriptors · Wavelet scattering

1 Introduction

Over the past few years, progress in emotion recognition has dealt with face tracking [1], voice analysis [2] and body gestures [3]. Facial expression is one of the most fundamental and spontaneous ways to identify a person's emotional state over the past decade. The unusual behaviour includes head movement, partially occluded

S. Vaijayanthi (✉) · J. Arunnehru
Department of Computer Science and Engineering, SRM Institute of Science and Technology, Vadapalani, Chennai, Tamil Nadu, India
e-mail: vaijayanthisekar@gmail.com

B. Pati et al. (eds.), *Proceedings of the 6th International Conference on Advance Computing and Intelligent Engineering*, Lecture Notes in Networks and Systems 428,
https://doi.org/10.1007/978-981-19-2225-1_37

face, temporal actions and facial micro-expressions. However, human facial expression emotions are challenging to determine the emotional states as facial expression characteristics are sensitive to external factors such as illumination conditions, light, pose variations and shadows. Researchers in this area are interested in enhancing the techniques to read frames from the video dataset and extract facial micro-expressions to improve machine prediction.

The primary approaches for facial expression recognition (FER) is (i) face component detection using Facial Action Coding System (FACS) [4]. (ii) Feature selection generally includes appearance-based features and geometric features [5]. (iii) The extracted feature vectors are passed through the machine learning algorithms to recognize the seven emotions, namely neutral, anger, fear, surprise, sad, disgust and happy.

Recently, many machine learning models [6, 7] have been emerging based on sentimental analysis and gained several advantages over traditional geometric emotion recognition techniques. This paper proposes a CNN-based image scattering transform with Haar and HFT to classify emotions based on appearance-based features. Initially, the input RGB frames are converted to grey frames, cropped, and the face is detected using the Haar cascade classifier and passed through HFT for identifying the facial regions using saliency map; later, these saliency vectors are passed to the convolution-based wavelet transform. Finally, the facial scattering features are fed to the four different machine learning algorithms for training and to predict the accuracy of the classifier from the MUG dataset for recognizing the seven facial emotions like anger, fear, disgust, surprise, joy, neutral and sad.

The paper is composed in the following manner: Sect. 2 provides the previous work done in facial-based emotion recognition. The detailed proposed workflow is explained in Sect. 3. Section 4 explains the classification approach. Section 5 describes the dataset used, performance metrics and experimental results on various machine learning algorithms on the MUG dataset, and finally, the conclusion and future work are summarized in Sect. 6.

2 Prior Research

This section aims to review the recent aspects of vision-based facial emotion recognition. Typically, facial emotion recognition involves detecting the face in the frame, extracting specific features and classifying the emotions [8]. Pre-processing [9] includes contrast adjustment, image scaling and supplementary enhancement, which helps boost facial expressions in-depth. Previously, many research articles focus on detecting AU occurrence, AU intensity and facial action points detection to classify emotions [10]. A brief survey regarding various FER methods involving visual pattern techniques such as geometric-feature-based, appearance-based, texture-based and hybrid facial image recognition techniques is discussed [11].

Anima et al. [12, 13] developed a self-organizing map describing the 26-dimensional facial point vector using the MMI dataset, especially the geometric features from eyebrow, eyelid, nose and lip are considered. The facial image is cropped and scaled with a nose as a midpoint, and other vital components of the face are incorporated physically. Machine learning [14] has become a significant source in today's world to recognize facial expression recognition with different levels of regularization and dropout with reduced overfitting. Recently, researchers [15] use Bessel transform with AdaBoost to protect the original frame's features and perceptual value in the deformed facial dataset to speed up the recognition accuracy of emotions.

From the above discussion, the recognition of facial emotions, visual information needs greater improvement in the field of emotion recognition. The solution for static facial regions is still preceding and is useful for discriminating between only a few subsets of expressions. Our method mainly focuses on identifying human emotions, in a static sequence by applying scattering features. These wavelet scattering visualizations, encode the magnitude and directional information by mapping the input RGB frame to grey colour space by the property of invariance and rotation. Finally, the feature vectors are passed through four different learning approaches to classify the emotions.

3 Proposed Approach

As seen in Fig. 1, initially, the frame is pre-processed for removing redundant features. This paper uses the MUG dataset to recognize facial expressions from static facial action sequences consisting of different emotions such as surprise, anger, fear, joy, neutral, sad and disgust. The input RGB frames are converted to grey frames, and the face is detected using the Haar cascade, cropped and passed through HFT for recognizing the facial regions through saliency map; later, these saliency vectors are passed to the convolution-based wavelet transform. Finally, the facial scattering features are fed to the machine learning algorithms like SVM, K-nearest neighbour, decision tree and naive Bayes for training and to predict the accuracy of the classifier.

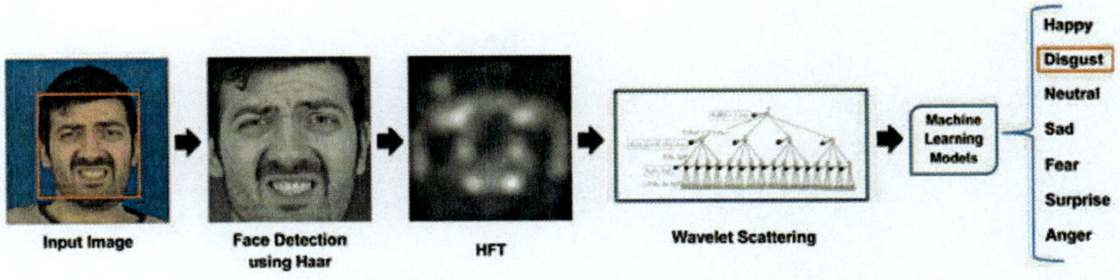

Fig. 1 Proposed approach for emotion recognition

3.1 Face Detection and Saliency Using Hyper-Complex Fourier Transform (HFT)

Haar cascade classifier is a dynamic approach used for object detection with a cascade function using various weight features and AdaBoost algorithm. A saliency map increases performance intensity by adding more feature maps such as colour and motion information in the pre-processing stage of feature extraction [16]. The hyper-complex transform matrix combines multiple feature maps to replace the normal Fourier transform. Here the input to the conventional discrete Fourier transform [17] is a real matrix represented as real numbers. This combination of features in a hyper-complex matrix forms a vector field for every element in the frame. Considering a hyper-complex matrix

$$f(n, m) = a + bi + cj + dk \tag{1}$$

The HFT of Eq. 1. in discrete version is expressed as:

$$F_H[x, y] = \frac{1}{\sqrt{MN}} \sum_{m=0}^{M-1} \sum_{n=0}^{N-1} e^{-\mu 2\pi ((\frac{my}{M})+(\frac{nx}{N}))} f(n, m) \tag{2}$$

From Eq. 2, $F_H[x, y]$ represents a hyper-complex matrix and $\mu^2 = -1$ is a pure quaternion function. Here, the absolute saliency map S_k chooses the optimal magnitude k_p. The optimal scale helps in prediction of the best saliency expressive region for entropy mapping.

$$k_p = \arg_k \min\{H(S_k)\} \tag{3}$$

where

$$H(x) = -\sum_{i=1}^{n} p_i \log p_i \tag{4}$$

The entropy x initially picks the pixel information in Eq. 4 from the histogram spatial geometry and Gaussian kernel is used to filter the traditional entropy of smoothed 2D signals. As a result, we get

$$H_{2D}(x) = H\{g_n * x\} \tag{5}$$

From Eq. 5, a low-pass Gaussian kernel g_n is obtained with scale S. By inheriting the border avoidance strategy, the area around the extracted feature is chosen by $\{S_k\}$ in HFT. The final parameter λ_k is expressed as,

$$\lambda_k = \sum \sum k(n, m).N(S_k(n, m)) \tag{6}$$

The k value in Eq. 6 refers to the 2D central Gaussian mask of equal size as s, $\sigma_w = w/4, \sigma_h = H/4, k(n, m) = 1.N(.)$ with normalized vector S. Therefore, from Eq. 3. the λ, k_p with reference to the entropy is given as

$$k_p = \arg_k \min\{\lambda_k^{-1} H_{2D}(S_k)\} \tag{7}$$

The normalized saliency vector k_p is then passed through the wavelet scattering transform for further process of classification of emotional instance before feature extraction phase.

3.2 Proposed Wavelet Scattering Transform

Wavelet scattering is a convolution image scattering decomposition technique used with explicit invariant transformations to perform translation and rotation. This paper advances Brunna and Mallat's [18] scattering coefficient's for facial feature representation. Initially, the features extracted from the Haar and HFT are passed as an input to the scattering coefficients to increase the frequency using multiple scales and orientation descriptors to recover the missing component in high frequencies.

The suggested wavelet scattering 2 frameworks have a 200 × 200 image size with a pixel scale of variance 150 and six rotations. The accurate information at various interval scales of HFT are obtained by invariance property. The primary meaning of wavelet transform is to compute the joint energy spectrum signals in the time-frequency domain, thus identifies both the frequency and time information associated with a distinct mode. Here the filter $\psi_{i,\beta}$ is constructed by scaling a filter ψ with 2^i and rotating $z \in \mathbb{R}^2$ with an angle β

$$\psi_{i,\beta}(z) = 2^{-2i}\psi(2^{-i}R_{\beta,z}) \tag{8}$$

In Eq. 8. R_z represents the z rotation and the resultant wavelet transform (f) at z is defined by

$$W_I f(z) = \left(\frac{f*\psi_{i,r}(z)}{f*\varphi_I(z)}\right)_{i<I,\beta\in\Gamma} \tag{9}$$

The $\varphi_{I(z)}$ in Eq. 9 implies low-pass frequency filter with the scattering transform. In this paper, the 2D Gaussian scaling filter act as the low-pass filter $\varphi_{I(z)}$, and thus, the wavelet coefficient amplitude is averaged as follows:

$$\left|f * \psi_{i,r}\right| * \varphi_I(z) \tag{10}$$

Convoluting the low-pass filter increases the invariance property distribution in high frequencies in Eq. 11. Hence, invariance and discriminate function in Eq. 12 reduces the variability of the coefficient by averaging the low-pass filtering and given as

$$\left|f * \psi_{i_1,r_1}\right| * \psi_{i_2,r_2} \tag{11}$$

where $i_1 < i_2, r_1, r_2 \in \Gamma$

$$||f * \psi_{i_1,r_1}| * \psi_{i_2,r_2}| * \varphi_I(z). \quad (12)$$

The wavelet coefficients are insensitive to local deformation by averaging their amplitudes. In general, scattering coefficients help to compute the cascading structure, which is analogous to a convolution network architecture.

4 Emotion Classification

4.1 K-Nearest Neighbour

The K-nearest neighbour is exclusively a statistical analysis and pattern recognition algorithm used in computer vision. This supervised lazy learning algorithm classifies the k wavelet features obtained through the votes of the neighbour pairs with high probability. The training phase saves the data and classifies the data based on the arrival of new data. If the value of k is 1, the assigned nearest neighbours will get a particular data value and vice versa when k is 0. The Euclidean distance (ED) function in Eq. 13 is a continuous distance measurement computed by calculating the approximate distance between any two pairs of points.

$$\text{ED} = \sqrt{\sum_{i=1}^{n} (x_i - y_i)^2} \quad (13)$$

4.2 Support Vector Machine (SVM)

SVM is an efficient supervised learning algorithm mainly explored for multi-class regression and classification tasks for recognizing visual patterns. The features obtained from wavelet scattering plots as a point in 'n' dimensional space and optimizing it with maximum margin with a minimum number of support vectors. Here the training set is not linearly separable, and hence, RBF kernel function is used in high-dimensional space to classify the suitable hyper-plane into seven classes.

4.3 Decision Tree (DT)

Decision tree is a supervised learning algorithm well known for its classification and regression tasks. The vital use of DT is to estimate the value of the facial emotions by learning the image wavelet features vectors inferred from the prior training data. Here roots represent the tree attributes, and the leaf nodes represent the corresponding child node of the tree. The training data is divided into subset and each sub-set obtain the value for individual attribute. The multi-class decision tree classifier uses information gain and Gini index for further classifying the human facial emotions more precisely.

4.4 *Naive Bayes (NB)*

Naive Bayes classifier [19], influenced by Bayes theorem, is most suitable for appearance-based human facial emotions. The pre-processed features from the MUG dataset containing static image sequence is passed through the probabilistic naive for classifying the seven emotions with k features. In Eq. 14, (x_i, i) is the frame pixel position, where $i = 1, 2...k$ and (y_j, j) implies the final emotional class instance and it is expressed as

$$Y_{nb} = \arg\max P(y_j) \prod_{i=1}^{n} P\left(\frac{x_i}{y_j}\right) \tag{14}$$

5 Experimental Results

The demonstrations are implemented in Windows 10 operating system with Intel Core i7 processor with 3.40 GHz using MATLAB 2019b. As explained in Sect. 3, the wavelet scattering feature vector extracted from the input frames is tested on different machine learning algorithms like SVM, K-NN, decision tree and naive Bayes to recognize the accuracy rate of seven different emotional states using the MUG dataset.

5.1 *MUG Dataset*

The multimedia understanding group (MUG) [20] is a facial expression dataset that contains static sequence frames aged between 20 and 35 years. Figure 2. shows the seven basic archetypical emotions like anger, disgust, fear, joy, neutral, sad and

Fig. 2 Sample frames of seven emotions from the MUG dataset

surprise expressed by 35 females and 51 males are used for this work. The recorded frames have a pixel size of 896 × 896.

5.2 Performance Metrics

The proposed approach uses an image scattering network with Haar and HFT with fivefold cross-validation strategy. Those extracted feature vectors are given to the K-NN, SVM, decision tree and naive Bayes to assess the statistical measures like accuracy (A), precision (P), recall (R) and F-measure performance, where tp is true positive, tn is true negative, fp is the false positive, fn is the false negative. Accuracy $(A) = \left[\frac{\text{tp+tn}}{\text{tp+fp+tn+fn}}\right]$ predicts the overall correctness of the facial emotions classified. Precision $(P) = \left[\frac{\text{tp}}{\text{tp+fp}}\right]$ gives the measure of perfection. Recognizing emotions in the exact way defined by Recall $(R) = \left[\frac{\text{tp}}{\text{tp+fn}}\right]$. F-measure $= 2\frac{P \times R}{P+R}$ gives the detailed mean of both precision and recall.

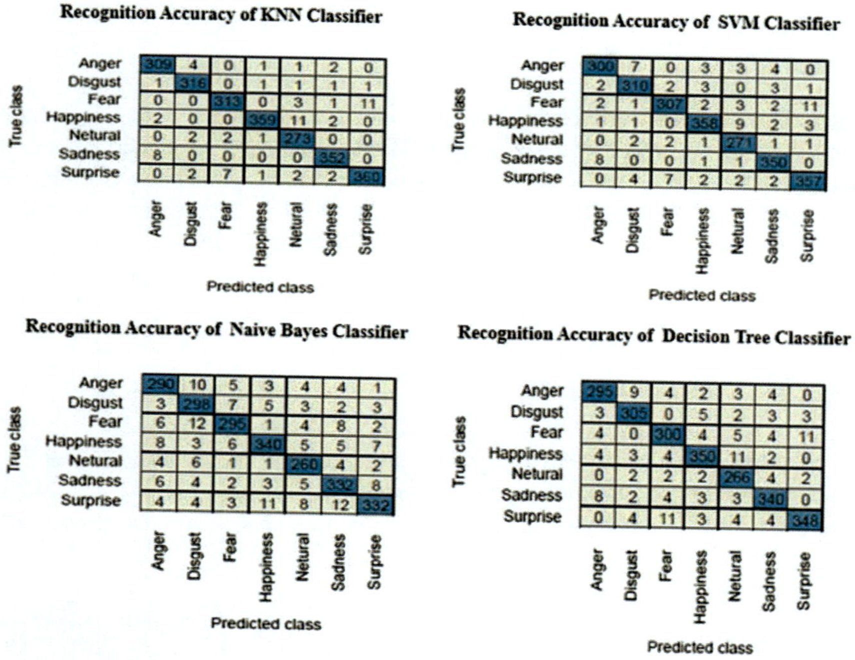

Fig. 3 Confusion matrix obtained for the K-NN, SVM , NB and DT classifier

5.3 Results and Discussion

The confusion matrix of the K-NN, SVM, NB and DT-based wavelet scattering two models on the MUG dataset is shown in Fig. 3, where the main diagonal denotes the instance correctly classified and most of the emotion classes like disgust, surprise and neutral are predicted well.The algorithms like K-NN, SVM, DT, NB performance measure are shown in Tables 1, 2, 3 and 4. The results show that K-NN gives an

Table 1 Performance measure of K-NN classifier

Class	Precision	Specificity	Recall	F-measure
Anger	0.9656	0.9946	0.9748	0.9702
Disgust	0.9753	0.9961	0.9844	0.9798
Fear	0.9721	0.9956	0.9543	0.9631
Happiness	0.9890	0.9980	0.9599	0.9742
Neutral	0.9381	0.9913	0.9820	0.9596
Sadness	0.9778	0.9960	0.9778	0.9778
Surprise	0.9677	0.9939	0.9626	0.9651
Mean (μ)	0.9694	0.9951	0.9708	0.9700

Table 2 Performance measure of SVM classifier

Class	Precision	Specificity	Recall	F-measure
Anger	0.9585	0.9936	0.9464	0.9524
Disgust	0.9538	0.9926	0.9657	0.9598
Fear	0.9654	0.9946	0.9360	0.9505
Happiness	0.9676	0.9939	0.9572	0.9624
Neutral	0.9377	0.9913	0.9748	0.9559
Sadness	0.9615	0.9930	0.9722	0.9669
Surprise	0.9571	0.9919	0.9545	0.9558
Mean (μ)	0.9574	0.9930	0.9581	0.9576

Table 3 Performance measure of Naive Bayes classifier

Class	Precision	Specificity	Recall	F-measure
Anger	0.9034	0.9848	0.9148	0.9091
Disgust	0.8843	0.9808	0.9283	0.9058
Fear	0.9248	0.9881	0.8994	0.9119
Happiness	0.9341	0.9879	0.9091	0.9214
Neutral	0.8997	0.9860	0.9353	0.9171
Sadness	0.9046	0.9824	0.9222	0.9133
Surprise	0.9352	0.9884	0.8877	0.9108
Mean (μ)	0.9123	0.9855	0.9138	0.9128

Table 4 Performance measure of decision tree classifier

Class	Precision	Specificity	Recall	F-measure
Anger	0.9395	0.9907	0.9306	0.9350
Disgust	0.9385	0.9902	0.9502	0.9443
Fear	0.9231	0.9876	0.9146	0.9188
Happiness	0.9485	0.9904	0.9358	0.9421
Neutral	0.9048	0.9865	0.9568	0.9301
Sadness	0.9418	0.9895	0.9444	0.9431
Surprise	0.9560	0.9919	0.9305	0.9431
Mean (μ)	0.9360	0.9895	0.9376	0.9367

overall recognition accuracy of 97% for the MUG dataset and 95.7% for SVM, and 93.7% for decision tree and 91.2% for naive Bayes, respectively.

From the observation, K-NN performs good in identifying the seven emotions when compared to SVM, DT and NB.

6 Conclusion

This paper proposes a novel method for recognizing seven emotions in MUG dataset using hyper-complex wavelet scattering network. The results show that K-NN gives an overall recognition accuracy of 97%, 95.7% for SVM and 93.7% for decision tree and 91.2% for naive Bayes, respectively. The different features from the Haar and HFT help to find the accuracy of quantitative metrics and increase the system's efficiency with a better recognition rate. From the experiment, it is concluded that the system could not be able to differentiate fear and sad with high accuracy. Our future research extends to recognize the various different emotional states by combining the visual body gestures with facial micro-expression recognition using real-time datasets.

References

1. Huang, Y., Chen, F., Lv, S., & Wang, X. (2019). *Facial expression recognition: A survey. Symmetry, 11*(10), 1189.
2. Vaijayanthi, S., & Arunnehru, J. (2021). Synthesis approach for emotion recognition from cepstral and pitch coefficients using machine learning. In *International conference on communication, computing and electronics systems* (pp. 515–528). Springer.
3. Arunnehru, J., & Kalaiselvi Geetha, M. (2017). Automatic human emotion recognition in surveillance video. In *Intelligent techniques in signal processing for multimedia security* (pp. 321–342). Springer.

4. Ekman, P., & Rosenberg, E. L. (1997). *What the face reveals: Basic and applied studies of spontaneous expression using the facial action coding system (FACS)*. Oxford University Press.
5. Youssif, A. A. A., & Asker, W. A. A. (2011). Automatic facial expression recognition system based on geometric and appearance features. *Computer and Information Science, 4*(2), 115.
6. Arunnehru, J., Nandhana Davi, A. K., Raghul Sharan, R., & Nambiar, P. G. (2019). Human pose estimation and activity classification using machine learning approach. In *International conference on soft computing and signal processing* (pp. 113–123). Springer.
7. Arunnehru, J., Kumar, A., & Verma, J. P. (2019). Early prediction of brain tumor classification using convolution neural networks. In *International conference on computational intelligence, security and internet of things* (pp. 16–25). Springer.
8. Michael Revina, I., & Sam Emmanuel, W. R. (2021). A survey on human face expression recognition techniques. *Journal of King Saud University-Computer and Information Sciences, 33*(6), 619–628.
9. Li, S., & Deng, W. (2020). Deep facial expression recognition: A survey. *IEEE Transactions on Affective Computing*.
10. Martinez, B., Valstar, M. F., Jiang, B., & Pantic, M. (2017). Automatic analysis of facial actions: A survey. *IEEE transactions on Affective Computing, 10*(3), 325–347.
11. Patel, K., Mehta, D., Mistry, C., Gupta, R., Tanwar, S., Kumar, N., & Alazab, M. (2020). Facial sentiment analysis using ai techniques: State-of-the-art, taxonomies, and challenges. *IEEE Access, 8*, 90495–90519.
12. Majumder, A., Behera, L., & Subramanian, V. K. (2014). Emotion recognition from geometric facial features using self-organizing map. *Pattern Recognition, 47*(3), 1282–1293.
13. Liu, X., Cheng, X., & Lee, K. (2020). Ga-svm-based facial emotion recognition using facial geometric features. *IEEE Sensors Journal, 21*(10), 11532–11542.
14. Geetha Manoharan, K., Nehru, J. A., & Balasubramanian, S. (2021). *Artificial intelligence and IoT: Smart convergence for eco-friendly topography* (Vol. 85). Springer Nature.
15. Owusu, E., Zhan, Y., & Mao, Q. R. (2014). A neural-adaboost based facial expression recognition system. *Expert Systems with Applications, 41*(7), 3383–3390.
16. Guo, C., Ma, Q., & Zhang, L. (2008). Spatio-temporal saliency detection using phase spectrum of quaternion fourier transform. In *2008 IEEE conference on computer vision and pattern recognition* (pp. 1–8). IEEE.
17. Ell, T. A., & Sangwine, S. J. (2006). Hypercomplex fourier transforms of color images. *IEEE Transactions on Image Processing, 16*(1), 22–35.
18. Bruna, J., & Mallat, S. (2013). Invariant scattering convolution networks. *IEEE Transactions on Pattern Analysis and Machine Intelligence, 35*(8), 1872–1886.
19. Noroozi, F., Kaminska, D., Corneanu, C., Sapinski, T., Escalera, S., & Anbarjafari, G. (2018). Survey on emotional body gesture recognition. *IEEE Transactions on Affective Computing*.
20. Aifanti, N., Papachristou, C., & Delopoulos, A. (2010). The mug facial expression database. In *11th international workshop on image analysis for multimedia interactive services WIAMIS 10* (pp. 1–4). IEEE.

Frequent Pattern Mining in Big Data Using Integrated Frequent Pattern (IFP) Growth Algorithm

Dinesh Komarasay, Mehul Gupta, Manojj Murugesan, J. Jenita Hermina, and M. Gokuldhev

Abstract Day-by-day, a huge volume of data are generated through electronic gadgets via recent technologies. The data are generated in different forms and thereby its complex in nature. Therefore, big data analytics techniques are introduced to analyse the complex dataset efficiently in different perspectives. Among the different perspectives, the finding of frequent pattern matching in the given dataset is addressed in this work. However, researchers introduced various algorithms for finding frequent pattern matching. But, the existing algorithms have less accuracy for predicting frequent pattern matching and take more retrieval time. In addition to that, the existing algorithms do not preprocess the dataset. Therefore, the main objective of the proposed work is to increase the accuracy by preprocessing the dataset and minimize the response time by parallel processing the dataset. In order to achieve the above objectives, this work introduces an integrated frequent pattern (IFP) growth algorithm to find the distributed frequent pattern effectively from the large dataset. Therefore, IFP growth algorithm deploys in a Hadoop platform. In Hadoop, the Hadoop distributed file system (HDFS) processes the dataset using a multiprocessor. Henceforth, the IFP growth algorithm improves the accuracy and also minimizes the prediction time. Thus, the proposed work chooses a supermarket dataset as a case study.

Keywords Big data · Modified frequent pattern · Hadoop · Online database

---

D. Komarasay (✉) · M. Gupta · M. Murugesan
Department of Computer Science and Engineering, Kongu Enginnering College, Perundurai, India
e-mail: dinesh.nova@gmail.com; dinesh.cse@kongu.edu

J. Jenita Hermina
Department of Computer Science and Engineering, Er. Perumal Manimekalai College of Engineering, Hosur, India

M. Gokuldhev
Vel Tech Rangarajan Dr. Sagunthala R&D Institute of Science and Technology, Chennai, India

B. Pati et al. (eds.), *Proceedings of the 6th International Conference on Advance Computing and Intelligent Engineering*, Lecture Notes in Networks and Systems 428,
https://doi.org/10.1007/978-981-19-2225-1_38

1 Introduction

Nowadays, most people are using electronic gadgets which generate a large volume of data. The data are generated from different sources such as media, sensors, online applications and so on [1]. Though the data generated in different sources, they are stored in different formats. Hence, it is difficult to handle and process the data with the existing technologies. In addition to that, day-by-day the rate of data generation increases into huge amounts. The huge generated data are stored in the cloud server [2]. Analysing the generated data is difficult because it has a huge amount of data. Hence, rule mining technique is used to analyse and produce the rules from which the association itemsets are identified [3]. The algorithm used in rule mining technique is the Apriori algorithm. Apriori algorithm finds the frequent itemset in the given set of items by scanning the whole dataset. It uses a candidate generation method to find the frequent itemset [4]. Hence, Apriori-based algorithm is extremely slow and requires more space as it scans the dataset many times to generate the candidate set [5].

In order to handle different varieties and huge volumes of data, big data analytics is introduced to analyse the data. Researchers introduced various algorithms for handling the complex data. In order to achieve the same functionality of finding the frequent itemset in a more efficient way, frequent pattern growth algorithm is most commonly used. This algorithm does not follow the candidate generation method. Instead of that, it represents the whole dataset in the tree-like form which maintains the association between the items [6]. The following section gives the motivation and overview of the proposed work.

1.1 Motivation and Overview of Proposed Work

From the state of the art, the existing algorithms do not preprocess the dataset. Therefore, the unwanted data or replicated data in the dataset will reduce the accuracy of the algorithms. Therefore, this work introduces integrated frequent pattern (IFP) growth algorithm to improve the accuracy of the algorithm by preprocessing the dataset. IFP algorithm predicts the frequent pattern itemset without generating the candidate keys. The proposed work has two phases, namely building an FP tree and mining the FP tree, which would be more effective than the conventional Apriori algorithm. Hadoop architecture is used to handle large amount of data in a distributed manner. Hadoop uses HDFS in which the entire data are broken into chunks and stored in the different nodes. Since the data are stored in the distributed nodes, map reduce framework is used for parallel processing of the data. Section 2 describes the literature review and followed by the proposed work.

2 Literature Review

Researchers introduced various algorithms for predicting the frequent pattern from the given dataset. From the state of the art, Jugendra Dongre proposed an approach to predict the buying pattern from the given dataset. This approach uses Apriori algorithm to find the association rules based on the minimum confidence. This approach analyses the transaction of supermarkets and predicts the frequent pattern mining. But, this approach does not support predicting the frequent pattern from large datasets [7]. Later, R. Agrawal et al. introduced rule-based approach to predict the frequent pattern from the large dataset. In this work, mining association rule technique finds the frequent itemset. In this technique, candidate itemsets are generated and compared with the minimum support. Thus, pruning technique is used to reduce the candidate itemset which reduces the number of candidate itemset. Though it supports a huge dataset, the generation of candidate sets is costlier [8]. So, J. Han et al. proposed a work called pattern mining approach to find the frequent pattern in minimum cost. The frequent pattern tree (FP tree) structure avoids candidate generation methods. This work focuses on the huge dataset to find frequent patterns. The considered large dataset is compressed into the small parts by divide and conquer method. Also, FP tree finds the frequent pattern without candidate generation such that it avoids more scans on the dataset. Hence, it is faster than the candidate generation method [9].

Similarly, William Cheung introduced a method called compressed and arranged transaction sequences trees (CATS) which works without candidate generation. Basically, CATS is the extension of the FP tree algorithm. CATS scans the dataset once rather than multiple scans like candidate generation. Once the tree is constructed, FrEquent/Large patterns mINing with CATS trEe (FELINE) method is used to find the frequent pattern in the itemset of the tree. The general concept behind FELINE is divide and conquer method. The dataset used in the work is generated by IBM QUEST [10]. But, CATS takes more time and space for scanning the database. Later, Akshita Bhandari et al. proposed a solution for minimizing the time for scanning the database and space for scanning the database. In this work, the existing Apriori algorithm is modified in such a way that it reduces the time and space usage. This is achieved by scanning the whole dataset initially and obtaining the items, minimum support and transaction ID's. Once the items and minimum support are obtained, the itemset is eliminated based on the minimum support. Instead of scanning again and again, self-join is done for finding the next candidate set. Therefore, memory space and time consumed are less in this work. Even though the work reduces the space and time, it does not focus on the uncertain data [11].

In order to find the frequent pattern in the stream of unclear data, Carson Kai-Sang Leung et al. proposed an algorithm called SUF-growth which effectively identifies the frequent pattern. In this work, the SUF tree is constructed from items in the tree which are arranged in the lexicographic order. Moreover, the tree is constructed with only one scan of the dataset. Hence it reduces the time. The stream of data is mapped to the tree. Once the tree is formed, the frequent pattern is found by the divide and conquer method. The dataset used is generated by IBM Almaden Research Center which contains nearly 1 million records [12]. In order to save more time, Ke Wang proposed a method to explore the tree in top down order. The tree is constructed by two scans. Once the tree is constricted, the mining technique is done by two pruning strategies such as TD-FP-Growth (M) and TD-FP-Growth (C) for minimum support and minimum confidence, respectively. The method generated 2051 rules in 5 s, this shows that this method executes faster than any other methods [13].

Similarly, Yaling Xun introduced to find frequent itemset. FiDoop algorithm is used in this approach to reduce the memory space using MapReduce model. Three MapReduce jobs are implemented in this approach. First, all possible frequent itemsets are identified in the first job, and the infrequent itemsets are removed in the second job. From the second job, the itemset is decomposed in the third job. The FIU tree is constructed in which the nonleaf nodes contain item-name and node-link. The leaf node contains the item-name and support. Different datasets like synthetic dataset and celestial spectral dataset are tested in this approach [14]. From the state of the art, researchers introduced various algorithms to find the frequent pattern matching from the dataset in less time and memory space. But the efficiency is lesser in the existing methods. Hence, the proposed work uses IPF growth algorithm to gain higher efficiency. The following section describes the proposed work in detail.

3 Proposed Work

In the proposed work, the dataset is collected from the online retailer shop to find the frequent pattern of buying by the customers. The collected data contains all the details of the transaction like products purchased, returned products and not delivered products. Among these, the data of the returned product might decrease the efficiency of the FP growth algorithm. Moreover, the collected dataset may also contain missing data which makes the dataset inconsistent. Therefore, this work proposes a new idea called integrated frequent pattern (IFP) growth algorithm in which removes the unnecessary and redundant data in the dataset through preprocessing. The preprocessed dataset is considered for predicting the frequent pattern in the dataset. The following section describes the integrated frequent pattern (IFP) growth algorithm to accurately predict the frequent pattern matching from the given dataset.

3.1 Integrated Frequent Pattern (IFP) Growth Algorithm

This work proposes integrated frequent pattern (IFP) growth algorithm to find the frequent pattern from the given dataset. IFP growth algorithm has two stages such as preprocessing and FP growth algorithm. Initially, IFP growth algorithm preprocesses the dataset to eliminate the unnecessary data in the dataset. After preprocessing, FP tree growth algorithm which generates the FP rules to determine the frequent item bought by the customers. The online retail dataset is obtained from the UCI repository, which contains nearly 55,000 records of transaction details of the customer. The dataset has details of successful and cancelled orders. The records are based on the UK-based online retail company which sells the gifts items. Transaction details of the customer like invoice number, product code, description about the product, quantity, date, price, customerID and country. Sometimes, the customer can cancel the order due to any reasons and thereby the invoice number prefixes with C denotes the cancelled order in the dataset. These records are removed because the cancelled order may affect the frequent pattern matching from the given dataset. From the literature, the missing values in the dataset are replaced by NA with the missing values may lead to the wrong prediction. Though the inconsistent data are preprocessed, the prediction result may not be efficient because of the negative descriptions about the product. Hence those records are removed using negative keywords like 'Wrong', 'Dislike', 'Damage', etc., and removed from the dataset. The transactions in the dataset are separated by the invoice number, this leads to the more number of records. In order to reduce the size of the dataset, all items purchased in the similar invoice numbers are grouped together as a single transaction as comma separated items. The preprocessed data are fed to FP growth algorithm.

The proposed work is to find the frequent pattern or item in the given dataset. Therefore, the frequent item has been computed in the given dataset. Initially, the frequency is calculated for all the products in the database. After that, the frequent patterns of the items are computed and the items sorted in descending order to minimize the searching time. The number of product lists will be shrinked based on the minimum support value. The minimum support values are customized by the administrator. This is achieved by eliminating the products with the frequency less than minimum support. This process of elimination is called pruning. The minimum support is calculated by choosing the minimum value of support count that generates in the first itemset. In this, the entire dataset is scanned once to find the frequency. The generated one itemset is mapped with the transaction ID to generate an ordered itemset. Here invoice number is considered as the transaction ID. The generated ordered itemset is passed as the input for constructing FP tree. Initially, the root of the FP tree is null, and the items in each transaction are linked with the root node in the FP tree along with the count of items. The count is incremented by 1 when there exists a direct link for the node with the required item. New node is generated when there is no direct link from the parent node and the new node is set with the count 1. After constructing the FP tree, conditional pattern base is identified for each item from the constructed FP tree. Conditional pattern base is the path from the starting node to the item taken at the instance. Once a conditional pattern base is generated, the conditional FP tree is constructed for each single itemset by summing the count of common items that occurs in the conditional pattern base. FP rules are generated by pairing the single itemset with the conditional FP tree (Table 1).

Table 1 Represents the sample dataset with cancelled transaction where C appends in cancelled transaction

Transactions	Items
T1	{C1, D1, A1}
T2	{A1, B1, C1}
T3	{E1, F1, A1, B1}
CT4	{C1, D1}
T5	{B1, C1}
CT6	{E1, F1}
CT7	{E1, D1, F1}
T8	{A1, B1}
T9	{D1, E1}
T10	{A1, C1, B1}
T11	{E1, A1}
T12	{A1, B1, C1}

3.2 *Illustration of IFP Growth Algorithm*

See Tables 1 and 2.

From the given dataset, a sample dataset is considered for illustration. From the dataset, the set of 12 transactions are taken into account which has eight fields such as invoice number, stock code, description, quantity, invoice date, unit price, customerID and country. Among the eight fields, this work considers only two fields such as invoice number as transaction ID and stock code as items.

Therefore, T4, T6 and T7 are removed from the sample dataset. The minimum support count is fixed as 2. Based on the minimum count, the items with the frequency less than the minimum support count are removed (Tables 3 and 4, Fig. 1).

The frequent pattern and IFP rule are different if the failed transactions are removed from the sample set. Thereby, this work achieves high accuracy of predicting the frequent pattern from the given database (Table 5).

Table 2 Represents the dataset with successful transactions

Transactions	Items
T1	{A1, C1, D1}
T2	{A1, B1, C1}
T3	{E1, F1, A1, B1}
T5	{B1, C1}
T8	{A1, B1}
T9	{D1, E1}
T10	{A1, B1, C1}
T11	{E1, A1}
T12	{A1, B1, C1}

Table 3 Table after the items are removed

Items	Frequency
A1	7
B1	6
C1	5
E1	3

Table 4 Ordered itemset of the transactions

Transactions	Ordered itemset
T1	{A1, C1}
T2	{A1, B1, C1}
T3	{A1, B1, E1}
T5	{B1, C1}
T8	{A1, B1}
T9	{E1}
T10	{A1, B1, C1}
T11	{A1, E1}
T12	{A1, B1, C1}

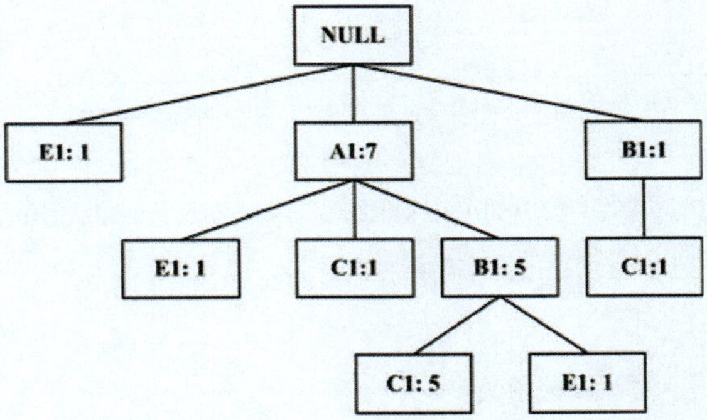

Fig. 1 IFP tree

4 Experimental Set-up and Performance Analysis

Figure 2 represents the execution time and accuracy of Apriori, FP growth and IFP growth algorithm. In IFP algorithm, execution time is less than the execution time of FP and Apriori algorithm because the failed transactions are removed from the dataset. Since the failed transactions are removed from the database, the size of the dataset is reduced, and the size of the FP tree is also minimized. Therefore, the IFP growth algorithm outperforms the other algorithms. In addition to that, the IFP growth algorithm achieves high accuracy since the incomplete transactions are removed in preprocessing. The failed transaction may repeat several time that result to the wrong

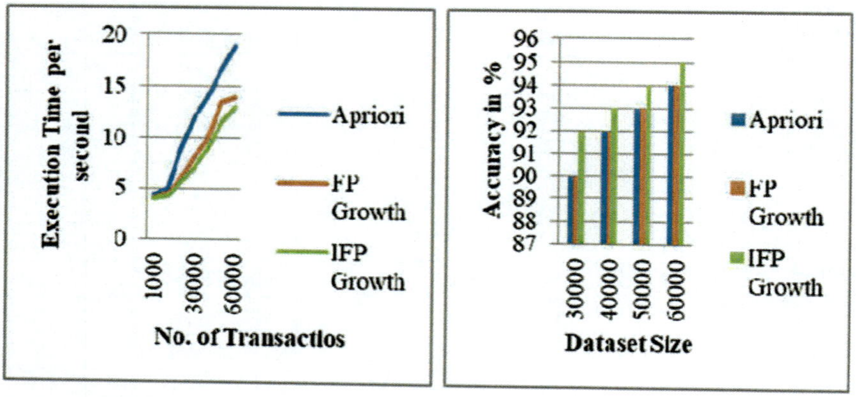

Fig. 2 Execution time and accuracy

Table 5 Conditional frequent pattern obtained from the IFP tree

Items	Conditional pattern base	Conditional frequent pattern	Rules
E1	{{A1: 1}, {A1, B1: 1}}	{A1: 2}	A1 -> E1 or E1-> A1
C1	{{A1: 1}, {A1, B1: 3}, {B1, 1}}		
B1	{A1: 5}	{A1: 5}	A1 -> B1 or B1 -> A1
A1			

From the conditional frequent pattern, rules are generated

prediction of frequent pattern. Therefore, IFP algorithm outperforms Apriori and FP growth algorithm.

5 Conclusion

The main objective is to improve the accuracy of predicting frequent patterns and to minimize the response time. In order to achieve the above objectives, this work introduces integrated frequent pattern (IFP) growth algorithm which eliminates the failed or incomplete transaction. By eliminating the failed transaction, IFP growth algorithm improves the accuracy of the predicting the frequent pattern. The size of the IFP tree is also reduced by eliminating the failed transaction, and FP rules are minimized. Thereby, the execution time for processing the dataset is also reduced in the proposed work. Moreover, the IFP growth algorithm is deployed in the Hadoop environment. The current work uses the pre-collected dataset to find the frequent pattern of buying by the customers. In future, the technique of Web crawling will be used to extract data from the Internet to find the frequent pattern of buying by the customers.

References

1. Sagiroglu, S., & Sinanc, D. (2013). Big data: A review. In *2013 International Conference on Collaboration Technologies and Systems (CTS)*, pp. 42–47. IEEE
2. Komarasamy, D., & Muthuswamy, V. (2018). Priority scheduling with consolidation based backfilling algorithm in cloud. *World Wide Web,21*(6), 1453–1471.
3. Solanki, S. K., & Patel, J. T. (2015). A survey on association rule mining. In *2015 Fifth International Conference on Advanced Computing & Communication Technologies*, pp. 212–216. IEEE
4. Aggarwal, C. C., Bhuiyan, M. A., & Al Hasan, M. (2014). Frequent pattern mining algorithms: A survey. In *Frequent Pattern Mining*, pp. 19–64. Springer.
5. Zhang, W., Liao, H., & Zhao, N. (2008). Research on the FP growth algorithm about association rule mining. In *2008 International Seminar on Business and Information Management*, Vol. 1, pp. 315–318. IEEE.
6. Pramudiono, I., & Kitsuregawa, M. (2003). Parallel FP-growth on PC cluster. In *Pacific-Asia Conference on Knowledge Discovery and Data Mining*, pp. 467–473. Springer.
7. Dongre, J., Prajapati, G. L., & Tokekar, S. V. (2014). The role of Apriori algorithm for finding the association rules in data mining. In *2014 International Conference on Issues and Challenges in Intelligent Computing Techniques (ICICT)*, pp. 657–660.
8. Panigrahi, C. R., Sarkar, J. L., Tiwary, M., Pati, B., & Mohapatra, P. (2019). DATALET: An approach to manage big volume of data in cyber foraged environment. *Journal of Parallel and Distributed Computing,131*, 14–28.
9. Han, J., Pei, J., Yin, Y., & Mao, R. (2004). Mining frequent patterns without candidate generation: A frequent-pattern tree approach. *Data Mining Knowledge Discovery in Data, 8*(1), 53–87.
10. Cheung, W. S., & Zaiane, O. R. (2003). Incremental mining of frequent patterns without candidate generation or support constraint. In *Proceedings of the Seventh International Database Engineering and Applications Symposium*, 2003. IEEE.
11. Bhandari, A., Gupta, A., & Das, D. (2015). Improvised apriori algorithm using frequent pattern tree for real time applications in data mining. *Procedia Computer Science, 46*, 644–651.
12. Leung, C. K.-S., & Hao, B. (2009). Mining of frequent itemsets from streams of uncertain data. In *2009 IEEE 25th International Conference on Data Engineering*. IEEE.
13. Wang, K., Tang, L., Han, J., Liu, J. (2002). Top down fp-growth for association rule mining. In *Pacific-Asia conference on knowledge discovery and data mining*, pp. 334–340. Springer.
14. Xun, Y., Zhang, J., & Qin, X. (2015). Fidoop: Parallel mining of frequent itemsets using mapreduce. *IEEE Transactions on Systems, Man, and Cybernetics: Systems, 46*(3), 313–325.

Image-Based Plant Seedling Classification Using Ensemble Learning

Deepak Mane, Kunal Shah, Rishikesh Solapure, Ranjeet Bidwe, and Saloni Shah

Abstract Agriculture is crucial for human survival and is a major economic engine across the world, particularly in emerging countries. Plant seed classification is a multi-class dataset with 5,539 pictures divided into 12 classes. We investigate various learning classifiers for the image-based multi-class problem in this study. We will start with a simple convolutional neural network (CNN) classifier model and work our way up to more complex options like support vector machines, and K-nearest neighbors. We will create an ensemble of classifiers to increase the current state-of-the-art accuracy. We will also investigate data preprocessing techniques like segmentation, masking, and feature engineering for an improvement in the overall precision. We will compare the performance as well as their impact on the final ensemble. To overcome this challenge, traditional techniques use complex convolution layer-based neural network architectures like Resnet and VGG-19. Though these techniques are effective, there is still scope for increasing accuracy. In this study, we propose a boosting ensemble-based strategy that employs a multilayer CNN model with a deep convolution layer that is boosted using the K-nearest neighbors lazily supervised learning technique. Although the fact that this combination is less complex than previous ways, it has obtained a higher accuracy of 99.90%.

Keywords Convolutional neural network · K-nearest neighbors · Ensemble learning · Agriculture

D. Mane
JSPM's Rajarshi Shahu College of Engineering, Pune, Maharashtra 411033, India

K. Shah (✉) · R. Solapure · R. Bidwe
SCTR's Pune Institute of Computer Technology, Pune, Maharashtra 411043, India
e-mail: kunaljshah03@gmail.com

S. Shah
KLE Technological University, Hubli, Karnataka 580031, India

B. Pati et al. (eds.), *Proceedings of the 6th International Conference on Advance Computing and Intelligent Engineering*, Lecture Notes in Networks and Systems 428,
https://doi.org/10.1007/978-981-19-2225-1_39

1 Introduction

People involved in agriculture still need to be able to classify and distinguish different plants and weeds, which takes a tremendous amount of effort eventually. Plant growth in the early season is critical for farm produce [1]. Effective weed control at the time of initial stage, according to research, is critical for higher production in some crops [2]. Herbicides having various widespread effects, for example, are used to control weeds chemically and have been popular in recent years because they can kill 90 to 99 percent of targeted weeds [3, 3]. Herbicides have several negative consequences on the biotic and abiotic environment, as well as the potential to affect human health. As a result, reducing the number of pesticides used in modern agriculture is a crucial step toward more sustainable farming. The classification job is connected to computer vision, which is centered upon Artificial Intelligence.

Machine learning can aid these personnel since it will reduce the amount of time and effort needed to identify plant seedlings. The capacity to do so efficiently can lead to greater agricultural yields and, overall, better environmental stewardship. As a result of being able to distinguish between plants and weeds quickly and accurately, this has the potential to have a favorable influence on agriculture.

CNN's accomplishments in image categorization began with the research of graphics processing units and the use of a trained model. In the agriculture domain, CNNs have recently been widely used to identify plant species [11]–12, weed detection [13], and the identification of plant diseases [14]. The authors of this work present an ensemble-based solution to this challenge. The final classifications are given by a KNN-based lazy learner model based on the features extracted from the earlier customized CNN. In contrast to the state-of-the-art models, it achieves superior results. The novelties of this approach include:

- In comparison to traditional architecture, the employment of ensemble techniques for plant seedling classification consists of fewer complicated models.
- The application of standout methods of image preprocessing including data augmentation, bilateral blurring, masking, and sharpening of the image.

The structure of the paper is given as: Sect. 2 contains background information and related literature, Sect. 3 contains information about the dataset and data preparation, Sect. 4 contains information about the suggested architecture, and Sect. 5 contains information about the findings obtained. In addition, Sect. 6 contains a comparison with the current models. Finally, Sect. 7 describes the conclusion and future scope.

2 Related Work

Machine learning algorithms for autonomous plant seedling classification have emerged as a key and encouraging research area for enhancing agricultural outcomes. Deep learning is a sort of machine learning that has sparked a lot of interest in a variety of fields. Convolutional neural networks (CNNs) are a type of deep neural network that is commonly used to evaluate visual images. In recent years, CNN's have made substantial advances in the field of computer vision. Furthermore, CNN's shown a great ability to acquire the most effective characteristics required for image categorization [9–11].

The authors of [15] supplied a dataset for weed or species spotting on the ground, as well as a baseline metric for researchers to use when comparing categorization results. In [16], the authors showed how a CNN could learn unsupervised feature representations for 44 distinct plant species with excellent accuracy.

For semantic segmentation, the authors in [17] proposed a 14-channel deep encoder-decoder CNN fed with other data, such as vegetation indices in order to solve the challenge of solely RGB data-based CNN-based segmentation of crops grown to distinguish sugar beet plants from weeds and backdrop.

The bagging ensemble approach was employed by the author in [18] to classify the CIFAR-10 dataset. For dimensionality reduction, the dataset was run through four distinct CNN models and the PCA method. The output of four CNN models was merged to achieve a precision of 93.99%, and the output of PCA was transmitted via the KNN model. With 94.03% accuracy, the output from KNN and combined CNN performed better. Primarily deep learning algorithms produced state-of-the-art results in different pattern classification problems [20–23]. However, recently many researchers tried to apply different amalgam-based approaches with deep CNN, i.e., CNN with KNN, CNN with Fuzzy, etc. [24–26] which produced better results than the traditional CNN model.

The authors presented a CNN model in their work [27]. Before being transmitted to the input layer, the seedling photos are all resized to 128 × 128 pixels. The Rectified Linear Units (ReLU) layers are linked to the convolutional layers. The global average pooling layer is used in the output layer to directly input the acquired feature maps into the feature vectors. Finally, to estimate the probabilities of each class label, a fully connected layer with n (number of classes) nodes and soft-max is created. The average test accuracy and recall for eight classes are almost 98%, with the precision and F1-score being about 99%. Various machine learning techniques are used in [4–7, 20] to simplify the problem while increasing its accuracy. We created a comparison Table V in our paper to show where our model stands in terms of accuracy.

3 Dataset and Data Processing

The dataset was created by looking into several periods of plant growth in the laboratory, as described in the paper [15]. The image database is intended for researchers that deal with plant classification, and it also serves as a presentation of data.

3.1 Dataset

There are 5,539 images in this dataset of crop and weed seedlings. Figure 1 depicts how the pictures have been divided into 12 distinct categories. The images in each class represent plants at various developmental stages and are represented as RGB images. The problem of overlapping plant leaves has been studied before [8], but they are small at the start of the growth of plants, thus the pictures were taken without them. In addition, the plants were grown in gravel-covered soil, which eliminates errors in pixel-based segmentation methods. Figure 1 shows a glimpse of the dataset.

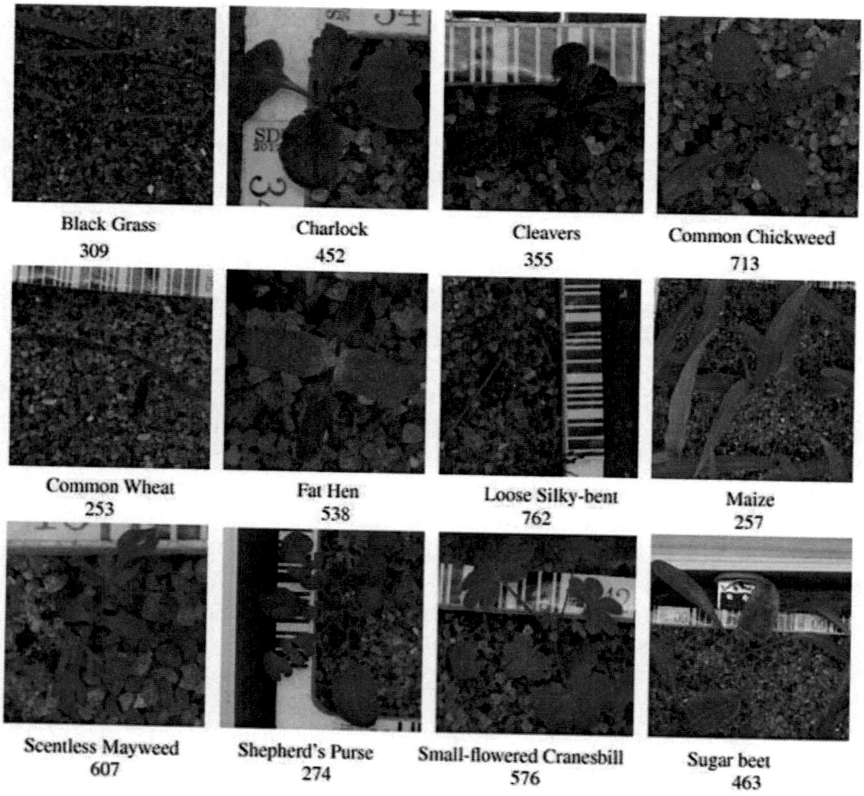

Fig. 1 Plant seedling dataset

3.2 *Data Preprocessing*

3.2.1 Data Augmentation and Bilateral Blurring

Data augmentation is the practice of creating new data to feed a machine learning model for training and testing. The more information your model has, the better it will perform. Another significance of this technique is to create an even dataset. It also helps to keep you from being too over-trained on the training set.

A bilateral filter is a non-linear option that preserves image edges while decreasing noise. It uses a weighted average of intensity values from adjacent pixels to replace each pixel's value. This ensures that the Euclidean distance between pixels remains razor-sharp so used a Gaussian weight distribution formula for this weighting.

$$(1/\mathrm{We}) * \text{Summation of}(Gs(||e - q||)Gr(|Ie - Iq|)Iq) \tag{1}$$

where

$(1/\mathrm{We})$: Normalization Factor
$(Gs(||e - q||)$: Space weight
$Gr(|Ie - Iq|)$: Range weight
Iq: Intensity at pixel q.

3.2.2 LAB Color Space and Segmentation

It uses three numbers to represent color: L* to represent perceived brightness, and a* and b* to represent the four primary hues of the human eye: red, green, blue, and yellow. There are things you cannot accomplish with RGB because LAB Color space is so expensive. Paper [19] claims that HSV is superior to LAB in terms of efficacy. However, after trying out various combinations, we discovered that LAB works out better than HSV for the proposed model. Figure 2 compares the HSV and LAB values for a sample image from a dataset used to classify plant seeds.

3.2.3 Masking and Sharpening

Figure 2 shows that the first LAB channel, which describes the hues from green to red, is significant for extracting plant information. The term "image sharpening" refers to any enhancing method that brings out the fine lines and features in a picture. Pictures that have been sharpened A sharper image is produced by increasing the amount (Fig. 3).

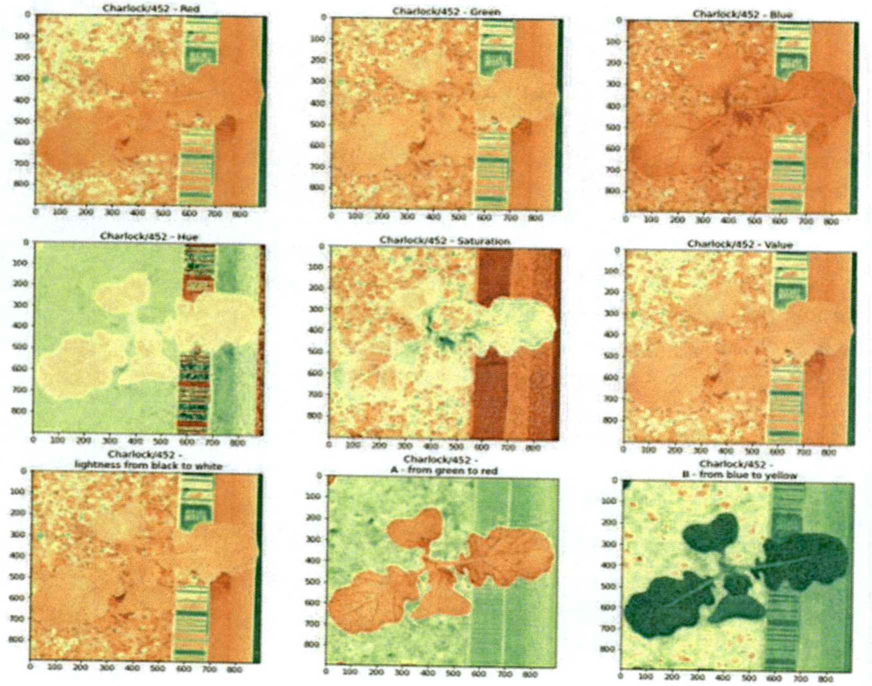

Fig. 2 Charlock 452.png HSV versus LAB images

Fig. 3 Masking of an image

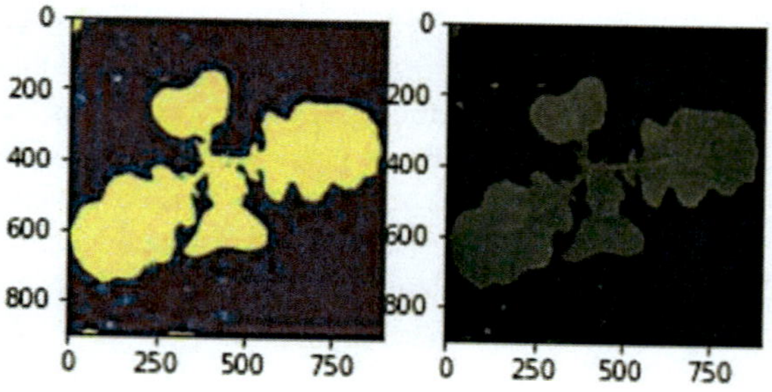

4 Proposed Ensemble Architecture

Neural networks are regarded as particularly powerful machine learning tools. We employ convolutional neural networks to extract features from images and then use these derived features as inputs to k-nearest neighbor classifiers in this study. The results show that features extracted using CNN improve the capabilities of KNNs when compared to running these algorithms on raw data and even outperform the neural network alone.

In this section, the author presents an ensemble-based framework for categorizing plant seedlings. A shallow CNN architecture is paired with a lazy learner approach of K closest neighbors to categorize the seedlings (KNN). The algorithms have an advantage in terms of faster convergence, quicker prediction, and are less likely to over fit due to their inherent lesser complexity.

4.1 Convolutional Neural Networks

For the multi-class images, we employed a simplified design of a standard convolution neural network as a feature extraction method. The network's input layer anticipates a LAB image. The input layer is fed with resized images (96 × 96 × 3). Three convolutional blocks are applied to the input image. Convolutional filters with 3 × 3 receptive fields are employed. A 2-D convolution layer function is included in each convolutional block (the number of filters changes between blocks). The activation function (nonlinearity operation) for the convolutional layers, as well as all hidden layers, is Rectified Linear Unit (ReLU), and spatial pooling is achieved using a max-pooling layer followed by a dropout layer. The formula for the ReLU activation function is $f(x) = \max(0, x)$. The hidden layer consists of 3 stages of learning layers. The filters used are all 3 × 3 kernel filters, with a total of 32, 64, and 64 filters for the various convolutional blocks. A pooling layer and a batch normalization layer follow each convolutional layer. A classifier block made up of two Fully Connected (FC) Dense layers completes the network (Fig. 4).

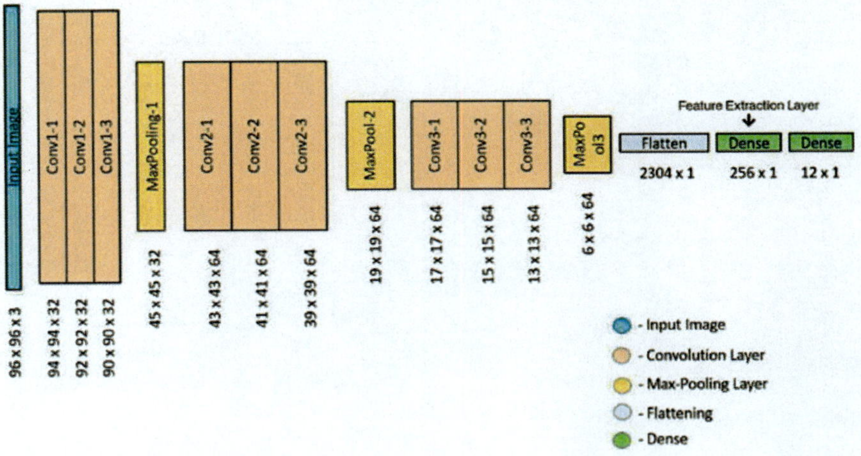

Fig. 4 CNN model architecture

4.2 K-fold Cross-Validation

A cross-validation technique is a randomized sampling method for testing machine learning algorithms on a small amount of data. The algorithm has only one parameter, k, which determines how many groups a given data sample, should be divided into. In our model, we observed that the value of $k = 5$ was an optimal solution to enhance the model accuracy the most. At the final fold, a validation accuracy of 99.29% was achieved (refer Figs. 6 and 7).

4.3 Ensembling

Ensemble learning is a generic machine learning meta method that aims to improve predictive performance by mixing predictions from many models. The features extracted from the Dense layer (Feature Extraction Layer), are the desired output of the CNN model which is then fed to the KNN model. Euclidean distance is being used in this model to classify plant seedling data (Fig. 5).

$$\text{Euclidean distance: } Dpq = Sqrt((p1 - q1)2 + (p2 - q2)2) \tag{2}$$

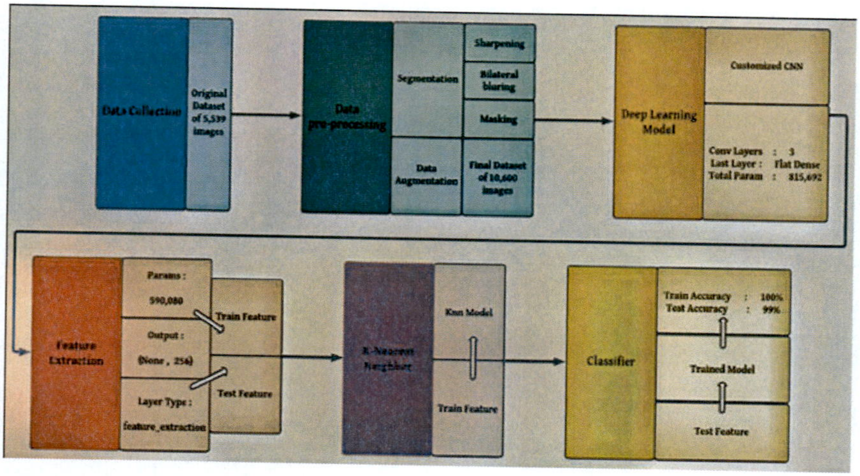

Fig. 5 Workflow of the model

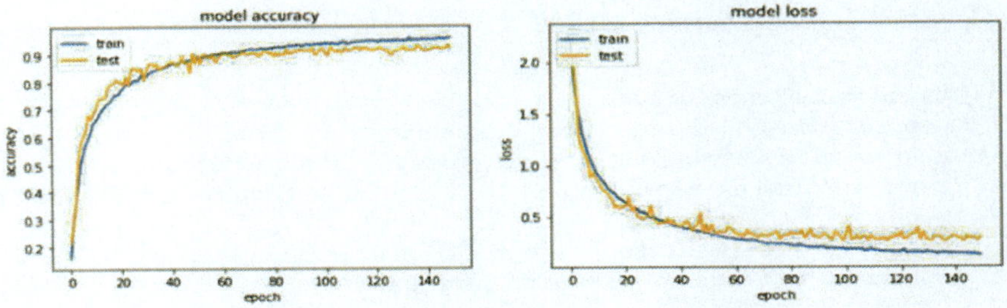

Fig. 6 A model trained for the 1st fold

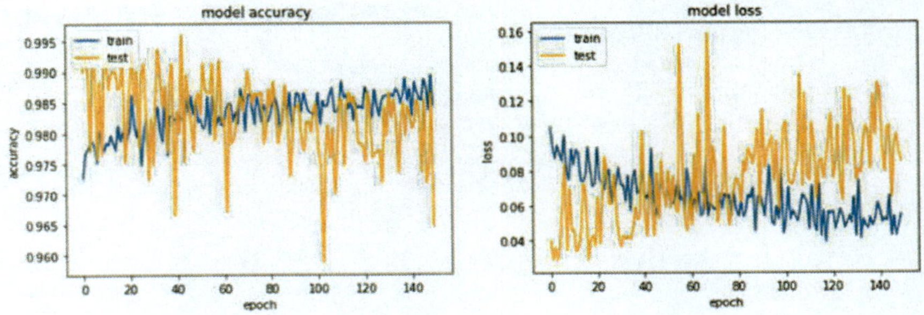

Fig. 7 Model trained for the final fold ($k = 5$)

Algorithm 1: CNN Working Model

For Training and Validation Dataset

```
skf=StratifiedKFold(n_splits=5, shuffle=True)
train, test = split (X, Y) {Splitting dataset into train and test}
for k values from 1 to 5: {K-fold cross validation}
split(train) into train and val dataset
for index in num_classes: {num_classes=no of species=12}
train_generator = ImageDataGenerator()
val_generator = ImageDataGenerator()
Cnn_model.fit(train_generator)
{Training Cnn_model}
Cnn_model.predict(val_generator)
{predicting the output using validation dataset}
Print performance matrix
```

For Testing Dataset

```
test_generator = ImageDataGenerator()
Cnn_model.predict(test_generator) {predicting for test dataset}
Print performance matrix
```

Algorithm 2. KNN Learning Model

```
inputs = Cnn_model.input
Outputs= Cnn_model.get_layer('feature_extraction'). output
Intermediate_layer_model= keras.models.Model(inputs, outputs)
{Here, feature extraction model is introduced to obtain features of
image}
Features = Intermediate_layer_model.predict(images)
{Obtained features for training purpose}
sc = StandardScaler()
Features = sc.fit_transform(Features)
{Transforming using StandardScaler() for fast computation}
import KNeighborsClassifier
Knn_model          =          KNeighborsClassifier(n_neighbors=5).
fit(traning_images ,target
{Training Knn model to improve performance}
Knn_model.predict(testing images)
Create an ensemble performance matrix and print it
```

5 Result and Analysis

5.1 Precision Metrics

The presented system's performance is measured using the mean accuracy, precision, recall, and F1-score.

$$\text{Average Accuracy: Total no. of correct samples/Total no. of samples} \tag{3}$$

$$\text{Average Precision: TruePositives}/(\text{TruePositives} + \text{FalsePositives}) \tag{4}$$

$$\text{Average Recall: TruePositives}/(\text{TruePositives} + \text{FalseNegatives}) \tag{5}$$

$$\text{Average F1 - score: }(2 * \text{Precision} * \text{Recall})/(\text{Precision} + \text{Recall}) \tag{6}$$

5.2 Obtained Results

The experiment used 10,600 photos of plant seedlings. The validation data is utilized to analyze and compare the results of each model. The dataset was divided as follows:

Training Set: 6787 images (shuffled for every fold), Validation Set: 1697 images (shuffled for every fold), Testing Set: 2116 images. A customized CNN model for seedling classification is shown in Table 1, with average validation results for 12 classes. Table 1 shows that for all species tested, model accuracy is 99.29%, precision, recall, and F1-score are all at 99.29% for 12 species. For 12 classes, the precision of the testing (accuracy) is 95.32%, while the F1-score, precision, and recall are all close to that mark (95.3%) (Figs. 6 and 7).

For the proposed technique, the confusion matrix of CNN model is shown in Table 2, while the confusion matrix of the proposed boosting ensemble technique is shown in Table 3. Table 3 shows the proposed ensemble technique's test results. The mean validation precision, recall, accuracy, and F1-score for 12 species are all around 100%. Finally, while testing for the 12 classes, the mean accuracy and F1-score were 99.906% and 99.906%, respectively, while the average recall and precision were 99.905% and 99.906%, respectively (Fig. 8).

Table 1 Classification report of CNN model

	Accuracy (%)	F1-score (%)	Precision (%)	Recall (%)
Train	99.2924	99.2929	99.2973	99.295
Test	95.32	95.301	95.315	95.307

Table 2 Confusion matrix of CNN for testing images

	SM	CW	CL	BG	SB	LSB	SP	MZ	CLV	CC	FH	SFC
SM	132	0	1	0	2	0	24	1	0	0	0	0
CW	0	161	0	0	0	0	0	0	0	0	0	0
CL	0	2	172	0	0	0	0	0	0	0	0	2
BG	0	0	0	179	1	0	0	0	1	2	0	0
SB	2	0	0	0	152	0		0	0	0	0	2
LSB	1	1	0	1	0	152	0	0	0	1	2	0
SP	31	0	0	0	0	1	127	0	2	0	0	1
MZ	0	0	0	0	0	0	0	159	0	0	1	0
CLV	0	0	0	2	0	1	0	0	154	3	0	1
CC	0	1	0	0	0	0	0	0	3	239	1	2
FH	0	0	0	0	1	0	0	0	0	0	226	0
SFC	0	2	0	0	0	0	0	0	0	0	0	161

Table 3 Final classification report of the ensemble model

	Accuracy (%)	F1-score (%)	Precision (%)	Recall (%)
Train	100	100	100	100
Test	99.905	99.906	99.905	99.906

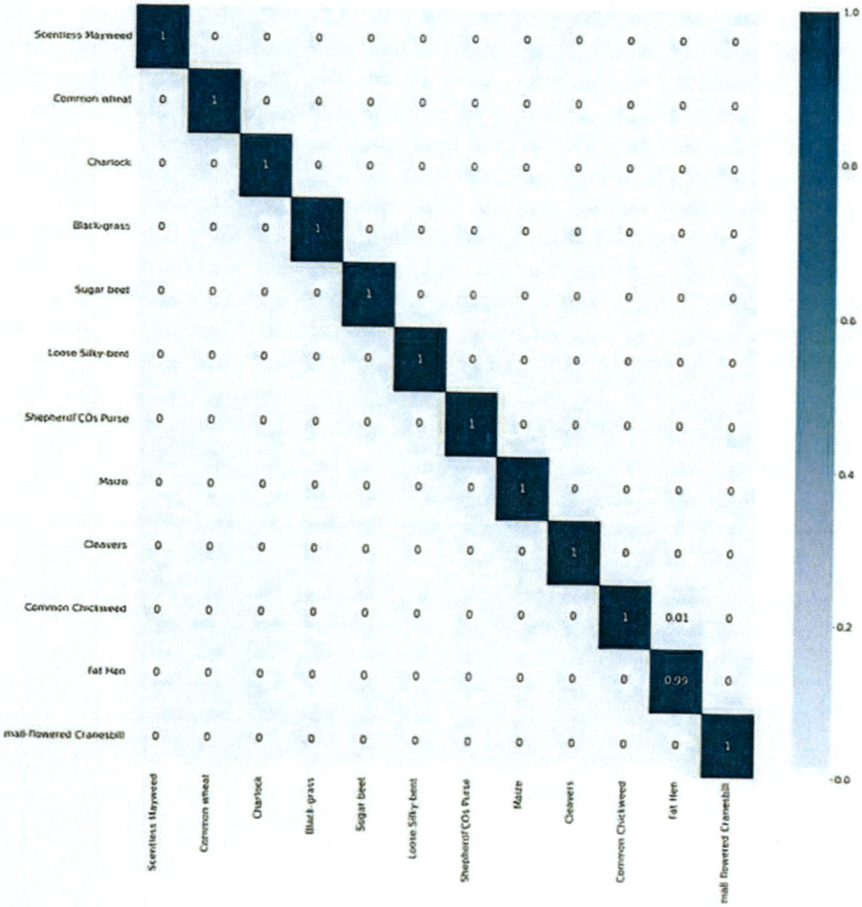

Fig. 8 Confusion matrix for the output of the ensemble model

6 Comparing Proposed Model with Existing Models

The authors of [5] compared various pre-trained models. All of the models that were utilized functioned wonderfully, with accuracy levels surpassing 90%. With a score of 97.08%, ResNet50 was by far the most accurate of the four. The comparison is depicted in Table 4.

Table 4 Result comparison of proposed system with existing/proposed algorithms

Year and ref.	Method	Accuracy %
(2018), [6]	CNN + OpenCV and background segmentation	92.6
(2019), [4]	Advanced CNN model	94.38
(2020), [20]	Transfer Learning ResNet (50) model	97.08
(2021), [5]	ImageNet + Transfer Learning VGG (19) model	97.54
(2021), [7]	Transfer learning EfficientNet-B1 model, a 7.8 million trainable parameters CNN of a mobile size	95.44
CNN	**Customized CNN model**	**95.32**
CNN + KNN	**Ensemble-based learning CNN + KNN model**	**99.9**

7 Conclusion and Future Scope

We came up with a method to help farmers improve crop yields while also improving the model's accuracy for better categorization. Implementation of novel data preprocessing techniques ascertained an increase in the overall precision of the model. Our boosting technique uses plant images from 12 distinct species as an input, constructs a model using CNN and k-nearest neighbors, and the model to identify the kind of (before unknown) features of plant seedlings. Overall, when trained on neural network features, the KNN outperformed the similar neural network-only structures. In particular, a testing accuracy of 99.90% is achieved using the proposed technique. Regardless of the network's depth, this trend tended to persist. Our proposed model has turned up with an accuracy of 99.90% and still holds up a future scope of reducing the time and space complexity, hence creating a machine/robot compatible system for practical use. This is because it is based on an ensemble-based learning model, which is an ensemble of comprehensive deep learning algorithms.

References

1. Lopez-Granados, F. (2011). Weed detection for site-specific weed management: Mapping and real-time approaches. *Weed Research, 51*, 1–11. https://doi.org/10.1111/j.1365-3180.2010.00829.x
2. Patel, T. U., Vihol, K. J., Thanki, J. D., Gudaghe, N. N., & Desai, L. J. (2018) Weed and nitrogen management in direct-seeded rice. *Indian Journal of Weed Science, 50*(4), 320–323. doi: https://doi.org/10.5958/0974-8164.2018.00069.2
3. Wang, H., Liu, W., Zhao, K., Yu, H., Zhang, J., & Wang, J. (2018). Evaluation of weed control efficacy and crop safety of the new HPPD-inhibiting herbicide-QYR301. *Science and Reports, 8*(7910), 1–11. https://doi.org/10.1038/s41598-018-26223-9
4. Bastiaans, L., Paolini, R., & Baumann, D. T. (2008). Focus on ecological weed management: What is hindering adoption? *Weed Research, 48*(6), 481–491. https://doi.org/10.1111/j.1365-3180.2008.00662.x
5. Heba A. E. (2019). Convolutional neural network architecture for plant seedling classification.

International Journal of Advanced Computer Science and Applications (IJACSA), 10(8), 319–325.

6. Hassan, E., Shams, M., Hikal, N. A., & Elmougy, S. (2021). Plant seedlings classification using transfer learning. In *2nd IEEE International Conference on Electronic Engineering*, pp. 1–7. Menoufia University.
7. Di, K., Omeiza, D., & Lubalo, N. (2018). *Deep convolutional neural network for plant seedlings classification*. arXiv:1811.08404
8. Martinson, O., & El-Gaya, O. (2021). An approach for weed detection using CNNs and transfer learning. In *Proceedings of the 54th Hawaii International Conference on System Sciences*, pp. 888–895.
9. Wang, A., Zhang, W., & Wei, X. (2019). A review on weed detection using ground-based machine vision and image processing techniques. *Computers and Electronics in Agriculture, 158*, 226–240. doi: https://doi.org/10.1016/j.compag.2019.02.005
10. Dhomne, A., Kumar, R., & Bhan, V. (2018). Gender recognition through face using deep learning. *Procedia Computer Science, 132*, 2–10.
11. Radovic, M., Adarkwa, O., & Wang, Q. (2017). Object recognition in aerial images using convolutional neural networks. *Journal of Imaging, 3*(21), 1–9.
12. Zayed, N. M., & Elnemr, H. A. (2019). *Intelligent systems for healthcare management and delivery*, pp. 101–147. doi: https://doi.org/10.4018/978-1-5225-7071-4.ch005
13. Bonnet, P., Goëau, H., Hang, S., Lasseck, M., Šulc, M., Malecot, V., Jauzein, P., Melet, J., You, C., & Joly, A. (2018). Plant identification: experts versus machines in the era of deep learning: Deep learning techniques challenge fora experts. In *Multimedia Tools and Applications for Environmental and Biodiversity Informatics. Multimedia Systems and Applications*. Springer, Cham, Chapter 8, pp. 131–149.
14. Qiu, Z., Chen, J., Zhao, Y., Zhu, S., He, Y., & Zhang, C. (2018). Variety identification of single rice seed using hyperspectral imaging combined with convolutional neural network. *Applied Sciences, 8*(2), 212, 1–12. doi:https://doi.org/10.3390/app8020212
15. Zhang, W., Hansen, M. F., Volonakis, T. N., Smith, M., Smith, L., Wilson, J., Ralston, G., Broad-bent, L., & Wright, G. (2018). Broad-Leaf weed detection in pasture. In *IEEE 3rd International Conference on Image, Vision and Computing (ICIVC)*, Chongqing, pp. 101–105.
16. Ferentinos, K. P. (2018). Deep learning models for plant disease detection and diagnosis. *Computers and Electronics in Agriculture, 145*, 311–318.
17. Giselsson, T., Jørgensen, R., Jensen, P., Dyrmann, M., & Midtiby, H. (2017). *A public image database for benchmark of plant seedling classification algorithms*, pp. 1–12. arXiv:1711.05458
18. Jeon, W.-S., & Rhee, S.-Y. (2017). Plant leaf recognition using a convolution neural network. *International Journal of Fuzzy Logic and Intelligent Systems, 17*(1), 26–34.
19. Milioto, A., Lottes, P., & Stachniss, C. (2018). *Real-time semantic segmentation of crop and weed for precision agriculture robots leveraging background knowledge in CNNs*, pp. 1–7. arXiv:1709.06764
20. Abouelnaga, Y., Ali, O. S., Rady, H., & Moustafa, M. (2016). CIFAR-10: KNN-based ensemble of classifiers. In *International Conference on Computational Science and Computational Intelligence (CSCI)*, pp. 1192–1195.
21. Bora, D. J., Gupta, A. K., Ahmad Khan, F. (2015). Comparing the performance of L*A*B* and HSV color spaces with respect to color image segmentation. *International Journal of Emerging Technology and Advanced Engineering, 5*(2), 192–203.
22. Mane, D. T., & Kulkarni, U. V. (2017). A survey on supervised convolutional neural network and its major applications. *International Journal of Rough Sets Data Analysis, 4*, 71–82.
23. Mane, D. T., & Kulkarni, U. V. (2018). Visualizing and understanding customized convolutional neural network for recognition of handwritten Marathi numerals. *Procedia Computer Science, 132*, 1123–1137.
24. Bhujbal A., & Mane, D. T. (2020). Vehicle type classification using deep learning. In *Soft Computing and Signal Processing. ICSCSP 2019. Advances in Intelligent Systems and Computing*, Vol. 1118, pp. 279–290. Springer. https://doi.org/10.1007/978-981-15-2475-2_26

25. Mane, D. T., Tapdiya, R., & Shinde, S. V. (2021). Handwritten Marathi numeral recognition using stacked ensemble neural network. *International Journal of Information Tecnology, 13*, 1993–1999. https://doi.org/10.1007/s41870-021-00723-w
26. Mane, D. T., Kumbharkar, P. B., Javheri, S. B., & Moorthy, R. (2022). Ensemble architecture for malware detection. In: *International Conference on Innovative Computing and Communications. Advances in Intelligent Systems and Computing*, vol. 1394, pp. 647–659. Springer. https://doi.org/10.1007/978-981-16-3071-2_53
27. Mane, D. T., & Uday, V. (2019). Kulkarni: A novel fuzzy convolutional neural network for recognition of handwritten Marathi numerals. *International Journal of High Performance Computing and Networking, 15*, 158–169.
28. Mane, D. T., Kumbharkar, P. B., Dhotre, P. S., Borde, S. (2021). Vehicle-type classification using customized fuzzy convolutional neural network. In *Data Engineering and Intelligent Computing. Advances in Intelligent Systems and Computing*, Vol. 1, pp. 419–429. Springer. https://doi.org/10.1007/978-981-16-0171-2_40
29. Gupta, K., Rani, R. (2020). Plant-seedling classification. Using transfer learning-based deep convolutional neural networks. *International Journal of Agricultural and Environmental Information Systems, 11*(4), 25–40.

Processing Kashmiri Text from a Document Image and Associated Challenges

Nawaz Ali Lone, Kaiser J. Giri, and Rumaan Bashir

Abstract Text processing in document images has evolved as the most prominent research area these days attracting a lot of interest from researchers across the globe. It encompasses a range of activities from simple text detection to text translation. However, there are a good number of languages for which very little or no research work has been done. Most of these languages lack basic resources to carry out research. Kashmiri language based on Perso-Arabic script is one among such languages, which has been discussed in this paper. This paper presents a brief introduction of text processing in the image domain, a description of the Kashmiri language and related issues for kashmiri text processing in the image domain.

Keywords Kashmiri · Image · Dataset

1 Introduction

The process of identifying text from a document image and understanding its content is the subject area of text processing in the document Image. Text processing in the document image is an important research area at present and plays a significant role in image/video analysis, image/video indexing, etc. It can be divided into various types according to the availability of data and type of writing as online versus offline and machine text/handwritten text. All of these types have certain inherent challenges. However, certain issues tend to happen because of the nature of the script followed.

Certain scripted texts such as those based on Arabic script, Devanagari script, Latin script, Chinese script have comparatively more challenges than Roman scripted text

N. Ali Lone · K. J. Giri (✉) · R. Bashir
Department of Computer Science, Islamic University of Science and Technology, Kashmir, India
e-mail: kaiser.giri@islamicuniversity.edu.in

N. Ali Lone
e-mail: lonenawazali@gmail.com

R. Bashir
e-mail: rumaan.bashir@islamicuniversity.edu.in

B. Pati et al. (eds.), *Proceedings of the 6th International Conference on Advance Computing and Intelligent Engineering*, Lecture Notes in Networks and Systems 428,
https://doi.org/10.1007/978-981-19-2225-1_40

which are mainly because of the cursive nature of such scripts creating connected words, characters changing shape based on position within a word, etc.

Present times throw an open challenge of digitization of written literature so as to access it electronically, which in turn enable archived material to be made usable to the society irrespective of the language of the material. While there is significant work related to major languages such as Roman scripted text, very less efforts have been made toward African scripts and Indian languages [1]. Unfortunately, no work has been done related to Kashmiri text processing in document images. The field if explored will help in a long way in research related to the Kashmiri language. Document image processing for the Kashmiri language can be exploited to create electronic documents of the books which are lying in various academic or administrative institutions of the erstwhile state. It can also be leveraged for the creation of digital research datasets.

In this paper, a brief about the Kashmiri language is presented. It also presents certain issues related to the processing of document images involving Kashmir text. The paper is organized as follows:

1. Section 2 provides an overview of the Kashmiri language.
2. Section 3 presents various issues concerning document image processing pertaining to Kashmiri text.
3. Section 4 presents the conclusion.
4. Section 5 finally presents the future work.

2 Brief About the Kashmiri Language

Kashmiri, also known as Koshur [2], is spoken in the erstwhile state of Jammu and Kashmir, an Indo-Aryan language belonging to the Dardic subgroup [3–5] of languages. Figure 1 below depicts the descent of the Kashmiri language as per Grierson.

The language got its name derived from the word "Kashmir" as the name of the region where it is mostly spoken. It is one of the languages written on Indian paper currency [6]. In addition to being an official language of Jammu and Kashmir [7], it is also one of the scheduled languages of the Indian Republic as per the eighth schedule of the Indian Constitution [8]. To promote the Kashmiri language, it has been made a mandatory subject up to higher secondary level since 2008 in the schools of Kashmir valley [9]. As per the 2011 census, there are approximately 7 million speakers in Jammu and Kashmir, mainly in the Kashmir Region, Chenab Valley, Neelam Valley, Leepa Valley, and the Haveli district [10–12].

The Kashmiri language is written using three orthographic systems: the Sharada script, the Devanagari script, and the Perso-Arabic script. However, these days,

Fig. 1 Kashmiri language tree

Fig. 2 Consonants

ب	[b]	ج	[d͡ʒ]	ر	[r]	ض	[z]	کھ	[kʰ]	ے	[j]
پ	[p]	چ	[t͡ʃ]	ڑ	[ɽ]	ط	[t]	گ	[g]		
پھ	[pʰ]	چھ	[t͡ʃʰ]	ز	[z]	ظ	[z]	ل	[l]		
ت	[t]	ح	[h]	ژ	[t͡s]	ع	[ø]	م	[m]		
تھ	[tʰ]	خ	[x], [kʰ]	ژھ	[t͡sʰ]	غ	[g]	ن	[n], [̃]		
ٹ	[ʈ]	د	[d]	س	[s]	ف	[f], [pʰ]	ں	[̃]		
ٹھ	[ʈʰ]	ڈ	[ɖ]	ش	[ʃ]	ق	[k]	و	[w]		
ث	[s]	ذ	[z]	ص	[s]	ک	[k]	ہ	[h]		

Fig. 3 Vowels

اَ	[ə]	اۄ	[o]
اٲ	[əː]	او	[oː]
اِ	[i]	اۆ	[ɔ]
ای	[iː]	اوا	[ɔː]
اؠ	[ɨ]	اےٚ	[e]
اٟ	[ɨː]	اے	[eː]
ا	[a]	اُ	[u]
آ	[aː]	اوٗ	[uː]

Roman script is commonly used for online communications. Kashmiri is now primarily written in Perso-Arabic and Devanagari (modified) scripts [13]. Kashmiri is one of the Perso-Arabic script languages, which represents all vowel sounds. The Kashmiri (Perso-Arabic) script [13] is a superset of Arabic because it contains 13 more letters than Arabic, which only has 28. Kashmiri has more consonants and vowels than Arabic and other languages. Figures 2 and 3 depict the character set of the Kashmiri language.

3 Related Issues

Processing Kashmiri text in the image domain is a challenging task because of many intricate features related to the nature of Perso-Arabic. Kashmiri text whether machine printed or handwritten throws challenges based on different features of each writing system. Various fundamental issues with text processing in general are text

Fig. 4 Connected/disconnected issue

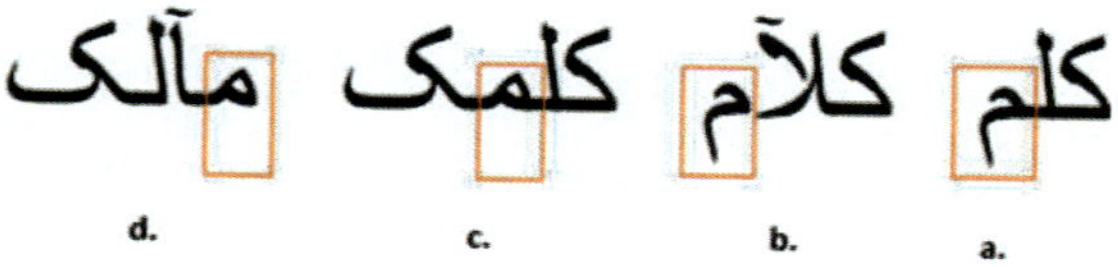

Fig. 5 Context-specific/free shape

localization [14], text verification [1], text detection [15], text segmentation [16], text recognition [17], and building end-to-end systems [17] irrespective of the script chosen. However, certain issues that may be specific to the Kashmiri language based on the Perso-Arabic script are discussed as under.

3.1 Connected/Disconnected Issue

Word formation in the Kashmiri language is somewhat trivial as some words have a dual visual profile as in the case of the word shown in Fig. 4 below, it is composed of four characters as in the last, two characters are connected, but in b, they are disconnected posing a challenge for double labeling of the single word and also there is the good number of words with such properties.

3.2 Context-Specific Shape/Free Shape

Certain characters in the Kashmiri language are sensitive with respect to the position in the word as they change their shape in different positions as is evident from Fig. 5 below, where the same character has four different structural representations based on different positions.

3.3 Looping to Start

As the Kashmiri language is written from right to left but with certain exceptions, some characters change their direction after certain plotting of the character as the word in Fig. 6 below shows here that the last character comes under the second

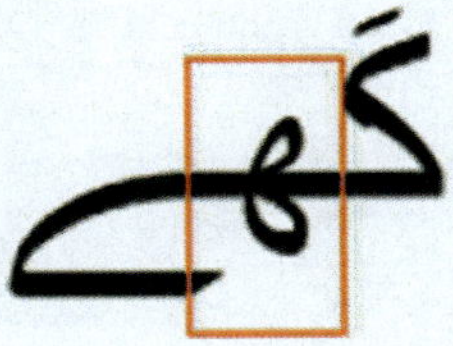

Fig. 6 Looping to start

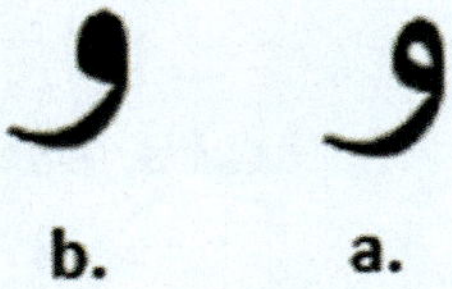

Fig. 7 Closed/open loops

character in vertical alignment thus posing a challenge to detect the character in vertical splitting.

3.4 Bi-directional Writing System

We follow a different direction to write numbers in the Kashmiri language, which needs to be taken into consideration during the identification process.

3.5 Closed/Open Loops

Certain characters create loops in their formation, but the loops are not visible in some font sizes, and also for the handwritten text, these loops get filled as in the given Fig. 7 representing the same character with a different visual appearance.

3.6 Ambiguity in Structure

There are certain words in the Kashmiri language that resemble the shape of head/dot/superscript/nukta of other characters which becomes hard to differentiate between a nukta and a character as in Fig. 8 below. Here in a, part of the character resembles the shape of the character in b.

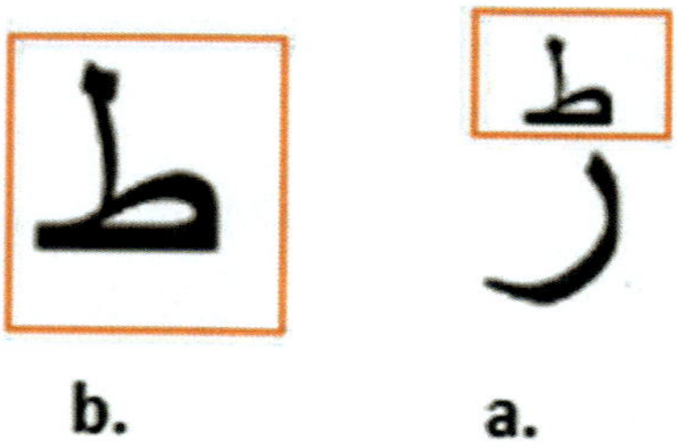

Fig. 8 Ambiguity in structure

3.7 *Lack of Dataset*

Though the above issues are inherent to certain languages, the main problem with Kashmir language processing in the image domain is the availability of the standard research dataset like Synth90k [18] which acts as the main hindrance in its research.

4 Future Work

Image processing involving Kashmiri text is in its infancy. There is no prominent work available in literature related to document image processing of the Kashmiri language. Therefore, there is a wide scope for undertaking research in this arena. The research work can be conducted related to different stages of document image processing either for machine printed or handwritten texts.

5 Conclusion

Kashmiri text processing in the image domain is an open research area, as yet no work has been published. Potential researchers need to resolve various issues in which certain have been presented/discussed above and also there are other issues as well which have not been discussed yet. As it through other challenges as well which are the result of a rich character set in comparison to other related languages such as Urdu, Arabic, and Persian.

References

1. Meshesha, M. (2008). *Recognition and retrieval from document image collections*. Ph. D. thesis.
2. Wikipedias, Kashmiri Language. Accessed November, 2021. https://en.wikipedia.org/wiki/Kashmiri_language
3. Kashmiri Language: Roots, Evolution and Affinity. Accessed 2021.

4. Kashmiri Language. Accessed 2021. http://heritagekashmir.blogspot.com/2007/04/kashmiri-language-roots-evolution-and.html
5. Koul, O. N. (2021). *Koshur: An Introduction to Spoken Kashmiri*. Accessed 2021. www.koshur.org/Kashmiri/introduction.html
6. Bashir, R., & Quadri, S. (2013, December). Identification of Kashmiri script in a bilingual document image. In *2013 IEEE Second International Conference on Image Information Processing (ICIIP-2013)*, pp. 575–579. IEEE.
7. The Jammu and Kashmir Official Languages Bill, 2020. Accessed June, 2021. http://164.100.47.4/BillsTexts/LSBillTexts/PassedLoksabha/124-C_2020_LS_Eng.pdf
8. Constitutional Provisions Relating to Eighth Schedule. Accessed June, 2021. https://www.mha.gov.in/sites/default/files/EighthSchedule_19052017.pdf
9. Kashmiri Made Compulsory Subject in Schools. Accessed November, 2020. https://www.oneindia.com/2008/11/01/kashmiri-made-compulsory-subject-in-schools-1225558278.html?story=3
10. Koshur: An Introduction to Spoken Kashmiri. Accessed November, 2020. https://www.koshur.org/
11. Snedden, C. (2015). *Understanding Kashmir and Kashmiris*. Oxford University Press.
12. Shakil, M. *Languages of Erstwhile State of Jammu Kashmir*, 10.
13. Kashmiri Alphabet, Pronunciation & Language. Accessed November, 2020. www.omniglot.com/writing/kashmiri.htm
14. Rainer, L., & Wernicke, A. (2002). Localizing and segmenting text in images and videos. *IEEE Transactions on Circuits and Systems for Video Technology, 12*(4), 256–268.
15. Lafferty, J. D., McCallum, A., & Pereira, F. C. N. (2001). Conditional random fields: Probabilistic models for segmenting and labeling sequence data. In *Proceedings of ICML*, pp. 282–289.
16. Wang, Y., Xie, H., Zha, Z.-J., Tian, Y., Fu, Z., & Zhang, Y. (2020). R-Net: A relationship network for efficient and accurate scene text detection. *IEEE Transactions on Multimedia.*
17. Von Ahn, L., Maurer, B., McMillen, C, Abraham, D., & Blum, M. (2008). Recaptcha: Human-based character recognition via web security measures. *Science, 321*(5895), 1465–1468.
18. Wang, K., Babenko, B., & Belongie, S. (2011). End-to-end scene text recognition. In *Proceedings of ICCV*, pp. 1457–1464.

In Plane and Out Of Plane Rotation Face Detection Model

Bibek Majumder and Sharmistha Bhattacharya

Abstract Face Detection is a famous topic in computer vision. Over the most few couple of years researchers have attempted to improve the performance of face detection algorithm in plane and out of plane rotation. In this paper, we propose a quick way to deal with face detection algorithm using support vector machine (SVM) and golden ratio. For performing this new algorithm, the main prerequisite is the preparation dataset in the front facing appearances to prepare SVM for skin filtering. In the proposed algorithm first we apply color histogram equalization (If the face detection algorithm is not able to detect any face) which can address the mistake of the skin filter then apply SVM for removing non-skin color, i.e., a skin filter machine is developed using SVM and lastly apply golden ratio for detecting the face region correctly. Proposed algorithm is compared on three datasets XM2VTS, FERET, and BioID with a high discovery rate not less than 95%. The experimental result shows the proposed algorithm not only runs comparatively fast but also gives an upgrade performance.

Keywords Face detection · Color histogram equalization · Support vector machine (SVM) · Facial landmark localization · Golden ratio

1 Introduction

Distinguishing human countenances is vital for face recognition [1, 2] and tracking. Consequently, we really wanted a strong and viable face detection algorithm which is comparatively fast. As of late many examination works resolved this issue, but distinguishing human faces is as yet a difficult issue in PC vision.

Deep learning is making significant advances in taking care of issues that have confined the best endeavors of the artificial intelligence community for a long

B. Majumder (✉) · S. Bhattacharya
Tripura University, Agartala, India
e-mail: bibekmajumder6@gmail.com

B. Pati et al. (eds.), *Proceedings of the 6th International Conference on Advance Computing and Intelligent Engineering*, Lecture Notes in Networks and Systems 428,
https://doi.org/10.1007/978-981-19-2225-1_41

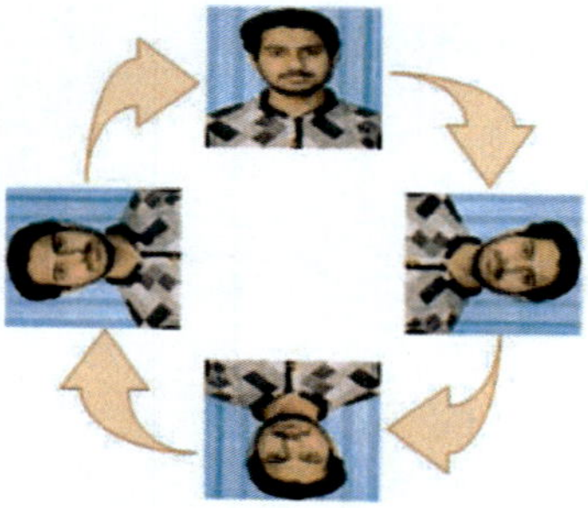

(a) In-plane rotation

(b) Out-of-plane rotation

Fig. 1 **a** In plane rotation and **b** out of plane rotation

time. It has demonstrated to be great at uncovering complex constructions in high-dimensional information and is thusly appropriate to lots of domains of science, business and government [3–6]. It resolves the issue of learning various leveled portrayals with a single algorithm or a few algorithms and has mainly beaten records in natural language processing, image recognition, semantic segmentation and many other real world scenarios. There are distinctive profound learning approaches like convolutional neural network (CNN), Stacked Auto encoder, and Deep Belief Network (DBN). CNN is generally utilized calculation in picture and face acknowledgment. CNN is a sort of counterfeit neural organizations that utilizes convolution procedure to remove the components from the input data to expand the quantity of provisions. Human face recognition in plane and out of plane turns exceptionally difficult point now in a day. In plane pivot a picture may addresses face turn from 0 to 360°, then again in out of plane revolution a face picture might contain just half nose, half mouth and one eye for example a revolution from −90 to 90° (Fig. 1).

Many existing writing [7–12] required an extremely enormous measure of face pictures to prepare the finder for each point. As of late, the neural organization becomes famous in view of its astounding presentation and learning capacity [13–16]. Other pixel-based calculations utilize the component of the skin tone in various shading spaces to recognize the face [17–20]. These techniques are exceptionally quick and robust to face rotation but are intolerant of environmental changes such as the lighting effect. Our focus is to develop the face detection algorithm in plane and out of plane rotation and execute the algorithm in a short span of time. Our algorithm is developed utilizing Color Histogram Equalization for noise reduction, SVM for skin sifting, and 5 point facial landmark detection for identifying faces.

Our paper is constructed as follows: In Sect. 2 we introduce the brief information about the notion and properties of SVM, Color Histogram Equalization, 5 point facial landmark detection, etc. Section 3 contains the algorithm and face detection procedure. In Sect. 4 we discuss about the comparative study between our developed algorithm and previously other defined algorithms. Lastly, the concluding remarks are discussed in Sect. 5.

2 Preliminaries

In this section, we discuss some of the research work which has been done by other researchers but are needed in the paper for further proceeding.

2.1 Color Histogram Equalization [21]

In this paper, we used color histogram equalization for removing small face candidates which can correct the error of the skin filter. Nonetheless, applying the technique on the Red, Green and Blue parts of an RGB picture might yield sensational changes in the picture's shading balance since the overall appropriations of the shading channels change because of applying the algorithm. However, if the picture is first changed over to another shading space, Lab tone space or HSL/HSV shading space specifically, then, at that point, the calculation can be applied to the luminance or worth divert without bringing about changes to the tint and immersion of the picture. In the proposed algorithm we apply color histogram adjustment in HSV color space. There are a few histogram balance strategies in 3D space. Trahanias and Venetsanopoulos applied histogram balance in 3-D shading space. However, it brings about "brightening" where the likelihood of splendid pixels are higher than that of dim ones. Han et al. proposed to utilize a new cdf characterized by the iso-luminance plane, which brings about uniform dark dispersion (Fig. 2).

2.2 Skin Filter Using Support Vector Machine (SVM)

Skin tone is a vital element of human faces. Handling tone is quicker than handling other facial provisions. Consequently, skin shading discovery is first performed on the info shading picture to diminish the computational intricacy. On account of the precision of skin shading identification influences the consequence of face location framework, picking an appropriate shading space for skin shading recognition is vital. Among various shading spaces, RGB shading space is delicate to the variety of force, and hence it is not adequate to utilize just RGB shading space to identify skin tone. Initial, a few skin and non-skin pixels are browsed the dataset. Then, at that point, a skin channel is prepared in Lab shading space by the support vector machine (SVM) [22, 23] to eliminate non-skin shading pixels. The hyper-boundaries of the skin channel are controlled by the preparation period of the SVM (Fig. 3).

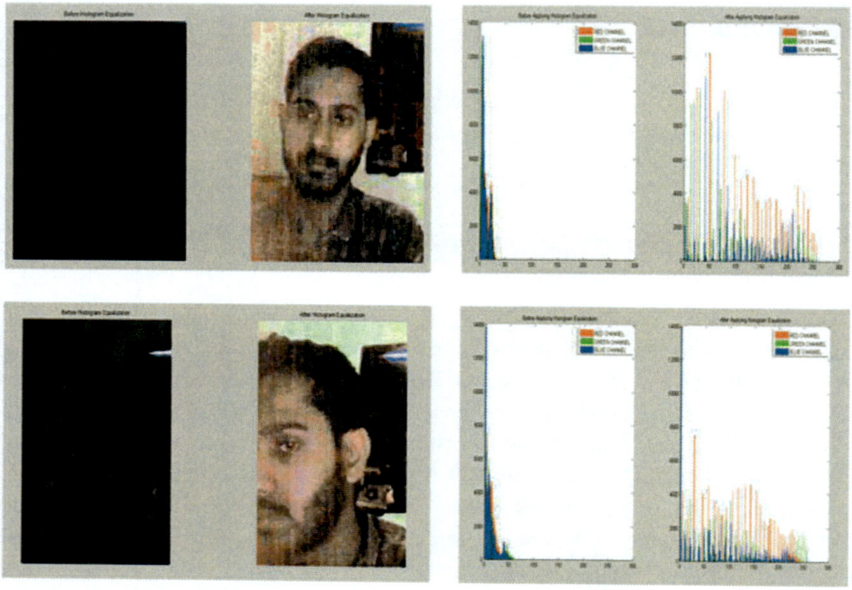

Fig. 2 Before and after applying color histogram equalization

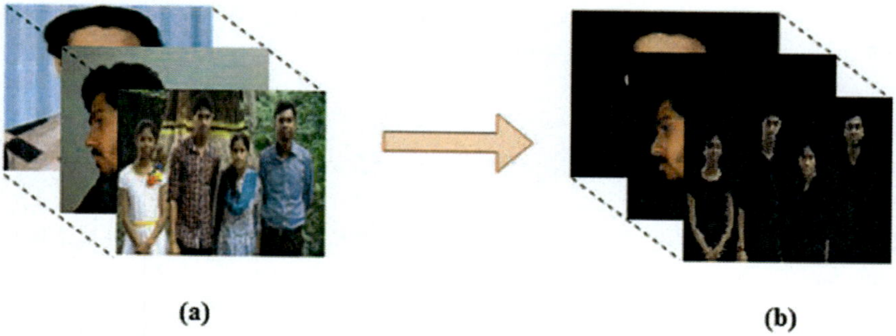

Fig. 3 **a** is the original image and **b** is the output of skin filtering

2.2.1 Point Facial Landmark Localization

Facial landmarks are standard reference points, for example, the inward and external corner of the eye gap where the eyelids meet. By and large the landmarks utilized in computational face analysis are basically the same as the physical soft issue landmarks utilized by doctors. The undertaking of consequently confining these landmarks is useful for different reasons. For example, a proficient gage of the head posture can be acquired with just a few landmarks. Additionally, facial landmarks can be utilized to adjust countenances to one another, which are important in an identification, arrangement and acknowledgment pipeline; better adjusted faces give better

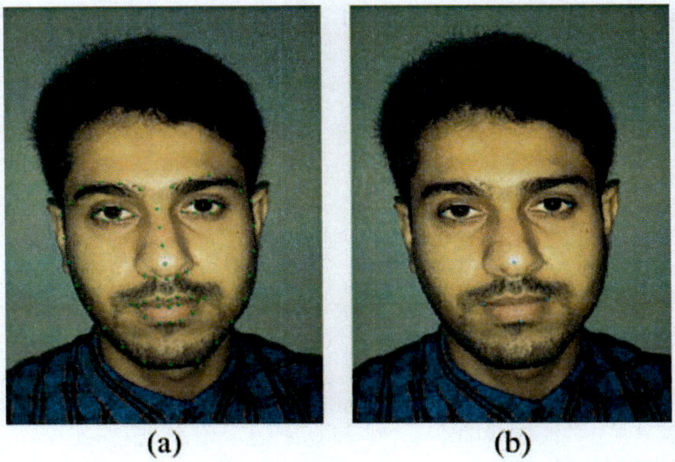

Fig.4 **a** Is the 68 point facial landmarks and **b** is the 5 point facial landmarks (after modification)

recognition results. Further, we can remove properties that have a nearby nature, for example, face credits (for example, shaggy eyebrows, skin tone, mustache), neighborhood descriptors or to prepare adaptable part-based locators. Some face data sets give facial landmarks, nonetheless, they need essentially somehow or another: for example, the pictures were gained under controlled conditions, are restricted to pretty much front facing sees or the assortment in identity is somewhat restricted.

We have collected freely available datasets namely, CUHK database, it contains three separate databases that are CUHK [24, 25] database, AR database, and XM2VTS database, the CUHK student database contains 188 image pairs, the AR database [26] contains 123 image pairs followed by XM2VTS [27] database which contains 295 images pairs. So, from here we managed to collect a total of 606 image database. We have used a total of 311 databases because here XM2VTS database is not a free one (Fig. 4).

2.3 *Face Definition*

The golden ratio [28], also known as the divine ratio, is a proportion or ratio defined by the number Phi, $\Phi = 1.618$. It has been utilized for a really long time in numerous fields, for example, Egyptians pyramids and Leonardo da Vinci's Mona Lisa [29, 30]. The golden ratio appears repeatedly in the physical proportions of the human body, particularly in the face. There are many facial parts or face components that form among one another based on the golden ratio, like the width and the height of the face. As per the golden ratio between the facial parts, we have these relations:

$$h/w = 1.618, \text{ and } d_{\text{eye}}/m = 1.618 \quad (1)$$

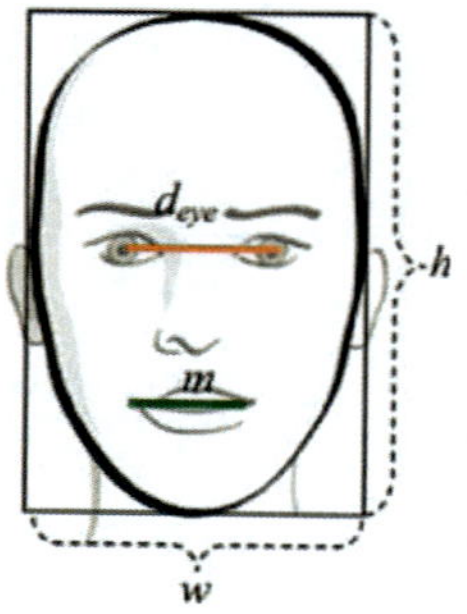

Fig. 5 Definition of the perfect face based on golden ratios between the elements of face

where h, w, d_{eye} and m are the height of face, the width of face, distance between the centers of eyes, and the width of mouth, respectively, as shown in Fig. 5. Based on statistical analysis of thousands of images from distinct races, we might also approximately expect that

$$w \approx 0.70 \times m + d_{\text{eye}} + 0.70 \times m \tag{2}$$

Using the second relation in (1), we can rewrite (2) in the following manner

$$\begin{aligned} w &\approx 0.70 \times d_{\text{eye}}/1.618 + d_{\text{eye}} + 0.70 \times d_{\text{eye}}/1.618 \\ &= 1.865 \times d_{\text{eye}} \end{aligned} \tag{3}$$

Also, from the first relation in (1) and using (3), we get

$$\begin{aligned} h/w &= 1.618, \text{ then } h = 1.618 \times w \\ &= 1.618 \times 1.865 \times d_{\text{eye}} = 3.018 \times d_{\text{eye}} \end{aligned} \tag{4}$$

Therefore, the width and height of the right human face can be estimated based on the distance between the eyes as

$$w = 1.865 \times d_{\text{eye}}, \text{ and } h = 3.018 \times d_{\text{eye}} \tag{5}$$

In this way, the golden ratio helps in assessing the face size according to the distance between the centers of eyes. Since, we have already calculated the nose tip pixel value using 5 points facial land mark detector, so take this point as a center of our face box. Thus, assuming the face is rotated then proposed method identify the turned face accurately.

The ideal face size determined utilizing (5) is displayed in Fig. 6 with the red square shape. In this specific circumstance (5) addresses a precise assessment of the human face size dependent on his/her between visual distance d_{eye} that is not the same as an individual to another. There are a few other facial parts having relations among one another closed to the brilliant proportion [30], which might be utilized in other fields, like 3-D facial models or facial highlights identification-based mathematical

Fig. 6 Correctly detect face region

data. In this work, we focus only on two of these relations (1) (i.e., height/width of face and inter-ocular distance/mouth width), which are significant for assessing the size of the human face.

3 Face Detection Frameworks

Our proposed face detection frameworks are presented in a following manner:

(a) Apply skin filter to remove unlikely face candidates and reduce the detection complexity. With them, the background region can be effectively removed.
(b) Apply 5 point facial landmark detector to detect the eye center, nose tip, and mouth corner points. If the face landmark not able to detect face features then apply color histogram equalization for removing the noise of face candidates which can correct the error of the skin filter.
(c) The centers of eyes in each face are manually labeled.
(d) Lastly apply the face golden ratio for detecting the face region correctly. For that we take nose tip as a center point of the face detector rectangle box.

Three publicly available datasets as well as a new challenging dataset are used in the evaluation of the face detection methods. They are the XM2VTS, FERET, and BioID. The centers of eyes in each face are manually labeled. Neither of face detection methods have been evaluated under these situations before nor the current common datasets consider such imaging conditions (i.e., headscarf occlusion).There are many people who wear headscarf and they use cameras; they buy an expensive camera (due to face detection technology included in this camera). The question

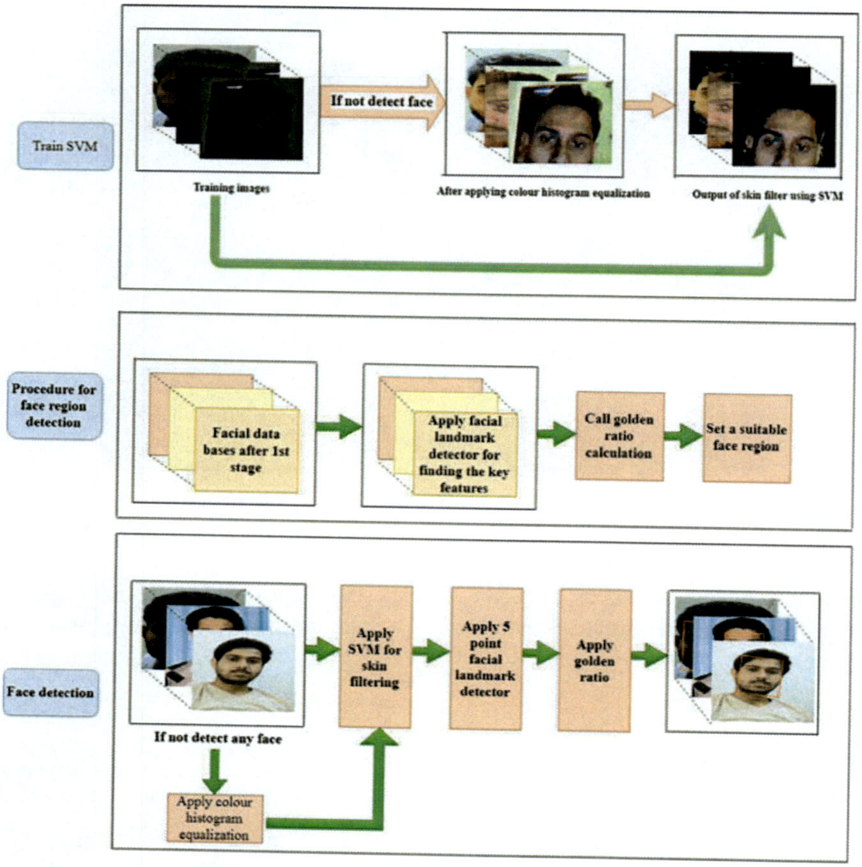

Fig. 7 Block diagram of proposed face detection algorithm

is this technology mature for them to buy it? The headscarf is a type of occlusion in the image, and it should be taken into account in evaluating the performance of face detection algorithms. Building the scarf dataset helps in highlighting the above question and studying the effect of headscarf occlusion on face detection performance (Fig. 7).

4 Results and Discussion

Figure 6 shows that (5) is a precise assessment for the human face determined dependent on the distance between eyes. Note that working out the ideal face (attracting Fig. 6) requires just the ground reality of eyes centers, which is done physically, and there is no compelling reason to assess the eyes positions in the distinguished face by

Table 1 Comparison of our model with other face detection methods on the four datasets

Method	XM2VTS	FERET	BioID	Scarf
	D. Rate (%)	D. Rate (%)	D. Rate (%)	D. Rate (%)
Our method	97.8	98.5	95 0.3	74 0.9
Chunlei Peng et al. [32]	95.5	96.3	95.18	67.9
Kienzle et al. [31]	83.9	62.3	40.6	49.2

eye discovery techniques. This guarantees that the proposed assessment measure is autonomous of the eyes position assessment. Consequently, the proposed assessment measure does not influence by imaging conditions like variety in light and impediment. While, the current proportion of Jesorsky et al. [11] influences such a great amount by these conditions since, it relies upon both the assessed eyes position and the ground reality of eyes. Since, it is notable that these imaging conditions make recognizing the eye position off base as well as a very difficult task. In the subsequent analysis, the proposed assessment measure is utilized to comp face location algorithm. To guarantee that the examination between various face location calculations is really reasonable, a similar assessment model is used (i.e., the proposed measure). Moreover, if at least two areas yield by the calculation fulfill the previously mentioned measure for same face, just one is considered as a right face and the others are considered false positives (FP).

The performance of our proposed algorithm is compared with the methods of Kienzle et al. [31] and Chunlei Peng et al. [32]. The consequences of this test are displayed in Table 1. The execution of proposed algorithm is compared on three datasets XM2VTS, FERET and BioID with a high discovery rate not less than 95%.While; there is a perceptible variety in the presentation of different techniques as indicated by the difficulties in the pictures.

5 Conclusion

Direct correlation of face detection techniques is a difficult task, specifically because there is not a clear definition for face and what the correct face detection is. In this paper, we presented a definition for the face based on the golden ratio among the components of the human face. Utilizing this definition, another assessment measure is proposed. For performing this new algorithm, the main prerequisite is the preparation dataset in the front facing appearances to prepare SVM for skin filtering. In the proposed algorithm first we apply color histogram equalization (If the face detection algorithm is not able to detect any face) which can address the mistake of the skin filter then apply SVM for removing non-skin color, i.e., a skin filter machine is developed using SVM and lastly apply golden ratio for detecting the face region correctly. The experimental result shows the proposed algorithm not only runs comparatively fast but also gives an upgrade performance. Since, the centers of eyes in each face are labeled manually, so this is the only drawback of our proposed method. In future if this issue can be addressed then a good face detection model can be proposed.

References

1. Bhattacharya (Halder), S., Majumder, B. (2019). Face recognition using fuzzy minimal structure oscillation in the wavelet domain. *Pattern Recognition and Image Analysis, 29*(1), 174–180.
2. Majumder, B., Bhattacharya (Halder), S. (2020). Face recognition based on human sketches using fuzzy minimal structure oscillation in the SIFT domain. In Bhattacharjee, A., Borgohain, S., Soni, B., Verma, G., Gao, X. Z. (Eds.), Machine learning, image processing, network security and data sciences. MIND 2020. Communications in computer and information science, vol. 1241. Springer.
3. Lal Sarkar, J., Majumder, A. (2021). gTour: Multiple itinerary recommendation engine for group of tourists. *Expert Systems with Applications.*
4. Lal Sarkar, J., Majumder, A. (2021). A new point-of-interest approach based on multi-itinerary recommendation engine. *Expert Systems with Applications, 181.*
5. Sarkar, J. L., Majumder, A., Panigrahi, C. R., Ramasamy, V., & Mall, R. (2021) TRIPTOUR: A multi-itinerary tourist recommendation engine based on POI visits interval. In *12th International Conference on Computing Communication and Networking Technologies (ICCCNT)*, pp. 1–7.
6. Sarkar, J. L., Majumder, A., Panigrahi, C. R., & Roy, S. (2020). MULTITOUR: A multiple itinerary tourists recommendation engine. *Electronic Commerce Research and Applications, 40.*
7. Viola, P., & Jones, M. J. (2004). Robust real-time face detection. *International Journal of Computer Vision, 57*(2), 137–154.
8. Negri, P., Clady, X., Hanif, S., & Prevost, L. (2008). A cascade of boosted generative and discriminative classifiers for vehicle detection. *EURASIP Journal on Advances in Signal Processing, 136.*
9. Liu, S., Dong, Y., Liu, W., & Zhao, J. (2012). Multi-view face detection based on cascade classifier and skin color. In *IEEE 2nd International Conference on Cloud Computing and Intelligence Systems*, vol. 1, pp. 56–60.
10. Fu, Y. J., & Li, J. W. (2010). Rotation invariant multi-view color face detection based on skin color and Adaboost algorithm. *Biomedical Engineering and Computer Science, IEEE*, 1–5.
11. Han, D., Kim, J., Ju, J., Lee, I., Cha, J., & Kim, J. (2014). Efficient and fast multi-view face detection based on feature transformation. *Advanced Communication Technology*, 682–686.
12. Liu, S., Don, Y., Liu, W., & Zhao, J. (2012). Multi-view face detection based on cascade classifier and skin color. *Cloud Computing and Intelligent Systems, IEEE, 1*, 56–60.
13. Yang, B., Yan, J., Lei, Z., & Li, S. Z. (2014). Aggregate channel features for multi-view face detection. In *International Joint Conference on Biometrics, IEEE*, pp. 1–8.
14. Farfade, S. S., Saberian, M., & Li, L. J. (2015). Multi-view face detection using deep convolutional neural networks. In *Proceedings of the 5th ACM on International Conference on Multimedia Retrieval*, pp. 643–650.
15. Zhang, K., Zhang, Z., Li, Z., & Qiao, Y. (2016). Joint face Detection and alignment using multitask cascaded convolutional networks. *IEEE Signal Processing Letters, 23*(10), 1499–1503.
16. Zuo, F. (2008). Cascaded face detection using neural network ensembles. *EURASIP Journal on Advances in Signal Processing*, 14.
17. Ghimire, D., & Lee, J. (2013). A robust face detection method based on skin color and edges. *Journal of Information Processing Systems, 9*(1), 141–156.

18. Ban, Y., Kim, S. K., Kim, S., Toh, K. A., & Lee, S. (2014). Face detection based on skin color likelihood. *Pattern Recognition, 47*(4), 1573–1585.
19. Cheddad, A., Condell, J., Curran, K., & Kevitt, P. M. (2009). A skin tone detection algorithm for an adaptive approach to steganography. *Signal Processing, 89*(12), 2465–2478.
20. Seshadrinathan, M., & Jezekiel, B. A. (2003). Pose invariant face detection. In 4th EURASIP Conference Focused on Video/Image Processing and Multimedia Communications, vol. 1, pp. 405–410.
21. Han, J.-H., Yang, S., & Lee, B.-U. (2011). A novel 3-D color histogram equalization method with uniform 1-D gray scale histogram. *IEEE Transaction on Image Processing, 20*(2), 506–512.
22. Chang, C. C., Lin, C. J. (2011). LIBSVM: A library for support vector machines. *ACM Transactions on Intelligent Systems and Technology, 2*(3), 27.
23. Chang, C. C., & Lin, C. J. (2011). LIBSVM: A library for support vector machines. *ACM Transactions on Intelligent Systems and Technology, 2*(3), 27.
24. Wang, X., Tang, X. (2009). Face photo-sketch synthesis and recognition. *IEEE Transactions on Pattern Analysis and Machine Intelligence, 31*(11), 1955–1967.
25. Zhang, W., Wang, X., & Tang, X. (2011). Coupled information-theoretic encoding for face photo-sketch recognition. In *Proceedings of IEEE Conference on Computer Vision and Pattern Recognition (CVPR).*
26. Panigrahi, C. R., Sarkar, J. L., & Pati, B. (2018). Transmission in mobile cloudlet systems with intermittent connectivity in emergency areas. *Digital Communications and Networks, 4*(1), 69–75.
27. Messer, K., Matas, J., Kittler, J., Luettin, J., & Maitre, G. (1999). XM2VTSDB: The extended of M2VTS database. In *Proceedings of International Conference on Audio- and Video-Based Person Authentication*, pp. 72–77.
28. Panigrahi, C. R., Sarkar, J. L., Tiwary, M., Pati, B., & Mohapatra, P. (2019). DATALET: An approach to manage big volume of data in cyber foraged environment. *Journal of Parallel and Distributed Computing, 131*, 14–28.
29. Walser, H. (2001). *The golden section.* Peter Hilton Trans. The Mathematical Association of America, ISBN: 0883855348
30. Sarkar, J. L., Panigrahi, C. R., Pati, B., Trivedi, R., & Debbarma, S. (2018). E2G: A game theory-based energy efficient transmission policy for mobile cloud computing. In *Progress in advanced computing and intelligent engineering*, pp. 677–684. Springer
31. Kienzle, W., Bakir, G., Franz, M., Scholkopf, B. (2005). Face detection efficient and rank deficient. In Weiss, Y. (Ed.), *Advances in neural information processing systems*, vol. 17, pp. 673–680. MIT Press.
32. Peng, C., Gao, X., Wang, N., Li, J. (2018). Face recognition from multiple stylistic sketches: Scenarios, datasets, and evaluation. *Pattern Recognition, 84*, 262–272.

Business Management and Sustainable Engineering

Using Machine Learning Techniques for Earthquake Prediction Through Student Learning Styles

Amr Mohamed El Koshiry, Mamdouh M. Gomaa, Entesar Hamed Ibraheem Eliwa, and Tarek Abd El-Hafeez

Abstract The Earthquake is an essential problem in human life, by using machine learning techniques in earthquake prediction, we can save humankind. Using the successful application of machine learning techniques indicates that it would be possible to use them to make accurate forecasts to avoid short-term earthquake damage. In this paper, with the first aim, we have applied seven machine learning techniques, namely, Artificial Neural Network (ANN), Decision Tree (DT), Logistic Regression, Random Forest Classification, Naïve Bayes (NB), K-Nearest Neighbors (KNN), and Support Vector Machine (SVM) to reach the best technique for prediction. The second aim used the methods of two learning styles, surface and deep learning, in training students with programming skills to use the seven techniques. Through two experimental groups, one of them used the method of surface learning (collective), and the other used the method of deep learning (individual).This is to determine the best learning style to teach students programming skills.

M. M. Gomaa · E. Hamed Ibraheem Eliwa · T. Abd El-Hafeez (✉)
Department of Computer Science, Faculty of Science, Minia University, EL-Minia, Egypt
e-mail: tarek@deraya.edu.eg; tarek@mu.edu.eg

M. M. Gomaa
e-mail: Mamdouh.gomaa@mu.edu.eg

E. Hamed Ibraheem Eliwa
e-mail: entesar.eliwa@mu.edu.eg; eheliwa@kfu.edu.sa

A. Mohamed El Koshiry
Faculty of Specific Education, Minia University, EL-Minia, Egypt
e-mail: al_koshiry@mu.edu.eg; Aalkoshiry@kfu.edu.sa

College of Education, King Faisal University, Al-Hofuf, Saudi Arabia

E. Hamed Ibraheem Eliwa
College of Science, King Faisal University, Al-Hofuf, Saudi Arabia

T. Abd El-Hafeez
Computer Science Unit, Deraya University, EL-Minia, Egypt

B. Pati et al. (eds.), *Proceedings of the 6th International Conference on Advance Computing and Intelligent Engineering*, Lecture Notes in Networks and Systems 428,
https://doi.org/10.1007/978-981-19-2225-1_42

Keywords Earthquake prediction · Artificial neural network (ANN) · Decision tree · K-nearest neighbors (KNN) · Logistic regression · Naive Bayes · Random forest · Support vector machine (SVM)

1 Introduction

The scientific concept of an earthquake and how occurred, which was ambiguous until the beginning of the twentieth century; where the emergence of what is named as seismology, which was able to provide answers to all vague questions related to this phenomenon. An earthquake is the sudden movement of the ground caused by the rapid release of energy that has accumulated along fault zones in the earth's crust. The earth's fundamental structure and composition are revealed by earthquakes through the study of waves that reflected from the interior of the earth.

For the time being, the earthquake warning system has already installed in many volcanic areas of the world that might increase the number survivor expectation. Many research outcomes gain more information about earthquake impacts and characteristics to the surrounding area. However, some machine learning techniques work result still has not provided accurate prediction, and sometimes rise up a false alarm because of the lack of the volume of data or the prediction method, this fact can make the problem of earthquake prediction critical to protect humankind [1]. Moreover, a good and reasonable prediction will provide opportunities to manage the emergency route path for the evacuation plan, which may reduce the injuries [2].

To provide datasets for prediction, we utilize the data collection from several earthquake and seismological repositories. The list of data resources for our research is as follows, the dataset of China and Japan countries. We are training some students of the instructional technology department on the skills of how to classify data by using two learning styles, surface and deep learning.

In this paper, we compare the performance of seven machine learning approaches, which are Artificial Neural Network (ANN), Decision Tree (DT), Logistic Regression, Random Forest Classification, Naïve Bayes (NB), K-Nearest Neighbors (KNN), and Support Vector Machine (SVM), to reach the best technique for prediction. The contribution of this paper is two aims:

(a) In earthquake predicting, a comparison between seven machine learning techniques may give light for a new approach. We made a comparison between techniques for earthquake prediction in Japan and China regions. Our method facilitates use of prediction and visualization that range within several years of seismic historical data, which is particularly helpful to classify how different machine learning performance could put light on our method of prediction. For this, our work can also adjust the size of data for better prediction. This is useful since the size of data sometimes influences the training and testing process for ultimate prediction. Other than that, we have flexibility in testing our results.

(b) Identify the best style of learning (surface—deep), which means the level of information processing helps to develop some learning outcomes for some students, skills to use seven machine learning techniques for Earthquake Prediction.

Students' learning styles are among the factors influencing the level of achievement of their learning outcomes, especially when developing their programming skills. The goal is to determine which students' favorite learning styles are most effective and to achieve outstanding learning outcomes. There is no doubt that the learning method is the method used by the learner to solve the problems faced during the educational situation; and what determines for him in the future his choice of one of these methods in achieving good learning outcomes in similar topics.

There are many global pedagogical attitudes that are concerned with students' learning styles, "The learning method affects the way an individual accepts information" [3].

2 Related Work

Several machine learning techniques are used in a lot of research on prediction field, especially in earthquake as a computer science application. In [4], a model of prediction was generated with maximum obtainable seismic features for training of an earthquake prediction model for Hindukush, Chile, and Southern California regions. The prediction model consists of (SVR) following by Hybrid Neural Networks and Enhanced Particle Swarm Optimization (EPSO). SVR-HNN prediction model trained and tested successfully for three regions.

In [5], the study shows earthquake occurrence in nonlinear and appears to be a random phenomenon, and it can be modeled using learning approaches of machine learning.

In [6], Rotation forest attained higher prediction performance with regard to seismic activity in Hindukush region, compared to decision trees, random forest, and rotboost. In [7], the study shows different damage grades due to earthquake in Nepal. The analysis indicates that random forest method is better than the neural network approach for building damage prediction.

In [8], the study tested seven machine learning techniques to classify the major earthquake events; which applied to a dataset collected from California. It shows that SVM, KNN, MLP, and Random Forest classify the higher number of outputs correctly. While KNN, Random Forest, and MLP were the best by producing the least false output (FP).

In the other side of our paper, little research of education, which applied to the methods of educating and training students as an experimental research in the field of learning styles. In [9], the authors examined in English courses, whether students' learning styles, thinking styles, and their attitudes studied their special specialization. They referred to the found negative relationship between thinking styles and deep learning style, and the study revealed that there was a positive relationship between achievement and attitudes. The results reported that students were equal in deep and surface learning styles.

In [10], the study aimed to describe the predictive ability of the big five factors and the learning environment for learning style. The results indicated that the deep learning style is prevalent among the sample, and learning environment dimension level was in order.

In [11], this study aimed at recognize the impact of augmented reality and the method of learning (surface-deep) in the development of the skills of designing educational web sites for students of educational technologies at the Faculty of Education. It indicated that there are statistically significant differences between the scores of the students of the two groups (deep—superficial). The results indicated that these differences are in favor of deep learning style. The study recommends the developing of positive attitudes toward the use of augmented reality and learning style during the study.

3 Machine Learning

The ability of machines to learn is called machine learning, which considers a subfield of Artificial Intelligence. A machine is built up using a certain techniques through which it can take its own decisions and then submit the result to the user. Machine learning techniques used in this paper for prediction. The prediction techniques have been used are Artificial Neural Network (ANN), Decision Tree Classification, K-Nearest Neighbors (KNN), Logistic Regression, Naive Bayes, Random Forest Classification, Support Vector Machine (SVM), and Kernel Support Vector Machine. In the next section, some state-of-the-art articles about these techniques that we have used them.

3.1 Artificial Neural Network (ANN)

ANN's are software emulators of the nervous system that simulate the way the human brain [12], and suitable for earthquake forecasting because of their mathematical operations and their ability to bear the error. In the study [13] it was shown that a neural network behaves as a model for the earthquake process, especially prediction tasks. Neural networks consist of neurons with an activation function. Figure 1 illustrates the A single layer artificial neurons which gives single output [14].

The Activation function of the neuron has the following form:

$$\sum_{i=1}^{n} x_i . w_i . \tag{1}$$

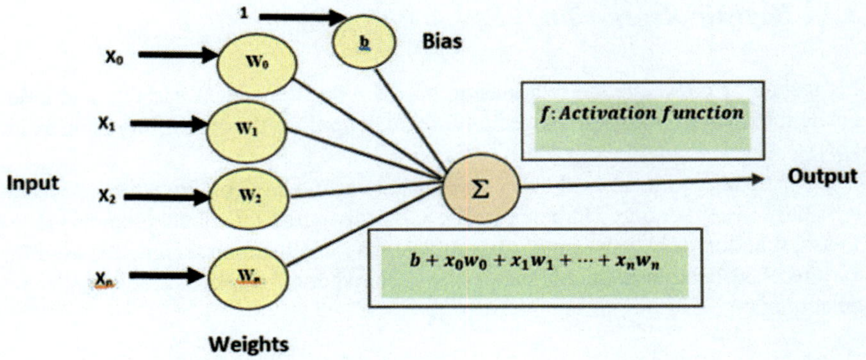

Fig. 1 A single layer artificial neurons

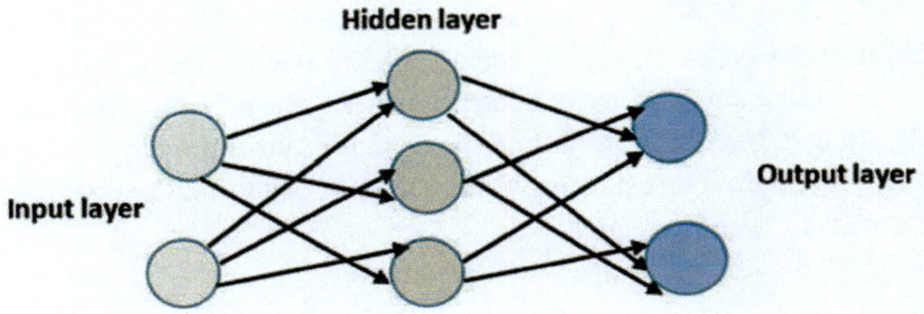

Fig. 2 Neural network

where x_i is the value of input, w_i is the corresponding weight. When simple neurons are connected, a complex function built.

Figure 2 showed an example of the neural network where the activation functions applied more than one to generate the final outputs of the whole networks. An error measured between the network's output and the target value reduced.

3.2 K-Nearest Neighbors (KNN)

KNN is a technique used for classification, and more frequently used for classification predictive problems. KNN based on the rule that case in point with the same properties exist nearness, where the class value of the unclassified instance is predicted by monitoring the class of its nearest neighbors. KNN finds the only amenable parameter in this technique, the K-nearest neighbor, to the query instance and classifies it by determining the most frequent class label [14–16].

3.3 Logistic Regression (LR)

LR technique considers the relationship between one dependent variable and independent variables. By using a logistic function, it predicts the probability of an event [17].

Logistic regression is used when the data in question has a binary output, so when it belongs to zero or one. Mathematically, a logistic regression model predicts P(Y = 1) as a function of X. It is one of the simplest machine learning techniques used for various classification problems such as cancer prediction, diabetes prediction, spam detection.

3.4 Naive Bayes (NB)

NB technique is based on applying Bayes' theorem with the assumption that all predictors are independent to each other. Bayesian classification assumes that the data belongs to a distinct class, and then the probability for the hypothesis to be true is calculated. The main interest in Bayesian technique is to find the next probability. Bayes' Theorem is stated as:

$$\mathrm{P(X|Y)} = \frac{\mathrm{P(Y|X).P(X)}}{\mathrm{P(Y)}}. \quad (2)$$

P(X) is the initial probability of event X, P(Y) is the next probability of event Y, P(X|Y) is the probability of event X given event Y, and P(Y|X) is the probability of event Y given event X [18].

3.5 Support Vector Machine (SVM)

SVM technique has two forms for binary classification problems. To break up any data, we define certain classes and depending on the complexity of the datasets, we define it as the linear or nonlinear classification [19, 20]. SVM is defined as a prediction machine, in which we search for a certain line or decision boundary called hyperplane, which separates out the datasets or classes easily, so it avoids the extra fit to the data. It uses the assumption space of a linear space into a high dimensional feature space. It is also capable of classifying the nonlinear data where it uses kernel functions.

3.6 Decision Tree (DT)

DT technique is a technique used in data classification and prediction problems. It has outcome was easy to understand, because the data set is represented in the form of a hierarchical tree, which consists of logical decisions. Each node represents an attribute to evaluate, the branches represent decision options on the given attribute, and each sheet represents a result. The result is in the form of a tree build, each interior node represents a dependent value of an attribute, and each sheet represents the decision for a particular class. The complication of the tree was controlled during the use of stopping criteria and pruning methods. However, the metrics used to measure the complications of the tree are the depth, the number of leaves, the number of nodes, and the number of attributes used [21].

3.7 Random Forest (RF)

RF technique based on majority voting of decision trees [22]. Different number of decision trees generated from the training dataset. The final decision was made by collecting the predictions obtained from the decision trees. Because multiple decision trees are tangled in random forest, compared to the use of single decision tree, the results are improved. Random forest focuses to improve the prediction accuracy and minimizes the overall error rate [23].

4 Learning Styles

Educational psychology defined the learning method as a key interface to describe individual differences in the context of learning. Different individuals learn in different ways, which are appropriately, noticeably related to different learning outcomes for them. This is what Backer emphasized when linking surface treatment with quantities without quality, and deep by quantity and quality together [24].

5 Performance Evaluation and Datasets

The dataset is separated into test and train sets within the ratio of 25 and 75%, respectively. The supervised ML classification approaches provided in Sect. 3 require training before prediction. For binary classification issues, there are several performance measures available. The accuracy of earthquake prediction is measured using the following criteria true positives (T_P), true negative (T_N), false positive (F_P), and false negative (F_N). The number of times the algorithm correctly predicted an earthquake and it happened, this is called T_P. T_N indicates that the number of times the algorithm predicted there would be no earthquake and exactly there was no earthquake. In the other hand, F_P means the number of occasions when the algorithm predicted an earthquake but there was no actual seismic activity, but when the number of times the algorithm correctly predicted there are no earthquake occurred and it happened, this is called F_N.

Another set of evaluation criteria is derived from the four indicators stated clearly and in detail before, Sensitivity and Specificity are two of the most often used statistical measurements. In the following equations, sensitivity S_n refers to the rate of actual positives predicted, while specificity S_p refers to the rate of actual negatives predicted.

$$S_n = \frac{T_p}{T_p + F_N}$$
$$S_P = \frac{T_N}{T_N + F_P}. \qquad (3)$$

6 Results

This section discusses the results yielded by the above discussed ML classifiers over unseen data of Japan and China country region. Figure 3 shows the results of earth-

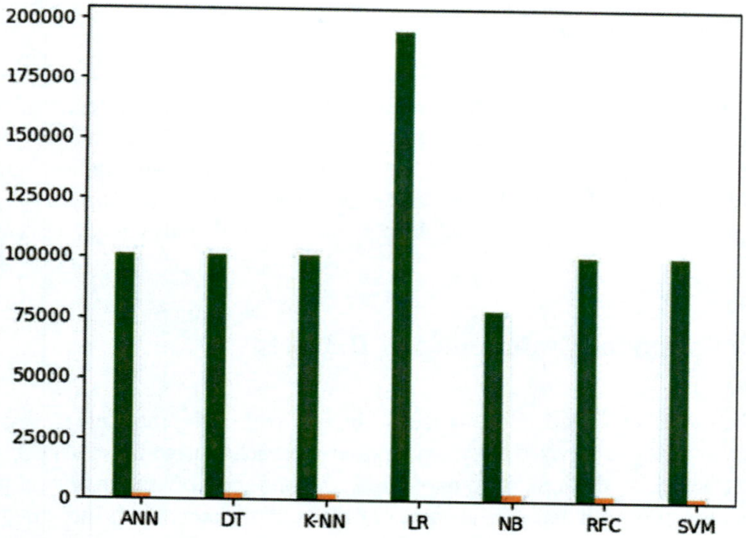

Fig. 3 Results of earthquake prediction false positive between algorithms

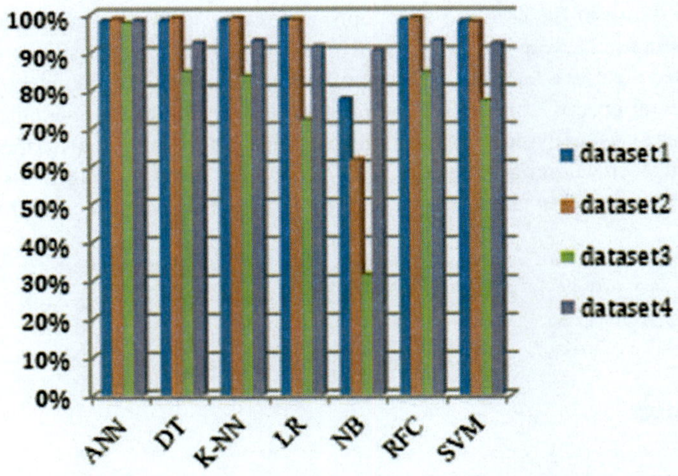

Fig. 4 Results of earthquake prediction performance between algorithms

quake prediction on the test dataset. The results produced by these ML techniques are different from each other in terms of the discussed evaluation criteria, yet quite encouraging over unseen data, considering the fact that there is no such robust system available for earthquake prediction.

From our results shown in Fig. 4, we can see that KNN is simple to understand, fast, and efficient but need to choose the number of neighbor's k. SVM is performant, not biased by outliers, not sensitive to overfitting, but it is not appropriate for nonlinear problems, not the best choice for large number of features. Naive Bayes is efficient, not biased by outliers, works on nonlinear problems, and a probabilistic approach but based on the assumption that features have the same statistical relevance. Decision Tree Classification is interpretability, no need for feature scaling, works on both linear / nonlinear problems, but in has poor results on too small datasets, overfitting can easily occur. Random Forest Classification is considered powerful and accurate, with good performance on many problems, including nonlinear, but it is not interpretability, overfitting can easily occur, but need to choose the number of trees.

7 Conclusion

Seven machine learning algorithms were employed to predict earthquakes in the sub ducting, Japan, and China regions, which are among the world's most seismically active. Every classifier that has been used yields somewhat different results. Although earthquake occurrence is thought to be nonlinear and random, the study reveals that it can be modeled using geophysical information from the seismic zone, as well as very advanced modeling and learning methodologies from machine learning. For the second aim of the study in education, we examined two experimental groups; one of them used a deep learning style with individual learning (8) students, another group used a surface learning style with traditional learning (13) students; the two experimental groups were trained on programing skilling to use machine learning techniques by using Python programming language. The results reported that students who used deep learning style had higher scores in programming skills, and the students were equal in classification datasets in the deep and surface approach.

Acknowledgements This work was supported by the Deanship of Scientific Research, King Faisal University, Saudi Arabia [**grant number NA000175**].

References

1. Murwantara, M., Yugopuspito, P., Hermawan,R. (2014). Comparison of machine learning performance for earthquake prediction in Indonesia using 30 years historical data. *Neuro-computing, 135*, 273–28.
2. Ruano, A., Madureira, G., Barros, O., Khosravani, H., Ruano, M., Ferreira, P. (2020, June). Seismic detection using support vector machines, *TELKOMNIKA Telecommunication, Computing, Electronics and Control, 18*, 1332.
3. Snyder, R. F. (2000). The relationship between learning styles/multiple intelligences and academic achievement of high school students. *High School Journal, 83*(2), 11–21.
4. Asim, K., Idris, A., Iqbal, T., Martínez-Álvarez, F. (2018). Earthquake prediction model using support vector regressor and hybrid neural networks, open access. *PLOS ONE, 13*, 1–22.
5. Asim, K., Martínez-Álvarez, F., Basit, A., & Iqbal, T. (2017). Earthquake magnitude prediction in Hindukush region using machine learning techniques. *Natural Hazards, 85*, 471–486.
6. Asim, K., Martínez-Álvarez, F., Idris, A., & Iqbal, T. (2016). Short term earthquake prediction in Hindukush Region using tree based ensemble learning. *IEEE, 1*, 365–370.
7. Chaurasia, K., Kanse, S., Yewale, A., Singh, V., Sharma, B., & Dattu, B. R. (2019). Predicting damage to buildings caused by earthquakes using machine learning techniques. *IEEE, 7*, 81–86.
8. Mallouhy, R., Abou Jaoude, C.,Guyeux, C., & Makhoul, A. (2019). Major earthquake event prediction using various machine-learning algorithms. *IEEE, 2*, 1–7.
9. Ismail, N., & Abdulwareth, S. (2012). The contributions of students learning styles, thinking styles and their attitudes towards studying their specializations in the English Language to their achievement at the college level. *Journal of the Faculty of Education, Zagazig University, 77*, 333–374.
10. Al-omari, E. (2018). *The big five personality factors and learning environment as predictors for the surface and deep learning styles among a sample of Yarmouk University Students*. PH.D Dissertation, Yarmouk University, pp. 1–151.
11. Hasan, A., & Al-Zahrany, A. (2019). The effect of augmented reality and learning style (deep-surface) in the development of web-design skills among educational technology students in the Faculty of Education of the University of Jeddah. *Journal of the Faculty of Education, Sohag University, 68*, 1563–1591.
12. Alves, E. I. (2006). Earthquake forecasting using neural networks: Results and future work. *Nonlinear Dynamics, 44*, 341–349.
13. Alves, E. I. (1995). A simple seismic computational model. *Memórias e Notícias* 37–42.
14. Gu, G., He, Q., Wang, B., & Dai, B. (2018). Comparison of machine learning techniques for the detection of the electricity theft. *IEEE, 2*, 116–119.
15. Li, A., & Li, K. (2009). KNN-based modeling and its application in aftershock prediction. In *2009 International Asia Symposium on Intelligent Interaction and Affective Computing, IEEE*, pp. 83–86.

16. Cover, T., Hart, P. (1967). Nearest neighbor pattern classification. *IEEE Transactions on Information Theory, 13*, 21–27.
17. Wahbeh, A., Al-Radaideh, Q., Al-Kabiand, M., et al. (2011). A comparison study between data mining tools over some classification methods. *International Journal of Advanced Computer Science and Applications, 35*, 18–26.
18. Greenlan, S. (1994). Alternative models for ordinal logistic regression. *Statistics in Medicine, 13*, 1665–1677.
19. Manning, C., Raghavan, P., Sch¨utze, H. (2010). Introduction to information retrieval. *Natural Language Engineering, 16*, 100–103.
20. Vapnik, V. (1995). *The nature of statistical learning theory*. Springer.
21. Cortes, C., & Vapnik, V. (1995). Support-vector networks. *Machine Learning, 20*, 273–297.
22. Rokach, L., Maimon, O. (2005). Decision trees. In *Data mining and knowledge discovery handbook*, pp. 165–192. Springer.
23. Breiman, L. (2001). Random forests. *Machine Learning, 45*, 5–32.
24. Asim, K., Idris, A., Alvarez, F., Iqbal, T. (2016). Short term earthquake prediction in Hindukush Region using tree based ensemble learning. In *Conference: International Conference on Frontiers of Information Technology (FIT)*, vol. 1, pp. 365–370.
25. Backer, R., & Yelich, S. (2002). *Comparison of learning styles and student achievement of aviation students, San Jose*. San Jose State University.

Role of Internet of Things in Global Business

Md. Shah Alam

Abstract The paper aims at presenting Internet of Things in global business relating to sustainable development with the emphasis on its prospects and problems. Secondary data have been used in conducting the paper. Internet of Things presently is at an infancy stage with some devices which emerges with the revolutionary change in information and telecommunication systems that facilitate the growth of trade and multinational companies during globalization. At this juncture, cloud computing starts using as one of the important transformations of information technology, where computer along with Internet provides many facilities with a view to achieving goals of both companies and end users. Production systems starting from 3D printing and high-performance computing to the Internet of Things, and industrial robots are facilitated through cloud computing. During the study, it is clarified that IOT can be applied in many areas as wearables, health relating activities, traffic monitoring, fleet management, agriculture, hospitability, smart grid and energy saving, water supply, and maintenance management. The study also points out some limitations of using IOT in case of control and reliability, compatibility and contacts, and lock-ins. Despite the limitations of using IOT, it has emerged as a boon of human civilization at the age of globalization. With the passes of time, pattern of business and its management style is changing fast. For sustainable development of a country in particular and the world in general, learning and implementation of IOT may be considered to be an alternative technique in the world.

Keywords Data analytics · Revolutionary change in information technology · Industrial Internet · Global business · Sustainable development

Md. Shah Alam (✉)
Department of Accounting and Information Systems, University of Rajshahi, Rajshahi 6205, Bangladesh
e-mail: shahmdalam@ru.ac.bd

B. Pati et al. (eds.), *Proceedings of the 6th International Conference on Advance Computing and Intelligent Engineering*, Lecture Notes in Networks and Systems 428,
https://doi.org/10.1007/978-981-19-2225-1_43

1 Introduction and Justification of the Study

Internet of Things abbreviated as IoT is a new concept relating to technical, social, and economic significance. It transforms the way where people work, live, and play with the everyday objects in case of consumable products and services, durable goods like cars and trucks, industrial goods and services, as well as sensors with Internet connectivity and powerful data analytic capabilities. The impact of IoT is considered to be important in case of anticipating as many as 100 billion connected IoT devices and more than $11 trillion global economic impact (www.Internet.society.org) [1].

1.1 Review of Related Works

In an article [2], it is clarified that IoT is also called industrial Internet which emphasizes on machine and devices capable of interaction with each other. Another article [3] points out the latency critical of IoT applications and analyzes their requirements in different organizations. Cheah et al. [4] mentioned that in an industry, there are thousands of sensors for different uses, viz pressure transmitters, flow meter, temperature transmitters, and so on. Wireless networks among the sensors, i.e., IoT, can help accomplish many day-to-day business activities with minimum cost. Jayavarghana et al. [5] identified that IoT wherein sensors and actuators combinedly shared in an environment in order to develop a common operating picture of measuring the environmental indicators in case of delicate ecologies and natural resources.

1.2 Motivation and Contribution of the Study

From the above-mentioned points, it is clarified that IoT is considered as an emerging technology that can be applied in different organizations with a view to achieving organizations goals and end users for cost minimization and profit maximization. Considering the importance of using IoT, the present study titled "role of **Internet of Things in global business**" has been undertaken.

2 Objectives of the Study

The objectives of the study are as follows:

(i) To understand the IoT;
(ii) To point out the possible applicable areas of IoT in global business;
(iii) To find out the limitations of IoT;
(iv) Lastly, to point out some suggestions.

3 Methodology of the Study

A study of this nature cannot be conducted without methodology. Data used in the study are secondary in nature. It has been collected from different journals, books, and websites. Only descriptive analysis has been used in conducting the study.

4 Global Business

Business is considered to be global when companies operate facilities in many countries around the world. Global business is different from international business as in case of international business, companies sell products worldwide but have facilities only in its home country. Management of global business includes project management, financial management, operations management, human resource management, management, and leadership which can ensure navigate any business environment (www.google.com) [6]. Strategy of a company which creates facilities doing business worldwide with a view to reaping maximum rewards of trading in a worldwide market. For profit maximization, global business follows a financial strategy in order to achieve the most cost-effective way of delivering goods and services to the required level of quality which refers to reduction of costs resulting in higher profits and better cash flows. IoT can help much in achieving profit maximization of global business.

5 IoT and Global Business

According to Brendan (www.techwireasia.com) [7], it is clarified that the IoT creates a heightened level of awareness and monitor business activities in a changing environment. The available devices of IoT stand at 100 billion devices and 11 trillion items in its capacity to connect. Global enterprises installing the connected smart devices and data analytics start implementing IoT enabled systems which represent 77% of corporation in Australia in one form or another.

It is seen in the outlook of Boston Consulting Group that global business will be spending amounted to US $285 billion on IoT services and products in manufacturing, transport, and utility industries. Bill Ruh, Chief Digital Officer of a US company (www.google.com) [6] says data analytics and machine connectivity may be considered to the next level productivity. The big data analytics and broadband communications may drive a significant awareness for taking efficient decision and bolster force multiplier in order to reducing overall cost of company functions.

IoT can facilitate managers to monitor fleet management and transportation systems introducing up-to-date alerts on things, viz traffic, road conditions, passenger demands, warranting a smooth, and stress-free transport system from beginning to end.

IoT creates facilities to business leaders in the world to improve asset utilization, reliability and productivity, analysis of production, plant conditions, staff, customer demand and unique market trends, and supply chain management of an organization.

IoT facilitates the global business on innovative functions like remote monitoring and control for product support simplicity and flexibility in a competitive environment.

6 IoT and Sustainable Development

IoT with connectivity of different devices and objects through Internet as smart in day-to-day operations of industries, institutions, households, and individuals for smooth functioning with a view to reducing costs and maximizing benefits. It is considered as a part of ICT which help facilitate an opportunity for the present and future in case of digitization in different organization. Its impact can be viewed as sustainable in water management, energy efficiency, environmental pollution management which is very much related to United Nations sustainable development goal. Accordingly, sustainable development can be said as the meeting of needs of present generation without sacrificing the ability of future generation.

7 Limitations of IoT

Limitations of IoT are considered from the viewpoints of breach of privacy, over reliance on technology, the loss of jobs, and security issues.

- **Breach of privacy**:
  If companies' information is stored on the Internet, question may arise who will control the information as there are more companies offering the facilities.
- **Overreliance of technology**:
  Users of IoT may be over reliance on technology. With the passes of time, it is seen that overreliance on technology for decision making could lead to devastation.
- **The loss of jobs**:
  Introducing more and more devices to IoT, there are more and more chances for automation in an organization which may lead to curtail of job opportunity.
- **Security issues**:
  As people are used to IoT more and more, information will remain on the Internet, there is chance that hacker may hack the information which may raise question for security issue.

8 Concluding Remarks

The study has discussed on "Internet of Things in global business." It is conducted on the basis of secondary data. The study has covered brief introduction of IoT and its impact on conducting day-to-day activities in industry, institutions, households, and individuals with a view to achieving sustainable development giving emphasis on cost minimization and maximization of benefits. During the study, it is seen that there are some limitations of IoT for its use, but despite the limitation, it may be said that IoT emerges in modern civilization as boon through which society at present is getting maximum benefits and in the future will get more diversity of benefits and when the world's development in general and a country's development in particular will be sustainable. Lastly, further research may be suggested to overcome the limitations of IOT in its application in global business for ensuring cost minimization and profit maximization.

References

1. www.internet.scociety.org
2. Lee, I., & Lee, K. (2015). The internet of things (IoT) Application, investments and challenges for enterprises. *Bus Horizons, 58*, 431–440.
3. Schulz, P., Matthe, M., Klissig, H., & Simsek, M. (2017). Latency critical IoT applications in 5G perspective on the design of radio interface and network architecture, IEEE xplore, ieee.org.
4. Cheah, W. Z., Jegathesan, J., & Loon, S. C. (2015). *International Journal of Computer Network Applications, 2*(1).
5. Jayavardhana, G., Buyya, R., Marusic, S., & Palaniswabri, M. (2013). *Future Generation Computer Systems, 29*(7), 1645–1660.
6. www.google.com
7. www.techwireasia.com

A Study on the Behaviours and Attitudes of Students in Online Learning at the Open University of Mauritius

Rubeena Doomun and Darelle van Greunen

Abstract The worldwide focus on education has led universities to opt for online learning since the COVID-19 pandemic. Although online learning is becoming the new normal, there is minimal research on learners' attitudes and behaviours towards online learning. The purpose of this paper is to explore the current behaviours, attitudes and learning experiences of students towards online learning through both students and academics perspectives. A mixed approach method was adopted, with a focus group discussion conducted via Zoom videoconferencing with eight academics and a survey-based questionnaire of 520 students via Google platform from the Open University of Mauritius. The main findings of this paper showed that learners need a sense of community between learners and instructors whereby there was a feeling of isolation. Hence, they value online discussions and opine that learning activities promote interaction with others. Making sure to study on a regular basis followed by attending online classes were the top characteristics of the learners. As for the academic perspective, low attendance was a major issue while rewards and teamwork were salient characteristics that came out from the findings. The paper concludes with limitations and suggestions for future work.

Keywords Online learning · Behaviours · Attitudes · Engagement · Interaction

1 Introduction

The education sector has experienced a paradigm shift from traditional classroom environments to online learning settings because of the unprecedented development of the COVID-19 pandemic. Lockdowns have been imposed in many countries which has led to the closure of university and college campuses, thereby rapidly switching

R. Doomun (✉)
Open University of Mauritius, Réduit, Mauritius
e-mail: r.doomun@open.ac.mu

D. van Greunen
Nelson Mandela University, Gqeberha, South Africa
e-mail: Darelle.vanGreunen@mandela.ac.za

B. Pati et al. (eds.), *Proceedings of the 6th International Conference on Advance Computing and Intelligent Engineering*, Lecture Notes in Networks and Systems 428,
https://doi.org/10.1007/978-981-19-2225-1_44

to online learning. Beginning in March 2020, Mauritius has practised strict lockdown measures significantly impacting the education sector. The Open University of Mauritius switched from blended learning to fully online mode because of the lockdown. Online learning was activated through the learning management system Moodle and the web videoconferencing tool Zoom. Guidelines and training sessions about how to use Moodle and Zoom for online teaching were implemented.

1.1 Motivation and Contribution

Academic institutions are facing a challenging time to adapt to changes while maintaining the effectiveness of lecture delivery. The contribution of this study is to evaluate the learner's attitudes and behaviours in online learning at the Open University of Mauritius. It was conducted during the COVID-19 pandemic.

In this study, we set out to answer the primary research question:

1. What are the experiences and characteristics of students in online learning?

There were two phases of this research study. First, the instrument, the Student Learning and Satisfaction in Online Learning Environments (SLS-OLE), was adapted after permission received from Dr. DiLoreto [7]. After validation of the adapted questionnaire, it was made available to students enrolled in an undergraduate program at the Open University of Mauritius. Next, a focus group discussion was carried out between the academics at the Open University of Mauritius to underscore the findings from the survey data.

Among the research in online learning, there is a scarcity of attitudinal studies exploring the attitudes and characteristics of the students in online learning especially among Mauritian university students. As such, this paper singles it out as a significant contribution to the current learning. The findings of this paper will be useful to universities to better engage students in online education delivery.

2 Literature Review

2.1 Online Learning

Online learning environments involve another pedagogy and different skills compared to traditional classrooms [5]. The type of interaction and engagement of learners in virtual learning environments is not similar to face-to-face settings causing learning outcomes to be set [9]. Online education in Mauritius has increased significantly after the COVID-19 outbreak. There was a paradigm shift in online learning due to the national lockdown. Educators had to align their pedagogical approach according to the new normal. In this pandemic period, the majority of the academic institutions have focused primarily on converting the educational materials into online content and have not changed the online teaching and delivery methods [25]. The Open University of Mauritius went through the same transformation while struggling to maintain the learning and engagement of its learners.

According to [18], there are two significant determinants to measure the effectiveness of online learning environments, namely instructor presence and interactive teaching style. Another study by [16] showed that learning and engagement can be improved by integrating virtual communities in online learning environments. [13] recommended that the students' say and perspective on online learning must be explored. Hence, to better understand students' perception of online learning, it is crucial to discover their attitudes and behaviours.

2.2 Community of Inquiry

The Community of Inquiry [8] is a well-defined framework that applies social constructivist ideas to the online learning environment whereby learning happens through the social presence, cognitive presence and teaching presence. Garrison et al. [8] found that "social presence marks a qualitative difference between a collaborative community of inquiry and a simple process of downloading information" (p. 96). Thus, online students do not experience a feeling of isolation in their learning, but rather they feel connected virtually. In another study, Jaggars et al. [11] defined certain characteristics for a quality course. Clearly written objectives, a well-organised content, interpersonal interaction and appropriate use of technology were the salient characteristics that a quality course should contain.

To date, there have been limited studies on learner's attitudes, behaviours and motivation in online learning. By elucidating the factors that affect the behaviour of learners in online learning, better ways of student engagement may be provided. Therefore, this paper aims to examine and identify the characteristics, attitudes and behaviours of students in online learning.

3 Methodology

This section illustrates the research design process. It also discusses data collection methods and data analysis techniques that were used during the research.

3.1 Research Model and Procedure

A mixed approach called the sequential quantitative–qualitative explanatory method has been adopted. Based on the approach of [10], the qualitative results can help to understand the quantitative results. The first part of the study involved a survey in the form of a questionnaire with the target audience, the students enrolled at the Open University of Mauritius. The study was restricted to learners from the Open University of Mauritius who were enrolled at the undergraduate level at the university at the time of the study. The questionnaire consisted of questions related to demographic and Likert-scale type of questions. The second part of the study consisted of a qualitative approach through a focus group discussion among academics at the Open University of Mauritius. It was used to determine the views on the student behaviour, motivation and engagement in online learning as well as to understand the academics' personal experience in conducting online lectures. Convenience purposive sampling was used to choose the participants. The participants agreed to join the focus group discussion by responding to the email sent to them. Permission was granted to both survey the students and carry out the focus group discussion at the Open University of Mauritius.

3.2 Research Context and Sample

The questionnaires were distributed electronically via Google Forms in December 2020, and a total of 520 responses were received after two weeks. According to [24], the population of Mauritius was 1,271,768 people in 2020, and hence, the adequate sample size is 385 at a 95% confidence level and 5% margin of error. 520 responses were collected, which is a 4.30% margin of error at a 95% confidence level at a 50% response distribution [17]. After the data was captured and cleaned, the statistical package SPSS was used for the analysis.

The sample size of the focus group discussion was six participants. The focus group discussion lasted for one hour, and it was recorded. Later on, it was transcribed verbatim. Furthermore, the transcribed data was anonymised whereby participants were given pseudonyms. Pseudonyms were devised into "P1", "P2", etc., for all participants. Finally, based on the thematic analysis, several themes emerged.

3.3 Instruments Used and Their Validation

The SLS-OLE instrument was referred to when designing the questionnaire. The questionnaire consists of certain constructs, namely course organisation/structure, learner interaction, student engagement, instructor presence, student satisfaction and perceived learning. Based on them, the student's satisfaction and perceived learning in online environments are assessed. Permission was granted by Dr. Melanie DiLoreto (2015) to adapt the SLS-OLE instrument for the purpose of the study. The questionnaire was then piloted on a sample of five members of the public where it was found to be viable.

On the other hand, the focus group was conducted online via the Web videoconferencing platform Zoom. Open-ended questions were posed to the participants. The latter were allowed to discuss and debate among themselves, while the researcher moderated the flow of the debate.

4 Data Findings and Analysis

4.1 *Quantitative Results*

The demographic profile of the respondent group is presented in Table 1. Most of the respondents (46.7%) were between 18 and 25 years of age. The gender split for the respondent group is female-dominated, with 71.3% of the respondents being female. All (100%) of the respondents studied in online or blended learning mode, and they were studying at the undergraduate level.

The respondents were asked several questions regarding online learning to determine their behaviour and attitudes. Questions regarding their motivation and interaction in online learning were also asked.

4.1.1 Field of Study

The top four majorities of the respondents were from the B.Sc. (Hons) Applied ICT with specialisation programme (n = 60) followed by B.Sc. (Hons) Human Resource Management programme (n = 56), then B.Sc. (Hons) Business Management programme (n = 54), followed by B.Ed. (Hons) Early Childhood and Education Care programme (n = 43) (Fig. 1).

Table 1 Demographic profile

	% of total	Total (n = 520)
Age group		
18–25	46.7	243
26–34	29.8	155
35–50	21.7	113
51–65	1.7	9
Gender		
Male	28.1	146
Female	71.3	371
Other	0.6	3
Learning mode		
Online learning	72.5	377
Blended learning	27.5	143

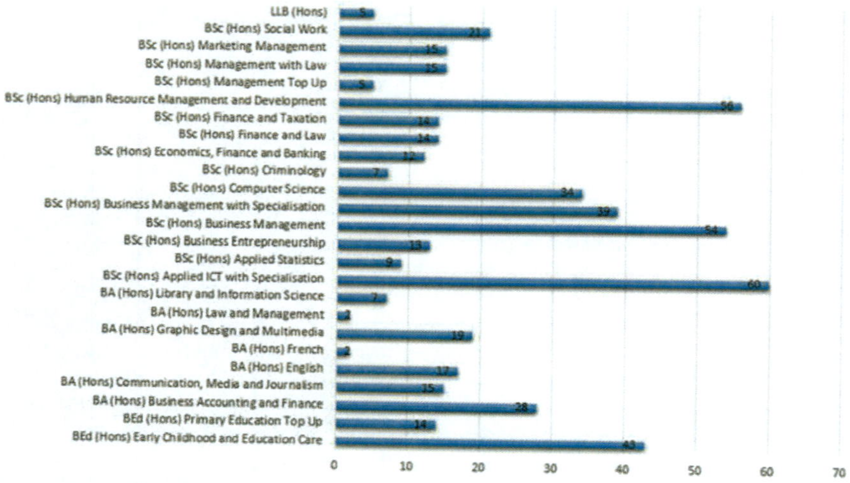

Fig. 1 Programme of study

4.1.2 Level of Study

The respondents were mostly in year 2 semester 1 ($n = 196$) in their level of study. Most of the courses were a three-year duration programme (Fig. 2).

Fig. 2 Year and semester of study

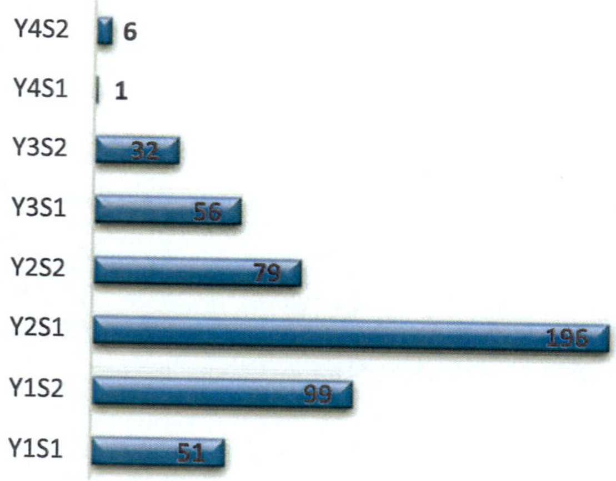

4.1.3 Motivation and Interaction of Students in Online Learning

Respondents were asked to indicate their motivation and interaction in online learning on a 5-point Likert scale ranging from strongly agree to strongly disagree. The variables were firstly treated as categorical and secondly as numerical in the measurement level.

Treating the variables as categorical in the measurement level

Considering the combined size of the agree and strongly agree proportions in the graph below, one can see that there are 87.2% ($n = 453$) of respondents indicating a need to have a sense of community between instructor and students in online learning, see Fig. 3. Furthermore, 72.7% ($n = 378$) of the respondents' value participation in online discussions while 48% ($n = 250$) of respondents express a feeling of isolation during online learning which underscores the fact that 66.2% ($n = 344$) respondents agree that they miss the social experience in an online class. About 54% ($n = 281$) of students indicated that they completed readings as assigned during the course.

In the same vein, the statement "participate in online discussions to earn an acceptable participation grade" scored 44% ($n = 229$). Likewise, 44.8% ($n = 233$) of the students rated affirmative to "learning activities promoted interaction with others" and 41.8% ($n = 217$) of students agreed that they received ongoing feedback from their classmates.

The reaction to having the opportunity to introduce oneself to others showed up slightly positive with 39.2% ($n = 204$) of students while 32.5% ($n = 169$) of students reacted negatively. Similarly, the response to get tasks done only when reminded to do so is slightly positive with 37.3% ($n = 194$) of students compared to 32.1% ($n = 167$) of negative responses. However, the question on "participate in online discussions to build friendships and help classmates" was stated on the neutral side with a score of 40% ($n = 208$) rather than on the positive with a score of 33.8% ($n = 176$).

Treating the variables as numerical in the measurement level

On average, having the opportunity to introduce oneself ($M = 2.98$, SD = 1.195) is indicated as the strongest behaviour, see Fig. 4. The overall tendency of the mean is towards 3 (neutral) except for three statements that tend to 2 (agree). These results carry the same information as the previous section in a more parsimonious way.

4.1.4 Behaviour, Thoughts and Feelings of Students in Online Learning

Respondents were asked to rate the characteristic on a 5-point Likert scale ranging from very characteristic to not at all characteristic. The variables were firstly treated as categorical and secondly as numerical in the measurement level.

	Strongly Agree	Agree	Neutral	Disagree	Strongly Disagree	Total
Want to have a sense of community between instructor and students	227	226	54	6	7	520
	43.7%	43.5%	10.4%	1.2%	1.3%	100.0%
Value participation in online discussions	132	246	87	36	19	520
	25.4%	47.3%	16.7%	6.9%	3.7%	100.0%
Feeling of isolation	139	111	125	115	30	520
	26.7%	21.3%	24.0%	22.1%	5.8%	100.0%
Miss the social experience with other students	236	108	89	64	23	520
	45.4%	20.8%	17.1%	12.3%	4.4%	100.0%
Get task done when reminded of	72	122	159	118	49	520
	13.8%	23.5%	30.6%	22.7%	9.4%	100.0%
Receive ongoing feedback from classmates.	55	162	149	100	54	520
	10.6%	31.2%	28.7%	19.2%	10.4%	100.0%
Complete readings as assigned during the course	78	203	149	67	23	520
	15.0%	39.0%	28.7%	12.9%	4.4%	100.0%
Participate in online discussions to earn an acceptable participation grade	49	180	172	91	28	520
	9.4%	34.6%	33.1%	17.5%	5.4%	100.0%
Participate in online discussions to build friendships and help classmates	39	137	208	102	34	520
	7.5%	26.3%	40.0%	19.6%	6.5%	100.0%
Learning activities promoted interaction with others	62	171	153	79	55	520
	11.9%	32.9%	29.4%	15.2%	10.6%	100.0%
Had the opportunity to introduce oneself to others	50	154	147	96	73	520
	9.6%	29.6%	28.3%	18.5%	14.0%	100.0%

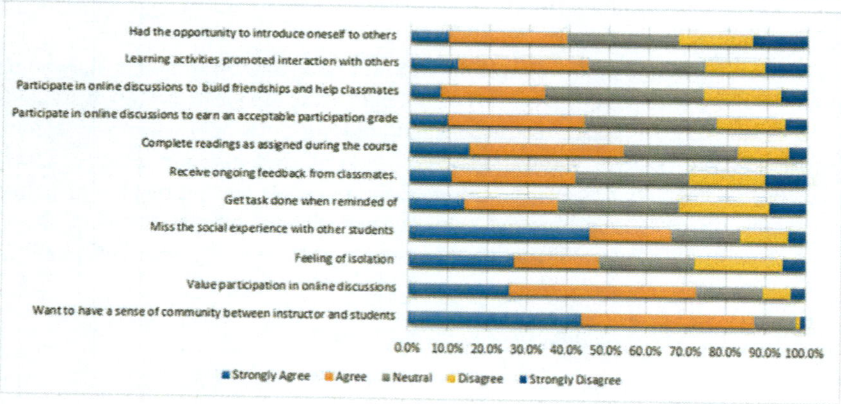

Fig. 3 Motivation and interaction in online learning: categorical measurement

Treating the variables as categorical in the measurement level

Considering the combined size of the very characteristic and characteristic proportions in the graph below, one can see that there are 62.1% ($n = 323$) of respondents indicating they attend all online classes, see Fig. 5. Likewise, 60% ($n = 312$) of respondents tend to make sure to study regularly while 65.4% ($n = 340$) of respondents find ways to make the course interesting.

Furthermore, a high percentage of 79% ($n = 411$) respondents has rated looking over notes/video lectures to understand the topic as their characteristic. Doing well

	Mean	N	Std. Deviation
Want to have a sense of community between instructor and students	1.73	520	0.798
Value participation in online discussions	2.16	520	1.001
Feeling of isolation	2.59	520	1.252
Miss the social experience with other students	2.10	520	1.228
Get the task done when reminded of	2.90	520	1.177
Receive ongoing feedback from classmates.	2.88	520	1.153
Complete readings as assigned during the course	2.53	520	1.037
Participate in online discussions to earn an acceptable participation grade	2.75	520	1.026
Participate in online discussions to build friendships and help classmates	2.91	520	1.008
Learning activities promoted interaction with others	2.80	520	1.158
Had the opportunity to introduce oneself to others	2.98	520	1.195

Fig. 4 Motivation and interaction in online learning: numerical measurement

	Very characteristic of me	Characteristic of me	Moderately characteristic of me	Not really characteristic of me	Not at all characteristic of me	Total
Attend all online classes	166	157	113	54	30	520
	31.9%	30.2%	21.7%	10.4%	5.8%	100.0%
Make sure study on a regular basis	113	199	134	56	18	520
	21.7%	38.3%	25.8%	10.8%	3.5%	100.0%
Put forth effort to complete the learning activities	152	222	97	35	14	520
	29.2%	42.7%	18.7%	6.7%	2.7%	100.0%
Looking over notes/video lectures to understand the topic	216	195	73	20	16	520
	41.5%	37.5%	14.0%	3.8%	3.1%	100.0%
Find ways to make the course interesting	144	196	113	43	24	520
	27.7%	37.7%	21.7%	8.3%	4.6%	100.0%
Getting a good grade	238	214	44	16	8	520
	45.8%	41.2%	8.5%	3.1%	1.5%	100.0%
Doing well in tests/quizzes	165	219	95	25	16	520
	31.7%	42.1%	18.3%	4.8%	3.1%	100.0%
Really desire to learn the material	221	208	56	23	12	520
	42.5%	40.0%	10.8%	4.4%	2.3%	100.0%

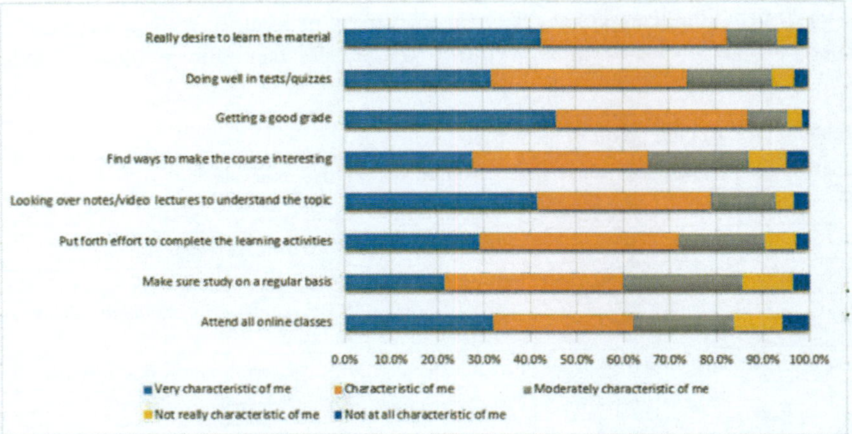

Fig. 5 Behaviour, thoughts and feelings of students in online learning: categorical measurement

	Mean	N	Std. Deviation
Attend all online classes	2.28	520	1.182
Make sure study on a regular basis	2.36	520	1.044
Put forth effort to complete the learning activities	2.11	520	0.990
Looking over notes/video lectures to understand the topic	1.89	520	0.989
Find ways to make the course interesting	2.24	520	1.088
Getting a good grade	1.73	520	0.857
Doing well in tests/quizzes	2.05	520	0.984
Really desire to learn the material	1.84	520	0.945

Fig. 6 Behaviour, thoughts and feelings of students in online learning: numerical measurement

in tests/quizzes and putting forth an effort to complete the learning activities is a highly rated characteristic of 73.8% ($n = 384$) and 71.9% ($n = 374$), respectively. Finally, two characteristics were rated very high, namely getting a good grade with 87% ($n = 452$) of respondents and really desiring to learn the material with 82.5% ($n = 429$) of respondents.

On average, making sure to study on a regular basis ($M = 2.36$, SD $= 1.044$) is graded as the most rated characteristic of students in online learning, followed closely by attending online classes ($M = 2.28$, SD $= 1.182$), see Fig. 6. The overall tendency of the mean is towards 2 (agree). These results carry the same information as the previous section in a more parsimonious way.

4.2 *Qualitative Results*

The profile of the participants is shown in Table 2. The participants were either full-time or part-time lecturers at the Open University of Mauritius. Each lecturer has taught at least one module in a specific programme in either online or blended mode.

Table 2 Demographic profile of participants

Participants	Gender	Age range	Occupation	Programme taught
P1	Male	26–34 years	Full-time lecturer	LLB
P2	Male	35–50 years	Full-time lecturer	B.Sc. (Hons) entrepreneurship
P3	Male	35–50 years	Full-time lecturer	B.Ed. (Hons) primary education B.Ed. (Hons) early childhood education and care
P4	Male	66 +	Part-time lecturer	B.Sc. (Hons) business management
P5	Female	26–34 years	Part-time lecturer	B.Sc. (Hons) applied ICT
P6	Male	35–50 years	Part-time lecturer	B.Sc. (Hons) graphic design

From the analysis of the qualitative data, the participants highlighted aspects of their experience in online teaching. The emerging themes of the study are discussed in the foregoing paragraphs.

4.2.1 Attendance

Low Online Attendance

First, it has been noted that the attendance in online classes is low. Participant P2 stated "The attendance if we compare it with a face-to-face session, it is not that much. I would say 50% since the sessions are recorded. So they don't attend it live because they know they will have it recorded." Another participant P4 agreed that the reason behind low attendance was the availability of the recordings for later consultation by stating "It's I attend the class or not, it's not a problem being given that it is distance learning so they can have the material from the website or whatever, and they can get ready for the exams." Participant P5 also mentioned that attendance tended to decrease in forthcoming sessions through this statement "But it's true from the first session, we had more students compared to the last sessions, maybe they were busy, etc."

Feel Isolated in the Online Class

The findings revealed a feeling of isolation when conducting an online class. Participant P4 highlighted that "But as a class, we have two-way traffic which personally, I didn't find, at times, I felt like I was speaking to a screen that's not giving any two-way traffic..." Similarly, participant P2 corroborates through this statement "Sometimes it feels like you are just talking alone". Participant P6 also agreed that it was a one-way communication only by stating "It was difficult for some participants to take important ideas and as agreed it is a one-way traffic, they just listen." Students were passive learners as illustrated by Participant P1 "they were passive learners... there was an inability to respond to it."

4.2.2 Participation of Learners

Members of the focus group were questioned about the participation of learners in an online class as well as the reasons behind why learners were not participating.

Ask Questions

Students were reluctant to ask questions or clarifications during an online class. Participant P2 explained "I try to ask a few questions about what they understood and encourage them to ask questions if they have any. But I don't think it worked." Similarly, participant P1 bemoaned that "I try to generally ask them the question, but I believe that eventually from a field of study with respect to education, it's quite difficult for them to get engaged about it". This situation persisted with Participant P4 who stated "Participation, online participation has been very, very deceptive. In the sense that to ask a question, nobody seemed to be there to answer...the participation on answering questions was quite, quite low. I should say that out of 11, 12 learners were logged in, only 1 or 2 make an effort to ask questions, and that's it. Nobody seemed to be interested in the questions that were put in." Participant P4 also pointed out that language is a barrier through the statement "At some point in time, I have noted that language is a problem. So I said you can ask me your question in French or Creole what matters is that you communicate to me. You tell me what the problem is and as far as possible I will be able to interact and solve the problem."

Personality Traits of Learners

Shyness, fear and lack of confidence were considered as the main reasons behind non- participation in the online class. Participant P3 acknowledged that "*many of the students don't want to participate because they are afraid of the reaction of other students.*" Participant P3 felt that students were shy declaring that "*why there is shyness, maybe that student has got a very low self-esteem to talk in front of other classmates in a group.*" The profile of the learners plays a role in their participation. Participant P3 stated that "*maybe the students, they do not have any previous experience or prerequisite knowledge as in the case of law. So they can't voice out or they may be shy about giving an answer that they think is more or less wrong.*" Adding further, Participant P3 recommended that "*If you already know the name of all your students and you already know those who are extroverts, you may know some of the students who do not hesitate to talk. So I think what we can do is that you ask those people who are not shy in the classroom to give their opinion of what they think that others can follow. Or you can ask someone else who is shy, you know, to just add a little something on what the other classmates have said.*"

On the other hand, participant P2 outlined that "*the difference is in a face-to-face session since they are in front of you there is a level of confidence. They will tend to participate. But in online since we don't see them, there is a lack of confidence where they don't ask questions or participate.*" Similarly, Participant P1 responded that "*There is this fear that you are behind the screen and you don't know really how to react...even if I drop a message in the chat box, I may have to wait for the whole session for them to respond, or actually I don't know if they are shy or if they are fearing what their answers are not good.*"

Learning Outcomes

There is high participation of learners due to well-defined learning objectives set. Participant P3 voiced out that "Students whether they participate or not participate, it depends on the type of activities that you have planned. What are your learning outcomes?" Adding further by mentioning "I think there was that high participation, because at the start of the session, I already set the expectation in terms of the learning...I think for me, this is the main reason why they are going to get engaged in the teaching and learning, it depends on whether the learning outcomes will have an impact on them or not."

4.2.3 Engagement of Learners

Participants were questioned about how they kept the learners engaged during an online class.

Use of Online Activities

Quiz and online activities helped to keep learners engaged as mentioned by Participant P6 "Task like quiz, online tutorial or any other materials which have been posted online, but some are very pleased to answer or to participate in these type of questions." Making use of discussion forums was also beneficial as reported by Participant P5 "I was using scrum boards, online scrum boards where I can set questions and then I will ask them questions. And then they have a time slot to give me that answer and correct it instantly in the class or after the quiz. So for engaging students, I was doing that every session...I was asking questions, sometimes we were replying in the Zoom chat box." However, Participant P2 responded negatively by saying "I tried to keep them engaged. After each topic I try to ask a few questions about what they understood and encourage them to ask questions if they have any. But I don't think it worked. I believe I should add other activities."

Peer Learning/Collaborative Work

Using a constructive approach for learning encourages participation as stated by participant P3 "I just introduce the concept and then just leave the floor to all of them and then they are free to give it like a type of brainstorming where each of them interrelates the concept to their own experience." Besides brainstorming, participants have tried making students work in groups. Participant P5 mentioned that "I group them four students in one group so that they can communicate between themselves if they have any queries etc. so that they can work together. I think this was one step I could do for them so that they can communicate and work." Adding further, "I think these interactions were good for programming compared to when we are in a class and everybody is staring at you and not knowing whether they are understanding or exchanging something."

4.2.4 Ways of Motivation

Participants were asked to comment on how to motivate the learners in online learning. They were asked to comment on the kind of motivation they should give to the learners so that they are able to participate in the class. Also, the kind of extrinsic or intrinsic motivation lecturers can provide to the learners.

Encouragement Through Feedback

From the participants' answers, it can be determined that feedback is considered key in the learning process. Participant P3 pointed out that "*feedback is very important and I think we should start by giving them positive feedback. We should not criticize them though, we see that there is something wrong, we have to praise them...it's my duty to praise him or her as a form of encouragement for later participation.*" Further adding "*they have got their self-esteem, so we have to praise them because if we praise them, let's say that guy has said something and it's wrong, but you can start by saying. What a little bit of information that has been given that is good and we don't criticize him for what is wrong. So in that case, he will have encouragement to participate.*"

In addition, for improvement and better performance, Participant P5 felt that "*I think when they know where their weaknesses are, they are more motivated to pursue what they are doing, I think so.*" Further stating that "*The most important thing in education, because it helps a student to adjust his or her performance. OK, this depends on receiving feedback. It helps them to reshape their performance to better achieve their goals in the future and keep them involved as well.*"

Rewards/Incentives

Many participants mentioned that rewards or incentives play an important role in motivating learners. Participant P1 suggested that "*a good way might have been by the means of incentives…if they do not have something to be rewarded in exchange of a commitment, they won't actually be engaged at all, so that's why I believe that maybe we could have a forum and tell them actually it will be used as a means to provide some grace marks or some other incentives...*" Further adding that "*when it comes to the field of law education, I simply believe that the best way once more remains a rewarding system by the means of marks.*" Similarly, Participant P2 mentioned, "*to award marks on attendance and to award marks on the activities online, they will be automatically motivated if marks are allocated.*" However, Participant P6 reported that "*It's one of the most challenging things as to how to motivate the students...one of the ways to motivate them, maybe can get them involved through compulsory attendance, quizzes so that they get marks after the lecture. So things that are compulsory will force them to attend class.*"

Teacher as a Role Model

The teacher itself can be a motivation to the learners. Participant P2 stated that "Yes the lecturer actually himself can be a motivating factor, just a plain example. For example, we all went to universities. If there is a really, really nice lecturer, if you really like someone, you will make sure that you attend his class." Participant P3 strengthened the argument that "They say they are just bearing him, but they don't participate because if they ask any question, they know he will be very, very harsh in the way that he is going to reply to them." Likewise, participant P5 pointed out that "I think for them to keep motivated, we need to be very attentive to their questions." In contrast, Participant P4 disagreed on the fact that teachers can be a motivational factor through this statement "You can be a role model for young students, those fresh comers but you cannot be a role model to adults. They are here just for the lecture and they go away. Whether they see you as a role model is very rare."

5 Discussion

Based on the findings of this study, most of the questions did reflect a positive characteristic in behaviour and attitude. Most of the students indicated that they want to get a good grade and they really desire to learn the material. Indeed, a student who is motivated extrinsically looks for external signs of worth and validity [19]. Usually, grades act as a validation that the students have achieved the course learning outcomes. However, external motivators like grades and rewards can undermine intrinsic motivation for a task [6]. Hence, grades do not act as the only or best motivation for learning.

Intrinsic motivation is an internal desire to learn something. Desiring to learn the material, putting forth the effort to complete the learning activities, looking over notes/video lectures to understand the topic and making sure to study on a regular basis emerge as positive indicators for student's learning. Vansteenkiste et al. [22] found out that students that were intrinsically motivated read material with more depth, scored higher grades, and were more persistent than extrinsically motivated students. Similarly, one theme that emerged was rewards which are linked to high motivation and positive results.

An important outcome was respondents finding ways to make the course interesting. The driving force behind intrinsic motivation is individual interest [1]. Scoring well on quizzes (the majority of students responded) provides positive and fast reinforcement of good study habits. Lecturers make use of online activities as a form of engagement with learners. By conducting online quizzes, they make sure that students have understood the topics that have been covered. Most of the students want to get good grades and need little incentive to do so. Hence, online quizzes act as instant gratification.

Moreover, the students and lecturers have conflicting views on attendance. Students responded that one of their characteristics is to attend all online classes. However, the lecturers indicated that online attendance was low most of the time. Online learning made attending classes convenient in terms of energy and travel time. However, good Internet connectivity is required to ensure effective learning. One of the reasons behind low attendance was the availability of the recordings on the learning platform for later viewing.

Another set of questions addressed the concept of interaction and participation of students in online learning. Students' responses showed they missed the social experience with other students in online classes. They feel isolated. This is in line with the findings of studies that have previously been investigated. Similarly, in previous studies [2, 14, 21], learners in this study felt disconnected in the online class. Furthermore, lecturers also felt lonesome with passive learners attending the online class. This might be alleviated by creating online discussion boards or using the live chat feature in the online course as students agreed that they value participating in online class discussions. To help build a sense of community, the instructor can provide a break the ice post to the students whereby they can introduce themselves. Students also agreed that they participate in online discussions to earn an acceptable participation grade. However, participating in online discussions to build friendships and help classmates was not supported by the students.

Learning activities promoted interaction with others. Peer learning and brainstorming encourage meaningful interaction. Social constructivist theories created by [23] and Bandura [3, 4] affirm that learning occurs through social interaction; that is, students will perform better when they work collaboratively with others. Students responded that they received ongoing feedback from their classmates. The lecturers pointed out that feedback acts as an encouragement to students. Constructive feedback acts as a motivational factor for students to perform well and improve. Muirhead [15] proposed several strategies to encourage student interaction in online learning. One of them is to offer positive feedback about student work which is in line with the findings from this study. Another positive result is that students do agree that getting reminders helps them to complete their tasks.

In online learning environments, the primary role of the instructor is to establish his presence and personality in the course instructions, discussions and learning activities [20]. Instructors can enhance online content and "engender a sense of caring by soliciting student feedback about the course and using that feedback to enhance the course" [12, p. 6]. One of the outcomes that came out of this study is the instructor's personality. However, there were mixed views that it can act as a motivational factor.

Finally, the last element that came up from this study is the personality traits of the students. Lecturers have noticed that students are reluctant to ask questions in an online class. Character traits like shyness, fear and lack of confidence made it difficult for students to ask questions or respond to lecturers in an online class. This suggests that the student's personality plays an important role in class participation and interaction. Having the opportunity to introduce oneself to others in an online class is a positive outcome that emerged from the survey. To encourage shy students to talk, lecturers can initiate the interaction with the extroverted students first, and then, the rest can simply follow the herd.

6 Limitation

The scope of the study was to focus on the Open University of Mauritius and on the attitudes of the learners rather than the technological aspects. As the study was conducted within the Mauritian context, future work has to be done to explore the relevance of the study in different contexts.

7 Conclusion

In this research, quantitative and qualitative methods have been combined to analyse students' learning experience from both students' and lecturers' dimensions. This study showed that students usually did miss interaction with the instructor and other students in an online class. Students did experience the feeling of being isolated. These two aspects may be overcome through a chat feature, where students can discuss freely on certain topics. Concerning their reasons for participating in online discussions, the participants were more keen in earning an acceptable participating grade than in forming friendships and helping with classmates. This can be due to the busy work schedules of students, and hence, the feasibility of forming friendships was restricted. To promote active participation in online discussions, lecturers can refer to the interplay between intrinsic/extrinsic motivation and Community of Inquiry when creating online courses. Finally, gamification can be a useful strategy to apply in online learning for behavioural change and engagement of students. As future work, artificial intelligence and virtual reality can be brought in to create a virtual classroom to enhance the interaction between instructors and students.

References

1. Alexander, P. A., Murphy, P. K., Woods, B. S., Duhon, K. E., & Parker, D. (1997). College instruction and concomitant changes in students' knowledge, interest, and strategy use: A study of domain learning. *Contemporary Education Psychology, 22*, 125–146.
2. Ali, S., Uppal, M. A., & Gulliver, S. R. (2018). A conceptual framework highlighting e-learning implementation barriers. *Information Technology & People, 31*(1), 156–180. https://doi.org/10.1108/ITP-10-2016-0246

3. Bandura, A., Ross, D., & Ross, S. A. (1961). Transmission of aggression through imitation of aggressive models. *Journal of Abnormal and Social Psychology, 63*(3), 575–582.
4. Bandura, A., Ross, D., & Ross, S. A. (1963). Imitation of film-mediated aggressive models. *Journal of Abnormal and Social Psychology, 66*(1), 3–11.
5. Boling, E. C., Hough, M., Krinsky, H., Saleem, H., & Stevens, M. (2012). Cutting the distance in distance education: Perspectives on what promotes positive, online learning experiences. *The Internet and Higher Education, 15*(2), 118–126. https://doi.org/10.1016/j.iheduc.2011.11.006
6. Deci, E. L., Koestner, R., & Ryan, R. M. (1999). A meta-analytic review of experiments examining the effects of extrinsic rewards on intrinsic motivation. *Psychological Bulletin, 125*, 627–668.
7. DiLoreto, M., & Gray, J. (2015, October). Student satisfaction and perceived learning in online courses: The development of the SLS-OLE. Paper presented at the Annual Meeting of the Online Learning Consortium, Orlando, FL.
8. Garrison, D. R., Anderson, T., & Archer, W. (2000). Critical inquiry in a text-based environment: Computer conferencing in higher education. *The Internet and Higher Education, 2*(2–3), 87–105.
9. Harris, R. A., & Nikitenko, G. O. (2014). Comparing online with brick and mortar course learning outcomes: An analysis of quantitative methods curriculum in public administration. *Teaching Public Administration, 32*(1), 95–107. https://doi.org/10.1177/0144739414523284
10. Ivankova, N., Creswell, J. W., & Stick, S. L. (2006). Using mixed-methods sequential explanatory design: From theory to practice. *Field Methods, 18*(3). https://doi.org/10.1177/1525822X05282260
11. Jaggars, S. S., & Xu, Di (2016). How do online course design features influence student performance? *Computers and Education, 95*, 270–284. https://doi.org/10.1016/j.compedu.2016.01.014
12. Jaggars, S. S., Edgecombe, N., & Stacey, G. W. (2013). Creating an effective online instructor presence. Report completed for the Community College Research Center, Teachers College, Columbia University, NY. Retrieved on May 10, 2021, from http://files.eric.ed.gov/fulltext/ED542146.pdf
13. Mailizar, M., Almanthari, A., Maulina, S., & Bruce, S. (2020). Secondary school mathematics teachers' views on E-learning implementation barriers during the COVID-19 pandemic: The Case of Indonesia. *Eurasia Journal of Mathematics, Science and Technology Education, 16*, em1860. https://doi.org/10.29333/ejmste/8240
14. McInnerney, J. M., & Roberts, T. S. (2004). Online learning: Social interaction and the creation of a sense of community. *Journal of Educational Technology & Society, 7*(3), 73–81. Retrieved on May 10, 2021, from https://www.jstor.org/stable/pdf/jeductechsoci.7.3.73.pdf
15. Muirhead, B. (2004). Encouraging interaction in online classes. *International Journal of Instructional Technology and Distance Learning, 1*(6), 45–50.
16. Panigrahi, R., Srivastava, P. R., & Sharma, D. (2018). Online learning: Adoption, continuance, and learning outcome—A review of literature. *International Journal of Information Management, 43*, 1–14. https://doi.org/10.1016/j.ijinfomgt.2018.05.005
17. Raosoft. (2004). *Sample Calculator*. Retrieved on March 30, 2021, from http://www.raosoft.com/samplesize.html
18. Richardson, J. C., & Swan, K. (2019). Examining social presence in online courses in relation to students' perceived learning and satisfaction. *Online Learning, 7*, 68–88.
19. Sansone, D., & Smith, J. L. (2000). Interest and self-regulation: The relation between having to and want to. In: Sansone & J. M Harackiewicz (Eds.), *Intrinsic and extrinsic motivation: The search for optimal motivation and performance* (pp. 341–372). San Diego, CA: Academic Press.
20. Shea, P., Li, C. S., & Pickett, A. (2006). A study of teaching presence and student sense of learning community in fully online and web-enhanced college courses. *The Internet and Higher Education, 9*(3), 175–190.

21. Szeto, E., & Cheng, A. Y. N. (2014). Towards a framework of interactions in a blended synchronous learning environment: What effects are there on students' social presence experience? *Interactive Learning Environments, 24*(3), 487–503. https://doi.org/10.1080/10494820.2014.881391
22. Vansteenkiste, M., Simons, J., Lens, W., Soenens, B., Matos, L., & Lacante, M. (2004). Less is sometimes more: Goal content matters. *Journal of Educational Psychology, 96*, 755–764.
23. Vygotsky, L. (1978). Interaction between learning and development. In: *Mind and society* (pp. 79–91). Cambridge, MA: Harvard University Press. (Reprinted from Readings on the development of children, pp. 29–36, by M. Gauvain & M. Cole, Eds., 1997 New York: W. H. Freeman and Co.).
24. Worldmeter. (2020). Mauritius population. Retrieved on December 10, 2020, from http://www.worldometers.info/world-population/mauritius-population/
25. Wu, Z. (2020). How a top Chinese university is responding to coronavirus. World Economic Forum. Retrieved on December 18, 2021, from https://www.weforum.org/agenda/2020/03/coronavirus-china-the-challenges-of-online-learning-for-universities/

Data-Driven Decision Management from a Dashboard Perspective

Nirmal Kumar Betchoo

Abstract Organisations are eager to use and develop decisions by managing data in the most efficient way. To become more competitive and better meet their current and future needs, data-rich information enables managers to prospect the market, use the information for analysis and make decisions that could keep their businesses one step ahead of the rest. The scarcity of such data creates gaps in research information for business intelligence. Although algorithms and artificial intelligence are introductory concepts of data-driven decision management, the dashboard can be a useful tool that organisations can use both to present useful data and to make good decisions effectively. This paper describes the concept of using and managing data for decision-making in areas other than information technology. Using a conceptual framework to describe the importance of data-driven decision management, it argues that the creation of dashboards could be applied effectively the Université des Mascareignes (UdM) where it could meet the needs of users and decision-makers willing to get a clear picture of the problem affecting them. Rather than just disseminating information, the dashboard could provide important information that improves decisions and actions that can be initiated more efficiently and productively.

Keywords Data-driven decision management · Benefits · Attributes · Application · Dashboard

1 Introduction

Data-driven decision management (DDDM) is defined as a corporate governance approach that values decisions that can be supported by verifiable data. The success of the data-based approach depends on the quality of the data collected and the effectiveness of its analysis and interpretation [1]

N. Kumar Betchoo (✉)
Department of Management, Faculty of Business and Management, Université Des Mascareignes, Beau Bassin-Rose Hill, Republic of Mauritius
e-mail: nbetchoo@udm.ac.mu

B. Pati et al. (eds.), *Proceedings of the 6th International Conference on Advance Computing and Intelligent Engineering*, Lecture Notes in Networks and Systems 428,
https://doi.org/10.1007/978-981-19-2225-1_45

In the early days of IT, an expert with a strong background in technology was often needed to extract data for informational purposes, as this person needed to understand how databases work. Today, business intelligence tools often require very little or no support from the IT department. Sales managers can customise dashboards to easily view the data they want to see and run custom reports [1]. Changes in the way data can be extracted and visualised allow business leaders with no technology experience to work with analytic tools and make data-driven decisions.

1.1 Conceptual Framework of the Paper

The structure of this paper follows from a brief description of the data-driven decision management (DDDM) process. Although this concept may be familiar to information and communication technology (ICT) specialists, little is known to the public beyond this field. The attributes of DDDM are highlighted with respect to its importance and innovation in decision-making. At the same time, the advantages of DDDM are explained since they can be effectively addressed to different organisations. Some practical information on the use of DDDM is then developed to show how it has been useful in a practical way. Finally, the dashboard is presented as one of the levers that encourage DDDM to meet the needs of an external stakeholder which could be provided by the Université des Mascareignes, a public research university in Mauritius.

1.2 Attributes of Data-Driven Organisation

There exist some attributes related to data-driven organisations that engage in. Firstly, a data-driven organisation can perform continuous testing that includes user testing, working directly with real customers to get direct feedback on any new features or products [2]. Secondly, a data-driven organisation needs a continuous improvement mindset. This requires careful analysis, the development of mathematical or statistical models and simulation [2]. Next, a data-driven organisation might be involved in predictive modelling, forecasting sales, stock prices or company earnings, but more importantly, feeding forecast errors and other learnings into models to improve them [2]. Finally, a data-driven organisation will almost certainly choose future options or actions using a set of weighted variables.

Such characteristics or attributes require a highly efficient, data-based culture, supported by modern tools to promote fluid innovation. Many companies excel in some areas, but few do well in all of them [3]. In a way, accelerating processing and data gains is a scissors effect. Companies with traditional mindsets and manual processes will be left behind, while those with easy access to data scientists and computing resources will reap a huge harvest.

1.3 Benefits of Data-Driven Organisation

There exist some advantages of DDDM. A data-driven approach to problem-solving improves an enterprise's communication and teamwork and develops performance metrics like revenue and margin. Data and evidence-based brainstorming help turn creative ideas into business transformation [4]. The use of measures and analytical methods specified at key points in your decision-making processes improves a company's accountability. An added benefit of using data is that a fact-based discussion tone helps teams work on issues with minimal personal or emotionally charged conversation [4].

Data collection and analysis has long played an important role in companies and organisations. Although data-driven decision-making has existed in businesses in one form or another for centuries, it is a truly modern phenomenon [5]. Today's largest and most successful organisations use data to their advantage when making high-impact business decisions.

1.4 Insights into Data-Driven Decision-Making

Software algorithms are changing the way people work in an increasing number of fields today, managing distributed human workers at scale [6]. In these work environments, human tasks are assigned, optimised and evaluated using algorithms and tracking data. There has been a positive impact of this data-driven algorithmic management on workers and work practices in relation to Uber and Lyft, a new ridesharing service.

One of their main findings was that algorithmic passenger allocation in Uber and Lyft automatically distributed countless ride requests to drivers in a matter of seconds. The rapid and frequent acceptance of assignments by the drivers ensured the efficiency of the service and maximised the number of passengers to obtain a fast service [6].

Although many managers have pledged their allegiance to the principles of data-driven decision-making, there is still no comprehensive approach to measuring a company's ability to leverage the potential, its own information assets; in other words, there is no measure of "data productivity" [7]. Inspired by the traditional OEE framework, there exist innovative perspectives where (a) data productivity is divided into data availability, quality and performance of the decision-making process using this data, and (b) technical and organisational factors have been used, helping companies to assess their current level of productivity and actions to improve it [7]. The model has been tested through three case studies, and the results obtained from its application reflect the expectations of business leaders who are accelerating the cultural change necessary to fully express the potential of Industry 4.0.

Data-driven decision-making might further be considered through cluster analysis methods, namely centroid, connectivity and density [8]. Drawing on traditional clustering techniques, there exists algorithmic extensions and innovative efforts to process dynamic, large-scale, representative, non-convex and consensus data in new business environments. Clustering could be integrated into the global solution in contexts of management assistance, collaborative business economy and decision support in health [8].

Two cases of data-driven management were analysed in the context of highly digitalised Danish universal well-being. The first case showed how data-based management was implemented in a hospital in Denmark. Danish hospitals and the Danish healthcare system in general have a long tradition of data-based processing supporting data-based management in Danish hospitals [9]. The second case investigated under what conditions it is possible to establish data-driven management in an environment without any tradition of data-driven decision-making [9]. Data-driven management has consequently helped improve the delivery of public social protection services to the long-term unemployed to help them find employment. The researcher concluded that one of the main findings is the need to create a delicate balance between top-down and bottom-up management, which is essential for data-driven management to work in practice.

1.5 Insights into Data-Driven Decision-Making

In a related research, detailed survey data on business practices and investments in information technology from 179 large publicly traded companies revealed that those adopting DDD have output and productivity between 5 and 6% higher than one would expect from other investments using information technology. Additionally, the relationship between DDD and performance also appears in other performance measures, such as asset utilisation, return on equity and market value [10].

Data-driven decision management might be undertaken as a way to gain a competitive advantage. A study from the MIT Center for Digital Business found that organisations motivated primarily by data-driven decision-making had 4% higher productivity rates and 6% higher profits [11].

The DDDM conceptual model was validated in a research with 456 usable responses from employees of different companies using different business analysis tools. The study found that data-driven culture has strongly influenced product and process innovation, making the company more competitive in the industry [12].

2 The Dashboard as a Key Element of Data-Driven Decision-Making

In today's organisation, all departments can benefit from business dashboards. Corporate dashboards are essential tools that help managers visualise the information they get from their data. Business intelligence tools are essential in this regard, as they help managers collect all their data in a centralised place and work there.

Corporate dashboards can be of many types, from analytics to operational to strategic, each of them tracks a certain type of metric and tells a specific data story [13].

There exist three types of dashboards for data-driven decision-making. They are as follows:

Strategic dashboards

High-level dashboards for executives and senior management to help them measure the performance of their strategies [14].

Analytical dashboards

Data-rich dashboards with filters and breakdowns for data analysts and business analysts to investigate various business questions posed by their bosses, such as executives and managers [14].

Operational dashboards

Key performance indicator (KPI) dashboards are used by operations managers to monitor the KPIs of their processes and services on a daily basis, primarily for compliance or the pursuit of excellence [14].

Dashboards can focus on particular tasks in a department and how they are performed or provide an overview of the current state of general activities. Working with modern corporate dashboards improves collaboration and communication within a team and between business units.

2.1 Creating a Dashboard for DDDM at the UdM

This paper considers the creation of a dashboard at the Université des Mascareignes (UdM), a public university in Mauritius. Data collection and use might be common in a university with regards to the various uses of such data. It is important to consider the relevance of using data and how this could be helpful in effective decision-making.

2.2 Practical Applications of the Dashboard

A local authority recently aimed at devising an appropriate value proposition to target and attract international institutions to set up in Mauritius. It wanted to collect information in the key following areas. (a) opportunities for linkages and collaboration with international institutions, either for curriculum development and student exchange, (b) current value proposition to link up with international universities and institutions, (c) developing new programmes as per industry needs and d) attracting international students.

To that end, the UdM was called to explore avenues of collaboration with international institutions. In the first instance, such form of collaboration depended a lot of data-driven decision-making. The main question was to find out where information existed and how this could be showcased to the incumbent so as to create a value proposition. It became important for the university to come up with data in such a situation.

An analytical dashboard could effectively address the key questions from which elaborate information might be developed afterwards. For two selected arguments, a dashboard is developed to provide sufficient data for decision-making. These relate to (a) opportunities for linkages and collaboration with international institutions, either for curriculum development and student exchange and, (b) current value proposition to link up with international universities and institutions.

2.3 *Dashboard One: Opportunities for Linkages and Collaboration with International Institutions*

The two dashboards each answer the key questions asked by the local authority. Figure 1 explains the main information that the UdM could provide with regards to opportunities for links. The first dashboard clearly illustrates the different affiliations over the years (top left), the university's new course offering today and in the near future (top right), the guiding principle of links and collaboration at the centre and the benefits proposed by the UdM within the framework of a possible exchange programme (Fig. 2).

The dashboard shown in Fig. 1 elicits an immediate response from the organisation seeking an effective response regarding cooperation and linkages. In a nutshell, it provides the basic data with both facts and figures for the incumbent. Rather than

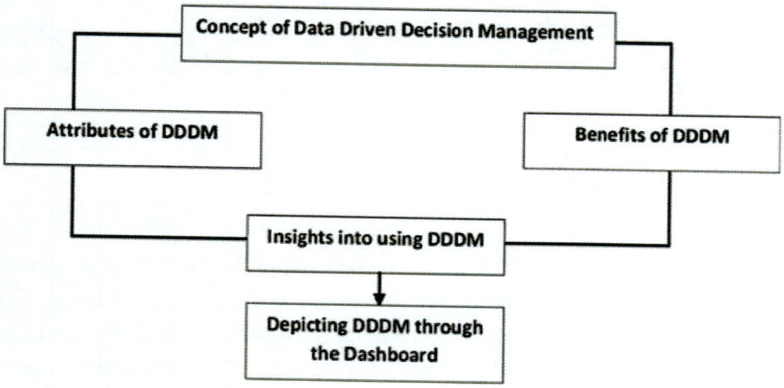

Fig. 1 Conceptual framework of the paper

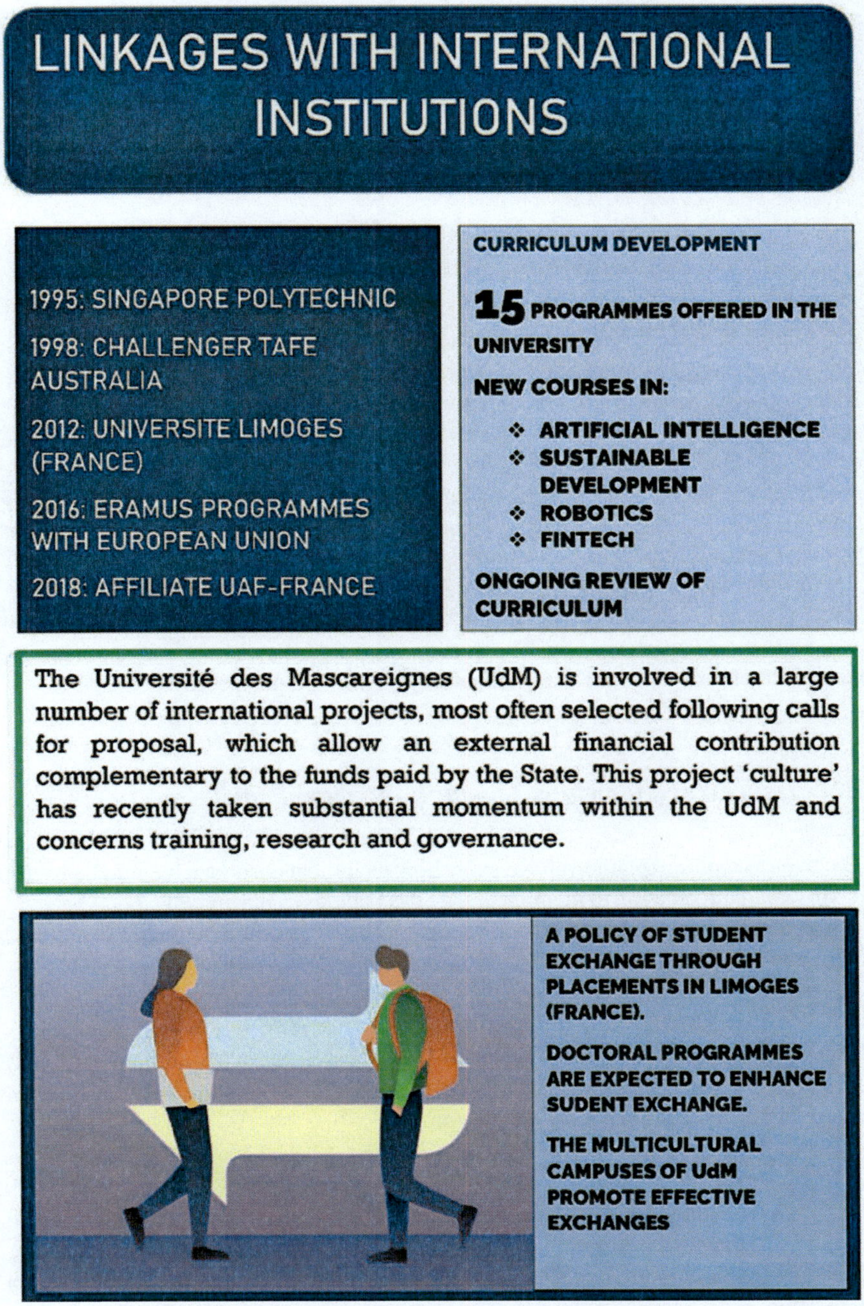

Fig. 2 Dashboard illustrating potential linkages with international institutions (Author's Source)

being a simple narrative of the key information requested, the dashboard provides a serious yet engaging overview of the possibilities for collaboration without being too detailed. With succinct but clear understanding, the dashboard already presents the initial documentation needed for a specific question.

2.4 Dashboard Two: Current Value Proposition to Link up with International Universities and Institutions

Regarding the expected value proposition of UdM, the second dashboard offers a unique presentation with enough information through the exploration of data mining. At the top right, the laboratory facilities are briefly mentioned, while at the top left, the academic profile of UdM staff is presented. In the centre on the left, the progression of doctoral students over the years is evoked, while in the centre on the right, the profile of graduates by department is highlighted. Bottom left, upcoming training courses reflecting current trends are mentioned while bottom right, the opening of two new schools at UdM are mentioned.

All of the data provided in Dashboard 2 illustrates the value proposition at a glance. There is no exaggeration of the facts but a simple presentation of rich data that will capture the needs of the public. The quality of information is important because the dashboard encapsulates the expected data from a single perspective (Fig. 3).

2.5 Improving Decisions Through the Dashboard

The two examples explained above show how analytical dashboards could be useful in providing information to incumbents. At the same time, dashboards might be good at decision-making from the university perspective. Certain examples are outlined below. A non-exhaustive list is briefly discussed.

Development of new courses

New programme development needs to be discussed after a situational analysis aiming to find out the attractiveness of a new course, its relevance as a new curriculum and the cost implications of such a programme.

Suppressing unattractive courses

Although the elimination of some courses might be inconvenient to academics, there might be a financial and human implication behind eliminating courses that have gone obsolete and redundant. There must be enough documentation to support such a decision.

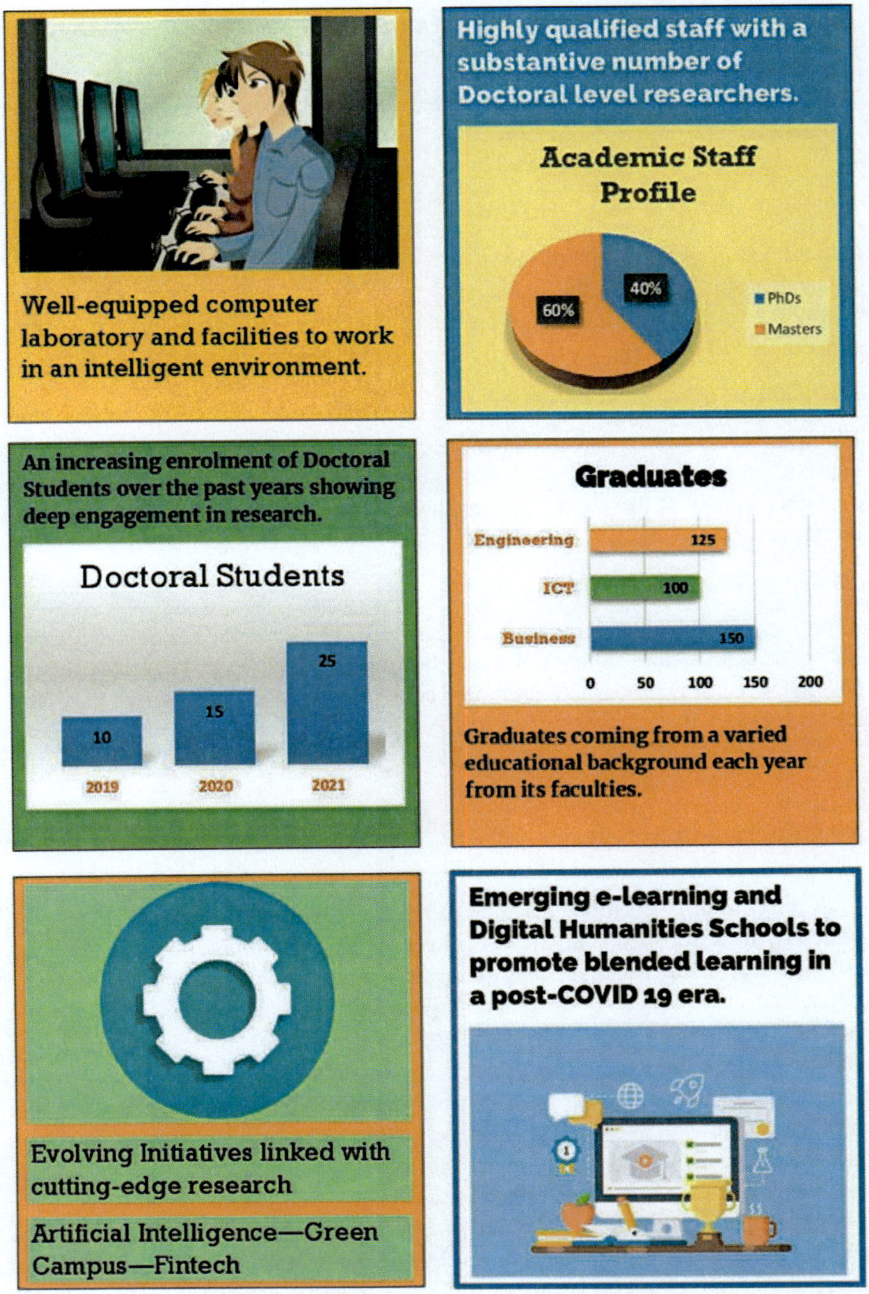

Fig. 3 Dashboard illustrating current value proposition to link up with international universities and institutions (Author's Source)

Forecasting potential avenues for research collaboration

Research collaboration is a well-sought strategy at the UdM. Yet, there needs to be a clear understanding of the key issue, avenues for collaboration for exchange programmes and dual collaboration. Here again, DDDM insights through a dashboard add value to the search for collaboration.

Improving student intake at the UdM

The attraction of more qualified students along with a diversity component could be an area of prospect for the UdM. Attracting doctoral students from other spheres willing to further research in the university could be a suitable decision-making process.

Recruitment strategy at the UdM

Recruitment strategy is a core human resource management concept. A situational analysis of existing staff as well as forecasting trends in recruitment could help develop capacity at the UdM. The dashboard might initially help find out gaps in recruitment.

Quality excellence at the UdM

Improving quality in an ongoing strategy at the UdM while, at the same time, compels staff to think in a more collaborative way how to reduce bottlenecks and ineffectiveness in the learning process. A dashboard could add data-rich information to address the problem.

3 Conclusion

Well-designed and developed dashboards are usually the result of careful planning guided by in-depth specification of requirements. They can provide insight, explanation, and a shared understanding of critical organisational information and then empower users to act on the information when and where it is needed. A dashboard basically represents a visual display of the most important information needed to achieve one or more goals, consolidated and organised on a single screen so that information can be monitored at a glance [15]. For a professional to get the best insight into the data and analyse it properly, they need to identify what to act on and streamline the workflow. Dashboards bring together data from multiple sources and combine it into a single interface for a detailed overview of the business while reducing reporting time [16]. The purpose of this paper was to briefly explain the importance of DDDM for organisations that are not necessarily IT-based. The relevance of the DDDM has been illustrated through the development of dashboards that can respond effectively to the strategic arguments of the university.

3.1 Future Directions

This study might be also influenced by some future directions. For instance, on being able to present data effectively, the university uses the dashboard as a means of conveying information to external audiences as well as developing the potential for using and exploiting data more competitively in the changing business environment today. The dashboard concept could also be integrated in the development of future university programmes with reference to the data available for study as well the interpolations and extrapolations that might be undertaken for effective decision-making.

References

1. Wigmore, I. (2021). Data-driven decision management, Techtarget.com.
2. Oreilly.com. (2020). What do we mean by data-driven, www.oreilly.com
3. Baker, M. (2020). Hallmarks of a data-driven business, Dell Technologies.
4. King, C. (2021). Five reasons data-driven management is a good idea, www.smartrecruiters.com
5. Stobiersky, T. (2019). The advantages of data-driven decision-making, Harvard Business School Online.
6. Lee, M., Kusbit, D., Metsky, E., & Dabbish, L. (2015, April). Working with machines: The impact of algorithmic and data-driven management on human workers. In *CHI '15: Proceedings of the 33rd Annual ACM Conference on Human Factors in Computing Systems* (pp. 1603–1612). https://doi.org/10.1145/2702123.2702548
7. Miragliotta, G., Sianesi, A., Convertini, E., & Distante, R. (2018). Data driven management in Industry 4.0: a method to measure data productivity. In *IFAC-Papers Online* (Vol. 51, Issue 11, pp. 19–24). ISSN 2405-8963, https://doi.org/10.1016/j.ifacol.2018.08.228
8. Sun, L., Chen, G., Hui Xiong, H., & Guo, C. (2017). Cluster analysis in data-driven management and decisions. *Journal of Management Science and Engineering, 2*(4), 227–251, ISSN 2096-2320, https://doi.org/10.3724/SP.J.1383.204011
9. Pedersen, J. (2019). Data-driven management in practice in the digital welfare state (Chap. 10, pp. 200–223). https://doi.org/10.4337/9781788112352.00014, Elgaronline.com.
10. Brynjolfsson, E., Hitt, M., & Heekyung H. (2011). Strength in numbers: How does data-driven decision making affect business performance? https://doi.org/10.2139/ssrn.1819486
11. Townsend, T. (2014). Data-driven decision making—The real advantage of big data, LinkedIn.com.
12. Chatterjee, S., Chaudhuri, R., & Demetris, V. (2021). Does data-driven culture impact business innovation and performance? *An Empirical Review Annals of Operations Research.* https://doi.org/10.1007/s10479-020-03887-z
13. Datapine. (2021). Business dashboard examples for every use case, www.datapine.com
14. Ubiq. (2020). What are the components of a successful dashboard? What are the components of a successful dashboard? Ubiq Business Intelligence.
15. Panigrahi, C. R., Tiwary, M., Pati, B., & Das, H. (2016) Big data and cyber foraging: Future scope and challenges. Techniques and environments for big data analysis. Studies in big data (Vol. 17). Springer.
16. Tamhankar, A. (2019). How dashboards help decision making, getting started media, https://medium.com

"Cultivating" Social Skills and Competence Through Digital Drama: A Case Study at Université des Mascareignes (Mauritius)

Neelam Pirbhai-Jetha and Shyama Ramsamy

Abstract Classroom scenarios have long shifted from their traditionality in the modern era of blended pedagogical strategies to accommodate the didactic reality of the mixed abilities setup. The shift from offline to online mode has undeniably necessitated a thorough examination of the curriculum and methodologies adapted to ensure effective teaching–learning through the introduction of interactive digital tools, platforms, and resources. The emergence of digital drama, which includes humanities and technology-based tools, in tertiary classroom settings establishes interactions between lecturers and "net generation" learners. Our objective is an attempt to cope with the challenges and constructive potentials to enhance their French language skills with drama as well as foster empathy and socially competent behaviours. In adopting a mixed research method, the paper aims at identifying and scrutinising the social skills developed, acquired and adopted by the sample student population at the Université des Mascareignes, Mauritius.

Keywords Digital drama · Net generation · French language · Mixed research · Social competence · Social skills · Empathy · Socially cognitive behaviour · Digital humanities · Education

1 Introduction

The logical application of technologically designed gadgets and accessories has paved their way into our everyday lives. Our Mauritian education system has been implementing technology-based tools and resources but has gained momentum particularly with the COVID-19 pandemic. The digitalisation of the teaching–learning approaches is being redefined accordingly. As it assesses current educational challenges, the Commonwealth of Learning (COL) endorses this innovative and adaptive teaching–learning practise. In the aftermath of achieving the United Nations' Sustainable Development Goals (SDGs), ICT-based curriculum designs,

N. Pirbhai-Jetha (✉) · S. Ramsamy
Centre for Digital Humanities, Université des Mascareignes, Beau Bassin-Rose Hill, Mauritius
e-mail: npirbhaijetha@udm.ac.mu

B. Pati et al. (eds.), *Proceedings of the 6th International Conference on Advance Computing and Intelligent Engineering*, Lecture Notes in Networks and Systems 428,
https://doi.org/10.1007/978-981-19-2225-1_46

teaching, and learning ensure "inclusive and equitable quality education and lifelong opportunities" [2].

Keeping the SDG 4 ("Quality Education") in mind, it is therefore from this perspective that the paper has opted for a mixed research methodology to understand the extent to which ICT-based learning and the introduction of digital drama (as a teaching tool), in the French course(s), can assist learners at the tertiary level in acquiring and cultivating the relevant social competence and skills.

Out of the three aspects of education—personal, economic and civic, David Ferrero, posits that Humanities Education plays a significant role in the personal and civic aspects of education [9]. According to him, the value of a school is defined "by whether it helps students earn credentials that will make them employable" and because the Humanities cannot be justified "in economic or credentialist terms" [4], it appears to have lost all meaning and value [4]. Many educationalists, linguists, and critics, including Kylene Beers and Robert E. Probst, support Ferrero's conceptualization of such competencies in their article "Literature as a 21st Century Skill": they relate the importance of studying Literature to "cultivate simple humanity" and develop certain life skills in students, so that they can become "ethical participants in our society" [1].

However, the economic purpose of schooling has borne an overbearing influence pushing the other two factors (personal and civic) into almost oblivion within the education policy. The emphasis on young learners' economic conditioning as a precursor to their entry into the country's employability sectors systematically deprives them of the holistic aspects of education. Kylene Beers and Robert E. Probst both argue that the education system, under pressure from funding agencies, prioritises "a race to the top," ignoring lack of respect, abuse, and the critical need to "produce humane and decent citizens" [1].

This paper is divided into three sections: The section titled "Theoretical Perspectives" underlines the significance of twenty-first century skills, particularly social competence, interpersonal, and intercultural skills and the focus is mainly on digital education, social competence and digital drama. The "Research Methodology" section sets the methods used in this research. "Findings and Analysis" expands the hypothesis which sustains that an introduction to digital drama enhances competencies so sought after in the workplace particularly empathy of the net-youth/digital natives and socially competent behaviours.

2 Theoretical Perspectives: An Overview

2.1 Digital Education

The adaptive use of technology and ICT-based teaching methodologies and practises within the classroom premise is to facilitate the learning-teaching process at the tertiary level. The digitalisation of the educational sector in Mauritius has become a pertinent tool for addressing the future needs of the young graduates' professional, holistic and social competencies in order for them to fit into the larger social context of their employability.

Digital society has changed the industrial society's mode of production towards innovation, new products, new lifestyles, and even new business models [24]. These innovative digital dynamics have dismantled the conventional precepts of the Mauritian tertiary education system, we are now enacting the role of being a "knowledge-net-society". Consequently, Friesen in his 2008 article titled "Critical theory: Ideology critique and the myths of e-learning" favours that such claims have ideological underpinnings which implicitly understands knowledge as a productive force [11]. As a result, Friesen [11] argues that such claims have ideological underpinnings that implicitly understand knowledge as a productive force [11]. A direct connection between knowledge, digitalisation and education has strengthened itself within the tertiary mindset of the Mauritian framework. Besides the cultivated digital competencies, it is imperative for us to cater for the other competencies holistically and socially.

2.2 *Social Competence*

Social Competence, a psychology lexicon, over the years, has gained an interdisciplinary significance. Amongst the definitions of social competence, Gresham et al. [12] defined social competence as the ability of digital natives to "establish and maintain satisfactory interpersonal relationships, gain peer acceptance, establish and maintain friendships, and terminate negative or pernicious interpersonal relationships" [12] whereas Desjarlais et al. [7] posit that social competence refers to "a person's ability to get along with other people and is one of important indicators for psychological well-being and development". Both statements strongly sustain the potential of social competence to generate and uphold relationships within the digital learners. According to existing literature and research project that by providing formal and informal social skills training, the school can become a potential optimal setting that fosters the development of social competence in students with behavioural problems [26]. A global overview about the curricula including those in the United States, reveals that most school curricula do not place achievement in social skills on a par with achievement in academic subjects [16]. It is further supported by claims that for students with emotional/behavioural disorders, social skills intervention is as critical as, if not more critical than, an academic curriculum that more effort should be devoted to improving current social skills training practises and to identifying more proven strategies [21, 23].

As one of the components of professional competence, the modern development of society necessitates the formation of students' primary social competencies [17]. Currently, the tertiary system must assess and revise its pedagogical practises that would allow people to be networked and social rather than individual, to meet ethical and value representations rather than isolating themselves and being free of values and objective "knowledge" [25]. Adapted cooperative learning, where young learners foster and practise not only group learning but most importantly, building rapports to get along with other learners and co-workers, later on, is a necessity amongst other social competences. Thus, digital drama-based teaching methods have been developed and implemented in order to adapt the French language and drama to assist learners in developing social aptitudes.

2.3 *Humanities Education: Digital Drama*

The language learning-teaching is a challenge in Mauritius, given that young learners face commendable difficulties during their learning apprenticeship. To enhance the French language learning and acquisition, the French teaching faculty needs to redesign, update and adapt the French curricula, pedagogies and technologies at all levels. Incorporating drama through digitalisation is undoubtedly a challenge, but it is also beneficial in the classroom setting. According to Kim Flintoff, in "Inter-facing: Drama, The Arts and I.C.T.", the uses of technology in Drama education have not been adequately explored and as such drama teachers have few models from which to develop their own practise [10].

The Western Australian Curriculum Framework (WACF), in its debate about the significance of the "Arts" in the educational frameworks and systems, acknowledges that drama, as one of the "arts" subjects in schools, is to "contribute to the development of an understanding of the physical, emotional, intellectual, aesthetic, social, moral and spiritual dimensions of human experience" as per the Curriculum Council of Western Australia (1998) (qtd in [10]. Furthermore, it posits that drama tends to "assist the expression and identity of individuals and groups through the recording and sharing of experiences and imagination" (Curriculum Council of Western Australia 1998), thus, students will benefit from the digitalisation of drama to reach their goals. Drama, in this way, serves as the resourceful, adapted and pedagogical tool to reach out to many learners to advance in both the academics and social competence acquisition.

To sum up, the learning-teaching of digital drama is as dynamic as the field itself and seeks innovative and technological tools to promote an overall academic and non-academic performance in cultivating workplace competencies and skills.

3 Methodological Approach

French Language is compulsory from pre-primary to secondary school (Cambridge O-level) in the public Mauritian education system. A close observation about the tertiary learners at the Université des Mascareignes (Mauritius) prompted this research as the learning trends for the French language elucidated some challenges and concerns. In fact, Université des Mascareignes enrols students mainly in the "technical" fields of ICT, Management and Engineering. Communication modules in English and French, such as presentation skills and technical writings, are compulsory for all students. Normally for the module "Renforcement de la langue française" (French Reinforcement Course), students must have a good notion of French language (at least B2 level), and the programme consists mainly of revision of French grammar/vocabulary and pronunciation. In August 2019, instead of the traditional teaching–learning of French language, which Mauritian students are used to (such as dictation, exercises), digital drama was introduced in class. In parallel, whilst the students were working on their drama texts, grammatical or syntax rules were explained. At the end of all sessions, a questionnaire was prepared on Google Forms and the link was sent by email to all 105 students to get their feedback on the learning of French with drama, their motivation and their opinions on the social skills they developed during this activity. The recommended sample of 83 (according to Raosoft quantitative calculator) was, however, not met since the questionnaire was sent after the last class, and only 48 students answered it. There are, anyways, several debates around the "correct" sample size in qualitative studies [8, 19]. Case studies seem to be the best methodological approach for this study: as "an empirical investigation" [13], it reflects the purpose of this observational and exploratory study, which is to collect and identify the social skills developed or used by students during class. Since, this is a mixed method research, and emphasis is mainly on its qualitative aspects, our sample size seems sufficient to come to certain conclusions. We will, therefore, focus on the following questions, translated and rephrased from French:

- Closed-ended questions of Likert type on group work and the social skills and competences that were developed during the activity, the group/class atmosphere, their attitudes during group work and any conflicts detected during group work,
- Open-ended questions on the conflicts and conflict resolution in the team.

4 Findings and Analysis

In this paper, we focused primarily on a few aspects of social skills. According to research done in the field of social competence, "numerous social competence models (of Cavell 1990; DuBois and Felner 1996; Rose-Kransor 1997) have identified social skills as the foundation upon which other facets of competence build upon" [18]. Rust and Huang posit that "behaviours categorised as social skills can contribute to social competence for, they produce the desired outcome for social interaction" [22]. As an ICT-enabled teaching pedagogy, digital drama aids in the development of requisite rapports, connections, and networking amongst learners. The constructivist and didactic mode promoted by digital drama within the selected sample sustains the paradigms of departing from the conventions and reductive methods towards a more adapted and inclusive approach to language learning and teaching. In this light, the paper seeks to achieve its stated goals by evaluating the effectiveness of digital drama, merging Arts, creativity and technology as a pedagogical tool and resource to upskill the learners' social skills and competencies, especially in developing empathy and socially competent behaviours.

4.1 Empathy

In Z. A. P. Del Prette and A. Del Prette's categorisation of social skills, some elements of empathy are "maintain eye contact", "approach another person", listen actively without interrupting, "adopt another's perspective", "express understanding", "encourage trust", "show willingness to help", "share the other's joy and fulfilment" [5]. In our study, about 15 closed-ended questions were asked to students on the social skills they acquired during the drama class. Empathy is the taking of personal responsibility (for another's distress, for instance), and it was found that sharing is part of the mindset. One of the Respondents/group members who answered our questionnaire mentioned that they had, for instance, a few issues concerning the choice of the costumes during the play, and since some did not have the costumes, [they] shared/lent clothes. When asked if they acquired "sharing" skills during the digital drama activity, about 75% chose the upper Likert scale of agreement as shown in the Table 1.

Helen Demetriou considers empathy as "the ability to reach to others" and, more importantly, as "a building block of social life" [6]. According to her, even if empathy has some genetic influences, research also reveals that the environment has a role to play and whether derived from nature or nurture, empathy is one of the key elements at the workplace.

Developing emotional skills is critical for the future in "the complex social dynamics of the pupils' world within the social conditions of learning in the classroom" [6]. Indeed, the importance of sharing and "giving help" bears a positive impact on "engagement" at the workplace and this "helping behaviour" that is "the willing devotion of time and attention to assist with the work of others", caters for employees' "well-being" [15]. Furthermore, as Rust and Huang posit in the "Feeling Economy", where artificial intelligence is taking over most specialists' tasks, there is a dire need of workers who are able to develop skills such as "judgement, creativity, intuition, emotion, empathy, and people skills—the things AI currently has more trouble doing" [22]. Working alone was not an option during the digital drama class activity, whether it was to prepare and enact the play, or to video record and video edit it. The following prerequisite skills were vital: team spirit, interpersonal communication, compromise and listening skills to be able to lead and work harmoniously and maturity and responsibility to respect the deadlines. Table 2, which tabulates the

Table 1 Skills acquired—sharing (response from our respondents)

1	2	3	4	5	6
0%	6%	0%	19%	35%	40%

Table 2 Social skills acquired (according to our respondents)

Skills and competences acquired:	0	1	2	3	4	5
Interpersonal communication	1(2.1%)	0 (0%)	3 (6.3%)	12 (25%)	13 (27.1%)	19 (39.6%)
Compromise	3 (6.3%)	1 (2.1%)	6 (12.5%)	15 (31.3%)	16 (33.3%)	7 (14.6%)
Sharing	0 (0%)	3 (6.3%)	0 (0%)	9 (18.8%)	17 (35.4%)	19 (39.6%)
Listening skills	0 (0%)	2 (4.2%)	5 (10.4%)	8 (16.7%)	17 (35.4%)	16 (33.3%)
Personality development	1(2.1%)	2 (4.2%)	1(2.1%)	12 (25%)	15 (31.3%)	17 (35.4%)
Teamwork/Team spirit	2 (4.2%)	2 (4.2%)	3 (6.3%)	10 (20.8%)	17 (35.4%)	14 (29.2%)
Maturity and responsibility	0 (0%)	0 (0%)	1 (2.1%)	7 (14.6%)	18 (37.5%)	22 (45.8%)
Leadership	0 (0%)	1 (2.1%)	4 (8.3%)	13 (27.1%)	11 (22.9%)	19 (39.6%)

responses of the students, provides a positive indication of the digital drama activity in gaining social skills and in being able to "co-exist" amongst peers [22].

4.2 *Socially Competent Behaviours*

The terms "social skills" and "social competence" are often used interchangeably, and "understood as conceptual equivalents" [5]. The distinctions between the two terms are not accepted by all researchers [5], but according to Z. A. P Del Prette and A. Del Prette, whilst social skills is a "descriptive construct" that lists and describes the behaviour and the different skills one must possess, social competence is an "evaluative construct of an individual's behaviour (thoughts, feelings, and actions)" [5]. In other terms, it is the "evaluation of the quality of this interaction" and "its effectiveness in terms of results" [5]. In our study, students were asked to rate their behaviour when confronted with someone who has been unable to complete a work or in helping members to integrate a team amongst others. Table 3 exemplifies the Respondents' level of self-awareness as measured by self-analysis of social behaviours cultivated during the digital drama class activity.

Moreover, when asked if they had encountered any challenges in the group, 22.9% replied positively. Following this question, 16 students responded and explained their experience in the group, the conflict they encountered and its resolution. The qualitative data below summarises some of the main disagreements they faced and two types of behaviours could be deduced.

The first type of behaviour is the recourse to complete negativity and lack of resilience when faced with social behaviours considered different or anti-social: difficulties in leading and motivating the members of the group, absenteeism and the refusal to deal with any clashes were observed; group members had the tendency to

Table 3 Self-awareness of social behaviours (according to our respondents)

	0	1	2	3	4	5
Group atmosphere	1 (2.1%)	0 (0%)	2 (4.2%)	16 (33.3%)	13 (27.1%)	16 (33.3%)
I help others integrate the group or I make them feel at ease in the group	0 (0%)	0 (0%)	4 (8.3%)	7 (14.6%)	13 (27.1%)	24 (50%)
I respect the opinion of others, even if I do not agree with them	0 (0%)	2 (4.2%)	2 (4.2%)	6 (12.5%)	14 (29.2%)	24 (50%)
Members of my group seek me for advice	1 (2.1%)	3 (6.3%)	3 (6.3%)	15 (31.3%)	12 (25%)	14 (29.2%)
Members of my group are at ease to discuss/work with me	0 (0%)	1 (2.1%)	0 (0%)	5 (10.4%)	18 (37.5%)	24 (50%)
It is not an issue for me to help a friend/colleague to finish a work because (s)he did not have the time to do it	1 (2.1%)	1 (2.1%)	3 (6.3%)	8 (16.7%)	22 (45.8%)	13 (27.1%)

delegate their work to the member who works more in the group; time management and lateness in submission of their assigned task was also a problematic issue. 1 respondent mentioned that the opinion of only one person counted; and the others could not say anything. So, "they preferred not to say anything". The lack of resilience could be seen in the wish to abandon or change group. Another respondent mentioned that members of the group had decided to discuss to solve a problem but it had not worked. So, they preferred not to deal with the problem, and just submitted the work.

The second type of behaviour perceived is the "compromise" type, where positivity is sought even within a negative situation. For instance, 1 student reported that their work remained "static" since "some members of the group did not participate; or they had different opinions". They decided to sit together, discuss and "assembled all ideas". Another list of obstacles mentioned by another student was "communication problems, coming on time to work together and completing the assigned task". However, s/he confirms that they "have been able to find solutions, encourage each other, accept the opinions of each and every one" and "have learnt to be patient and not to get angry".

There were also many students who found that working in groups was positive and commented about the "friendly" atmosphere in which they evolved. 1 student added that working with international students was a great experience, and s/he learnt many things. According to some students this activity (drama class) has also helped them to exchange and find more ideas whilst thinking together. "Cooperation" was one of the recurrent words used.

It is to be noted that students were allowed to choose their own members for their group for this activity, and there were no imposed groupings. Since they were Year-I students, they had just recently met and did not know one another well. However, conflicts are inherent to any relationship, but communication, especially in cases where discussion sessions and sharing of opinions are mentioned, seems to have a positive impact on the group growth and has led to "productive management" [14]. Indeed, as seen in some cases, there were anger management, negotiations and discussions, and a compromise has been reached. As stated by 1 Respondent, "we had discussion sessions in our group to resolve all problems, and we could find solutions together". Nevertheless, some attitudes can be damaging as in the case of students with a domineering personality and dishonest attitude. For instance, 1 respondent mentioned that in his/her group:

> the problem was that one of the students had a domineering personality and was refusing the good ideas of others. Then he/she was taking credit of the work that others did.

Teamwork means mutual interdependence and can become unproductive when "conflicts turn into gridlocked conflicts" [14]. This is the case where some students disengaged completely from the activity, and completed the work merely for the sake of completing it. This unresolved conflict shows a lack of emotional intelligence skills in a couple of groups.

5 Conclusion and Recommendations

To sum up, the purpose of this study is to determine whether digital drama can serve to remediate difficulties in social skills [3] of university students enabling them gain social competence skills, useful at the workplace. Using the detailed explanation of social skills and social competence proposed by Del Prette & Del Prette [5], we can, therefore, conclude with a representation of the skills acquired by most students during the digital drama activity in the Fig. 1.

In fact, there seems to be a dire need for social skills work sessions in the education system. If we look at the functional outcomes of social competence, we realise that it is "largely built upon social problem-solving research" [18], and problem-solving is not an inborn quality, but must be taught. Theoretical notions of managing people and emotional intelligence are certainly taught in class to Marketing and Human Resource Management students. We, therefore, recommend that students must be coached in social competence and engaged in real-life simulations, where they can put into practice the knowledge of these skills. Consequently, research shows that even if "social competence is difficult to operationalize", students who have been "trained in social skills, as compared to untrained peers, exhibit greater social knowledge and greater social proficiency" [18]. They enjoy "peer acceptance and other positive functional outcomes (e.g. making friends, achieving social goals, etc.)" [18].

The research further suggests that social learning such as working collaboratively develops empathy and conflict resolution aptitudes and they are key skills to be learnt. Even if a majority of existing literature review focusses on using drama to teach social skills to children with special needs, the feasibility of introducing digital drama in mainstream education can, therefore, be undertaken.

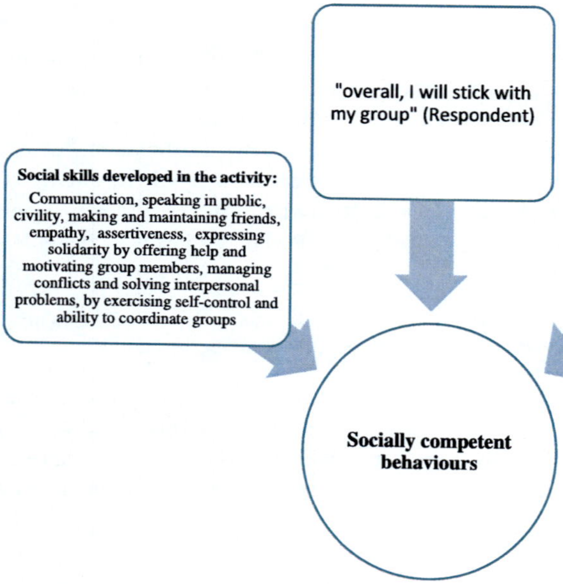

Fig. 1 Socially competent behaviour during the digital drama class (created by Pirbhai-Jetha and Ramsamy, using the explanations of Del Prette & Del Prette [5])

References

1. Beers, K., & Probst, R. E. (2011). Literature as a 21st century skill. *Language Arts Journal of Michigan, 26*(2). https://doi.org/10.9707/2168-149X.1792
2. Commonwealth of Learning: Open Educational Resources (2021). Retrieved on September 11, 2021, from https://www.col.org/programmes/open-educational-resources
3. Cruz, R. E., Lian, M.-C.J., & Morreau, L. E. (1998). The effects of creative drama on social and oral language skills of children with learning disabilities. *Youth Theatre Journal, 12*(1), 89–95. https://doi.org/10.1080/08929092.1998.10012498
4. Dass, R. (2014). Literature and the 21st century learner. *Procedia—Social and Behavioral Sciences* 289–298.
5. Del Prette, Z. A. P., & Del Prette, A. (2021). *Social Competence and Social Skills: A Theoretical and Practical Guide*, Springer, Switzerland.
6. Demetriou, H. (2018). *Empathy, Emotion and Education.* Palgrave-Macmillan, London.
7. Desjarlais, M., & Willoughby, T. (2010). A longitudinal study of the relation between adolescent boys' and girls' computer use with friends and friendship quality: Support for the social compensation or the rich-get-richer hypothesis? *Computers in Human Behavior, 26*(5), 896–905.
8. Dworkin, S. L. (2012). Sample size policy for qualitative studies using in-depth interviews. *Archives of Sexual Behaviors, 41*, 1319–1320. https://doi.org/10.1007/s10508-012-0016-6
9. Ferrero, D. J. (2011). The humanities: Why such a hard sell? *Educational Leadership, 68*(6), 22–26.
10. Flintoff, K. (2002). Interfacing: Drama, the arts and ICT. Unpublished paper.
11. Friesen, N. (2008). Critical theory: Ideology critique and the myths of e-learning. *Ubiquity, 2*, 1–13.
12. Gresham, F. M., Sugai, G., & Horner, R. H. (2001). Interpreting outcomes of social skills training for students with high-incidence disabilities. *Exceptional Children, 67*(3), 331–344.

13. Hancock, D. R., & Algozzine, B. (2006). *Doing case study research—A practical guide for beginning researcher*. Teachers College Press.
14. Hocker, J. L., & Wilmot, W. W. (2018). *Interpersonal conflict*. McGraw-Hill Education.
15. Hoeven, C. L., & Zoonen, W. V. (2020). Helping others and feeling engaged in the context of workplace flexibility: The importance of communication control. *International Journal of Business Communication*, 1–22. https://doi.org/10.1177/2329488419898799
16. Janney, R., & Snell, M. (2006) *Social Relationships and Peer Support*. Brookes, Baltimore.
17. Kalashnikova, L. A., & Kostina, E. A. (2016). Psychological and pedagogical approaches to the concept of social maturity of personality in foreign and domestic science. *Bulletin of the Novosibirsk State Pedagogical University*, (4), 18–28. https://doi.org/10.15293/2226-3365.1604.02
18. Kowert, R. (2015). *Video games and social competence*. Routledge.
19. Malterud, K., Siersma, V. D., & Guassora, A. D. (2015). Sample size in qualitative interview studies: Guided by information power. *Qualitative Health Research*.https://doi.org/10.1177/1049732315617444
20. Panigrahi, C. R., Tiwari, M., Pati, B., & Prasath, R. (2014). Malware detection in big data using fast pattern matching: A hadoop based comparison on GPU. In *Mining Intelligence and Knowledge Exploration* (pp. 407–416). Springer.
21. Regan, K. S. (2009). Improving the way we think about students with emotional and/or behavioral disorders. *Teaching Exceptional Children, 41*(5), 60–65.
22. Rust, R. T., & Huang, M. H. (2021). *The Feeling Economy—How Artificial Intelligence is Creating the Era of Empathy*. Springer, Palgrave Macmillan, Switzerland.
23. Sugai, G., & Horner, R. R. (2006). A Promising Approach for Expanding and Sustaining School-Wide Positive Behavior Support. *School Psychology Review,* 35(2), 245–259.
24. Tapscott, D., & Williams, A. D. (2006). *Wikinomics: How mass collaboration changes everything*. Portfolio.
25. Tynjala, P. (2013). Toward a 3-P model of workplace learning: A literature review. *Vacations and Learning, 6*(1), 11–36.
26. Walker, H. M., Colvin, G., & Ramsey, E. (1995). *Antisocial Behavior in School: Strategies and Best Practices*. Brooks/Cole, Pacific Grove (CA).

Organizing for Social Media Marketing: A Case of Conglomerates in Mauritius

Swaleha Peeroo

Abstract Conglomerates in Mauritius are leveraging social media as a marketing tool. However, studies on the implementation of social media marketing by companies are limited. This paper aims to explore the internal organization of businesses in Mauritius to use social media as a strategic marketing tool. A qualitative research design has been developed to gather data from Chief Communication Officers and Chief Marketing Officers of major conglomerates in Mauritius. Semi-structured interviews were conducted. This study shows that conglomerates either manage their social media marketing in-house or they hire an agency to manage their social media accounts. To guide behaviors of employees, all conglomerates have developed a social media policy. Social media marketing requires specific skills and competencies and has given rise to new job descriptions. This study adds to the body of knowledge of social media marketing by providing insights about the approaches conglomerates have adopted to leverage social media.

Keywords Social media marketing · Social media · Content management

1 Introduction

Social media are an undisputed essential element of our daily lives. Social media platforms are increasingly being used by businesses to connect various stakeholders—users, businesses, organizations, customers, and the society [1]. As at February 2021, active global social media users amounted to 4.2 billion with Facebook remaining the most popular social networking site among the global audience with more than 2.85 billion monthly active users [2]. Facebook is also the most commonly used social media platform among marketers worldwide. According to a global survey, 93% of responding social media marketers used the network to promote their business, while another 78% did so via Instagram [3].

S. Peeroo (✉)
Université des Mascareignes, Pamplemousses, Mauritius
e-mail: speeroo@udm.ac.mu

B. Pati et al. (eds.), *Proceedings of the 6th International Conference on Advance Computing and Intelligent Engineering*, Lecture Notes in Networks and Systems 428,
https://doi.org/10.1007/978-981-19-2225-1_47

The role of social media has progressively evolved from being a single marketing tool to that of a key element of strategies of businesses [4]. Increasingly, marketers strategically leverage social media marketing to achieve superior competitive edge and performance [5]. Despite the increasing importance of social media within the marketing realm, there is scant knowledge as to how businesses organize themselves to implement social media marketing.

To fill this gap in the literature, this paper aims to explore the internal organization of businesses in Mauritius to use social media as a strategic marketing tool. This research is important as it will guide social media managers in implementing their social media plan. Once social media accounts are created, managers need to know what content to post, when to post content, how and when to answer to customer queries, how to handle customer complaints, and respond to comments in a way to provide positive customer experience. Social media marketing needs to be managed.

A qualitative research design has been developed to gather data from Chief Communication Officers and Chief Marketing Officers of major conglomerates in Mauritius. Semi-structured interviews were conducted to answer the following research question: How do conglomerates in Mauritius organize for implementing social media marketing strategies?

This paper contributes to the literature as it provides knowledge on the internal organization of conglomerates for implementing social media marketing. This paper informs about people and tasks involved and internal factors to be considered when using social media. This paper also provides knowledge on how conglomerates in Mauritius monitor the use of social media for marketing. This paper will provide marketing managers practical guidelines on how to manage social media platforms to reach desired marketing outputs.

The paper provides an overview of social media marketing, then sets out to explain the methodology used. The findings are then presented and discussed. The managerial implications are provided in the conclusion part as well as the limitations of the research and avenues for future research.

2 Literature Review

Having gained wide acceptance, social media have radically transformed the marketing landscape [6, 7]. Users and customers sharing similar opinions, tastes, and preferences connect with brands on social media platforms [8]. Considered as a major dynamic within the marketing realm, social media are the prime channels through which users interact and engage with a brand or an organization [9].

2.1 Social Media Marketing

Social media marketing refers to "the utilization of social media technologies, channels, and software to create, communicate, deliver, and exchange offerings that have value for an organization's stakeholders" [10]. The growth of social media has altered the balance of power between the company and the customer in terms of control of a shared reality and the customer's ability to create a brand narrative [11]. Quinton advocated a shift from a relational orientation dealing with one-to-one communication to an interactional orientation focusing on the various relationships based on sharing in online-branded communities [11].

Customers join corporate social media platforms where they give their consent to receive information and commercial messages as member of the online social media communities [12]. Social media communities enable marketers to engage with consumers on a one-to-one level and give real-time information about brands [13]. Moreover, social media marketing activities enable marketers to learn about their audience from the information obtained both through consumer-to-consumer interactions and consumer-to-brand interactions [14].

Managing social media marketing in the fast evolving digital age is very complex [15], and marketers need to manage and adapt to new challenges, organizational, and philosophical changes pertaining to the role of employees and departments for managing social media marketing. Responsibilities of social media marketers and roles of stakeholders within the social media realm need to be considered within a holistic and interdisciplinary framework for conceptualizing strategic social media marketing [14].

2.2 Strategic Social Media Marketing

As an interdisciplinary and cross-functional concept, strategic social media marketing involves organizational decisions about the scope, culture, structure, and governance, to achieve organizational goals by creating value for stakeholders through the use of social media platforms [14]. Felix et al. developed a strategic social media marketing framework along four main dimensions: scope, culture, structure, and governance [14].

The scope of social media marketing comprises of how businesses use social media marketing for communication with one or few stakeholders or for all stakeholders with the aim to collaborate. Defenders use social media marketing as a one-way communication tool, while explorers seek a genuine social media marketing collaboration through interactions with various stakeholders [14].

The social media culture ranges from conservatism which adopts a traditional mass advertising approach to social media marketing, whereas within a modernism culture, social media marketing is open, permeable, and flexible [14].

The social media marketing structure encapsulates how social media marketing activities are assigned in the organization in terms of organization and departmentalization. Within a centralized hierarchical approach, social media marketing is the responsibility of one department or an individual. In a decentralized approach, social media marketing is structured within a network where all employees are responsible for its execution [14].

The governance of social media marketing deals with the issue of how businesses establish guidelines and rules for social media marketing within the company. Under an autocratic governance, there are precise rules on who can interact on corporate social media platforms, while anarchy characterizes a situation of absence of rules and guidelines for social media marketing [14].

To effectively navigate the dynamic and complex arena of social media marketing, Felix et al. advocate cross-functional collaborations to decide about the four dimensions [14] while going through the social media management process.

2.3 *Social Media Management Process*

Managing social media marketing is a complex task which requires managers to address challenges linked to the fast-evolving digital arena [16]. To assist managers in developing and using social media as a tool for marketing, a framework consisting of four dimensions relating to activities managers execute when implementing social media marketing [17] has been developed. The social media management process consists of four actions: messaging/projecting, monitoring, assessing, and responding [17].

The first step is messaging/projecting—managers need to decide what content to post online and how often, with the aim to create awareness, build image, foster interaction with customers, and build relationships [17]. The second step is monitoring—scanning the environment and closely monitoring the online conversations to gain insights into the needs of customers to enhance the customer experience by providing meaningful content. It also involves gathering competitive intelligence [17]. The third phase is assessing the data collected during the monitoring phase to decide upon the appropriate response. Information can be evaluated by examining performance indicators such as shares, likes, and followers or by carrying out sentiment analysis of the comments posted by customers [17]. The fourth step is responding, which involves taking actions. Existing procedures will guide employees to determine how and when they should communicate and interact with the customer, which may generate creativity online for the employees [14]. The strategic focus of the business, i.e., it is marketing scope, culture, structure, and governance will determine how the business adopts this four-phase process [18].

3 Methodology

This paper aims to explore the internal organization of businesses in Mauritius to use social media for strategic marketing. A qualitative methodology was used to examine the use of social media platforms by conglomerates in Mauritius.

In a qualitative research using an exploratory research design, it is appropriate to conduct qualitative interviews as a data collection method [19]. The qualitative research interview tries to understand the world from the interviewee's point of view, to uncover the meaning of their experiences, to reveal their lived world before any scientific explanations [20]. Since it is important to capture personal insights to answer the research questions, interviewing is the right method for gathering data in this research. Furthermore, interviews are suitable since the aim is to understand the internal organization of conglomerates for implementing social media marketing.

Semi-structured interview is defined as an interview with the aim of gaining descriptions of the life world of the subject in order to construe the meaning of the described phenomena [20]. Semi-structured interviews were conducted to gather data from Chief Communication Officers and Chief Marketing Officers of the six major conglomerates in Mauritius. Each interviews lasted for about 45 min and was transcribed and analyzed using NVivo 12. Semi-structured interviews are appropriate for this research as questions can be prepared before the actual interview and allow the interviewees to express their views in their own words.

Thematic analysis was carried out following the procedures outlined by Braun and Clark [21] based on the themes identified in the literature review. To gain insights from the transcripts, the constant comparative method was used to identify similarities and differences [22]. Firstly, open coding was initially applied to the data. Then, after carefully scrutinizing and comparing data, axial coding, which involves linking the identified categories, was then performed to attempt to identify any patterns or trends in the phenomenon under study.

4 Findings and Discussion

Semi-structured interviews were carried out to gain an understanding of the internal organization of conglomerates in Mauritius for implementing social media marketing strategies. The preliminary findings show that conglomerates in Mauritius have widely adopted social media. The social media platforms currently used are Facebook, LinkedIn, Instagram, YouTube, and Twitter. Facebook, LinkedIn, and Instagram are the most popular platforms in the business world in Mauritius.

All the participants in this study have embarked on the social media bandwagon as they had no choice than to create their pages as illustrated by the two participants:

> Social media are unavoidable; its a must be. (P2)

Creating social media pages have been a strategic move for the conglomerates as they could foresee the affordances of the social media platforms for their business.

> We must be where the audience is. The audience is online, so when you know that you cannot not be on social media. (P1)

> Because it is the easiest way to communicate with our audiences and promote the brand, our values and what we do. (P3)

This study shows that businesses have started using social media platforms because of the numerous advantages that social media offer. All participants agree that harnessing social media was not an option; they had to use these communication platforms. As their audiences are using social media, the conglomerates created their social media platforms to be where their audiences are. The participants also joined social media as the business world is becoming digital, and it is perceived to be the logical evolution of all businesses.

4.1 *Organizing for Social Media*

Findings show that to manage social media marketing, conglomerates have either recruited staff specifically to manage their social media accounts or have hired agencies to help them with their social media accounts. Though the interviewees are all executives of the largest conglomerates in Mauritius, this study reveals that 50% of the interviewees has opted for the handling their social media accounts by external agencies, but they collaborate closely with the agencies as mentioned by one of the participants.

> We do our editorial plan management jointly with the agency, we also draft all the answers and the agency post our answers. (P3)

The other 50% of conglomerates have set up full-fledged departments and hired staff to implement social media marketing as stated by one of the participants.

> We have a team that are totally focused on the digital. So, how do we work, we have a calendar, for example, we decide to publish three posts per week, so on the calendar, we have a template; we will work on the date; then, we will work on the topics linked to the marketing which are our product, which market we are targeting, we have a segmentation, and then, the graphic designer will work on the visuals because visuals are key. (P4)

Social media marketing requires specific skills and competencies, and businesses have set up their team accordingly as illustrated below.

> I think that what is beautiful about social media, about brands taking to social media, it has given rise to a whole pool of new competencies. Community manager is now a profession; graphic designers, photographers, videographers have all started becoming professionals now. So, it has created a whole pool of creative talents and jobs actually, which is very good. (P2)

Findings of this study, whereby some business resort to the services of external agencies to manage their social media platforms are similar to previous research. These authors claimed that managers were intimidated by the social media space and perceived that they lacked the knowledge and expertise to navigate effectively through the social media and the related technological sphere [23].

4.2 Social Media Policy

To implement social media marketing, all informants agreed that having rules and regulations about the use of social media platforms by the company was very important. All conglomerates in the study have set up their social media policy as illustrated below:

> Yes, we do have our own social media policy. We have to make them aware, and we have to educate them; it is purely recommendations, awareness, and education. (P5)

> Yes, it is part of our Corporate Communication Policy. The social media policy, it is twofold. One is about our corporate pages and how we communicate, and one is about our employees and the behavior they should take on social media as employees. (P6)

From the comments quoted above, it can be seen that the managers have felt the need to create awareness about opportunities and dangers that social media may pose to their organization. Hence, these organizations have created a social media policy to guide their employees as to the behavior to be adopted. They have provided training to educate their employees on the guidelines for using social media. Conglomerates in Mauritius have more of an autocratic governance, as there are rules on who can interact on corporate social media platforms [14]. Creating and implementing a social media policy are one element of social media marketing governance [14].

4.3 Social Media Content Management

For effective social media marketing, content is key. Who gets to decide what will be said on social media platforms of organizations depends on how content is managed within that organization. Informants revealed that content management is either done in-house, or for those organizations hiring agencies, it is jointly done by the marketing or communication team with the agency.

> We do that jointly with the agency; we draft all the answers, and the agency posts our answers. (P3)

> So, how do we work, we have a calendar, for example, we decide to publish three posts per week, so on the calendar, we have a template, we will work on the date, then, we will work on the topics linked to the marketing which are our product, which market we are targeting, we have a segmentation, and then, the graphic designer will work on the visuals depending on what we are publishing, we will decide whether we are going to boost, and boost with a targeting community behind it. (P6)

Conglomerates in Mauritius use scripted texts to respond to queries of users. Scripted texts allow the organizations respond to customers in a professional manner.

> We have templates of how to reply. For example, we have answers which I have; we have drafted for all sorts of FAQs. All FAQs have its answers. For example, for consumer finance, we are talking to the B2C market, so often they are using creole and often when there is credit behind, money brings negative emotions, sometimes they are quite brutal. And my staff being young persons, they have to stick to what we have decided before. Never be brutal on our site, never be aggressive, stay firm. We use scripted texts. (P6)

Managers need to decide on the desired content, the frequency of contents, ensure contents are relevant and meaningful to the brand and also decide which contents should be published on which corporate social media platform [24].

4.4 Social Media Monitoring

Managers need to monitor social media activity to be aware of what is being said about their brands online. Whether social media are being managed in-house or through an agency, all participants monitor their social media activity as illustrated below:

> Yes, we have a monthly report about our social media pages and corporate Web site *sent to us by our agency, but we also have direct access if we want to check the progress. (P3)*

> We measure our impact every week. On the back office, Facebook provides us with analytics same for LinkedIn, and from the Web site, we connect all our platforms, and we have Google Analytics which give us the segment, the number of visits, the amount of time they spent on the Web site, what they were looking at; we drop an email to all those who have spent more than X number of seconds on a subject. So it is a powerful tool (P4).

The conglomerates monitor the social media activity also by reading comments posted on social media platforms and blogs by customers. They also gather competitive intelligence by watching competitors' activities on social media platforms. Monitoring social media is a valuable source of primary research which managers can leverage to improve their performance and competitive positioning [25, 26].

5 Conclusion

This paper aims to explore the internal organization of businesses in Mauritius to use social media as a strategic marketing tool. Conglomerates in Mauritius either manage their social media marketing in-house or hire agencies to help them. All conglomerates have established a social media policy to guide behaviors of employees. It was also found that social media marketing requires specific skills and competencies and has given rise to new job descriptions. Contents are created to boost interactions with customers, and social media activities are tracked through various forms of reporting, which is then used to guide future decisions.

This study adds to the body of knowledge of social media marketing by providing insights about the approaches conglomerates have adopted to leverage social media.

This paper provides managers with some useful guidelines to implement social media marketing within their organization in a strategic perspective. Managers can gain an understanding of the various issues which needs to be considered when establishing an online social presence based on the organization's scope, culture, structure, and governance. It also provides guidance to managers about the steps to be followed to implement social media marketing.

Being an exploratory study, this research has some limitations as only six conglomerates participated in the study, and therefore, generalizing the findings of this study to other contexts is not recommended.

Future research could explore how businesses in different sectors of different sizes have organized themselves to implement social media marketing. This study has focused on the large conglomerates, which have a sizeable budget for marketing. Smaller organizations may not have sufficient funding to hire an agency and so would most probably manage their own social media accounts.

References

1. Mention, A.-L., Barlatier, P.-J., & Josserand, E. (2019). Using social media to leverage and develop dynamic capabilities for innovation. *Technological Forecasting and Social Change, 144*, 242–250. https://doi.org/10.1016/j.techfore.2019.03.003
2. Social Media—Statistics & Facts|Statista. https://www.statista.com/topics/1164/social-networks/. Last accessed 2021/09/11.
3. Social media platforms used by marketers 2021|Statista. https://www.statista.com/statistics/259379/social-media-platforms-used-by-marketers-worldwide/. Last accessed 2021/09/11.
4. Li, F., Larimo, J., & Leonidou, L. C. (2021). Social media marketing strategy: Definition, conceptualization, taxonomy, validation, and future agenda. *Journal of the Academy of Marketing Science, 49*, 51–70. https://doi.org/10.1007/S11747-020-00733-3
5. Lamberton, C., & Stephen, A. T. (2016). A thematic exploration of digital, social media, and mobile marketing: Research evolution from 2000 to 2015 and an agenda for future inquiry. *80*, 146–172. https://doi.org/10.1509/JM.15.0415
6. Arnaboldi, M., & Coget, J.-F. (2016). Social media and business. *Organizational Dynamics, 45*, 47–54. https://doi.org/10.1016/j.orgdyn.2015.12.006
7. Jurgens, M., Berthon, P., Edelman, L., & Pitt, L. (2016). Social media revolutions: The influence of secondary stakeholders. *Business Horizons, 59*, 129–136. https://doi.org/10.1016/j.bushor.2015.11.010
8. Kozinets, R. V. (2002). The field behind the screen: Using netnography for marketing research in online communities. *Journal of Marketing Research, 39*, 61–72. https://doi.org/10.1509/jmkr.39.1.61.18935
9. Carvalho, A., & Fernandes, T. (2018). Understanding customer brand engagement with virtual social communities: A comprehensive model of drivers, outcomes and moderators. *Journal of Marketing Theory and Practice, 26*, 23–37. https://doi.org/10.1080/10696679.2017.1389241
10. Tuten, T. L. (2020). *Social media marketing*. Sage.
11. Quinton, S. (2013). The community brand paradigm: A response to brand management's dilemma in the digital era. *29*, 912–932. https://doi.org/10.1080/0267257X.2012.729072

12. Habibi, M. R., Laroche, M., & Richard, M. O. (2014). The roles of brand community and community engagement in building brand trust on social media. *Computer Human Behavior, 37*, 152–161. https://doi.org/10.1016/j.chb.2014.04.016
13. Bazi, S., Filieri, R., & Gorton, M. (2020). Customers' motivation to engage with luxury brands on social media. *Journal of Business Research, 112*, 223–235. https://doi.org/10.1016/j.jbusres.2020.02.032
14. Felix, R., Rauschnabel, P. A., & Hinsch, C. (2017). Elements of strategic social media marketing: A holistic framework. *Journal of Business Research, 70*, 118–126. https://doi.org/10.1016/j.jbusres.2016.05.001
15. de Swaan Arons, M., van den Driest, F., & Weed, K. (2014). The ultimate marketing machine. *Harvard Business Review, 92*, 54–63.
16. Parsons, A. L., & Lepkowska-White, E. (2018). Social media marketing management: A conceptual framework. *17*, 81–95. https://doi.org/10.1080/15332861.2018.1433910
17. Parsons, A. L., & Lepkowska-White, E. (2018). Social media marketing management: A conceptual framework. *Journal of Internet Commerce, 17*, 81–95. https://doi.org/10.1080/15332861.2018.1433910
18. Lepkowska-White, E., Parsons, A., & Berg, W. (2019). Social media marketing management: An application to small restaurants in the US. *International Journal of Culture, Tourism Hospitality Research, 13*, 321–345. https://doi.org/10.1108/IJCTHR-06-2019-0103
19. Bryman, A., & Bell, E. (2007). *Business research methods*. Oxford University Press.
20. Kvale, S., & Brinkmann, S. (2009). *InterViews: Learning the craft of qualitative research interviewing*. Sage Publications.
21. Braun, V., & Clarke, V. (2006). Using thematic analysis in psychology. *Qualitative Research in Psychology, 3*, 77–101. https://doi.org/10.1191/1478088706QP063OA
22. Miles, M. B., & Huberman, A. M. (1984). *Qualitative data analysis: Handout. A source. New methods*. SAGE Publication Inc.
23. Durkin, M., McGowan, P., & McKeown, N. (2013). Exploring social media adoption in small to medium-sized enterprises in Ireland. *Journal of Small Business Enterprise Development, 20*, 716–734.
24. Fox, G., & Longart, P. (2016) Electronic word-of-mouth: Successful communication strategies for restaurants. *Tourism Hospitality Management, 22*, 211–223. https://doi.org/10.20867/THM.22.2.5
25. Kwok, L., & Yu, B. (2013). Spreading social media messages on facebook: An analysis of restaurant business-to-consumer communications. *Cornell Hospitality Quarterly, 54*, 84–94. https://doi.org/10.1177/1938965512458360
26. Malthouse, E. C., Haenlein, M., Skiera, B., Wege, E., & Zhang, M. (2013). Managing customer relationships in the social media era: Introducing the social CRM house. *Journal of Interactive Marketing, 27*, 270–280. https://doi.org/10.1016/j.intmar.2013.09.008

The Transformation in Higher Education: From Face-to-Face to Online Distance Learning and Ultimately to Blended Mode: A Case Study at the University of Mascareignes (UdM), Mauritius

Bushra Khoyratty

Abstract Distance learning is currently of national interest in this period of confinement following the Covid-19 pandemic. In Mauritius, for example, the decision to suspend classes by the government is to ensure pedagogical continuity with online learning and teaching on the platforms of the Université des Mascareignes or via other platforms of e-learning. This article examines the measures taken by the authorities to ensure the quality of learning in terms of educational and didactic effectiveness and efficiency. This prompts us to take a new look at the issue of distance learning (DL), from the point of view of educational policies, and on the side of innovative approaches and emerging practices such as hybrid learning.

Keywords Distance learning · Educational continuity · Health crisis · Covid-19 · Hybrid learning

1 Introduction

Education is deemed to be the sector most affected by the Covid-19 pandemic as it has affected the entire world population. In Mauritius, as in most countries, lacking a prior strategy to deal with this unexpected situation, had to think about the continuity of learning. The duration of this situation forced us to think about the continuity of learning for all learners by using online education, social media, television, radio, and other online educational offerings. This is how several alternative methods have been developed and the implementation of digital platforms such as Zoom, Google Classroom, Office 365's, and Ms. Team have been used by several educational and tertiary establishments, among others. The objectives of this research are to know the effects of distance education on the development of autonomy in students. Other objectives stand out, such as the types of constraints teachers faced in terms of

B. Khoyratty (✉)
Department of Software Engineering, Pamplemousses, Mauritius
e-mail: bkhoyratty@udm.ac.mu

B. Pati et al. (eds.), *Proceedings of the 6th International Conference on Advance Computing and Intelligent Engineering*, Lecture Notes in Networks and Systems 428,
https://doi.org/10.1007/978-981-19-2225-1_48

infrastructure and mastery of computer tools. Has DL been carried out effectively in terms of support? The evaluations made during confinement really reflect the performance of the students while considering the inequalities between the different social classes. At the Université des Mascareignes (Udm), we want to adopt a so-called hybrid system according to Covid-19, can we continue e-learning without even having a good reform in teaching and learning and how to ensure its proper implementation?

This paper begins with the explosion observed in the educational uses of technology and pedagogical questioning imposed by distance education at the Universite des Mascareignes during the pandemic Covid-19, 2021.It analyzes the specificities of the act of distance education and the difficulties inherent with it. The paper also explains the pedagogical and didactic practices put in place by lecturers, the difficulties they encountered in doing so, and the positive lessons learnt from this period.

The rest of the paper is divided into seven sections. Section 2 reviews the background study of the existing system, while Sect. 3 discusses some of the related works. Section 4 presents the methodology used in this study, including the search strategy and the selection process. On the other hand, Sect. 5 presents the findings of the study and discusses them in detail, while Sect. 6 concludes with a summary of the research outcomes. Finally, Sect. 7 consists of a list of recommendations and possible future work.

2 Background Study

It is interesting to note that all these new forms of online learning, known as e-learning, aim to learn at your own pace on your computer, tablet, or smartphone. The learner, being responsible for his learning, must be able to develop cognitive autonomy based on the principle of "learning to learn" [1]. It is both "a goal and a means of acquiring the ability to communicate and learn on one's own. The learner develops competent skills, no one can do it for him, and the teacher must resist this temptation. It is only an aid in this learning." [1]. With the evolution of new information and communication technologies (NICT), it becomes possible to integrate in distance learning the human relationship continuously, the interaction and the dialogue supported, and especially by the asynchronous communication [2].

However, at the Université des Mascareignes, Mauritius, higher education places a major emphasis on the new learning framework that promotes the empowerment of the various actors involved, through the distance learning platform (moodle) already present on the university's Web site. This platform offers free access to teaching content, from simply uploading courses to videoconferencing through virtual rooms, and promotes collaborative learning. The many features available to it rely on the ability of students and teachers to adapt to this new form of e-learning in a changing landscape, whereby the classic methods of teaching and learning are no longer the same within the university community. To develop the hybrid teaching method in our educational ecosystem, important actions and projects have been undertaken. On the technological and pedagogical level, there are finally challenges to be anticipated for the migration of conventional face-to-face learning to hybrid learning.

In these times of health crisis, Udm has switched to a distance education mode. It was the use of digital tools that made it possible to avoid a complete interruption of this pedagogical relationship between students and teachers. These tools have been implemented thanks to the considerable effort of all actors, who have mobilized massively to ensure educational continuity including students, teachers, academics, and central services.

The use of online communication tools to motivate and engage students in a pandemic context should not be restricted to this exceptional situation. The return to the classroom should benefit students from new teaching strategies that will allow students to continue learning with digital technology. Hence, the proposed solution of this post-Covid-19 educational turmoil is to maintain a balance between online and offline classes (hybrid solution).

3 Literature Review

3.1 Distance Education: Definition

UNESCO defines "Distance learning, also called e-learning, as a mode of teaching, provided by an institution that does not involve physical presence of the teacher and students" [3]. It is aimed at a wide audience and touches various fields. According to the Association Francaise de Normalisation (AFNOR), distance learning is defined as "a mode of learning designed to allow individuals to train without moving to the training site and without the physical presence of a trainer. Distance learning is a concept of openness and distant" [4]. Peraya [5] emphasizes that "Distance learning dissociates the teaching/learning process in time and space, it appears from the outset as deferred training and consequently, it must necessarily be conceived and implemented as media training." In our view, distance education or distance learning is like a device composed of a set of material, technical, human, and pedagogical resources set up to provide teaching to individuals who are distant [6].

3.2 Historical Background of Distance Learning

Distance learning is not a new concept as it began in the mid-nineteenth century. There have been significant developments in the use of printing, post-delivered media, video cassettes, radio and television broadcasting, and computer-aided or multimedia teaching to interactive media, including the Internet [7, 8].

The first generation is characterized by correspondence courses. The individual pedagogical approach used in this mode of learning has several shortcomings including the slowness of the postal communication system and the possibility that mail arrives late or is lost along the way. Interaction in this mode of learning was low, and dropouts were high [9].

The second generation is marked by the introduction of a greater variety of media, such as television and radio. Printing is no longer the only educational medium; educational materials are also delivered as audio and tapes [7]. Isolated learner support schemes have been established by tutors and regional resource centers to help learners. Correction of work done by correspondents was carried out through telephone calls between learners and tutors [7].

The third generation is characterized by computer-aided instruction and the use of multimedia. By the 1980s, all distance courses began using the educational systems based on multimedia materials such as tutorials, hypermedia, and intelligent tutors. Telecommunication services such as telephone, fax, and emails provide a means of communication to facilitate interaction between learner and tutor [7].

The fourth generation is specified using interactive media ad Internet which creates a more interactive learning environment and facilitates communication between learner and educational institution and even more, by reinforcing the relationship between tutors and encourage pairs interaction. The use of basic tools, such as discussion forums, email, and Web page consultation or web conference, allows learners to exploit interactive technologies for collaborative learning [10]. These types of interactions enhanced by the evolution of technology can be synchronous or asynchronous.

3.3 Course Typology

In recent years, the diffusion of the use of digital technologies in face-to-face or distance learning has allowed the emergence of various ways of teaching models or courses. Since 2012, Frank Mayadas, Gary Miller, and John Sener have proposed a typology of the different types of courses. This typology is updated over time by its authors, based on comments made by the community of professionals and experts in the field of online training [11–13]. To Sener's seven types of course, another type put forward by Michael Power and his colleagues in recent years [14, 15], namely the online hybrid courses also called (Blended learning). Thus, it is possible to identify eight types of courses:

1. In class or face-to-face: courses offered in a classroom, in different formats (presentation, workshop, laboratory, etc.) which can use ICT tools for simulations, or creations of projects [13].
2. Augmented face-to-face [13]: courses offered face-to-face, but in which students can be invited to participate in online learning or assessment activities through a platform (LMS) or a Web site.

3. In class with online extension [13]: courses offered face-to-face, but accessible online through a videoconferencing system.
4. Hybrids [13]: courses in which a significant number of face-to-face sessions are replaced by online learning sessions or activities, synchronously or asynchronously.
5. Asynchronous online [13, 16]: courses in which all the content and activities are made available to students who choose the moment to carry them out and possibly communicate with an instructor or peers by sending messages.
6. Synchronous online [13, 16]: courses in which all the content and activities are made available to students who must participate in certain teaching or exchanges at different times and according to technological means provided by the device.
7. Online hybrids (Blended) [14, 15]: online course combining synchronous and asynchronous modalities.
8. Flexible, co-modal or HyFlex [13]: courses in which students can choose between online (synchronous or asynchronous) or face-to-face monitoring mode. This choice can be made during the course.

3.4 Benefits of Hybrid Learning

Akhtar et al. [17] stipulated that blended learning provides greater opportunities to link teaching materials and merge them into one interconnected unit. This link and connections would facilitate the learning process by focusing on learning facts rather than concepts because concepts are broader in their meanings as well as their relationships and connections [17].

Blended learning requires building on a new foundation to develop curricula that allow learners to better understand what they are learning so that their learning achieves meaning, value, and the desired benefit. Rao [18] added that combined learning has come to serve as a link between traditional learning and online learning, thus having many advantages that outweigh the advantages of traditional learning and online learning. E-learning if they are tracked separately, as it is not limited to the wall of the classroom, but it stays in touch with its learners even after the end of the classroom lessons to achieve continued effectiveness even with the presence of many learners. One of the biggest advantages of blended learning is that it addresses a uniqueness of every aspect of learning methods; there is no educational method or strategy because each educational situation presents specific characteristics in light of the many variables imposed on it [19]. The real challenge is therefore the possibility of differentiating these methods and strategies and choosing the most appropriate and compatible with the objectives of each educational situation separately [19].

3.5 Challenges in Implementing Hybrid Modalities

Blended learning still poses some challenges for teachers and students although research tells us that it has many benefits. Blended learning can have technical drawbacks if it is not designed and implemented properly because it relies on technical resources or equipment [19].

Kaur [20] mentioned that the need for all students to be online at the same time, the lack of student and teacher needs for sophisticated workstations and high-speed connection, lack of technical skills, and resources needed to interact are the downsides of blended learning.

Lieberman [21] recognized the difficulty of implementing a radically new educational strategy in most US schools during a pandemic. He said much of the problem can be attributed to a lack of financial resources for the education sector, and school budgets have continued to be tightened as the pandemic has weighed on state and local finances. Li and Lalani [22] argued that there are still many challenges to overcome in implementing blended learning, especially with regard to access. They found that some students who do not have stable access to infrastructure find it difficult to engage in virtual learning. They cited a survey conducted by the Organization for Economic Co-operation and Development (OECD) which found that while 95% of students in Switzerland, Norway, and Austria has access to a computer for their schoolwork; only, 34% of students in Indonesia has access which demonstrates the economic disparities between countries.

Garcia and Weiss identified the limitations of implementing hybrid systems. The idea of effective learning is beautiful, but it has not yet been proven [23]. They focused on the issue of teacher training and access to resources as determining factors for successful use of online and blended education. They also noted that the design and development of educational tools intended for large-scale deployment are subject to severe time restrictions. In addition, planning and designing effective education for the COVID-19 era, when teachers and school districts lacked a systematic framework to adapt to what they were doing, is a major challenge [23].

Rasheed et al. analyzed the issues facing students, teachers, and institutions in achieving blended learning. They suggested that difficulties with self-regulation and the use of learning technologies are the main challenges for students. They designated five general categories for the challenges encountered: (1) self-regulation challenges (SRC), (2) technological literacy and competency challenges (TLCC), (3) student isolation challenges (SIC), (4) technological sufficiency challenges (TSC), (5) technological complexity challenges (TCC). Gilmour [24] added resistance as another factor to this list [25].

3.6 Engagement in Student Learning and Participation

Engagement refers to the investment and effort of the learner in the learning task [25]. In some blended education settings, student engagement may be weakened compared to a traditional teaching setting due to reduced face-to-face interactions, social support, and feedback [26]. However, student engagement is considered a fundamental variable in learning and linked to performance or other learning factors [26]. In hybrid or distance education devices, it is possible to measure an observable aspect of student engagement through their participation. Indeed, participation, measured quantitatively, turns out to be an indicator of student engagement in courses offered online [27–29]. A study by Huang [30] shows, for example, a link between the learning style of students and their participation. The authors measured online participation quantitatively that is they recorded the digital traces of students: The number of files opened and pages read, and the time spent on the platform. Therefore, this measure appears interesting and complementary to others when we seek to characterize the variables related to the performance of students in a hybrid system.

4 Methodology

4.1 Research Approach

This study uses an exploratory mixed method research design process which consists of both quantitative and qualitative methods. An online questionnaire using Google Forms was designed for this purpose. A survey was set up on June 4, 2021, and it focused its activities on studying the educational uses of digital technology during this period of confinement. The study respondents were drawn from 57 teachers who are representatives of all the players in the educational community in Udm. The report began by mentioning the impact made by the educational uses of digital technology and the pedagogical issues imposed by distance education. It then analyzed the specificities of distance education and the difficulties inherent with it in such a situation. Finally, it is devoted to an analysis of the pedagogical and didactic practices that had been effectively put in place by the lecturers, the difficulties they encountered in doing so, and the positive lessons that could be drawn from this period.

Table 1 Gender, year of service by gender and age group

		Number of years of experience			Age group			
	Sex (%)	Mean	Minimum	Maximum	18–25 (%)	26–35 (%)	36–45 (%)	+ 45 (%)
Female	26.3	19	1.5	28	2.6	13.2	31.6	52.6
Male	73.7	19	1.5	30				

Table 2 Distribution of lecturers by discipline

Field of studies	Number of lecturers
Accounting	6
Mathematics and information Technology	17
Engineering sustainable development	24
English and French communication	3
Marketing and banking	7

5 Key Results and Discussion

5.1 *Lecturers' Demography*

Respondents were predominantly male, 73.7%. For this survey, regardless of the reason for the difference in the responses, it should be noted that the sample does not represent a gender-balanced grouping. Teachers, on average, have 19.8 years of seniority in the National Education with a maximum of 30 years of experience. It may be important to keep in mind that our respondents were not quite young people. Indeed, 52.6% of respondents was over 45 years of age (Tables 1 and 2).

5.2 *Methods of Exercising Educational Continuity During the Covid-19 Pandemic*

Based on the analysis conducted, it was found that at the start of the second confinement (March 2021), it was through telephone or mobile phone, and the social network WhatsApp, was mainly used to compensate for communication problems. Then, the synchronous virtual classroom took over [31]. Email was one of the most widely used methods. 60.5% of respondents said they had done remote works (example) in the form of a virtual classroom. So, the preferred tool by all lecturers remained the synchronous virtual classroom. It maintained the relationship with the students and the pedagogical team to ensure continuous follow-up. This result was interesting because it showed the preference for a static teaching method that involved exchanges, engagement, and participation [32].

The virtual classroom had played a vital role in student engagement and perseverance, restoring personal work pace through peer interactions. The dropouts were few in the groups that were able to have regular meetings. The lecturers even testified the fact that attendance was much more important and regular in synchronous class than in face-to-face.

The contemporary era was marked by the development of telecommunications, distance education networks and platforms, and the virtual campus. This virtual space is defined by Daniel Peraya [33] as "a unique environment integrating different functions of information, communication (synchronous or asynchronous), collaboration, management, and learning."

5.3 *Online Teaching Software Tools Used/Implemented*

The mostly used application during the lockdown period, at the time of this questionnaire's response, was Microsoft Teams. Despite of the strong presence of the distance learning platform Moodle, at UDM, Microsoft Teams proved to be more user-friendly and easy access as compared to other online applications like Easyclass, Zoom, and Google Classroom. The main reasons being there were easy sharing of materials; students could chat; and lecturers had no problem in organizing their online sessions through calendars. Prior to security reasons, lecturers were encouraged to shift to the new platform Moodle.

A study conducted by the firm McKinsey [34] focusing on reliable and secure platforms demonstrated that there is a strong relationship between security and adoption of platforms; this aspect was researched during the Covid-19 pandemic, whereby McKinsey financed teaching activities in university.

5.4 *Adjusting Teaching and Learning Practices*

Barzman et al. [35] pointed out that the digital transition had influenced higher education by modifying content, tools, and teaching methods. Indeed, the respective roles of teachers and students, as well as the development of training content, are and will be transformed by teachers [36].

More than 52.6% of lecturers declared that they had been led to modify their teaching and didactic practices. The lecturers said they had to rely initially on the work that had been carried out in class face-to-face such as consolidation and revision work, then gradually introduce the complexity of new pedagogical concepts. The main changes and challenges encountered by lecturers were as follows:

i. 73.7% of lecturers offered students a weekly work plan to maintain peer contact.

ii. 84.2% of lecturers requested individual work (one-to-one interaction was encouraged). In this case, lecturers had to customize and adapt themselves with practical activities that were either easy to understand or shown on video and simulations.

iii. A good majority of the lecturers (92.1%) had to keep the existing groups of students due to lack of time and having very busy schedules.

iv. 71.1% of lecturers respected the timeslot from posted timetables (which lacks flexibility).

v. 21.1% of lecturers said they chose a regular weekly schedule but different from the normal schedule.

vi. 58% of lecturers believed that the implementation of the educational continuity system could have a positive impact on the autonomy of some students.

vii. Only, 6% believed that this had not influenced the autonomy of any student. Peraya [37] found, in the case of distance education, that learners were most often technophile and relatively expert, developing autonomy and self-direction of their learning by placing themselves at the center of the learning process. In the French context [38], it underlined that ICTs have changed the relationship between teacher and students (faster and more efficient) [37].

viii. 50% of lecturers said they did not use a common digital working space to make materials available to students. The main reason stated by lecturers was as follows: "We do not have a digital workspace with scheduling and academic management."

ix. 84.2% of lecturers requested individual work. In this case, lecturers had to content themselves with practical activities that were simply described, shown on video or, more rarely, simulated. While some lecturers had sought new solutions, others have changed their practice by returning to teaching that they themselves qualify as more "traditional," more frontal, reduce work complexity, less oral work on the part of the students, more written interactions (chat), no group work, no individual feedback on student activities or a control assessment instead of a formative assessment.

x. 57.9% of lecturers favored collective work productions to encourage mutual help and collaborative practices among learners.

xi. 76.3% of teachers continued with email or chat exchanges promoting student self-expression and personalization. Lecturers therefore adequately identify what constituted added value for their teaching in ordinary, hybrid, or remote conditions. But, their projects were conditioned on the need to accustom students to distance and asynchronous education to develop their autonomy and prepared them in the event of re-confinement.

xii. 77.3% of lecturers said they assessed student learning using homework in a limited timeframe. Some had used multiple-choice questionnaires (MCQs) to gain insight into the level of comprehension.

xiii. Regarding changes in practices, whether they were of didactic or pedagogical in nature, these had led lecturers to set up new working methods like individual comments, self-correcting exercises, synchronized lessons, and the interactive whiteboard, and it appears that the students appreciated the modalities put in place.

5.5 *Collaboration and Institutional Support*

57.9% of lecturers felt that they lacked some form of support during this period of confinement. It is the lack of support from upper management that is most often cited. Several teachers regret a lack of coordination and harmonization within the teaching teams. A significant number also mentioned having suffered from a lack of financial support and tools within the teaching teams.

6 Conclusion

Although Covid-19 is seen as the big disruptor, it has propelled UdM in the direction of a technology-integrated education system. The use of digital technology has been able to generate a natural solution to the pedagogical constraints in the sense that lecturers have been able to bring together “virtually” and would ensure “pedagogical continuity” capable of meeting the requirements of university programs.

In this regard, particular attention should be paid in the future, on the one hand, to the training of lecturers in the educational uses of digital technology, both at the level of their initial training and that of their continuity training and, on the other hand, to the training of both administrative and educational teams involving widespread social distancing. Digital technology cannot constitute a miracle solution to the constraints emerging from unexpected crisis situations, but the best preparation for the best possible coordination of the school’s actors can help to guarantee the benefit of its effects.

7 Suggestions and Recommendations

According to the World Economic Forum, in response to COVID-19, universities and higher education schools have rapidly implemented e-learning solutions. These transformations have been carried out urgently and each actor adapted to the situation according to its own resources. Concretely, most of the courses offered in face-to-face sessions have been delivered online. Now, we can use the most effective crisis-recovery strategies as the basis for long-term improvements in the following ways:

- Top management support and commitment to understand the basic meaning of online learning in the educational field.

- It is essential that university establishments have a complete digital workspace with the design of timetables and management of university life.
- The university could gain in teaching efficiency through agreements with Internet service providers and equipment suppliers.
- Encourage lecturers to learn and familiarize themselves with modern teaching methods.
- The need to prepare courses adapted to the Mauritian reality to upgrade the scientific level of teaching and learning.
- Prepare and record video clips in advance and post them onto platforms.
- Avoid overloading students and review the pace of work to be submitted (number and deadline) and sending the corrected work with appropriate feedback.

It is likely that after these urgent transformations, hybrid models of education will become the norm within institutions. However, many universities and higher educational schools are all asking the same question: Will hybridization be for the benefit of students?

This paper recommends future research verification on the effectiveness of hybridization in one of the courses at Udm and to examine the predictive factors that lead to student success in a hybrid environment.

References

1. Barbot, M. (2000). Les auto-apprentissages, Clé International, coll. *Didactique des langues étrangères, Paris.*
2. Chambers, M. (2002). The efficacy and ethics of using digital multimedia for educational purposes. In *The convergence of distance and conventional education* (pp. 17–28). Routledge.
3. UNESCO (1987). *Glossary of educational technology terms.* UNESCO.
4. Hantem, A., Iscae-Rabat, A., & Économiques, A. D. J. D. E. (2020). *Les conditions de l'enseignement à distance pendant le confinement dû au COVID19: Cas de l'enseignement supérieur au Maroc.*
5. Peraya, D. (2006). La formation à distance : un dispositif de formation et de communication médiatisées. Une approche des processus de médiatisation et de médiation. *Calidoscópio, 4*(3), 200–2004.
6. Ouerfelli, T., & Gharbi, K. (2008). *Le dispositif d'enseignement à distance à l'université de Bahreïn practiques et attentes des enseignants.* Retrieved from https://isdm.univ-tln.fr/PDF/isdm32/isdm32-gharbi.pdf
7. Marot, J. C., & Darnige, A. (1996). *La téléformation.* Presses universitaires de France.
8. Marchand, L., & Loisier, J. (2004). *Pratiques d'apprentissage en ligne.* Chenelière Éducation.
9. Glikman, V. (2002). *Des cours par correspondance au "e-learning" [From correspondence courses to "e-learning"].* Presses Universitaires de France.
10. Peraya, D., Viens, J., & Karsenti, T. (2002). Introduction: Formation des enseignants à l'intégration pédagogique des TIC: Esquisse historique des fondements, des recherches et des pratiques. *Revue des sciences de l'éducation, 28*(2), 243–264.
11. Coswatte, S. (2014). Updated e-learning definitions. *Online Learning Consortium.*
12. Blandin, B. (2004). Historique de la formation ouverte et à distance. *Actualité de la formation permanente (Paris-la-Défense), 189*, 69–71.
13. Sener, J. (2015). Updated e-learning definitions. *Online Learning Consortium.*
14. Vaughan, N., & Power, M. (2010). Blended online learning design: Shaken not stirred.

15. Lakhal, S., & Power, M. (2016). La formation en ligne en enseignement postsecondaire au Québec: proposition d'une typologie de cours. In *ACTES DU COLLOQUE CIRTA 2016 «Dépassons nos frontières»* (p. 104).
16. Gérin-Lajoie, S., & Potvin, C. (2011). Évolution de la formation à distance dans une université bimodale. *Distances et savoirs, 9*(3), 349–374.
17. Akhtar, R., Hassan, H., & Saidalvi, A. (2020). The effects of ESL student's attitude on academic writing apprehensions and academic writing challenges. *International Journal of Psychosocial Rehabilitation, 24*(05), 5404–5412.
18. Rao, V. (2019). Blended learning: A new hybrid teaching methodology. *Online Submission, 3*(13).
19. Alsarayreh, R. (2020). Using blended learning during COVID-19: The perceptions of school teachers in Jordan. *Cypriot Journal of Educational Sciences, 15*(6), 1544–1556.
20. Kaur, M. (2013). Blended learning-its challenges and future. *Procedia-Social and Behavioral Sciences, 93*, 612–617.
21. Lieberman, M. (2021). How hybrid learning is (and is not) working during COVID-19: 6 case studies.
22. Li, C., & Lalani, F. (2020). The COVID-19 pandemic has changed education forever. This is how. In *World economic forum* (Vol. 29).
23. García, E., & Weiss, E. (2020). COVID-19 and student performance, equity, and US education policy: Lessons from pre-pandemic research to inform relief, recovery, and rebuilding. *Economic Policy Institute.*
24. Gilmour, A. (2020). Getting the blend right: Inspiring student learning through blended learning.
25. Rasheed, R. A., Kamsin, A., & Abdullah, N. A. (2020). Challenges in the online component of blended learning: A systematic review. *Computers & Education, 144*, 103701.
26. Manwaring, K. C., Larsen, R., Graham, C. R., Henrie, C. R., & Halverson, L. R. (2017). Investigating student engagement in blended learning settings using experience sampling and structural equation modeling. *The Internet and Higher Education, 35*, 21–33.
27. Sun, J. C. Y., & Rueda, R. (2012). Situational interest, computer self-efficacy and self-regulation: Their impact on student engagement in distance education. *British Journal of Educational Technology, 43*(2), 191–204.
28. Baragash, R. S., & Al-Samarraie, H. (2018). Blended learning: Investigating the influence of engagement in multiple learning delivery modes on students' performance. *Telematics and Informatics, 35*(7), 2082–2098.
29. Cacciamani, S., Cesareni, D., Martini, F., Ferrini, T., & Fujita, N. (2012). Influence of participation, facilitator styles, and metacognitive reflection on knowledge building in online university courses. *Computers & Education, 58*(3), 874–884.
30. Huang, E. Y., Lin, S. W., & Huang, T. K. (2012). What type of learning style leads to online participation in the mixed-mode e-learning environment? A study of software usage instruction. *Computers & Education, 58*(1), 338–349.
31. Hrastinski, S. (2008). Asynchronous and synchronous e-learning. *Educause Quarterly, 31*(4), 51–55.
32. Paivandi, S., & Streitwieser, B. (2014). *Internationalisation of higher education and global mobility.* Symposium Books (p. 320); *Revue française de pédagogie. Recherches en éducation, 187*, 150–153.
33. Peraya, D., & Deschryver, N. (2003). Staf 17-Réalisation d'un dispositif de formation entièrement ou partiellement à distance. *La formation à distance, un cadre de référence. Tecfa-Document de cours*
34. Sneader, K., & Singhal, S. (2020). Beyond coronavirus: The path to the next normal. *McKinsey & Company, 5.*
35. Barzman, M., Gerphagnon, M., & Mora, O. (2020). *La transition numérique dans la recherche et l'enseignement supérieur à l'horizon 2040* (p. 146). éditions Quae.
36. Lassassi, M., Lounici, N., Sami, L., Tidjani, C., & Benguerna, M. (2020). Université et enseignants face au COVID 19: L'épreuve de l'enseignement a distance en Algérie. *les cahiers du cread, 36*(3), 397–424.

37. Peraya, D. (2011). Un regard sur la «distance», vue de la «présence». *Distances et savoirs, 9*(3), 445–452.
38. Gremmo, M. J., & Kellner, C. (2011). Pratiques pédagogiques et usages des TIC: enseigner à l'université, un impensé? *TIC et métiers de l'enseignement supérieur. Émergences, transformations*, 35–52.

Moving on with Digital Innovation: The Always-Connected Customer

Normada Bheekharry

Abstract With the evolution of the Internet of Things (IoT), marketing has moved from a product-centric marketing approach (Marketing 1.0) to a customer-centric era (Marketing 2.0) further to a period where value-driven approaches were prioritized (Marketing 3.0) to ultimately focus on Marketing 4.0. This new era of marketing has seen customers to be more participative, more curious of something that is new and different, and businesses are focusing on machine learning (ML) to understand customer's psychology. Traditionally, marketing was considered to be the synthesis of the four main elements of the marketing mix to satisfy customer needs, and there was a one-way communication: business to consumers (B2C). However, two decades after the twenty-first century, consumer experiences, information management, and predictive analytics are main factors to determine future trends. Practitioners argue that Marketing 3.0 is similar to Marketing 2.0 where customers' involvement and participation contribute to the overall business marketing process, but what is new is emotional marketing where individuals' needs and aspirations are also considered. With the advancement in information communications technology (ICT) and Web 3.0, the emergence of social media platforms has transformed the communication channels in marketing. Social media is an important marketing communication tool where customers share their experiences among themselves and businesses as well. Consumers' shared experiences, information, and knowledge management together with predictive analytics are the main factors that are used in Marketing 4.0 to predict future consumer buying behavior. The aim of this paper is to go beyond the acceptance of ICT and understand its contribution in the future of marketing.

Keywords Always-connected marketing · Marketing 4.0 · AI in marketing

N. Bheekharry (✉)
Université des Mascareignes, Beau Plan, Pamplemousses, Mauritius
e-mail: nbheekharry@udm.ac.mu

B. Pati et al. (eds.), *Proceedings of the 6th International Conference on Advance Computing and Intelligent Engineering*, Lecture Notes in Networks and Systems 428,
https://doi.org/10.1007/978-981-19-2225-1_49

1 Introduction

We are in the connected customer era. Digitalization has empowered customers to demand fast access to personalized content on their terms. To stay in competition and evolve in the digital transformation trend, companies are concentrating on customer engagement and customer experience to enhance customer value. Customers want more options and more content at all times, on any device. Companies must deliver more tailored information, faster, and across platforms to meet the changing demand of customers. Google, Amazon, Netflix, Spotify are associating customer interactions from numerous online platforms: Twitter, Instagram, and Pinterest and tapping customers' insight to better understand customer psychology and market to them better. Insights can be tapped from individuals of from collective audience interactions enabling marketers optimize segmentation, positioning, marketing, promotion, and pricing strategies. According to Berman and Kesterson-Townes [1], social viewing, distracted viewing, and on-demand viewing are becoming more popular among connected customer, and this concept was developed further by Malchenko et al. [2].

Malchenko et al. [2] suggest that depending on the function the customer and the purpose of the online transaction, there are five forms of exchange to be considered: Content consumption and development are the first two steps in the content creation process, (3) making a purchase of a goods or service, (4) providing a service or product, (5) privacy-related behavior [2]. It has also been brought forward by the same authors that digital exchanges are nonlinear where the customers can either be the receiver or sender, and the object of online transaction can either be: (1) searching and/or sharing of information on a particular product or service, (2) searching and/or sharing personal data. In the fast-changing digital marketplace, all of these sorts of trade provide a setting for the development of consumers' digital competencies. We should also not forget that content consumption leads to interactions with a range of sources, including eWOM exposure and evaluation, in the digital consumer journey [3].

Customers' connectedness has led to the creation of a sharing and collective business culture. The role of the customer as the end user of products and services has drastically changed: customers no longer buy products and services through digital marketplace but also to produce them and provide services on their own [4]. There is a shift in the role of the consumer from client to manufacturer/producer/intermediary.

The aim of this paper is to have a broad view of how technology is contributing to the change in customer psychology and is organized in the following sections. The first section is the background study. It covers areas on Internet of Things. The second section shows the evolution of marketing from Marketing 1.0 to Marketing 4.0. Section 3 outlines the implication of Web. The next section gives the characteristics of the always-connected customer, followed by Managerial Implication and Conclusion.

2 Background Study

To better understand the principles of digital marketing, practitioners and academics should first understand the influence of customer connectedness. The effective application and adoption of information technology (IT) have reshaped the business landscape. Innovation and evolution in IT have given rise to the Internet of Things (IoT) described by The Cluster of European Research Projects on the Internet of Things (CERP-IoT) [5] as: 'an integrated part of future Internet and is a dynamic global network infrastructure with self-configuring capabilities based on standard and interoperable communication protocols where physical and virtual "things" have identities, physical attributes, and virtual personalities and use intelligent interfaces, and are seamlessly integrated into the information network'. The evolution of the Internet of Things has stretched to an interactive network a Web of platforms for connected devices and objects forming a Web of "everything." Several studies and researchers have forecasted that by 2020 and the coming years that there will be more than 50 billion connections and the global economy will be driven by connecting the unconnected: people to people (P2P), machine to people (M2P), machine-to-machine (M2M) via the Internet of Everything.

The Internet of Everything can be defined as the networked connection of people, process, data, and things to create value [6]. The higher the level of connectedness the higher the value which can be realized by connecting the unconnected. Marketers must embrace the trend toward a more horizontal, inclusive, and social business environment. The market is broadening its appeal. Social media breaks down geographical and demographic barriers, allowing people to interact and communicate while also allowing businesses to collaborate and create. Customers are shifting their focus from vertical to horizontal. They are growing increasingly cautious of brand marketing communications and depending on the f-factor instead (friends, families, fans, and followers). Finally, the buying process for customers is becoming more social than it was previously. When it comes to making decisions, customers are paying more attention to their social network. They are looking for suggestions and feedback.

Researchers and marketing practitioners have drawn our attention to the changing role of the consumers prior to innovation in IT and at the same time the evolution in marketing concepts, fundamentals, philosophies, and practices. Jara et al. [7], emphasized the change in customers' requirements [7]. Customers are no longer looking for a product to satisfy their basic needs but want to be an integral part of the product and product decision-making. Furthermore, customers want to have the power to participate and interact with the product. They want their "voices" to be heard. Consequently, the technological improvement, in both information and communication, has been a key driver in enabling customers to share their experiences and more importantly to check whether the product satisfies what it promises. Marketing and the Internet are moving around customers, empowering their interaction with products while considering the values from users and offering them more reliable and up to date information.

3 Evolution of Marketing

The term marketing has long been disputed by different marketing theorists. Many authors have defined marketing and related concepts such as a marketing orientation. Two standard definitions of marketing include:

> Marketing as the activity, set of institutions, and processes for creating, communicating, delivering, and exchanging offerings that have value for customers, clients, partners, and society at large. (American Marketing Association 2013)

> Marketing is the management process, which identifies, anticipates, and supplies customer requirements efficiently and profitably. (Chartered Institute of Marketing)

These definitions of marketing are based on the concept of the marketing exchange, in which the supplier has a product to sell, and the customer is prepared to pay a price for the product [8] and to be competitive organizations have to understand the needs and wants of customers and deliver a better service compared to competitors. Over the past few years, marketers have recognized the importance of customer value and endorsed the concept of relationship marketing. Rowley [8] emphasizes that relationship marketing identifies that the core of marketing is the relationship between the organization and the customer, which may extend over many transactions and several years [8]. This leads to a different perspective on marketing. For instance, in traditional marketing, the focus is on the benefits from each individual transaction; in relationship marketing, the focus shifts to the value of the relationship and concepts such as customer lifetime value. The emphasis in marketing communication and service delivery concerns maintaining the relationship with the customer. The primary role in marketing is to satisfy customers [9, 10] have claimed that for an organization to be successful it must relate "all its thinking to the customer's needs." [9, 10]. Focusing on the root of these definitions, we conclude that both professional marketing associations place the customer at the heart of marketing activities and to deliver customer satisfaction. An equation can be developed between organization resources, capabilities, and know-how in terms of knowledge either about the customer or technical skills. Therefore, the more the organization knows about the needs and demands of the market the more it can prepare itself to deliver better goods and services and be competitive. Hence, we can conclude that there is a relationship between connectedness and value. The more the connectedness the more the value: **value = Connected**. Additionally, for the equation to stand good, different processes should be aligned, for example, the search of information, setting of marketing objectives, analyzing the firm's capabilities, understanding consumer behavior, and the decision-making unit, devising strategies, segmenting, targeting and positioning, implementing the marketing mix that is product, price, place, promotion, physical evidence, people and processes and controlling the whole operations. Traditional marketing practices have brought success to many organizations throughout the world; however, with research and development in the information technology, particularly, with the emergence of the Internet, major changes have contributed to marketing. The next section will consider the characteristics of Web 2.0 and its impact on marketing.

3.1 The Web 2.0

One feature of the marketplace is its increasingly digitization. Woon et al. [11] outline that technological changes have made traditional marketing more efficient and effective in reaching and selling to markets [11]. Web 2.0 is a powerful marketing tool and is pivotal to share ideas and create customer values. The notion of value and value proposition is central to marketing. Kotler [12] defines value from perspective of the customer, and he posits that customer value is the result of customer's assessment in weighing the bundle of benefits against the bundle of cost they expect to incur in evaluating, obtaining, and using the product or service [12]. One of the main roles of marketing is to understand and explain the value an intended consumer derives from the product or service. Digital media has created an unprecedented capacity to generate and capture customer demand through a variety of forms and through different channels. Web 2.0 technologies have three distinctive characteristics: collaboration, participation, and communication, and they are being used by customers and firms for different purposes, such as: information gathering and sharing. Some authors mention that organizations are using Web 2.0 technologies to improve decision cycle times, organizational effectiveness and innovation [13]. Sharma and Baoku [14] state that "Web 2.0 provides benefits by delivering access to collaboration and allows information to be spread more efficiently" [14]. Palacios-Marques [15] relates that Web 2.0 adoption might help organization to generate and disseminate market intelligence, and at the same time, they have the potential to increase firm's innovativeness by fostering knowledge creation and sharing [15]. Web 2.0 is also defined as Web pages that use a two-way streams of communication between users allowing them to socialize online and to share their own user generator content [16–18].

Examples of popular Web 2.0 applications which allow online user-generated content sharing or social media interactions include: file-sharing sites (Flickr for photo sharing: Eason [19]), blogs (e.g., Blogger.com: Thackeray et al. [18]), wikis (e.g., Wikipedia: Kennedy et al. [20]), and social networking sites (e.g., Facebook: Kennedy et al. [16]; Twitter: Lefebvre [20]) [16, 20]. Since Web 2.0 technologies are easy to use, customers and employees are using them for different purposes. We believe that thanks to these characteristics customers are sharing on the Web more information about their experiences with products and services. Therefore, firms tend to acquire market intelligence using Web 2.0 technologies. Additionally, it has been argued that Internet blog narratives can be used to determine one firm's competitive position [21]. Finally, Web 2.0 has the potential to ease the dissemination of market intelligence within the firm.

4 Power Shifts in Marketing

The field of marketing has kept on changing over decades, and with technology innovations, there has been a 360° transformation; however, understanding customer needs, creating, developing, and delivering value remain the core of marketing.

4.1 *Marketing 1.0*

Marketing 1.0 has be defined as a product-centric approach, and it dates back to the Industrial Age. At this stage, marketing was focused around sales and not the needs and wants of target customers.

4.2 *Marketing 2.0*

A radical change in the concept of marketing occurred with the innovation of IT and ease of communication via the Internet. Consumers started to be well informed and product value defined by them. This approach is defined by the customer-centric era which is mainly based on the information age.

Marketing research is conducted to better understand the needs and wants of the customers, and strategies are devised on how to satisfy customers better and more efficiently than competitors. The emergence of customer relationship management, relationship marketing, brand management, and social responsibility is geared toward customer satisfaction and loyalty also transpire at this stage.

4.3 *Marketing 3.0*

Kotler [22] defines marketing 3.0 as the human-centric era where values are driven by customers [22]. Jara et al. [7] stated that customers are treated as humans who are active, anxious, and creative, and the marketer's role lies in the understanding of human side of the customer [7]. Access to easy and cheap communication through the Internet has enabled customers to engage in the development of product and share their ideas and opinions. Customers are more aware and sensible about social and environment issues. Development in Web technologies has enabled customers to be involved in social networks, and user-generated content (UGC) provides valuable information for marketers to better understand customer's psychology and online behavior. In Marketing 3.0, emphasis is laid on data management, data warehousing, big data, big data analytics, predictive marketing, and marketing automation.

We are in an era where customers' participation and engagement are considered important in the creation, definition, and delivery of value, and this age is known as Marketing 3.0 [23]. Customers actively share their knowledge, anxiety, desire, and creativity along social networks. Hence, customer experience (CX) is shared through blogs, videos, and chats which validates, confirms, and checks the experiences of other customers with the product. As a result, assuring the brand satisfies what it is offering. A major integration of users with Internet through social networks has been observed, and social virtual environments are presenting a direct consequence over the products with platforms of opinions and experiences. The main drawback, however, is that the information gathered from the Internet is mainly based from CX and is not always the integration of opinions and justifications of the manufacturer or vendor.

4.4 Marketing 4.0

Marketing 4.0 concerns "a marketing approach that combines online and offline interaction between companies and consumers," [13]. It is the era where artificial intelligence (AI) is used to enhance productivity of other technologies. According to Fucci and Dumitrescu [24], AI has strengthened connectivity to improve the customer interaction process [24]. Industry 4.0 has accelerated the transition of marketing 3.0 to marketing 4.0 [25]. In marketing 3.0, concepts such as big data, social media, corporate social responsibility, and e-commerce were in the development stage and early growth stage; however, with Marketing 4.0, the connectedness between the customer and the business has become stronger [26]. Marketing 4.0 aims to better understand how business can better use and implement predictive marketing hence to better predict future trends with consumer trends, data management, and advanced analytics [27] (Table 1).

Analyzing the different opinions and views of researchers on the evolution of marketing, we conclude that movement of marketing from industrialization to the digital era, the basic concept of marketing remains the same. That is the core of marketing remains the customer and the four marketing concepts and inter-nested. We should not forget that marketing is about dealing with a constantly changing industry, and in order to comprehend cutting-edge marketing, we must first comprehend how the market has evolved in recent years. The Internet offers access to limitless connectivity and interactivity for not only corporations but individuals also, and this has reshaped the entire configuration of marketing practice; however, the core remained the same that is creating value and engaging in customer satisfaction.

Table 1 Evolution of marketing from origin

Concept	Era	Characteristics	Focus
Marketing 1.0	Industrial age	• Focused on selling products • Considering the needs and wants from the target market	Product-centric
Marketing 2.0	Information and communication age	• Exhaustive search and research about the needs and wants • Emphasis to discover new target markets • Consumers started to be well informed and compare several values offerings of similar products	Customer-centric approach
Marketing 3.0	Value-driven marketing	• Success of Internet • Emergence of the new generation of social networking sites • Users are able to leave opinions, have numerous connections to other site members • The user experiences and opinions over the products are used as marketing tools	Human-centric
Marketing 4.0	Connectedness	• The user and interaction with the product offers the capabilities for coalescence of the previous three generations of marketing. Customers are considered as individuals with feelings	Holistic and spiritual marketing

5 Always-On Customer

A review of the research has shown that marketing 4.0 revolves around the integration of four main pillars: people, data, things, and process. Marketing 4.0 has reinforced the creation of a reputation platform to assess consumer experiences and opinions with this new layer of capabilities to engage in and engage with products. The value is directed related to the level of connectedness experience by the customer. The always-on customer is connected 24/7 over technological devices, for example, digital television, mobile, social media. To better understand customers and their values, bearing in mind that with Marketing 3.0 and 4.0, customers have inculcated several identities through multi-online that is different email addresses, different accounts in different social platforms, and offline that is through mobile technologies.

5.1 People

In IoE, human beings will be able to connect to the Internet in innumerable ways. Most human beings join the Internet through their use of gadgets (such as PCs, tablets, TVs, and smartphones) and social networks (such as Facebook, Twitter, LinkedIn, and Pinterest). As the Internet evolves toward IoE, we will be related in extra relevant and valuable ways. According to Gartner, people themselves will turn out to be nodes on the Internet, with both static records and a constantly emitting endeavor system.

Rather than simply reporting raw data, linked things will soon envisage higher-level information again to machines, computers, and people for in addition comparison and choice making. This transformation from information to facts in IoE is necessary due to the fact it will permit us to make faster, greater intelligent decisions, as well as manage our environment extra effectively.

5.2 Data

With IoT, gadgets typically gather facts and move it over the Internet to a central source, the place it is analyzed and processed. As the capabilities of things related to the Internet proceed to advance, they will emerge as extra shrewd by means of combining facts into more useful information. Rather than just reporting raw data, related matters will soon distribute higher-level data again to machines, computers, and humans for further comparison and decision-making. This transformation from data to information in IoE is necessary because it will enable us to make faster, extra smart decisions, as properly as manipulate our surroundings greater effectively.

5.3 Things

This element is made up of physical objects such as sensors, consumer devices, and agency assets that are connected to both the Internet and each other. In IoE, these matters will feed greater data, become context-aware, and furnish greater experiential information to help people and machines make greater applicable and valuable decisions. Examples of "things" in IoE include clever sensors built into structures like bridges and disposable sensors that will be positioned on everyday items such as milk cartons.

5.4 *Process*

Process plays an essential position in how each of these entities—people, data, and things—works with the others to supply cost in the related world of IoE. With the right process, connections emerge as applicable and add cost due to the fact the right data are delivered to the right individual at the right time in the appropriate way.

6 Emerging Trends

The digital market is in continuous transition, and we can see the hints and emerging trends if we look hard enough. A new type of client is growing globally, one that will soon become the majority—young, urban, middle-class people with significant mobility and connectivity. While mature markets contend with an aging population, emerging economies benefit from a demographic dividend in the form of a younger, more productive population. They are not only young, but they are rapidly relocating to cities and adopting a big-city lifestyle. The bulk of them are middle class or upper middle class, with a significant amount of disposable income. They want to achieve better goals, experience nicer things, and emulate those who have risen from a lower socioeconomic standing. They want to achieve larger goals, have more luxurious experiences, and copy the habits of persons in higher social classes. They are a compelling market for marketers to seek because of these characteristics. However, what sets this new sort of customer apart from previous markets is their tendency to be mobile. They travel a lot, commute frequently, and live life at a faster pace. Everything must be instantaneous and time-saving. When they are interested in something they watch on television, they use their mobile devices to look for it. They examine pricing and quality online before deciding whether to buy something in store. They can make purchases anywhere and at any time using a variety of devices since they are digital natives. Despite their Internet prowess, they prefer to engage in physical connectedness. When connecting with brands, they value high-touch interaction. They are also quite social; they converse with one another and have faith in one another. In fact, they have more faith in their circle of friends and family than they do in organizations and brands. In a nutshell, they are always connected.

Many mainstream theories and major assumptions about customer, product, and brand management have been called into question as a result of connectivity. Connectivity lowers the cost of connection between businesses, employees, channel partners, customers, and other stakeholders. As a result, the hurdles to accessing new markets are lower; concurrent product development is possible, and brand building takes less time.

Various examples of how connectivity has quickly disrupted long-established businesses with ostensibly significant entry barriers have been recognized. Amazon created havoc on retailers and, later, the publishing business. Similarly, Netflix alongside the likes of Hulu has disrupted brick-and-mortar video rental establishments, as well as satellite and cable TV providers. Spotify and Apple Music have transformed the way music is distributed in a similar way. Connectivity also alters our perceptions of competitors and customers. Collaboration with competitors and customer co-creation is now essential. It is no longer a zero-sum game when it comes to competition. Customers are no longer passive recipients of a company's segmentation, targeting, or positioning efforts. Connectivity has accelerated market dynamics to the point where it is nearly difficult for a corporation to compete on its own or rely solely on internal resources. A corporation must accept the fact that in order to succeed, it must interact with third parties and even include customers.

7 Managerial and Marketing Implication

Customer acquisition is of upmost importance in the dynamic digital era. We can add: drive increased sales among identified leads, increase general brand awareness, acquire new customers, increase number of named contacts, create, and deliver personalized experiences to customers, improve post-purchase customer loyalty/repurchase, shift brand positioning, and retain existing customers, as top digital marketing objectives. Marketing functions that cultivate two key aspects of adaptability—foresight and flexible planning—are better positioned to sense and respond to rapidly changing customer needs. Marketing professionals who manage loyalty and customer connections must focus on two important tasks—listening and planning—to feel and respond to changing client preferences and demands, in order to promote adaptability.

No. 1: Generate customer foresight

Marketing managers should focus in generating customer foresight. Analyzing and forward-looking customer insight detects long-term shifts in customer needs. This can be applied to current and potential customers. To generate customer foresight, the following need to be done:

(i) Diversify data sources. Traditional behavioral and attitude data may not be accurate enough to capture altering consumer preferences or predict future customer behavior. Secondary sources including employee/frontline staff feedback, topic experts, and customer advisory boards should also be considered. Changes in conduct and attitude are put to the test. Some modifications made in response to disruption have a higher chance of sticking than others.

(ii) Pressure test behavioral and attitudinal shifts. Changes in conduct and attitude are put to the test. Some modifications made in response to disruption have a higher chance of sticking than others. If, for example, changes in customer behavior appear to be an acceleration of a trend that started before the disruption, it is likely that at least some of the consumers who are experimenting with new behaviors will not revert to their old ones.

No. 2 Make flexible and contingency planning a priority

Adaptive strategy has shown to be notably important during the COVID-19 interruption and continues to be so. To capitalize on changing consumer needs, marketing leaders must constantly reassess changes in client needs, shift current resources, and build important capabilities. Tactics to be followed:

(i) Customers should be re-segmented based on their changing needs. Revisit segmentation more frequently, especially during periods of uncertainty, to account for changes in customer needs, product fit, channel fit, and content fit.
(ii) Plan scenarios that are tailored to the needs of your customers. Scenario planning in adaptable firms goes beyond anticipating the impact on business performance; it also predicts how consumer demands will change in different situations and the marketing capabilities required to meet those changing requirements.
(iii) Change to a continuous, zero-based project prioritization system. Begin with a clean slate. Instead of trying to reduce an existing list of projects, justify each one from scratch. Starting from scratch necessitates retesting and validating assumptions, offering a lens through which to effectively reallocate resources.
(iv) Adaptability is essential for staying competitive in a world where client needs are always changing. Even in the most difficult times, the marketing function may find ways to develop the firm by modifying how they listen to customers and planning for changing client preferences.

No. 3 Facing challenges to execution

Marketing practitioners must plan, execute, and track their actions with a strong digital marketing focus that is from managing client acquisition and retention innovative technologies should be integrated to promote personalization. All marketing strategies, budgets, and key performance indicators are increasingly influenced by how digital marketing objectives are aligned to business goals. The Gartner 2021 Digital Marketing Survey confirms this effect and gives insight into digital marketing leaders' goals, priorities, and methods. It is important to keep in mind that this marketing development reflects fundamental change in client involvement choices, technological advancements, and other disruptive influences both internal and external to the organization.

No. 4 New technologies and customization

To fit with their customer acquisition and retention goals, digital marketing experts are growing their usage of new technologies like as artificial intelligence (AI) and machine learning (ML). While 84 percent of marketers think that AI/ML improves their capacity to provide real time, tailored experiences to clients, adoption is still low. Personalization projects that need a major additional investment raise concerns regarding the magnitude, pace, and surety of reward, and good outcomes are not always dependent on increasing personalization technology investments. With readily available data, information, and current skill and technology, a lot may be accomplished. Nonetheless, AI and machine learning will become more important tools for interpreting quick variations in client demand and making more correct judgments.

8 Conclusion

The digital disruption brought about by the evolution in Internet's growth has caused both scholars and marketing practitioners to reconsider the different consumer behavior models brought forward by researchers throughout the world. The Internet has had a huge impact on society and has changed the way people communicate [28]. The rise of social media has given rise to a new dimension in the shape of online communities of customers, who are today's marketing target groups. In this age of digitization, which has revolutionized the way individuals associate and connect with one another, the customer channels illustrated in earlier hierarchical models are insufficient [29]. The AIDA framework was the first model that was used to depict consumer behavior and most extensively utilized in various studies connected to advertising, as well as a popular model used by practitioners. However, in this digital phase, customers do not have time to assess advertising messages in the pre-purchase stage or marketed items in the post-purchase stage in the contemporary environment, where information flows at the speed of light. They have a penchant for seeking counsel from Internet sources. It encourages businesses to produce information that is trustworthy and worthy of community consideration. This emphasizes the significance of upgrading the customer journey throughout this period of connection. Customers' preferences and brand growth should be more concentrated in this age of the Internet, when individuals are socially linked with one another, beyond the confines of increased buy alone [30] (Diagram 1).

In this over-connected marketplace, managers and marketing managers should not only focus on customer relationship management but consumer relationship management. The customers trust the consumers, and this element of trust is decisive in the adoption or purchase of a product and brand integrity. Consequently, trust will

Stage 1: Product Identification
Stage 2: Check through social platforms and manufacturer
Stage 3: Construction and verification of the brand
Stage 4: Integrity and reputation reinforced on customer experience
Stage 5: Purchase
Stage 6: Establishment of trust relationship among the consumer of the brand

Diagram 1 Simple consumer purchase behavior framework in the digital era (Based on own elaboration of literature)

promote and create value. Hence, we can conclude that the higher the connectedness the higher the value creation.

References

1. Berman, S. J., & Kesterson-Townes, L. (2012). Connecting with the digital customer of the future. *Strategy & Leadership, 40*(6), 29–35.
2. Malchenko, Y., Gogua, M., Golovacheva, K., Smirnova, M., & Alkanova, O. (2020). A critical review of digital capability frameworks: A consumer perspective. *Digital Policy, Regulation and Governance.*
3. Rosario, A. B., de Valck, K., & Sotgiu, F. (2019). Conceptualizing the electronic word-of-mouth process: What we know and need to know about eWOM creation, exposure, and evaluation. *Journal of the Academy of Marketing Science, 48*(3), 422–448.
4. Kim, N. L., & Jin, B. E. (2020). Why buy new when one can share? Exploring collaborative consumption motivations for consumer goods. *International Journal of Consumer Studies, 44*(2), 122–130.
5. CERP-IoT (2009, September 19). *Internet of Things: Strategic research roadmap.* Retrieved from http://www.internet-of-thingsresearch.eu/pdf/IoT_Cluster_Strategic_Research_Agenda_2009.pdf. Accessed on December 13, 2019.
6. Sudharshan, D. (2020). Internet of everything (IoE). *Marketing in Customer Technology Environments*, 161–208.https://doi.org/10.1108/978-1-83909-600-620201008
7. Jara, A. J., Parra, M. C., & Skarmeta, A. F. (2012). Marketing 4.0: A new value added to the marketing through the ınternet of things. In *2012 Sixth international conference on innovative.*
8. Rowley, J. (2003). Information marketing: Seven questions. *Library Management, 24*(1/2), 13–19.
9. Witkowski, T. H. (2005). Sources of immoderation and proportion in marketing thought. *Marketing Theory, 5*(2), 221–231.
10. Woodall, T. (2007). New marketing, improved marketing, apocryphal marketing: Is one marketing concept enough? *European Journal of Marketing, 41*(11/12), 1284–1296.
11. Woon, K. C., Mathew, S., Christopher, W., & Lui, V. (2010). B2B emarketplace: An e-marketing framework for B2B commerce. *Marketing Intelligence & Planning, 28*(3), 310–329.
12. Kotler, P., Kartajaya, H., & Setiawan, I. (2010). *Welcome to marketing 3.0.* Wiley. https://doi.org/10.1002/9781118257883.ch1
13. Lim, X., Ren, X., & Zheng, X. (2015). Management of competition among sellers and its performance implications for business-to-business electronic platforms: Dynamic analysis by VAR model. *Nankai Business Review International, 6*(2), 199–222.
14. Sharma, G., & Baoku, L. (2013). Customer satisfaction in Web 2.0 and information technology development. *Information Technology & People, 26*(4), 347–367.
15. Palacios-Marqués, D., Cortés-Grao, R., Carral, C. L. (2013). Outstanding knowledge competences and web 2.0 practices for developing successful e-learning project management. *International Journal of Project Management, 31*(1).
16. Lefebvre, R. C. (2007). The new technology: The consumer as participant rather than target audience. *Social Marketing Quarterly, 13*(3), 31–42.
17. OECD (2007). *Participative web and user-created content: web 2.0, Wikis, and social Net255 working, organisation for economic co-operation and development*, Paris. Available at www.256biac.org/members/iccp/mtg/2008-06-seoulmin/9307031E.pdf. Accessed 14 January 2021.
18. Thackeray, R., Neiger, B. L., Hanson, C. L., McKenzie, J. F. (2008). enhancing promotional strategies within social marketing programs: Use of web 2.0 social media. *Health Promotion Practice, 9*(4), 338–343.
19. Eason, J. A. (2007). Online social networks: How travelers can benefit from using interactive sites. *Healthcare Traveler, 15*(5), 18–27.

20. Kennedy, G., Dalgarno, B., Gray, K., Judd, T., Waycott, J., Bennett, S., Maton, K., Krause, K.-L., Bishop, A., Chang, R., & Churchward, A. (2007). *The net generation are not big users of web 2.0 technologies: Preliminary findings.*
21. Crotts, J. C., Mason, P. R., & Davis, B. (2009). Measuring guest satisfaction and competitive position in the hospitality and tourism industry: An application of stance-shift analysis to travel blog narratives. *Journal of Travel Research, 48*(2), 139–151.
22. Kotler, P. (2011). Reinventing marketing to manage the environmental imperative. *Journal of Marketing, 75*(4), 132–135.
23. Kotler, P., Kartajaya, H., & Setiawan, I. (2017). *Marketing 4.0: Moving from traditional to digital.* Wiley.
24. Fucui, M., & Dumitrescu, L. (2018). From marketing 1.0 to marketing 4.0—The evolution of the marketing concept in the context of the 21st century. *International Conference Knowledge-Based Organization, 24*(2), 43–48.
25. Taş, B., & Şeker, ŞE. (2017). Nöropazarlama ve yönetim bilişim sistemleri. *YBS Ansiklopedi, 4*(2), 12–17.
26. Soyak, S., & Soyak, A. (2018). Sanayi 4.0 döneminde pazarlamada dijital arayışlar ve bazıküresel şirketlerden örnekler, 5. Uluslararası Multidisipliner Çalışmalar Kongresi, 2–3 November, Antalya (pp. 26–38).
27. Vassileva, B. (2017). Marketing 4.0: How technologies transform marketing organization. *Obuda University E-Bulletin, 7*(1), 47–56.
28. Cappo, J. (2003). *The future of advertising: New media, new clients, new consumers in the post-television age* (pp. 189–193). McGraw-Hill.
29. Wijaya, B. S. (2011). New model of hierarchy of effects in advertising. *ISSIT, 1*(1), 5–15.
30. Wijaya, B. S. (2012). The development of hierarchy of effects model in advertising. *International Research Journal of Business Studies, 5*(1). https://doi.org/10.21632/irjbs.5.1.73-85

Quality Estimation of Change-Point Detection by the Signals Ratio Algorithm for Random Processes

Elena N. Benderskaya

Abstract The change-point detection method based on decision-making statistics is considered. The average number of delay steps of the change-point detection and the average number of delay steps of detecting the returns of the statistical properties to the initial values are constructed as the mathematical models for the algorithm based on decision-making statistics with the signals ratio equations. The mathematical models for the average number of delay steps of detecting the returns of the statistical properties to the initial values are proposed for the first time. Proposed mathematical models for the average number of delay steps of the change-point detection have no limitation on the values of the algorithm's parameters, unlike already known ones. To ensure that the change-point detection in real-time systems will be done with a delay not exceeding a given value, it is critical that the obtained models already allow choosing the algorithm's parameters with the worst-case orientation.

Keywords Change-point detection · The moments of the random processes properties changes · Signals ratio algorithm · Quality measures · Decision-making statistics · Quality estimating · Fault detection and diagnosis

1 Introduction

Many problems formalizing in the different practical applications lead us to the well-known statistical task dealing with the changes in statistical properties of the random processes or the change-point detection task. The task of detecting the moments of changing the properties of the random processes emerges in many practical applications—beginning with such as seismic activity analysis [1], analysis of the water flow activities for the flood events prediction [2–5], technological processes monitoring [6, 7], and ending with the analysis of the user's behavior, monitoring computer networks activity [8, 9].

E. N. Benderskaya (✉)
Peter the Great St. Petersburg Polytechnic University, Polytechnicheskaya, 29, 195251 St. Petersburg, Russia
e-mail: helen.bend@gmail.com

B. Pati et al. (eds.), *Proceedings of the 6th International Conference on Advance Computing and Intelligent Engineering*, Lecture Notes in Networks and Systems 428,
https://doi.org/10.1007/978-981-19-2225-1_50

1.1 Motivation

Although there are many new methods of processing stochastic signals, including natural computing and fuzzy logic [10–12], interest in the classical approaches based on decision-making statistics remains generally excellent [13–16]. At the same time, the focus is on determining the parameters of the algorithms that provide the best quality of the detection change-point.

The false detection probability and the time delay of the detection are used to measure the algorithms' quality to compare the different change-point detecting algorithms. Many of the proposed approaches are based on a priori information about the processes parameters before and after the change-point to estimate the quality of the algorithms. In real applications, this information as the information about the distribution of change-point appearances is often unavailable, and the modes of the primary statistics accumulations are limited [13, 14, 16, 17].

The paper is devoted to the quality estimation of the algorithm constructed on the signals ratio principle. The algorithm features are simple decision-making statistics, the possibility to detect positive changes and negative changes in the parameters of the stochastic processes, the opportunity to catch the recovering of the parameters (the returns of the statistical properties to the initial values) by the same decision-making statistics.

1.2 The Mathematical Base of the Signals Ratio Algorithm

Decision-making statistic of the change-point (CP) detection is defined as signals ratio [18]:

$$G_{n+1} = \frac{S_{n+1}}{R_{n+1}}, \tag{1}$$

where

$$S_{n+1} = (1-\alpha)S_n + \alpha e_{n+1} \tag{2}$$

$$R_{n+1} = (1-\alpha)R_n + \alpha|e_{n+1}| \tag{3}$$

$$e_{n+1} = (x_{n+1} - \mu_1) \tag{4}$$

α smoothing coefficient, $0 < \alpha < 1$

x_{n+1} is the value of the stochastic process (SP) at the time moment $(n+1)$

μ_1 is the mean value of the SP before CP.

Based on (1)–(4), the decision-making rules can be represented as follows [18]. Let the hypothesis H^1 corresponds to the case of "change-point" (H^{1+}—positive CP as "mean changes," H^{1-}—negative CP as "mean changes") and the hypothesis H^0 to its absence; the hypothesis H^3 corresponds to the case of "change-point" as "variance changes" and the hypothesis H^2 to its absence, then:

$$H^0: h_1 \leq G_{n+1} \leq h_2 \tag{5}$$

$$H^{1+}: G_{n+1} > h_2 \tag{6}$$

$$H^{1-}: G_{n+1} < h_1 \tag{7}$$

$$H^2: R_{n+1} \leq h_3 \tag{8}$$

$$H^3: R_{n+1} > h_3 \tag{9}$$

Let us consider the case when the SP is represented by the normally distributed process with parameters μ_i (SP mean) and σ_i (SP variance), where $i = 1$—parameter values SP before CP, $i = 2$—after CP. The decision-making statistic G_n is used to detect CP as "mean changes," and the decision-making statistic R_{n+1} is used to detect CP as "variance changes;" since in most cases, it is necessary to detect precisely these kinds of mismatch errors.

In the case of CP with $\mu_1 < \mu_2$, the value of the function G_{n+1} is going to 1, and while $\mu_1 > \mu_2$, the value of the function G_n is going to $-$ 1. In the case of CP is absent $M[G_{n+1}] = 0$, therefore, if we use h_1, h_2 and $-1 < h_1 \leq G_{n+1} \leq h_2 < 1$, we can detect not only the fact of CP but their character too. The additional feature of the considered decision-making rules (5)–(9) is the ability to detect CP disappearance (returning a random process to its original parameters) with the step number fixation. It is essential for the mane practical application dealing with fault detection and diagnosis.

1.3 Contribution

In the case of the decision-making statistic (1), the probability of a false detection CP can be calculated with the mathematical models already obtained in [17, 18]. The proposed [17, 18] models can be used only if the smoothing coefficient α is smaller than 0.175. The same limitation is to use the proposed mathematical models for the average number of delay steps of the change-point detection [17, 18]. The case of the statistical properties returns to the initial values was not considered. Therefore, it is vital to complete the mathematical models by new models without limitation on the value of the smoothing coefficient and by the models for the average number of delay steps of detecting the recovering of the parameters of the random processes. Due to the absence of restrictions on the range of possible values, we will have a more powerful combination of algorithm parameters for its tuning to obtain the best quality detecting.

2 Mathematical Models of the Average Time of Delays in Detection of Change-Point

Let us consider instead of signals ratio (1) the follow equation.

$$Q_{n+1}^{j} = S_{n+1} - h_j R_{n+1} \tag{10}$$

where $j = 1, 2$.

The Eq. (10) is the signal difference with the parameters $h_{j,}$ and it may be used as a decision-making statistic to detect CP as "mean changes," then taking into account (5)–(9), the decision-making rules:

$$H^0\colon \begin{cases} Q_{n+1}^1 \geq 0 \\ Q_{n+1}^2 \leq 0 \end{cases} \tag{11}$$

$$H^{1+}\colon Q_{n+1}^2 > 0 \tag{12}$$

$$H^{1-}\colon Q_{n+1}^1 < 0 \tag{13}$$

Let us consider the case of positive CP in detail:

$$Q_{n+1}^2 = S_{n+1} - h_2 R_{n+1},\ Q_{n+1}^2 > 0 \tag{14}$$

In the case of CP is absent:

$$Q_{n+1} = (1-\alpha)^n z_1 + \alpha \sum_{i=1}^{n} (1-\alpha)^{n-i} z_{i+1} \tag{15}$$

$$M\left[Q_{n+1}^2\right] = -2h_2\sigma_1/\sqrt{2\pi} \tag{16}$$

$$z_i = e_i - h_2|e_i| \tag{17}$$

$$M[z_i] = -2h_2\sigma_1/\sqrt{2\pi} \tag{18}$$

Let $t = n_0 + 1$ is the moment of CP:

$$e_n = \begin{cases} e_n\,, & n < n_0 + 1 \\ \tilde{e}_n, & n \geq n_0 + 1 \end{cases} \tag{19}$$

where $e_n \in N(0, 1)$, $\tilde{e}_n \in N(\mu_2, \sigma_2)$, then taking into account (15)–(19), we obtain the follow equations:

$$Q_{n+1} = (1-\alpha)^n z_1 + \alpha \sum_{i=1}^{n_0} (1-\alpha)^{n-i} z_{i+1} + \alpha \sum_{i=n_0+1}^{n} (1-\alpha)^{n-i} \tilde{z}_{i+1} \tag{20}$$

$$M\left[Q_{n+1}^2\right] = (1-\alpha)^{n-n_0}(M[z_i] - M\left[\tilde{z}_i\right]) + M\left[\tilde{z}_i\right] \tag{21}$$

$$z_i = e_i - h_2|e_i|, \tilde{z}_i = \tilde{e}_i - h_2|\tilde{e}_i| \tag{22}$$

For the characteristic estimation of the z_i while the CP of the sequences $\{e_i\}$, it is assumed that $\tilde{e}_n \in N(\mu_2, \sigma_1)$, where $\mu_2 = r\sigma_1$, $\mu_2 > 0$, $r = 1, 2...$—the level of the "mean changes," then we obtain.

$$M[z_i] = 2\sigma_1(r/2 - (h_2/\sqrt{2\pi})\exp(-r/2) - h_2 r\Phi(r)) \tag{23}$$

where

$$\Phi(r) = (1/\sqrt{2\pi})\int_0^r \exp(-r/2)\mathrm{d}r \tag{24}$$

The detection delay time of the algorithm depends on the following condition.

$$M[Q_{n+1}^2] = 0 \tag{25}$$

Taking into account (17), (18), (21), (23), (24) and the first term of the approximate representation of the cumulative normal distribution function

$$F(r) = 1/(\sigma_1\sqrt{2\pi})\int_0^r \exp(-(r-\mu_2)^2/2\sigma_1^2)\mathrm{d}r \tag{26}$$

and the relation $\Phi(r) = F(r) - 1/2$, we obtain.

$$M[z_i] = \sigma_1 r(1-h_2), \tag{27}$$

$$\begin{aligned} M\left[Q_{n+1}^2\right] &= (1-\alpha)^{n-n_0} W + \sigma_1 r(1-h_2) \\ W &= ((-2h_2\sigma_1/\sqrt{2\pi}) - \sigma_1 r(1-h_2)) \end{aligned} \tag{28}$$

On the base of (28), we obtain the average detection delay time in the case of "mean changes" as follow

$$\overline{N}_{\text{det CP\_r}} = \left]\frac{\ln\left[\frac{r(1-h_2)}{r(1-h_2)+2h_2/\sqrt{2\pi}}\right]}{\ln(1-\alpha)}\right[- 1, \tag{29}$$

where $]X[$—the nearest whole, more than X.

In the case of CP as "variance changes," the decision rules (8) and (9) help to detect CP. Therefore, the average detection delay time may be considered based on a modified form of the (3) obtained by analogy with the (20) and (21). We have to find the time moment when

$$M\left[R_{n+1}\right] = h_3 R_0 \tag{30}$$

As $\tilde{e}_n \in N(0, \sigma_2)$, where $\sigma_2 = (1-u)\sigma_1$, $u > 1$, $u = 2, 3...$—the level of the "variance changes," then we obtain.

$$M\left[R_{n+1}\right] = (1-\alpha)^{n-n_0}(M[|e_i|] - M\left[|\tilde{e}_i|\right]) + M\left[|\tilde{e}|_i\right] \tag{31}$$

$$M[|e_i|] = 2\sigma_1/\sqrt{2\pi} \tag{32}$$

$$R_0 = \sigma_1/\sqrt{2\pi} \tag{33}$$

$$M\left[|\tilde{e}|_i\right] = 2\sigma_1(u+1)/\sqrt{2\pi} \tag{34}$$

$$\overline{N}_{\text{det CP\_u}} = \left]\frac{\ln[(u+1-h_3)/u]}{\ln(1-\alpha)}\right[- 1 \tag{35}$$

Comparing (35) with the equation from [18] leads us to the conclusion that the proposed approach to obtain the new mathematical model for the average time of delays in detection of change-point in the case of "mean changes" (29) is correct. A mathematical model for the average time of delays in detection of change-point in the case of "variance changes" obtained in [18] by another method is the same as obtained by our approach (35).

3 Mathematical Models of the Average Time of Delays in Detection of Change-Point Disappear

In many practical applications, it is essential not only to detect the CP but also to detect the disappearances of the CP. There is a case of multiple CP and recovering of SP. The moment of the detecting the disappearances of the CP may be calculated if we take into account that

$$Q_{n+1}^2 < 0, n+1 > N_{\det \mathrm{CP\_r}} \text{ in the case of "mean changes"} \tag{36}$$

$$R_{n+1} < h_3 R_0, n+1 > N_{\det \mathrm{CP\_u}}, \text{ in the case "variance changes"} \tag{37}$$

Let $t = n_0 + 1$ is the moment of CP, and $T - 1$—is the duration of CP, then:

$$\begin{aligned} Q_{n+1}^2 &= (1-\alpha)^n(e_1 - h_2|e_1|) + \alpha \sum_{i=1}^{n_0} (1-\alpha)^{n-i}(e_i - h_2|e_i|) + \\ &\alpha \sum_{i=n_0+1}^{n_0+T-1} (1-\alpha)^{n-i}(\tilde{e}_i - h_2|\tilde{e}_i|) + \alpha \sum_{i=n_0+T}^{n} (1-\alpha)^{n-i}(e_i - h_2|e_i|), \end{aligned} \tag{38}$$

$$M\left[Q_{n+1}^2\right] = (1-\alpha)^{n-n_0}(1-(1-\alpha)^{-T})(M[z_i] - M\left[\tilde{z}_i\right]) + M\left[\tilde{z}_i\right], \tag{39}$$

$$\overline{N}_{\det \mathrm{CP\_r}} = n - n_0 - T = \frac{\ln(M[z_i]/((M[z_i] - M\left[\tilde{z}_i\right](1-(1-\alpha)^T)))}{\ln(1-\alpha)}, \tag{40}$$

where $\tilde{e}_n \in N(\mu_2, \sigma_2)$,$\mu_2 = r\sigma_1$, $z_i = e_i - h_2|e_i|$, $\tilde{z}_i = \tilde{e}_i - h_2|\tilde{e}_i|$.

Having substituted in (39) by Eqs. (15)–(18), we obtain the average time of delays in detection of change-point disappear:

$$\overline{N}_{\det \mathrm{CP\_dis\_r}} = \left]\frac{\ln\left[\frac{2h_2/\sqrt{2\pi}}{r(1-h_2)+(2h_2/\sqrt{2\pi})(1-(1-\alpha)^T)}\right]}{\ln(1-\alpha)}\right[-1. \tag{41}$$

The same transformations may be done in the case of the "variance changes" CP disappearance. With the analogy to (34) and (36), we have obtained the equation for, and for it mathematical expectation, then by substitution to (37), we have equations:

$$\overline{N}_{\det \mathrm{CP\_dis\_u}} = \left]\frac{\ln\left[(1-h_3)/(u((1-\alpha)^T - 1))\right]}{\ln(1-\alpha)}\right[-1. \tag{42}$$

4 Conclusion

Verifying the reliability of the proposed mathematical models of the evaluation of the CP detection delay was carried out by simulation with the developed simulation model. The dependence of the average CP detection delay from its appearance for different parameters (thresholds and smoothing coefficient) of the decision-making statistic at varying levels of false detection was investigated. The comparison of the proposed analytical model with the simulation dependencies demonstrated that investigated mathematical models provide the estimations of time delay for the worst case. The simulation experiments were fulfilled in a wide range of conditions and with different values of CP "mean change" and "variance change."

The analysis of the simulation experiments showed that the analytical models reflect the nature of the investigated dependences entirely. In addition, the developed models (29) and (35) provide an estimation for the worst case for CP dealing with the "changes in the mathematical expectation," and with the "changes in the dispersion," therefore, these models are applicable for tuning the algorithm based on signals ratio principle. Moreover, the mathematical models (41) and (42) estimate the average time of delays in detecting change-point disappear in the case of multiple CP in the SP.

The new mathematical models provide a more extensive combination of algorithm parameters due to the absence of restrictions on the range of possible values.

Further work is the design of a new hybrid algorithm with the combinations of the various decision-making statistics with the different principles calculation of mismatch between actually observed signals and the expected model.

References

1. Lykou, R., Tsaklidis, G., & Papadimitriou, E. (2020). Change point analysis on the Corinth Gulf (Greece) seismicity. *Physica A: Statistical Mechanics and its Applications, 541*, 123630.
2. Alaya, M. A. B., Ternynck, C., Dabo-Niang, S., Chebana, F., & Ouarda, T. B. M. J. (2020). Change point detection of flood events using a functional data framework. *Advanced Water Resources, 137*, 103522.
3. Guo, L.-P., et al. (2018). Trend and change-point analysis of streamflow and sediment discharge of the Gongshui River in China during the last 60 years. *Water, 10*.
4. Akbari, S., & Reddy, M. J. (2020). Non-stationarity analysis of flood flows using copula based change-point detection method: Application to case study of Godavari river basin. *Science of the Total Environment, 718*, 134894.
5. Getahun, Y. S., Li, M.-H., & Pun, I.-F. (2021). Trend and change-point detection analyses of rainfall and temperature over the Awash River basin of Ethiopia. *Heliyon, 7*, e08024.
6. Basseville, M., & Nikiforov, I. V. (1993). *Detection of abrupt changes: Theory and application.* Prentice Hall.
7. Pan, Y., & Zheng, Z. (2020). Bayesian online change point detection method for process monitoring. In *2020 Chinese control and decision conference (CCDC)*, 3389–3393.
8. Skaperas, S., Mamatas, L., & Chorti, A. (2020). Real-time algorithms for the detection of changes in the variance of video content popularity. *IEEE Access, 8*, 30445–30457.

9. Kalus, D., Muma, M., & Zoubir, A. M. (2015). Distributed robust change point detection for autoregressive processes with an application to distributed voice activity detection. In *2015 IEEE international conference on acoustics, speech and signal processing (ICASSP)*, pp. 3906–3910.
10. Ageev, E. V., & Benderskaya, E. N. (2014). Review of natural computing: important trends. St. Petersburg. Polytechnic University. *Journal of Computer Science Telecommunication Control System*, 9–22.
11. Benderskaya, E. N. (2014). Nonlinear dynamics as a part of soft computing systems: Novel approach to design of data mining systems. *Advances in Intelligent Systems and Computing, 287*.
12. Lu, K.-P., & Chang, S.-T. (2016). Detecting change-points for shifts in mean and variance using fuzzy classification maximum likelihood change-point algorithms. *Journal of Computational and Applied Mathematics, 308*, 447–463.
13. Alippi, C., Boracchi, G., & Roveri, M. (2017). Hierarchical change-detection tests. *IEEE Transactions of Neural Networks Learning System, 28*, 246–258.
14. Padilla, O. H. M., Athey, A., Reinhart, A., & Scott, J. G. (2019). Sequential nonparametric tests for a change in distribution: An application to detecting radiological anomalies. *Journal of American Statistical Association, 114*, 514–528.
15. Zhao, K., et al. (2019). Detecting change-point, trend, and seasonality in satellite time series data to track abrupt changes and nonlinear dynamics: A Bayesian ensemble algorithm. *Remote Sensing of Environment, 232*, 111181.
16. Amaya-Gómez, R., Bastidas-Arteaga, E., Schoefs, F., Muñoz, F., & Sánchez-Silva, M. (2020). A condition-based dynamic segmentation of large systems using a changepoints algorithm: A corroding pipeline case. *Structural Safety, 84*, 101912.
17. Benderskaya, E. N., Kolesnikov, D. N., & Pahomova, V. I. (2000). *Functional diagnostics of control systems*. SPbGTU.
18. Kalishev, O. N. (1988). A method of diagnosing measuring channels in the light of their history. *Automation and Remote Control, 49*, 795–801.

Energy Efficient Street Lighting: A GIS Approach

Kazi Amrin Kabir, Pragna Labani Sikdar, and Parag Kumar Guha Thakurta

Abstract An efficient procedure to obtain information on lighting levels, uniformity, energy consumption, and energy classes regarding street lighting through GIS is proposed in this paper. It shows the illuminated street along with its plot in terms of map data. The proposed methodology captures a nocturnal image of an illuminated street and subsequently, it determines the values of the average illuminance and the electrical power consumption on each street. The map is explored into a GIS which can provide the amount of power consumption installed along with corresponding energy classes from the street's length. This proposed work highlights on installation of illuminance of the street lights as much as needed to cover the pavement area with least cost over the existing work.

Keywords Street lighting · Energy consumption · Illuminance · Energy class · GIS

1 Introduction

The development of a smart city is used to simplify the life-cycle of the citizens in terms of various aspects. Various physical devices connected to the Internet of Things (IoT) is integrated with information and communication technology (ICT) to provide the efficient city operations and services related to the citizens [1]. The interaction between citizens and the government becomes transparent with reduced resource consumption [2]. A computer-based tool known as geographic information system (GIS) [3] is used for smart city management. A GIS allows the interactive queries from user [4]. The spatial and geographic data are stored and analyzed. Subsequently, the results of these operations in terms of maps are shared [5].

Street lighting has a significant influence on smart city as it used to public safety and security [6]. It reduces the number of accidents and the night crime. In order to prepare a report on conducting the energy audit of a smart city, GIS tool is used

K. A. Kabir (✉) · P. L. Sikdar · P. K. Guha Thakurta
Department of Computer Science and Engineering, NIT Durgapur, Durgapur, West Bengal 713209, India
e-mail: kak.20cs1108@phd.nitdgp.ac.in

B. Pati et al. (eds.), *Proceedings of the 6th International Conference on Advance Computing and Intelligent Engineering*, Lecture Notes in Networks and Systems 428,
https://doi.org/10.1007/978-981-19-2225-1_51

for mapping of the street lighting system components onto the geotagging of photo documentation [5]. This georeferenced data format can help to work on the specific street as it can show the installation of lighting levels and appropriate luminaries [7].

An efficient procedure to obtain information on lighting levels, uniformity, energy consumption, and energy classes regarding street lighting through GIS is proposed in this paper. It shows the illuminated street along with its plot in terms of map data. It ascertains the different levels in energy efficient installations of lighting in the streets incorporated into GIS. This proposed methodology captures a nocturnal image of an illuminated street and subsequently, it determines the values of the average illuminance and the electrical power consumption on each street. Next, the map is explored into a GIS which is based on that connected street of the nocturnal image captured earlier. The proposed GIS-based framework can provide the amount of power consumption installed along with corresponding energy classes from the street's length. This proposed work highlights on installation of illuminance of the street lights as much as needed to cover the pavement area with least cost over the existing work.

The organization of this paper is as follows: the literature survey is discussed in Sect. 2. System model is presented in Sect. 3. In Sect. 4, the proposed methodology is discussed. In Sect. 5, various results and its analysis are shown. The paper is concluded in Sect. 6.

2 Literature Survey

An efficient installation of street lights on a road to improve the energy efficiency is suggested from various dimensions in different works. In part-night lighting for the street lights, the costs and emissions are limited to parts of the night. Here, the lights are switched off after midnight, so that the energy consumption can improve [8]. ICT exploits the data obtained from street lights for improving government infrastructures and public welfare [9]. The article in [10] presents an intelligent lighting in a smart city. The use of LED lamps [11] is becoming more for saving cost in the natural environment [12, 13]. High-quality LED lighting improves the quality of life in the city by making the streets brighter and safer [14]. More lifetime by LED lamps over high pressure sodium (HPS) is used for energy saving in street lighting systems [15] which reduces the maintenance cost [16] accordingly. The brightness of the streetlights is efficiently adjusted by an adaptive lighting system [17]. Another work on IoT [18]-based street lighting provides energy efficiency in a city by using ICT [19]. A free and open source lighting software known as Dialux [7] is used to calculate the values of various lighting parameters for installing street light.

The work in [20] introduces a GIS approach for night-time lighting evaluation and planning. A GIS is used to capture, manipulate, store, analyze, and display all types of spatial or geographic data for designing an energy efficient street lighting [21]. In order to assess lighting standards and guidelines, the capability of GIS is discussed in [22]. By using aerial imagery of cities, the lighting levels are measured in [7]. The safety assessment related to traffic management [23] is dependent on the urbanization through an efficient artificial lightening. Hence, in order to obtain better energy efficiency in a smart city, a method for appropriate fitting of streetlights on roads through GIS is discussed in the proposed work. In this work, it has already highlighted the same patterns of luminaries can be applied on the similar kinds of roads.

3 System Model

In order to measure the lighting levels, uniformity, energy consumption, and energy classes in street lighting, a GIS-based framework is developed in this paper. Before discussing the procedure, it is necessary to introduce different lighting parameters along with other essential factors related to the proposed work. These are discussed in next.

3.1 Camera Standardization

The details of a nocturnal image captured by a digital camera are various information regarding the pixels of the image, International Organization for Standardization (ISO) value, camera' aperture, focal length, and exposure time. It is transformed into a photometer. However, the camera cannot determine luminance value. This luminance value from a color model red, green, and blue (RGB) image [7] is determined by the following.

$$L = \left(Y \times f_s^2\right)/(K \times t \times S_{\mathrm{ISO}}) \quad (1)$$

In (1), f_s denotes the camera' aperture, the exposure time (t) considered in seconds, the ISO value is S_{ISO} and K denotes the calibration constant. By [7], the value of photonic luminance (Y) is obtained as follows.

$$Y = 0.2126.\mathrm{R} + 0.7152.\mathrm{G} + 0.0722.\mathrm{B} \quad (2)$$

3.2 Important Terminologies

For ease of further discussions, the following parameters are defined.

- **Definition 1**: The brightness emitted from the light source is measured by luminous flux.

- **Definition 2**: The illuminance (E) is estimated by the received luminous flux per unit of surface and calculated in terms of lux (lx).
- **Definition 3**: The level of illuminance on a horizontal work plane is denoted by an average horizontal illuminance.
- **Definition 4**: The minimum horizontal illuminance is required to maintain throughout the entire interior region.
- **Definition 5**: It is defined as the level of illuminance on a vertical work plan is denoted by minimum vertical illuminance.
- **Definition 6**: The minimum semi-cylindrical illuminance is the average value for vertical illuminance beamed from all directions around a single point (on a vertical axis).
- **Definition 7**: Luminance (L) is the emitted luminous flux within a given angle and is measured by candela per square meter.
- **Definition 8**: The average illuminance or luminance (E_{av} or L_{av}) is defined as the minimum value of E and L, respectively. This value of E_{av} represents the average illuminance in a pavement roadway. We can also find the average luminance value for each pixel by using [7] as follows.

$$L_{av} = q_0 \times E_{av} \tag{3}$$

  In (3), the unit of L_{av} and E_{av} are considered as lx and cd/m^2, respectively. The value of average luminance coefficient (q_0) is used as 0.07 cd/m^2/lx if the reflection properties of the pavement are unknown.
- **Definition 9**: The uniformity (U_0) is defined as $U_0 = L_{min}/L_{av}$ and $U_0 = E_{min}/E_{av}$ for luminance and illuminance, respectively.

3.3 *Luminaries by Using Illuminance Classes*

The criteria for street lighting are selected depending upon the class of road being lit. A class or a sub class is dependent on the factors like users' speed, traffic volume on the road, and difficulty of the navigational task [24]. The guideline from International Commission on Illumination (CIE) is used to discriminate different road users such as pedestrians and motorists [25]. In this paper, we have focused on the pedestrian (P) class where street light becomes "ON" by sensing a pedestrian. The values of lighting parameters, such as average horizontal illuminance, minimum horizontal illuminance, and additional requirements, for P class as recommended in CIE 115:2010 [26] are used in the proposed work.

3.4 *Energy Efficient Street Lighting Installation*

From [6], the installation of energy efficient lighting is expressed by the following.

$$e(N) = \sum_{n=0}^{N} \times P_{\max} \times \phi \times T \tag{4}$$

In (4), after N discrete timesteps, $e(N)$ denotes the energy consumed by streetlight, where, N and n denote the upper limit and the lower limit of discrete-timesteps. The $P_{\max}$ denotes the maximum rated power of streetlight. The duration of illuminance output is denoted by the product $(\phi \times T)$ for timestep n.

3.5 *Energy Classes for Street Lighting Installation*

Depending on the lamp' efficiency, the energy classification for installing the street lights is used by a criterion (*SLEEC*) [7]. The *SLEEC* indicator as power density indicator $[D_p]$ is a photometric measurement in terms of illuminance or luminance as follows.

$$S_x = 1/e(N) \tag{5}$$

In (5), the energy efficiency $e(N)$ is required for the street lighting installation; and depending on E or L, the type of luminous parameter is denoted by X. We can also define S_x as D_p. The classification of energy efficiency from energy class A–G [7] as per S_x is used for the proposed work.

3.6 *QGIS Software*

In our work, the QGIS software [6] is utilized to analyze and edit the spatial information of a map. The different layers in terms of georeferenced aspects are shown for associated street lighting installation. As a result, we can estimate the energy efficiency corresponding to the energy classes which is consumed by luminaries.

4 Proposed Methodology

In order to obtain information on lighting levels, uniformity, energy consumption, and energy classes regarding street lighting, an efficient GIS-based framework is proposed here. The illuminated street along with its plot in terms of map data can be shown by this work. Initially, a nocturnal image of an illuminated street is captured by using a digital camera with its pixels, ISO value, and aperture. Here, the illumination of a larger road area corresponds to a single image acquired by digital camera can be analyzed by its both quality and quantity. Next, the average illuminance value and

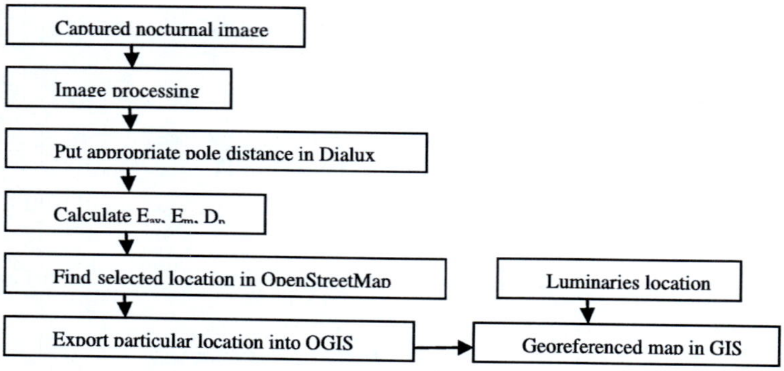

Fig. 1 Flowchart of the proposed methodology

minimum illuminance value by using appropriate pole distance are determined in Dialux software. Subsequently, the electrical power consumption on each street and the corresponding energy classes depending on the length of the road or the pavement area of the road are determined also. Afterward, a map of the captured illuminated street from OpenStreetMap is obtained and then, the selected area is exported into QGIS succeeding georeferenced the map image. It exhibits the captured area of the image corresponding with the connected roads. Subsequently, the luminaire locations, power consumption of the lighting systems, and energy classes of these road sections are incorporated into GIS. The entire procedure of the proposed GIS-based framework is described by the flow chart shown in Fig. 1.

5 Results and Analysis

The street lighting installations on a road with appropriate power LED bulbs are focused and improved their energy efficiency by efficient utilization of energy. For this purpose, an area name Golapbag near to the Burdwan University in Burdwan, West Bengal, India (Latitude: 23.45301, Longitude: 87.95956) is selected. Hence, a nocturnal image of this area as shown in Fig. 2 is captured by a digital camera having 4000 × 3000 Pixels, ISO 3062, f/1.7 aperture, 4.73 mm focal length, and 1/17 s exposure time. As per IRC [16] specification, we can know for carriageway breadth, the highest permissible vehicle breadth is 2.44 m, and the suitable side clearance for single lane road is 0.68 m on both sides. The double lane with no kerbs on sides is 7 m.

The map, of that nocturnal street and its corresponding connected streets are obtained from www.openstreetmap.org, then exports it into the QGIS. Figure 3a shows the map obtained from OpenStreetMap as well as incorporated into QGIS in Fig. 3b.

Fig. 2 Captured nocturnal image

In our proposed work, the Dialux is used for lighting design to carry out a normative road calculation with all the features like uniformity, luminance value, threshold values of illuminance, etc. Hence, in the result, we first work for pedestrian with the P1 class. Here, the two-way street has 7 m width and has 30 m distance between two consecutive light poles. For the simulation purpose, we have used 103 W LED light. In this calculation, the values of the parameters like threshold value, average illuminance (E_{av}), minimum illuminance (E_{min}), and the energy consumption are obtained as 8, 21.95 lx, 11.92, and 412 kW/year, respectively. In order to make a comparison, we have used same power in a different class *P*2 that does not satisfy with the same pole distance. Here, if we consider the pole distance as 44 m, then we can obtain the threshold value is 11, the average illuminance (E_{av}), minimum illuminance (E_{min}), and the energy consumption are 14.96 lx, 2.53 lx, and 412 kW/year, respectively. The captured area in the georeferenced format based on these parameter values are shown in Fig. 4a, b. By this GIS-based framework, the street lights of a particular street are highlighted which is our main interest to work.

Table 1 shows the pavement area shown in Fig. 4 with illuminance value in lux and D_p value also. In Table 1, the energy classes are determined also. From Table 1, it is observed that the top and bottom streets are parallel which are numbered as 1 and 4. Similarly, other two parallel streets numbered as 2 and 3 are positioned in left and right. Street number 5 is a straight street. Hence, the pavement areas of the streets 2 and 3 are greater than others as the lengths of those streets are longer. So, the LED lamp with 103 W power is suitable for those streets. Similarly, 51 and 88 W LED lamps are suitable for the parallel streets 1 and 4 as well straight street 5, respectively. This proposed work highlights on installation of illuminance of the street lights as much as needed to cover the pavement area with least cost which is beyond scope of the existing work [7]. The parameters necessary to install a streetlight are satisfied in the range of Dialux software. An effort has been taken to fit the similar pattern of street lights for both the parallel streets.

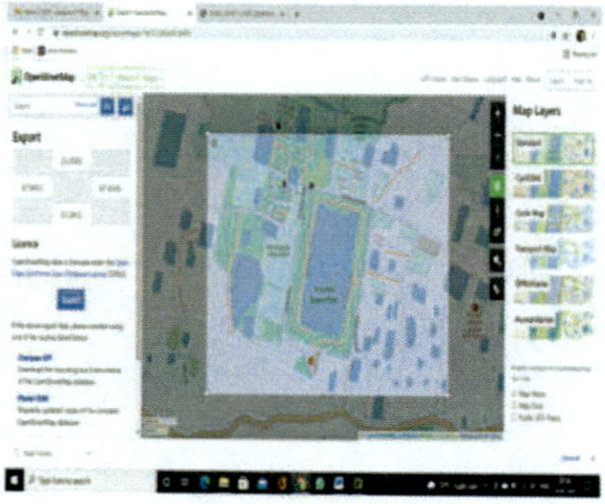

(a) Map from OpenStreetMap

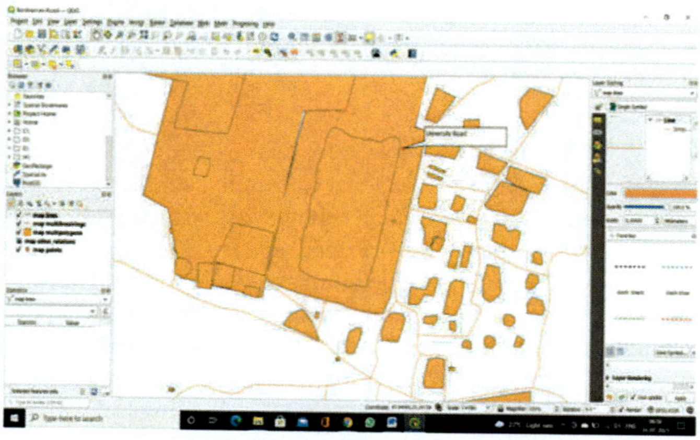

(b)Export the selected area into the QGIS

Fig. 3 **a** Map from OpenStreetMap. **b** Export the selected area into the QGIS

6 Conclusions

A GIS-based framework in this paper is proposed to achieve energy efficient street lighting. The spatial data related to the street lighting is gathered and analyzed for the infrastructural development of a smart city. The calibration of camera is used to estimate the parameters of an image. Appropriate luminaries by using illuminance classes and the pedestrian class have been considered here. The energy efficiency of street light installation has been highlighted in terms of SLEEC indicator. In order to install street lights, the associated layers in terms of georeferenced attributes are shown by QGIS software. It can suggest the power requirement of the LED lamp according to the street conditions. In the future, a further enhancement of proposed work can be done by suitable machine learning algorithm as a powerful tool for increasing the quality of the urban environment.

(a) Aerial image

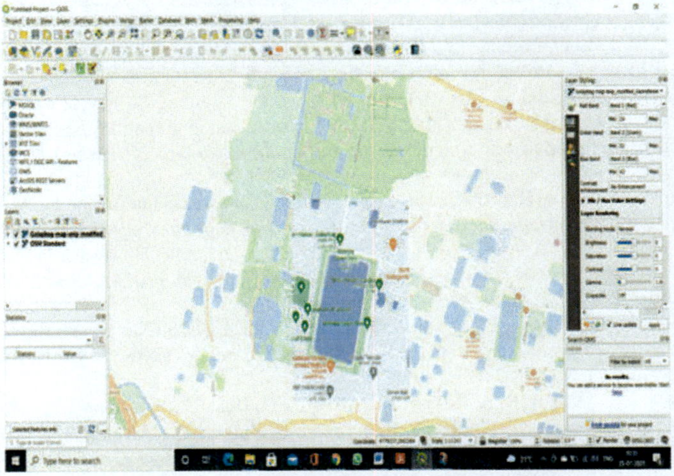

(b) Georeferenced view

Fig. 4 **a** Aerial image. **b** Georeferenced view.

Table 1 Lighting parameter values with energy class for street light installation obtained by the proposed work

Street	Power (W)	Pavement area (m^2)	Average illuminance E_{ave} (lux)	D_p (W/lx/m^2)	Energy class	Consumption (kWh/year)
1,4 (parallel top–bottom)	51 W	168	15	0.020	B	204
2,3 (parallel left–right)	103 W	210	21.95	0.022	B	412
5 (straight)	88 W	196	22.33	0.020	B	352

References

1. Cohen, B. (2015). *The 3 generations of smart cities. Inside the development of the technology driven city.*
2. Peris-Ortiz, M., Bennett, D. R., Yábar, Diana, P.-B. *Sustainable smart cities: Creating spaces for technological, social and business development*. Springer. ISBN: 9783319408958.
3. Clarke, K. C. (1986). Advances in geographic information systems. *Computers, Environment and Urban Systems, 10*(3–4), 175–184.
4. Pati, B., Panigrahi, C. R., & Sarkar, J. L. (2018). CETM: A conflict-free energy efficient transmission policy in mobile cloud computing. *International Journal of Communication Networks and Distributed System, 20*(2), 129–142.
5. Maliene, V., Grigonis, V., Palevičius, V., & Griffiths, S. (2011). Geographic information system: Old principles with new capabilities. *Urban Design International, 16*(1), 1–6.
6. Dizon, E., & Pranggono, B. (2021). Smart streetlights in smart city: A case study of Sheffield. *Journal of Ambient Intelligence and Humanized Computing, 24*, 1–6.
7. Rabaza, O., Gómez-Lorente, D., Pérez-Ocón, F., & Peña-García, A. (2016). A simple and accurate model for the design of public lighting with energy efficiency functions based on regression analysis. *Energy, 15*(107), 831–842.
8. Panigrahi, C. R., Sarkar, J. L., & Pati, B. (2018). Transmission in mobile cloudlet systems with intermittent connectivity in emergency areas. *Digital Communications and Networks, 4*(1), 69–75.
9. Wüller, D., & Gabele, H. (2007). The usage of digital cameras as luminance meters. In *Digital photography III. International society for optics and photonics*, 2007 Feb 20 (Vol. 6502, p. 65020U).
10. Zanella, A., Bui, N., Castellani, A., Vangelista, L., & Zorzi, M. (2014). Internet of things for smart cities. *IEEE Internet of Things Journal, 1*(1), 22–32.
11. Bachanek, K. H., Tundys, B., Wiśniewski, T., Puzio, E., & Maroušková, A. (2021). Intelligent street lighting in a smart city concepts—A direction to energy saving in cities: An overview and case study. *Energies, 14*(11), 3018.
12. Panigrahi, C. R., Verma, R. K., Sarkar, J. L., & Pati, B. (2018). Energy-aware issues for handling big data in mobile cloud computing. In G. Skourletopoulos, G. Mastorakis, C. Mavromoustakis, C. Dobre & E. Pallis (Eds.), *Mobile big data. Lecture notes on data engineering and communications technologies* (Vol. 10).
13. Fryc, I., Bisegna, F., & Tabaka, P. (2017). Lighting of recreation grounds as a source of sky glow—The influence of luminaire type on this phenomenon. In *2017 IEEE international conference on environment and electrical engineering and 2017 IEEE industrial and commercial power systems Europe (EEEIC/I&CPS Europe)*, 2017 Jun 6 (pp. 1–4). IEEE.
14. Panigrahi, C.R., Pati, M.T.B., Sarkar, J.L. (2015). EEOA: Improving energy efficiency of mobile cloudlets using efficient offloading approach. In *Proceedings of 9th IEEE international conference on advanced networks and telecommunications systems (IEEE ANTS)* (pp. 1–6).
15. Czyżewski, D. (2010.)Monitoring of the lighting conditions of a street illuminated with road lights equipped with LEDs. *Przegląd Elektrotechniczny, 86*(10), 170–172.
16. Leccese, F., Salvadori, G., & Rocca, M. (2017). Critical analysis of the energy performance indicators for road lighting systems in historical towns of central Italy. *Energy, 1*(138), 616–628.
17. Pati, B., Sarkar, J. L., & Panigrahi, C. R. (2017). ECS: An energy-efficient approach to select cluster-head in wireless sensor networks. *Arabian Journal for Science and Engineering, 42*, 669–676.
18. Lau, S. P., Merrett, G. V., Weddell, A. S., & White, N. M. (2015). A traffic-aware street lighting scheme for smart cities using autonomous networked sensors. *Computers & Electrical Engineering, 1*(45), 192–207.
19. Talari, S., Shafie-Khah, M., Siano, P., Loia, V., Tommasetti, A., & Catalão, J. P. (2017). A review of smart cities based on the internet of things concept. *Energies, 10*(4), 421.
20. Feng, X., & Murray, A. T. (2018). Spatial analytics for enhancing street light coverage of public spaces. *Leukos, 14*(1), 13–23.

21. Panigrahi, C. R., Sarkar, J. L., Pati, B., & Bakshi, S. (2016). E3M: An energy efficient emergency management system using mobile cloud computing. In *IEEE international conference on advanced networks and telecommunications systems, ANTS'16* (pp. 1–6).
22. Murray, A. T., & Feng, X. (2016). Public street lighting service standard assessment and achievement. *Socio-Economic Planning Sciences, 1*(53), 14–22.
23. Tagliabue, L. C., Cecconi, F. R., Moretti, N., Rinaldi, S., Bellagente, P., & Ciribini, A. L. C. (2020). Security assessment of urban areas through a GIS-based analysis of lighting data generated by IoT sensors. *Applied Sciences, 10*(6), 2174. https://doi.org/10.3390/app10062174
24. Panigrahi, C. R., Sarkar, J. L., Pati, B., & Verma, R. K. (2016). CQS: A conflict-free query scheduling approach in wireless sensor networks. In *3rd international conference on recent advances in information technology (RAIT)* (pp. 13–18).
25. Fotios, S. (2019). *A review of design recommendations for P-class road lighting in European and CIE documents—Part 1: Parameters for choosing a lighting class.* Oct 4, 2019.
26. Commission Internationale de l'Éclairage. (2010). Lighting of roads for motor and pedestrian traffic. CIE 115:2010. Vienna: CIE.

Algorithms and Emerging Computing

ANFIS Coupled Genetic Algorithm Modelling for MIMO Optimization of Flat Plate Heat Sink

S. Prasanna Devi, S. Manivannan, and J. Arunnehru

Abstract This paper introduces a new hybrid approach that uses the output of Taguchi design of experiment (DOE) matrix to train an adaptive neuro-fuzzy inference system (ANFIS) model, and the rules of the ANFIS model are used to perform multiobjective optimization. The proposed approach is applied to optimize the geometry parameters of a flat plate heat sink such as the heat sink is length, width, fin height, base height, fin thickness, and number of fins and the multiobjective functions are minimization of the thermal resistance and emitted radiations of the heat sink. Also, the trained ANFIS model is used to predict the performance of the multiple outputs given the combination of input parameters. The multiple responses such as emitted radiations and thermal resistance are optimized by 22.18% with respect to the original design. Also, this method serves to reduce the search space for a given problem. Using the reduced search space, genetic algorithm (GA) was used at the second level of optimization to further improve the performance by upto 34.12% compared to the original design. Thus, the proposed Taguchi-based ANFIS modelling can either be used as a simple and standalone approach for optimization or can be combined with GA in the next level for any kind of MIMO parameter optimization problems.

Keywords Taguchi-based ANFIS modelling · DOE · ANFIS · Flat plate heat sink · Genetic algorithm · Multiobjective optimization · Prediction · Emitted radiations · Thermal resistance · ANFIS with GA · Geometry optimization · MIMO optimization

S. P. Devi (✉) · J. Arunnehru
Department of Computer Science and Engineering, SRM Institute of Science and Technology, Vadapalani, Chennai 26, Tamilnadu, India
e-mail: prasannasiva11@gmail.com

S. Manivannan
Department of Electrical and Electronics Engineering, Dr MGR Educational and Research Institute University, Chennai 95, Tamil Nadu, India
e-mail: manivannan.s@drmgrdu.ac.in

B. Pati et al. (eds.), *Proceedings of the 6th International Conference on Advance Computing and Intelligent Engineering*, Lecture Notes in Networks and Systems 428,
https://doi.org/10.1007/978-981-19-2225-1_52

NOMENCLATURE

A	Fin height, mm
B	Base height (or) base thickness of the heat sink, mm
D	Fin thickness, mm
L	Heat sink length, mm
N	Number of fins
R_{sin}	Thermal resistance of the heat sink, 0 K/W
W	Heat sink width, mm

1 Introduction

Taguchi method (TM) is used to get complete information through a few. It is widely used for experiments to identify optimal process parameters within the given search area [1], most of the parameter optimization models. A comprehensive survey paper on the applications of Taguchi and grey—Taguchi methods for the parameter optimization methods for various applications was undergone in [2].

1.1 Motivation

Gupta et al. [3] proposed a robust approach by integrating a fuzzy logic system with a Taguchi method to identify optimal machining constraints for CNC turning section. Inturn fuzzy-based TMs help in developing the idea behind product design [3]. Ho et al. [4] enhanced the prediction of ANFIS by proposing a hybrid Taguchi-genetic learning algorithm (HTGLA) based on surface roughness concept by implementing end milling technique. The suggested HTGLA determines the most appropriate membership functions when applying to ANFIS. Our proposed methodology applies the concept in vice versa. That is, Taguchi-based ANFIS output is applied to genetic algorithm. A similar model is also used for the team-level service climate in GSD projects by [5]. In our proposed method, the vice versa is used. That is, Taguchi-based ANFIS output is applied to genetic algorithm. Sutono et al. [6] suggested a fuzzy-based Taguchi method to evaluate the design parameters of a car using an optimal combination of users perceptron, where the inputs are judgemental parameters and not quantitative parameters. Since ANFIS modelling is a popular model to predict the input–output relationship between variables, many researchers have used this method for predicting responses like [7].

1.2 Contribution

In this research, we have extended the research of [8], here the optimization technique involves multiobjective flat plate heat sink with the concepts of Taguchi parametric design using experiments, and the responses were predicted using multiple linear regression analysis (MLRA). The input factors for optimization are the length and width of the heat sink, the height of the fin, the height of the base, the number of fins and the thickness of the fin along with their chosen levels, as shown in Table 1 are used in our study. The six input parameters considered in this research help to generate the L27 Taguchi orthogonal array along with the responses obtained for thermal resistance, and emitted radiations as shown in Table 2. The problem for optimizing heat sink is used as an example to illustrate the proposed solution. The original heat sink had the design parameters as given in Table 1. To optimize it, [9] used Taguchi-based grey relational analysis (GRA). The Taguchi responses, electromagnetic radiations were obtained by Flotherm software which helps to simulate the Ansoft High-Frequency structure Simulator (HFSS) software and obtain the thermal resistance Tables 2 and 3 are taken from [8] for our research. In this paper, Sect. 2 illustrates fuzzy modelling of the traditional DOE, Sect. 3 explains ANFIS prediction, and Sect. 4 explains genetic algorithm for optimization. Section 5 describes results and discussion, and Sect. 6 describes conclusion and future work.

Table 1 Design parameters of the original heat sink

Method	Input parameters							Multiple responses studied	
	L	W	a	b	D	N	Response evaluation method	Radiations emitted	$R_{\sin}$
Actual design	90	70	30	6	1.2	17	Simulation using HFSS and Ansoft	16.12	0.2368

Table 2 Taguchi DOE Levels

Design parameters	Level 1 (low)	Level 2 (medium)	Level 3 (high)
Length of the heat sink (L) mm	70	80	90
Width of the heat sink (W) mm	70	80	90
Fin height (a) mm	10	20	30
Base height (b) mm	4	6	8
Fin thickness (d) mm	0.8	1	1.2
No. of fins (N)	10	20	30

Table 3 Responses obtained for L27 array

Exp. No.	Multiple inputs						Multiple outputs	
	L	*W*	*A*	*B*	*D*	*N*	Radiations emitted	$R_{\sin}$
1	70	70	10	4	0.8	10	11.303	0.88197
2	70	70	10	4	1	20	11.379	0.43256
3	70	70	10	4	1.2	30	12.93	0.29334
4	70	80	20	6	0.8	10	12.144	0.70237
5	70	80	20	6	1	20	11.94	0.32302
6	70	80	20	6	1.2	30	13.371	0.2122
7	70	90	30	8	0.8	10	14.852	0.64549
8	70	90	30	8	1	20	15.364	0.29914
9	70	90	30	8	1.2	30	15.702	0.19126
10	80	70	20	8	0.8	20	14.207	0.26968
11	80	70	20	8	1	30	14.525	0.17947
12	80	70	20	8	1.2	10	14.43	0.55137
13	80	80	30	4	0.8	20	12.626	0.27109
14	80	80	30	4	1	30	13.327	0.17736
15	80	80	30	4	1.2	10	12.122	0.52874
16	80	90	10	6	0.8	20	15.29	0.44891
17	80	90	10	6	1	30	15.65	0.29952
18	80	90	10	6	1.2	10	15.02	0.81851
19	90	70	30	6	0.8	30	16.716	0.16362
20	90	70	30	6	1	10	15.738	0.45558
21	90	70	30	6	1.2	20	16.406	0.20293
22	90	80	10	8	0.8	30	15.96	0.24145
23	90	80	10	8	1	10	15.02	0.65008
24	90	80	10	8	1.2	20	15.73	0.65009
25	90	90	20	4	0.8	30	15.746	0.21291
26	90	90	20	4	1	10	15.06	0.59193
27	90	90	20	4	1.2	20	15.47	0.28774

2 Fuzzy Modelling of Taguchi DOE

The design parameters of L27 matrix are converted into fuzzy levels using the triangular membership function as shown in Table 4, and fuzzification of the two output responses of the L27 array is also done as shown in Table 5. The triangular membership function was chosen since it is commonly used in practical applications and simplifies the computational process [10].

From the fuzzified input/output ranges of the Taguchi L27 matrix, the ANFIS model is created using Mamdani architecture as shown in Fig. 1 and the corresponding rules generated are listed in Table 6. To construct ANFIS model, the geometry

Table 4 Fuzzification of input parameters

Design parameter	Level 1 (low)	Level 2 (medium)	Level 3 (high)
Length of the heat sink (*L*) mm	65–75	75–85	85–95
Width of the heat sink (*W*) mm	65–75	75–85	85–95
No. of fins (*N*)	5–15	15–25	25–35
Fin height (*a*) mm	5–15	15–25	25–35
Base height (*b*) mm	3–5	5–7	7–9
Fin thickness (*d*) mm	0.6-.09	0.9-1.2	1.2-1.5

Table 5 Fuzzification of output responses

Output parameters	OBJ	Level 1 (low)	Level 2 (medium)	Level 3 (high)
$R_{\sin}$	Min	0.1–0.3	0.3–0.6	0.6–0.9
Emitted radiations	Min	11–13	13–15	15–17

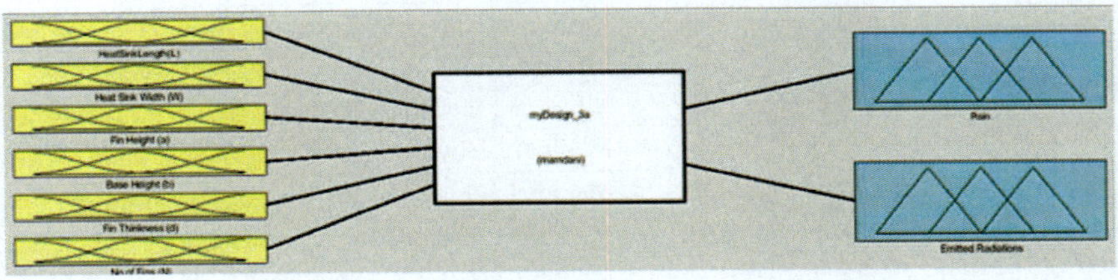

Fig. 1 Architecture of ANFIS

parameters of heat sink like its length, width, fin height, base height, fin thickness and number of fins were considered as inputs, and thermal resistance and emitted radiations were considered as outputs data for the modelling of growth factors. The triangular membership function was considered for both input and output variables with the centroid of the function to be the same as the level of the factor specified in Table 2.

The rules of the L27 array are interpreted as follows: If input1 = $x1$ & input2 = $x2$ & ... & input6 = $x6$, then output1 = $y1$ & output2 = $y2$.

Since we intend to do a multiobjective optimization function where both the output columns are minimization of thermal resistance and emitted radiations, respectively, we look for a rule that achieves this objective. In this way out of the L27 experiments. Thus, the experiments corresponding to rule 3 and rule 13 combination yield the expected output. Hence, the inputs corresponding to those experiments are chosen as optimal input parameters. To verify this, the rule 3 is mapped to Table 3 and the experiment 3 is chosen as the best optimal result. The input/output level of experiment 3 is replaced with fuzzy values set in Tables 4 and 5, respectively. The resultant

Table 6 Fuzzy rules of the L27 array

SI	Inputs						Outputs	
	L	*W*	*a*	*b*	*D*	*N*	R_{sin}	Emitted radia-tions
1	Low	Low	Low	Low	Low	Low	High	Low
2	Low	Low	Low	Low	Medium	Medium	Medium	Low
3	Low	Low	Low	Low	High	High	Low	Low
4	Low	Medium	Medium	Medium	Low	Low	High	Low
5	Low	Medium	Medium	Medium	Medium	Medium	Medium	Low
6	Low	Medium	Medium	Medium	High	High	Low	Medium
7	Low	High	High	High	Low	Low	High	Medium
8	Low	High	High	High	Medium	Medium	Low	High
9	Low	High	High	High	High	High	Low	High
10	Medium	Low	Medium	High	Low	Medium	Low	Medium
11	Medium	Low	Medium	High	Medium	High	Low	Medium
12	Medium	Low	Medium	High	High	Low	Medium	Medium
13	Medium	Medium	High	Low	Low	Medium	Low	Low
14	Medium	Medium	High	Low	Medium	High	Low	Medium
15	Medium	Medium	High	Low	High	LowLow	Medium	Low
16	Medium	High	Low	Medium	Low	Medium	Medium	High
17	Medium	High	Low	Medium	Medium	High	Low	High
18	Medium	High	Low	Medium	High	Low	High	High
19	High	Low	High	Medium	Low	High	Low	High
20	High	Low	High	Medium	Medium	Low	Medium	High
21	High	Low	High	Medium	High	Medium	Low	High
22	High	Medium	Low	High	Low	High	Low	High
23	High	Medium	Low	High	Medium	Low	High	High
24	High	Medium	Low	High	High	Medium	High	High
25	High	High	Medium	Low	Low	High	Low	High
26	High	High	Medium	Low	Medium	Low	Medium	High
27	High	High	Medium	Low	High	Medium	Low	High

MIMO optimization output is given in Table 7, and the input range is replaced with the centroid value of the input. The corresponding output is calculated using the predicted values of the model as described in the next section.

Table 7 MIMO optimization for fuzzy input

Design parameters	L	W	a	b	d	N	R_{sin}	Emitted raditions
Optimized rule 3	Medium	Medium	High	Low	Low	Medium	Low	Low
Range of fuzzy inputs	75–85	75–85	25–35	3–5	0.6–0.9	15–25	–	–
Centroid of inputs	80	80	30	4	0.75	20	-	–

3 ANFIS Prediction of MIMO

In this paper, ANFIS model developed is not only used to derive an optimal output, but also used for prediction of the multiple responses. The results of the ANFIS prediction for the L27 array shown in Table 3 are shown in Table 8, and the MAPE of actual vs predicted values of the emitted radiations and thermal resistance are 0.034058 and 0.193617, respectively. This error is better than the MAPE of MLRA prediction reported by [8].

The proposed MIMO ANFIS model is considered to be an effective approach for both optimization and prediction of geometry parameters. Using this prediction model, the output of the optimized input specified in Table 5 is calculated and is given in Table 9.

If equal weightage is given to each of the chosen objective functions, the net function is formulated as given below and the improvement in performance obtained by the proposed approach is approximately 22.18%. The performance comparison of the Taguchi-based ANFIS model with the actual design proposed by [8] is shown in Table 10.

$$Z = 0.5 \times y(1) + 0.5 \times y(2) \tag{1}$$

where emitted radiations, $y(1) = -5.17 + 0.127 \times L + 0.0584 \times W + 0.0254 \times a + 0.440 \times b + 0.649 \times d + 0.0458 \times N$,
R_{sin}, $y(2) = 1.13 - 0.00292 \times L + 0.00203 \times W - 0.00990 \times a + 0.0000 \times b - 0.028 \times d - 0.0214 \times N$
where Z denotes the weighted average of the two responses studied. R_{sin} and emitted radiations are taken from the linear regression equation given by [8].

Table 8 Predicted values of responses for the L27 array using ANFIS prediction

Exp. No.	Emitted radiations	R_{sin}
1.	12.456	0.6895
2.	12.456	0.4798
3.	12.461	0.2742
4.	12.456	0.6895
5.	12.456	0.4798
6.	13.999	0.2742
7.	14	0.6889
8.	15.511	0.2732
9.	15.511	0.2735
10.	14	0.2729
11.	14	0.2731
12.	14	0.48
13.	12.456	0.2722
14.	13.999	0.2731
15.	12.461	0.48
16.	15.511	0.48
17.	15.511	0.2739
18.	15.511	0.6887
19.	15.511	0.2731
20.	15.511	0.4799
21.	15.511	0.2735
22.	15.511	0.2731
23.	15.511	0.6891
24.	15.511	0.6887
25.	15.511	0.2732
26.	15.511	0.48
27.	15.511	0.2735

Table 9 ANFIS prediction of multiple outputs for the given input

Design parameters	L	W	a	b	d	N	Predicted R_{sin}	Predicted emitted radiations
Centroid of inputs	80	80	30	4	0.75	20	0.2722	12.4558

Table 10 Performance evaluation of Taguchi-based ANFIS modelling

Method	Inputs						Output using analytical model		Z
	L	*W*	*a*	*B*	*D*	*N*	Emitted radiations	R_{sin}	
Actual design [8]	90	70	30	6	1.2	17	16.12	0.2368	8.1784
Taguchi-based ANFIS modelling approach	80	80	30	4	0.75	20	12.4558	0.2722	6.364

4 Multilevel Optimization Using Genetic Algorithm

Second-level optimization was performed using genetic algorithm for the reduced search space obtained in Table 5. The objective function used is shown in Eq. 1, and the results of GA optimization with the constraints as the limits for inputs are given in Table 5. A genetic algorithm is predominantly used for global optimization in parameter optimization problems like optimization of bend stress proposed in [11], and Taguchi method hybridized with GA for the optimization of heat sink geometry is reported in [12]. The GA tuning parameters used in [12] such as crossover rate = 0.85, choice of roulette wheel crossover mechanism and mutation rate = 0.1 are chosen for the restricted space GA implementation. In GA, initial population is selected randomly, whereas our proposed approach has given a combination of variables for a given search space. Hence, instead of selecting the initial population at random for GA, the search space is restricted by the proposed approach. Table 11 illustrates the results of multilevel optimization which combines the results of ANFIS to GA.

5 Results and Discussions

It is observed from Table 12 that the proposed approach of Taguchi-based ANFIS modelling coupled with GA produces a performance improvement of 34.12% compared to the original design. It produces significant performance improvement of 10.64% compared to the results produced by Taguchi-based GRA. The methods compared for the study of multiobjective optimization are ranked based on their performance (*Z*) produced, and it is found that the proposed approach is an effective multiobjective optimization problem compared to those papers that were reported in the literature so far for the MIMO multiobjective optimization of a flat plate heat sink. It is a well-known challenge that searching for an optimal value in a huge variable space with acceptable solution quality is very high. Rather, if the search space is

Table 11 Multilevel optimization using GA

Method	Optimal input combinations						Response evaluation method	Multiple responses studied		Z	Rank
	L	W	A	b	D	N		Radiations emitted	R_{sin}		
Solution using GA for the problem [9]	70.03	78.05	21.4	4.16	1.13	27	Unconstrained search space	12.61	0.263	6.4365	4
Taguchi-based ANFIS modelling + GA	65	65	27	3	0.645	28	Restricted space search	10.5884	0.1876	5.388	1

Table 12 Comparison of results

Method	Optimal input combinations						Response evaluation method	Multiple responses studied		Z	Rank
	L	W	a	b	D	N		Radiations emitted	R_{sin}		
Actual design	90	70	30	6	1.2	17	Simulation	16.12	0.2368	8.1784	5
Taguchi GRA [9]	70	90	30	4	0.8	30	MLRA	11.9	0.19	6.02965	2
Taguchi-based ANFIS modelling	80	80	30	4	0.75	20	ANFIS	12.4558	0.2722	6.364	3
Taguchi-based ANFIS modelling + GA	65	65	27	3	0.645	28	Restricted space search	10.5884	0.1876	5.388	1
Solution using GA for the problem [9]	70.03	78.05	21.4	4.16	1.13	27	Unconstrained search space	12.61	0.263	6.4365	4

reduced by means of some search reduction technique, it would ease the process of moving towards the best optimal solution. Hence, we have used the ANFIS algorithm as a search reduction technique, and further as a next level of optimization, GA is used to obtain the optimal solution within the restricted solution space. The performance of the proposed algorithm was tested on the heat sink geometry optimization problem, and the results shown in Table 12 show that reduced solution space not only improves quality of solution but also reduces the processing time of the search algorithm.

6 Conclusion and Future Work

An hybrid optimization model is proposed in this paper. When limited time is available, the Taguchi experiments are simulated and modelled with ANFIS to optimize the MIMO parameter optimization problem. Further, to enhance the results of Taguchi-based ANFIS modelling, a second-level optimization is performed using genetic algorithm to obtain the global optimized multiple responses with the constrained input space given by the Taguchi-based ANFIS modelling approach. The proposed method can be applied for any MIMO parameter optimization problem.

References

1. Ross, P. J., & Ross, P. J. (1988). *Taguchi techniques for quality engineering: loss function, orthogonal experiments, parameter and tolerance design*. Number TS156 R12. McGraw-Hill.
2. Rajarasalnath, S., Balasubramanian, K., & Rajeswari, N. (2015). Multi-objective optimization problems in taguchi parameter design–A literature review. *Applied Mechanics and Materials, 813*, 1188–1192.
3. Gupta, A., Singh, H., & Aggarwal, A. (2011). Taguchi-fuzzy multi output optimization (MOO) in high speed cnc turning of aisi p-20 tool steel. *Expert Systems with Applications, 38*(6), 6822–6828.
4. Ho, W.-H., Tsai, J.-T., Lin, B.-T., & Chou, J.-H. (2009). Adaptive network-based fuzzy inference system for prediction of surface roughness in end milling process using hybrid taguchi-genetic learning algorithm. *Expert Systems with Applications, 36*(2), 3216–3222.
5. Sangaiah, A. K., Thangavelu, A. K., Gao, X. Z., Anbazhagan, N., & Saleem Durai, M. (2015). An anfis approach for evaluation of team-level service climate in gsd projects using taguchi-genetic learning algorithm. *Applied Soft Computing, 30*, 628–635.
6. Sutono, S. B., Abdul-Rashid, S. H., Aoyama, H., & Taha, Z. (2016). Fuzzy-based taguchi method for multi-response optimization of product form design in kansei engineering: a case study on car form design. *Journal of Advanced Mechanical Design, Systems, and Manufacturing, 10*(9), JAMDSM0108–JAMDSM0108.
7. Shivakoti, I., Kibria, G., Pradhan, P. M., Pradhan, B. B., & Sharma, A. (2019). Anfis based prediction and parametric analysis during turning operation of stainless steel 202. *Materials and Manufacturing Processes, 34*(1), 112–121.
8. Manivannan, S., Arumugam, R., Paramasivam, S., Salil, P., & Subba Rao, B. (2010). HFSS simulation, experimental investigation and optimisation of heat sink EMI. *IET Power Electronics, 3*(6), 881–891.

9. Manivannan, S., Prasanna Devi, S., Arumugam, R., & Sudharsan, N. M. (2011). Multi-objective optimization of flat plate heat sink using taguchi-based grey relational analysis. *The International Journal of Advanced Manufacturing Technology, 52*(5–8), 739–749.
10. Yeh, C.-H., & Deng, H. (2004). A practical approach to fuzzy utilities comparison in fuzzy multicriteria analysis. *International Journal of Approximate Reasoning, 35*(2), 179–194.
11. Vaghela, P. A., & Prajapati, J. M. (2019). Hybridization of taguchi and genetic algorithm to minimize iteration for optimization of solution. *MethodsX, 6*, 230–238.
12. Manivannan, S., Prasanna Devi, S., & Arumugam, R. (2011). Optimization of flat plate heat sink using genetic algorithm. In *2011 1st international conference on electrical energy systems* (pp. 78–81). IEEE.

Solving Fuzzy Quadratic Programming Problems by Fuzzy Neural Network

G. Selvaraj and L. Jarina Banu

Abstract A new fuzzy energy function for fuzzy quadratic programming problems is constructed using fuzzy norm. Based on the fuzzy energy function, a new fuzzy neural network is developed for solving fuzzy quadratic programming problems numerically in which all or some parameters are fuzzy. The stability of the proposed fuzzy neural network is established, and numerical examples are demonstrated to substantiate the significance of the proposed fuzzy neural network.

Keywords Fuzzy quadratic programming problem · Fuzzy neural network · Approximate optimum solution.

1 Introduction

An optimization problem is a mathematical programming problem in which a function of several variables, called objective function is to be optimized subject to constraints of these variables. It has many applications in the fields of engineering, technology, applied sciences and management sciences. Most of the optimization problems [1] are solved numerically by classical iterative methods. The iterative procedure of a classical method is required more long computation time for solving optimization problems. Alternatively, an artificial neural network or a neural network (NN) can be used to acquire a solution of the optimization problem swiftly since it is a computational mathematical structure that is stimulated by observed processes in the natural network of biological neurons which consists of simple computational highly interconnected units, called neurons. In the past few years, many researchers have applied various types of NNs for solving several classes of constrained opti-

G. Selvaraj (✉)
Department of Mathematics, Ramco Institute of Technology,
Rajapalayam, Viruthunagar, Tamil Nadu, India
e-mail: gselvaraj89@gmail.com

L. J. Banu
Department of Mathematics, Ayya Nadar Janaki Ammal College, Sivakasi, Tamil Nadu, India

B. Pati et al. (eds.), *Proceedings of the 6th International Conference on Advance Computing and Intelligent Engineering*, Lecture Notes in Networks and Systems 428,
https://doi.org/10.1007/978-981-19-2225-1_53

mization problems efficiently. Tank and Hopfield [2] initiated solving optimization problems using neural network, and also, they proposed a neural network with an LP circuit for solving linear programming problems. After then, many researchers have worked on NN implementation to solve LP problems in which they transform the given problem into dynamical systems. Kennedy and Chua [3] developed an improved neural network model of Tank and Hopfield's model for solving linear programming problems with guaranteed convergence. Followed by them, Maa and Shanblatt [4] presented two-phase network to solve linear programming problems. Xia [5] and Malek and Yari [6] solved linear programming problem with its dual using the two layers of neurons with their separate networks. Yinzhen et al.[7] developed a linear neural network based on penalization for solving the fuzzy shortest path problem by converting the fuzzy shortest path problem into crisp shortest path problem. Effati and Baymani [8] solved convex nonlinear programming problems by a new nonlinear neural network. The stability analysis problem for neural networks with time delay has attracted much attention, and many sufficient conditions have been proposed to guarantee the stability of neural networks [9].

Using three layers of neurons, Selvaraj and Pandian [10] solved fuzzy linear programming problems without converting into crisp linear programming problem. Effati et al. [11] proposed two fuzzy neural networks for solving two types of problems which are fuzzy shortest path problems and fuzzy maximum flow problems. Malek and Alipour [12] developed a recurrent modal with no parameter setting to solve linear programming (LP) and quadratic programming (QP) problems. Effati et al. [13] presented a neural network for solving linear and quadratic programming problems. Ghasabi et al. [14] stated the simulation aspect of neural network to solve linear and quadratic programming problems. Gao and Liao [15] developed a new NN model for LP and QP problems in real time by presenting some new vectors. Yang et al. [16] presented a new NN model to solve QP problems which is having inequality and equality constraints. He et al. [17] proposed a NN model for solving convex bilevel QP problems. Nazemi [18] presented a capable NN to solve strictly convex QP problems having general linear constraints. Miao et al.[19] developed the finite time dual NNs using a newly defined activation function for solving QP problems. Using delayed projection neural network, Sha et al. [20] solved QP problems with having inequality and equality constraints. Liu and Zhou [21] proposed a single-layer recurrent NN to solve non-smooth convex optimization problems. Recently, Swati Tyagi and Syed Abbas [22] have proposed a fractional order projection NN for solving QP problems.

Based on the concepts of decision making in the fuzzy environment developed by Bellman and Zadeh [23], many researchers have developed several methods for solving fuzzy optimization problems. The study of optimization problems in a fuzzy environment known as fuzzy optimization was first coined by Zimmerman [24]. Yinzhen et al. [7] obtained a linear neural network based on penalization for solving the fuzzy shortest path problem, by converting the problem into crisp shortest path problem which is a representation of LP model. Pehlivan and Apaydin [25] solved fuzzy LP problems by both simplex method and neural network approach, and also, they compared the results by converting the fuzzy LP problems into crisp LP problems. Selvaraj and Pandian [10, 26] proposed crisp and fuzzy NNs to optimize fuzzy linear programming problems numerically.

This paper is organized as follows: In Sect. 2, few concepts and results related to real intervals and fuzzy sets are presented. A fuzzy neural approach for solving fully fuzzy QP problems is discussed, and a fuzzy energy function for fuzzy QP problems using fuzzy norm is defined, and also, a fuzzy NN is constructed to optimize fuzzy QP problems numerically in Sect. 3. In Sect. 4, numerical examples are shown for understanding the proposed fuzzy NN approach, and finally, the conclusion is given in Sect. 5.

2 Preliminaries

The following definitions and results are used in this paper which can be found in [10, 27–30]. Let D denotes the set of all closed bounded intervals on the real line R. That is,

$$D = \{[a, b] : a \le b, a, b \in R\}$$

Definition 1 Let $A = [a, b]$ and $B = [c, d]$ be in D, then

- $A \oplus B = [a + c, b + d]$.
- $A \ominus B = [a - d, b - c]$.
- $kA = [ka, kb]$ if k is a positive real number.
- $kA = [kb, ka]$ if k is a negative real number.
- $A \otimes B = [p, q]$, where $p = \min\{ac, ad, bc, bd\}$ and $q = \max\{ac, ad, bc, bd\}$.
- $A \le B$ if $a \le c$ and $b \le d$.
- $A = B$ if $a = c$ and $b = d$.

Definition 2 Let X be a classical non-empty set and a set $A \subseteq X$. A fuzzy set $\tilde{A}$ with the membership function $\mu(x)$ is defined as

$$\tilde{A} = \{(x, \mu(x)) : x \in X\}$$

where, $\mu(x)$ is a function from X to $[0, 1]$.

Definition 3 The α-cut of a fuzzy set $\tilde{A}$ where $\alpha \in [0, 1]$ is defined as the set ${}^{\alpha}\tilde{A} = \{x : \mu_{\tilde{A}}(x) \ge \alpha, x \in X\}$. It is also denoted by the closed interval $[A^l, A^r]$, where A^l and A^r are the left and right values of ${}^{\alpha}\tilde{A}$.

Definition 4 A real fuzzy number $\tilde{a} = (a_1, a_2, a_3, a_4)$ is a fuzzy set from the real line R with the membership function $\mu_{\tilde{a}}(x)$ satisfying the following conditions:

- $\mu_{\tilde{a}}(x)$ is a continuous mapping from R to the closed interval $[0, 1]$.
- $\mu_{\tilde{a}}(x) = 1$ at $x \in [a_2, a_3]$.
- $\mu_{\tilde{a}}(x) = 0$ at $x \in (-\infty, a_1] \cup [a_4, \infty)$.
- $\mu_{\tilde{a}}(x)$ is strictly decreasing and continuous on $[a_3, a_4]$.
- $\mu_{\tilde{a}}(x)$ is strictly increasing and continuous on $[a_1, a_2]$.

Definition 5 A fuzzy number $\tilde{a}$ is a trapezoidal fuzzy number denoted by (a_1, a_2, a_3, a_4), where a_1,a_2,a_3,a_4 are real numbers and its membership function $\mu_{\tilde{a}}(x)$ is given below.

$$\mu_{\tilde{a}}(x) = \begin{cases} (x - a_1)/(a_2 - a_1) & a_1 \leq x \leq a_2 \\ 1 & a_2 \leq x \leq a_3 \\ (a_4 - x)/(a_4 - a_3) & a_3 \leq x \leq a_4 \\ 0 & \text{otherwise.} \end{cases}$$

Definition 6 Let $\tilde{a} = (a_1, a_2, a_3, a_4)$ and $\tilde{b} = (b_1, b_2, b_3, b_4)$ be trapezoidal fuzzy numbers. Then,

- $\tilde{a} = \tilde{b}$ iff $a_i = b_i$, $i = 1, 2, 3, 4$
- $\tilde{a} \leq \tilde{b}$ iff $a_i \leq b_i$, $i = 1, 2, 3, 4$
- $\tilde{a} \geq 0$ iff $a_1 \geq 0$

Definition 7 The α-cut of a fuzzy number $\tilde{a} = (a_1, a_2, a_3, a_4)$, where $\alpha \in [0, 1]$ is given as ${}^{\alpha}\tilde{a} = [a_1 + \alpha(a_2 - a_1), a_4 - \alpha(a_4 - a_3)]$.

Remark If $a_2 = a_3$ in $\tilde{a} = (a_1, a_2, a_3, a_4)$, then $\tilde{a}$ becomes $\tilde{a} = (a_1, a_2, a_4)$ known as triangular fuzzy number.

Theorem 1 *Let x^* be an equilibrium point for the system $\frac{dx(t)}{dt} = f(x(t))$. If $V : R^n \to R$ is a continuously differentiable function such that*

- $V(x^*) = 0$ *and* $V(x) > 0$ *for all* $x \neq x^*$,
- $V(x) \to \infty$ *when* $\|x\| \to \infty$, *and*
- $\frac{dV(x)}{dt} < 0$ *for all* $x \neq x^*$,

then $x = x^*$ is globally asymptotically stable.

Definition 8 Any scalar function $V(x)$ that satisfies all the requirements of Theorem 1 is called a Lyapunov function for the equilibrium state.

Definition 9 Let $\tilde{a} = (\tilde{a_1}, \tilde{a_2}, ..., \tilde{a_n})$ be a fuzzy vector which consists of fuzzy numbers. Then the fuzzy norm of $\tilde{a}$, $\|\tilde{a}\|_f$ is defined as

$$\|\tilde{a}\|_f = (\|\tilde{a}\|_l, \|\tilde{a}\|_r) = (\|\tilde{a}^l\|, \|\tilde{a}^r\|)$$

where $\|\tilde{a}\|_l$ and $\|\tilde{a}\|_r$ are Euclidean norms of the lower and upper values of the fuzzy vector $\tilde{a}$, $a^l = (a_1^l, a_2^l, ..., a_n^l)$ and $a^u = (a_1^u, a_2^u, ..., a_n^u)$, respectively.

Theorem 2 *Let $\tilde{f}$ be a fuzzy differentiable function of $x \in \Omega \subseteq R$. If x is a point of local minimum, then $\tilde{\nabla}\tilde{f}(x) = \tilde{0}$.*

3 Fuzzy Neural Network

Fully fuzzy QP problem is a unique type of the fuzzy optimization problem in which the objective is a fuzzy quadratic function and constraints are fuzzy linear functions. The mathematical form fully fuzzy QP is given as follows:

(*P*) Maximize $\sum_{j=1}^{n}\sum_{i=1}^{n} \tilde{a}_{ij} \otimes \tilde{x}_i \otimes \tilde{x}_j + \sum_{j=1}^{n} \tilde{c}_j \otimes \tilde{x}_j$

subject to

$$\sum_{j=1}^{n} \tilde{d}_{ij} \otimes \tilde{x}_j \approx \tilde{b}_i, \text{ for all } i = 1, 2, \ldots, m$$

$$\tilde{x}_j \succeq \tilde{0}, \text{ for all } j = 1, 2, \ldots, n$$

where for all i and j ,the values $\tilde{a}_{ij}, \tilde{d}_{ij}, \tilde{c}_j, \tilde{x}_j, \tilde{b}_i \in F(\mathbb{R})$, a set of all triangular fuzzy numbers.

Now, for solving the fully FQP problem, we construct a new fuzzy energy function based on fuzzy norm introduced in Effati et al. [11] and then, we propose a new fuzzy neural network based on the newly defined fuzzy energy function.

Now, we define a new fuzzy energy function for solving the problem (*P*) using fuzzy norm as follows:

$$\tilde{E}(\tilde{x}) = \sum_{j=1}^{n}\sum_{i=1}^{m} \tilde{a}_{ij} \otimes \tilde{x}_i \otimes \tilde{x}_j + \sum_{j=1}^{n} \tilde{c}_j \otimes \tilde{x}_j + \frac{\lambda}{2}\sum_{i=1}^{m} \left\| \sum_{j=1}^{n} \tilde{d}_{ij} \otimes \tilde{x}_j - \tilde{b}_i \right\|_f^2, \tilde{x}_j \succeq \tilde{0}. \tag{1}$$

For minimizing the fuzzy energy function stated in (1), the following governing fuzzy differential equation is obtained using the conjugate gradient learning algorithm

$$\frac{\tilde{d}\tilde{x}}{dt} = -\tilde{\nabla}\tilde{E}(\tilde{x}), \tilde{x} \in F(\mathbb{R}) \tag{2}$$

where $\tilde{\nabla}\tilde{E}(\tilde{x})$ is the gradient of the fuzzy energy function $\tilde{E}(\tilde{x})$.

The $\alpha-$cut of $\tilde{E}(\tilde{x})$ is given by ${}^{\alpha}\tilde{E}(\tilde{x}) = [\tilde{E}(\tilde{x})^l, \tilde{E}(\tilde{x})^r]$.

$\tilde{\nabla}\tilde{E}(\tilde{x}) = \left(\frac{\tilde{\partial}\tilde{E}(\tilde{x})}{\partial\tilde{x}_1}, \frac{\tilde{\partial}\tilde{E}(\tilde{x})}{\partial\tilde{x}_2}, \cdots, \frac{\tilde{\partial}\tilde{E}(\tilde{x})}{\partial\tilde{x}_n}\right)$

where $\frac{\tilde{\partial}\tilde{E}(\tilde{x})}{\partial\tilde{x}_i} = \left[\frac{\tilde{\partial}\tilde{E}^l(\tilde{x})}{\partial\tilde{x}_i}, \frac{\tilde{\partial}\tilde{E}^r(\tilde{x})}{\partial\tilde{x}_i}\right]$ for all $\alpha \in [0, 1]$.

From (2), the following dynamic system of fuzzy differential equations can be obtained.

$$\frac{\tilde{d}\tilde{x}_k}{dt} = -\sum_{j=1}^{n} \tilde{a}_{jk} \otimes \tilde{x}_i - (\tilde{c}_k) - k\sum_{i=1}^{m} (\tilde{d}_{ik})^T \otimes \left(\sum_{j=1}^{n} \tilde{d}_{ij} \otimes \tilde{x}_j - \tilde{b}_i\right), k = 1, 2, \ldots, n \tag{3}$$

For computing purpose, the above fuzzy network can be represented by the following pair of two crisp networks using the $\alpha-$cut.

$$\left.\begin{array}{l} \frac{d^l\tilde{x}_k}{dt} = -\sum_{j=1}^{n}\tilde{a}_{jk}^l \otimes \tilde{x}_i^l - (\tilde{c}_k^l) - \lambda\sum_{i=1}^{m}(\tilde{d}_{ik}^l)^T \otimes \left(\sum_{j=1}^{n}\tilde{d}_{ij}^l \otimes \tilde{x}_j^l - \tilde{b}_i^l\right) \\ \frac{d^r\tilde{x}_k}{dt} = -\sum_{j=1}^{n}\tilde{a}_{jk}^r \otimes \tilde{x}_i^r - (\tilde{c}_k^r) - \lambda\sum_{i=1}^{m}(\tilde{d}_{ik}^r)^T \otimes \left(\sum_{j=1}^{n}\tilde{d}_{ij}^r \otimes \tilde{x}_j^r - \tilde{b}_i^r\right) \end{array}\right\} k = 1, 2, \ldots, n \quad (4)$$

where $\frac{d^l\tilde{x}_k}{dt}$ and $\frac{d^r\tilde{x}_k}{dt}$ are the left and right derivatives of $\tilde{x}_k$ with respect to t By solving the dynamic system (4) for each value of α in [0, 1], the minima for the fuzzy energy function $\tilde{E}(\tilde{x})$ can be determined.

Now, we establish the stability of the proposed fuzzy neural network (4) which is used to solve the problem (*P*).

Theorem 3 *The fuzzy energy function $\tilde{E}(\tilde{x})$ for the problem (P) is a Lyapunov function for the fuzzy neural network (4). Also, the fuzzy NN is globally asymptotically stable.*

Proof For every $\lambda > 0$ and $\tilde{x}$, $\tilde{E}(\tilde{x})$ is positive and for all $\alpha \in [0, 1]$, we have

$$\begin{aligned} \frac{d^r}{dt}(\tilde{E}(\tilde{x})) &= \left(-\sum_{j=1}^{n}\tilde{a}_{jk}^r \otimes \tilde{x}_i^r - (\tilde{c}_k^r) - \lambda\sum_{i=1}^{m}(\tilde{d}_{ik}^r)^T \otimes \left(\sum_{j=1}^{n}\tilde{d}_{ij}^r \otimes \tilde{x}_j^r - \tilde{b}_i^r\right)\right)\frac{d^r}{dt}(\tilde{x}_k) \\ &= \left(-\frac{d^r}{dt}(\tilde{x}_k)\right)\left(\frac{d^r}{dt}(\tilde{x}_k)\right) = -\left(\frac{d^r}{dt}(\tilde{x}_k)\right)^2 < 0. \end{aligned}$$

Since $\frac{d^l}{dt}(\tilde{E}(\tilde{x})) \leq \frac{d^r}{dt}(\tilde{E}(\tilde{x}))$ for all $\alpha \in [0, 1]$, we have

$$\frac{\tilde{d}}{dt}(\tilde{E}(\tilde{x})) = \left[\frac{d^l}{dt}(\tilde{E}(\tilde{x})), \frac{d^r}{dt}(\tilde{E}(\tilde{x}))\right] < 0, \text{ for all } \alpha \in [0, 1].$$

Therefore, $\frac{\tilde{d}}{dt}(\tilde{E}(\tilde{x})) < 0$ and the function $\tilde{E}(\tilde{x})$ is a Lyapunov function for the system (4). Thus, the proposed fuzzy NN (4) is globally asymptotically stable by Theorem 1.

Now, the following theorem is established a relation between the equilibrium point of the fuzzy NN (4), that is, the minima for the fuzzy energy function $\tilde{E}(\tilde{x})$ and the optimum of the fully fuzzy QP problem (*P*).

Theorem 4 *If the problem (P) has an optimal solution and for any $\lambda > 0$, $\tilde{x}^*$ is the equilibrium point of the fuzzy neural network (4), then $\tilde{x}^*$ will be the optimal solution of the problem (P).*

Proof Now, since $\tilde{x}^*$ is the equilibrium point for the fuzzy neural networks (4) for any $\lambda > 0$, we must have $\tilde{\nabla}\tilde{E}(\tilde{x}^*) \approx \tilde{0}$ and $\tilde{x}^*$ is a feasible solution to the problem (*P*).

Now, from the definition of the energy function and for any feasible $\tilde{x}^{\circ}$ to the problem (*P*), we have

$$\tilde{E}(\tilde{x}^{\circ}) \approx \sum_{j=1}^{n}\sum_{i=1}^{m} \tilde{a}_{ij} \otimes \tilde{x}_i^{\circ} \otimes \tilde{x}_j^{\circ} + \sum_{j=1}^{n} \tilde{c}_j \otimes \tilde{x}_j^{\circ}. \tag{5}$$

Now, from (5) and since $\tilde{x}^*$ is a feasible solution to the problem (*P*) and $\tilde{E}(\tilde{x}^*)$ is minimum, the problem (*P*) attains its minimum at $\tilde{x}^*$.
Hence the theorem.

Remark From Theorems 3 and 4, we assure that the proposed fuzzy neural network will converge to the optimum solution of the problem (*P*).

4 Numerical Examples

In this section, we present different numerical examples to show the computational efficiency of the proposed fuzzy neural network for solving fully fuzzy QP problems in which all or some parameters are fuzzy.

Example 1 Consider the following fully FQP problem
Maximize $(0.5, 1, 2) \otimes \tilde{x}_1^2 \oplus (1, 2, 3) \otimes \tilde{x}_1 \otimes \tilde{x}_2 \oplus (2, 3, 4) \otimes \tilde{x}_2^2 \oplus (2, 4, 7) \otimes \tilde{x}_1 \oplus (2, 5, 9) \otimes \tilde{x}_2 \oplus (3, 6, 8) \otimes \tilde{x}_3$
subject to

$$(1, 1, 3) \otimes \tilde{x}_1 \oplus (1, 2, 3) \otimes \tilde{x}_2 \approx (1, 3, 9)$$

$$(2, 4, 4) \otimes \tilde{x}_1 \oplus (3, 5, 6) \otimes \tilde{x}_3 \approx (2, 6, 9)$$

$$\tilde{x}_1, \tilde{x}_2, \tilde{x}_3 \succeq \tilde{0}$$

Using $\alpha-$cut, the problem can be written as follows:
Maximize $[0.5 + 0.5\alpha, 2 - \alpha] \otimes \left[\tilde{x}_1^{l^2}, \tilde{x}_1^{r^2}\right] \oplus [1 + \alpha, 3 - \alpha] \otimes [\tilde{x}_1^l, \tilde{x}_1^r] \otimes [\tilde{x}_2^l, \tilde{x}_2^r]$ $\oplus [2 + \alpha, 4 - \alpha] \otimes \left[\tilde{x}_2^{l^2}, \tilde{x}_2^{r^2}\right] \oplus [2 + 2\alpha, 7 - 3\alpha] \otimes [\tilde{x}_1^l, \tilde{x}_1^r] \oplus [2 + 3\alpha, 9 - 4\alpha] \otimes [\tilde{x}_2^l, \tilde{x}_2^r] \oplus [3 + 3\alpha, 8 - 2\alpha] \otimes [\tilde{x}_3^l, \tilde{x}_3^r]$
subject to

$$[1, 3 - 2\alpha] \otimes [\tilde{x}_1^l, \tilde{x}_1^r] \oplus [1 + \alpha, 3 - \alpha] \otimes [\tilde{x}_2^l, \tilde{x}_2^r] \approx [1 + 2\alpha, 9 - 6\alpha]$$

$$[2 + 2\alpha, 4] \otimes [\tilde{x}_1^l, \tilde{x}_1^r] \oplus [3 + 2\alpha, 6 - \alpha] \otimes [\tilde{x}_3^l, \tilde{x}_3^r] \approx [2 + 4\alpha, 9 - 3\alpha]$$

Table 1 Left and right values of $\widetilde{x}_1, \widetilde{x}_2$ and $\widetilde{x}_3$

α	$\widetilde{x}_1^l$	$\widetilde{x}_2^l$	$\widetilde{x}_3^l$	$\widetilde{x}_1^r$	$\widetilde{x}_2^r$	$\widetilde{x}_3^r$
0	0.9944	0	0	2.2502	0.7455	0
0.1	1.0909	0.0816	0	2.1760	0.7817	0
0.2	1.1666	0.1761	0	2.1009	0.8205	0
0.3	1.2304	0.2653	0	2.0260	0.8513	0
0.4	1.2855	0.3486	0	1.9505	0.8726	0
0.5	1.3330	0.4259	0	1.8753	0.8839	0
0.6	1.3745	0.4973	0	1.8006	0.8838	0
0.7	1.4112	0.5633	0	1.7247	0.8698	0
0.8	1.4438	0.6244	0	1.6497	0.8423	0
0.9	1.4729	0.6808	0	1.5745	0.7941	0
1	1.4993	0.7443	0	1.4993	0.7443	0

$$\widetilde{x}_1^l, \widetilde{x}_2^l, \widetilde{x}_3^l \succeq \widetilde{0}, \alpha \in [0, 1]$$

Now, for each $\alpha \in [0, 1]$ and using the proposed neural network, we obtain the left and right values of ${}^{\alpha}\widetilde{x}_1, {}^{\alpha}\widetilde{x}_2$ and ${}^{\alpha}\widetilde{x}_3$ are+1 given below:

From Table 1, we can obtain that the optimal solution of the given problem is $\widetilde{x}_1 = (0.9944, 1.4993, 2.2502)$, $\widetilde{x}_2 = (0, 0.7433, 0.7445)$ and $\widetilde{x}_3 = (0, 0, 0)$. The solution graph of the network for the problem is shown in Fig. 1.

Now, the convergence of the proposed network at $\alpha = 0$ and $\alpha = 1$ for all the values of $\widetilde{x}_1^l, \widetilde{x}_2^l, \widetilde{x}_3^l, \widetilde{x}_1^r, \widetilde{x}_2^r, \widetilde{x}_3^r$ is shown in Figs. 2 and 3, respectively.

5 Conclusion

In this paper, we have discussed a numerical method of solving fully FQP problems using fuzzy neural network. Based on fuzzy norm in Effati et al. [11], we have defined a new fuzzy energy function for the FQP problem. Then, we have developed a new fuzzy neural network based on $\alpha-$cut for determining an approximate optimal solution of the FQP problem using the proposed fuzzy energy function. The convergence of the proposed fuzzy neural network and the result that the equilibrium point of the fuzzy energy function for a FQP problem is an optimal solution of the FQP problem have been established. The fuzzy neural network is uncertain and so, it differs from all other existing neural networks for solving FQP problems. The solution procedure for solving FQP problems is given, and the importance of the proposed network is demonstrated through numerical examples.

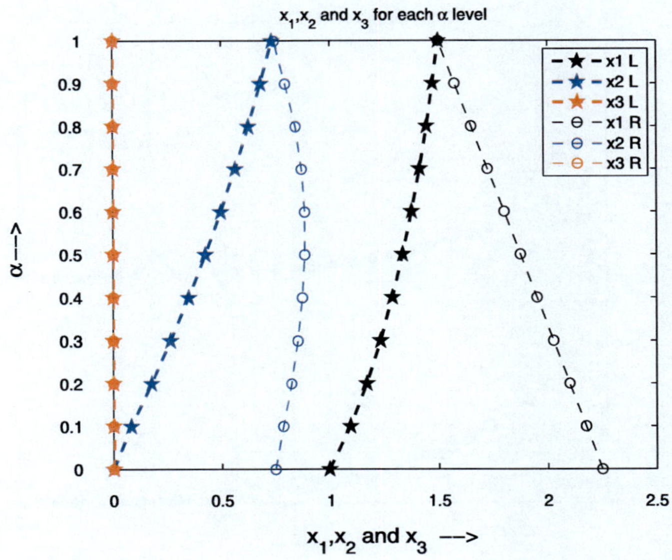

Fig. 1 Solution graph for different α levels

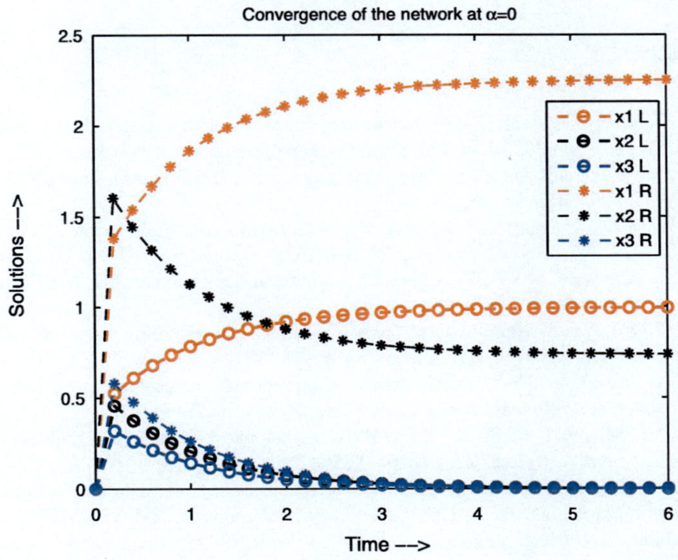

Fig. 2 Convergence of the network when $\alpha = 0$

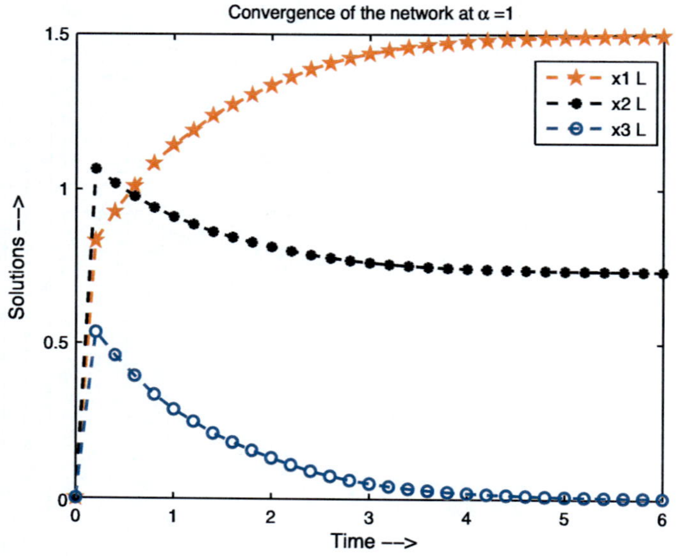

Fig. 3 Convergence of the network when $\alpha = 1$

References

1. Chong, E. K., & Zak, S. H. (2013). *An introduction to optimization*. John Wiley and Sons Inc.
2. Tank, D. W., & Hopfield, J. J. (1986). Simple neural optimization networks: An A/D converter, signal decision circuit and a linear programming circuit. *IEEE Transactions on Circuits and Systems, 33*, 533–541.
3. Kennedy, M. P., & Chua, L. O. (1998). Neural networks for nonlinear programming. *IEEE Transactions on Circuits and Systems, 35*, 554–562.
4. Maa, C., & Shanbaltt, M. (1992). A two-phase optimization neural network. *IEEE Transactions on Neural Networks, 3*, 1003–1009.
5. Xia, Y. (1996). A new neural network for solving linear programming problems and its applications. *IEEE Transactions on Neural Network, 7*, 525–529.
6. Malek, A., & Yari, A. (2005). Primal-dual solution for the linear programming problems using neural networks. *Applied Mathematics and Computation, 167*, 198–211.
7. Yinzhen, L., Mitsuo, G., & Kenichi, I. (1996). Solving fuzzy shortest path problems by neural network. *Computer Engineering, 31*, 861–865.
8. Effati, S., & Baymani, M. (2005). A new nonlinear neural network for solving convex nonlinear programming problems. *Applied Mathematics andComputation, 168*, 1370–1379.
9. Jarina Banu, L., Balasubramaniam, P., & Ratnavelu, K. (2015). Robust stability analysis for discrete-time uncertain neural networks with leakage time-varying delay. *Neurocomputing, 151*, 808–816.
10. Selvaraj, G., & Pandian, P. (2014). A new crisp neural network for solving fuzzy linear programming problems. *Global Journal of Pure and Applied Mathematics, 10*, 685–695.
11. Effati, S., Pakdaman, M., & Ranjbar, M. (2011). A new fuzzy neural network model for solving fuzzy linear programming problems and its applications. *Neural Computation and Application, 20*, 1285–1294.

12. Malek, A., & Alipour, M. (2007). Numerical solution for linear and quadratic programming problems using a recurrent neural network. *Applied Mathematics and Computation, 192*, 27–39.
13. Effati, S., & Nazemi, A. R. (2006). Neural network models and its applications for solving linear and quadratic programming problems. *Applied Mathematics and Computation, 172*, 305–331.
14. Ghasabi-Oskoei, H., Malek, A., & Ahmadi, A. (2007). Novel artificial neural network with simulation aspects for solving linear and quadratic programming problems. *Computers and Mathematics with Applications, 53*, 1439–1454.
15. Gao, X., & Liao, L. Z. (2010). A new one-layer neural network for linear and quadratic Programming. *IEEE Transactions on Neural Networks, 21*, 918–929.
16. Yang, Y., Cao, J., Xu, X., Hu, M., & Gao, Y. (2014). A new neural network for solving quadratic programming problems with equality and inequality constraints. *Mathemetics and Computers in Simulation, 101*, 103–112.
17. He, X., Li, C., Huang, T., & Li, C. (2014). Neural network for solving convex quadratic bi-level programming problems. *Neural Networks, 51*, 17–25.
18. Nazemi, A. (2014). A neural network model for solving convex quadratic programming problems with some applications. *Engineering Applications of Artificial Intelligence, 32*, 54–62.
19. Miao, P., Shen, X., & Xia, X. (2014). Finite time dual network with a tunable activation function for solving quadratic programming problems and its applications. *Neurocomputing, 143*, 80–89.
20. Sha, C., Zhao, H., & Ren, F. (2015). A new delayed projection neural network for solving quadratic programming problems with equality and inequatlity constraints. *Neurocomputing, 168*, 1164–1172.
21. Liu, X., & Zhou, M. (2016). A one-layer recurrent neural network for non-smooth convex optimization subject to linear inequality constraints. *Chaos Solitons and Fractals, 87*, 39–46.
22. Tyagi, S., & Abbas, S. (2017). Stability and Synchronization of delayed fractional-order projection neural network with piecewise constant argument of mixed type. *Tbilisi Mathematical Journal, 10*, 57–74.
23. Bellman, R. E., & Zadeh, L. (1970). A decision-making in a fuzzy environment. *Management Science, 17*, 141–161.
24. Zimmerman, H. J. (1983). Fuzzy mathematical programming. Computional. *Operations Research, 10*, 291–298.
25. Pehlivan, N. Y., & Apaydin, A. (2005). Artificial neural networks approach to fuzzy linear programming. *Selncuk Journal of Application Mathematics, 6*, 9–26.
26. Selvaraj, G., & Pandian, P. (2016). A fuzzy neural network for fully fuzzy linear programming problems. *International Journal Of Mathematical Modelling and Numerical Optimization, 7*, 231–243.
27. Zadeh, L. A. (1965). Fuzzy sets. *Information and Control, 8*, 338–353.
28. Boyd, S., & Vandenberghe, L. (2004). *Convex optimization*. Cambridge University Press.
29. Madan, M. G., Jin, L., & Homa, N. (2003). *Static and dynamical neural networks: From fundamentals to advanced theory*. John Wiley & Sons Inc.
30. Mansoori, A., Effati, S., & Eshaghnezhad, M. (2018). A neural network to solve quadratic programming problems with fuzzy parameters. *Fuzzy Optimization and Decision Making, 17*, 75–101.

Maximum Loadability Analysis of Particle Swarm Optimization Technique for Consideration of Stability by Continuation Power Flow

P. K. Dhal

Abstract The loadability of a large power system network is investigated in this paper, and a particle swarm optimization (PSO)-based technique is proposed to evaluate the optimum position and setup of the FACTS device to optimize loading margin and voltage stability. Voltage stability is one of the most problematic aspects of power system functioning. There are numerous methods for determining the voltage stability of a system. The calculation of the margin from the present point of operation to the MLP is one of these approaches. The maximum load capacity is determined using this approach. It is to increase the voltage stability of power network in critical operating circumstances; it is recommended for STATCOM usage. The search for the optimal solution planning is expressed as an objective function toward the least real-power loss in the particle swarm optimization technique. An optimization challenge may be seen as the difficulty of identifying the maximum load point. The proposed system's performance is assessed in a range of operational settings, including without STATCOM and with PSO-optimized STATCOM.

Keywords Maximum loadability · Particle swarm optimization · Continuation power flow · STATCOM

1 Introduction

Stability is a fundamental consideration in a power system network. To establish stability, a variety of methods are employed. The best way to examine the system is to keep the power flowing. The system's maximum load capacity is governed by CPF. It is used to strengthen and enhance the power grid. In this study, the maximum load capability of the FACTS device is studied using MATPOWER. STATCOM is used to increase the system's load capacity and stability. The STATCOM's key advantage is its quick responsiveness to system power flow. The topmost loadability control

P. K. Dhal (✉)
Department of Electrical and Electronics Engineering, Vel Tech Rangarajan Dr, Sagunthala R&D Institute of Science and Technology, Chennai, India
e-mail: pradyumna.dhal@rediffmail.com

B. Pati et al. (eds.), *Proceedings of the 6th International Conference on Advance Computing and Intelligent Engineering*, Lecture Notes in Networks and Systems 428,
https://doi.org/10.1007/978-981-19-2225-1_54

is the highest overall network load that the system can handle devoid of violating network values such as bus voltage limits, generator watt and wattles power limits, and transmission line power transfer limits. A nonlinear optimization approach was used to solve the problem of determining the topmost loadability limit. Addressing the feasibility and robustness of the improvement moth flame optimization (IMFO) algorithm in dealing with optimal power flow (OPF) issues [1]. In contrast to previous techniques, the IMFO algorithm is able to find accurate and better optimum flow solutions. If the convergence properties of the IMFO approach are compared with those of others, it is clear that IMFO is the best way to discover the best possible convergence flow solution. Using the famous GA and PSO hybrid optimization technique [2], the optimum load shed is focused, as is the system voltage profile optimization resulting in stable fast voltage stabilization indicators. In [3], the issue of where these devices should be positioned and which size the voltage stability margin (VSM) should be increased to a particular level is discussed in a metaheuristic manner. A combination of the continuous power flow technique (PFT) to calculate the topmost loadability of the power system (PS) with the PSO idea was developed to find the better positioning and size of a group of static synchronous compensator units. Ajjarapu and Christy [4] show how to discover solutions for the energy flow that start at a baseload and lead to the stability limit of the system's steady-state voltage in order to determine maximum power system loadability. Irisarri et al. [5] offer a nonlinear optimization (NLO) interior point (IP). The implementation of pure primal–dual and primal–dual IP methods with a predictor corrector is discussed in further depth [6]. The newly constructed evolutionary PSO is employed to address this optimization problem. Two testing systems (Ward-Hale 6-bus) and the suggested technique are described (IEEE 14-bus). The results are compared to the often-used continuous flow method. Shunmugalatha and Mary Raja Slochanal [7] utilize one technique to assess stability of voltage and maximum load capacity (MLL). The optimal generation cost for MLL may be expressed in a power system as a two-phase optimization issue, MLL computation and optimal production costs for MLL determination. Arya et al. [8] describe how reactive variables can be rescheduled so that the loadability margin at the current point of operation can be accepted. As an optimization technique, coordinated PSO-based aggregation (CAPSO) was utilized. In order to enhance the load margin, two goal functions were selected. The first aim is to minimize the wattles power loss overall, while the second objective is to minimize the wattles power loss (PL) over time. In [9], the method DEPSO is used to calculate the highest loadability level of a power system. Software from Mat Power has been used for the IEEE 30 node network and IEEE 118 node networks. In comparison with other evolutionary algorithms such as DE and multi-agency hybrid PSO, the analysis of the DEPSO method is compared. Saravanan et al. [10] discuss the use of PSO to determine the optimum place to put flexible FACTS (Flexible ACTS) devices with the lowest installation cost and increase system loadability. In the most critical and N-1 operating circumstances, a genetic algorithm-based SVC optimization technology is introduced [11] to enhance the VS of power network. In relations of maximal peor-reactive control boundary (PRCB), high load tensions at the critical operational places, less actual power losses, and lowest SVC device cost, the SVC placement was structured as a multi-objective optimization problem (MOOP). A nonlinear architectural control for the static VAR compensator is discussed in order to enhance VS [12] with SVC. The SVC system is described in a third-order dynamic, nonlinear way. A nonlinear controller is created using the direct feedback linearization (DFL) approach. Time and bifurcation analysis on a three-bus power system assesses the controls' effectivity for improving voltage stability. Rath et al. [13] offer a technique for measuring FACTS shunt controllers' location and size using fluorescent logic and a real genetic algorithm. Using the saddle-node fork distance, voltage, and capacity of the STATCOM controller, a fluctuating performance index is constructed. In order to optimize the location and configuration of FACTS devices to enhance VS, [14] proposed a PSO based. In order to optimize the placement and configuration of SVC as a FACTS device, the particles swarm optimization approach is utilized. This paper [15] is aimed at identifying the UPFC in electricity systems. The technique suggested is based on continuous research in the UPFC with OPF power injection model. Nagalakshmi et al. [16] present a method for advance device to optimize the loadability and differential development on the deregulated energy markets in transmission systems utilizing the particle swarm optimization and diversion models. A mixed discrete continuous nonlinear optimization issue is defined by [17] as a challenge of maximizing power network load capacity (PSLC) by installing two forms of the FACTS, namely the SCV and the TCSC. In order to enhance loadability, current electrical networks [18] need be increased. A mixed discreet, continuous nonlinear improvement is the ideal regions for network enhancement. In [19] proposes, a SCGS technique is adopted to overcome local optima and [20] suggested that CPSO technique is able to provide reliable convergence. The organization of this paper is the first section contributed motivation and contribution. The second section highlighted the continuation power flow analysis. The third section is the maximum loading point formulation. The fourth section is the overview of the PSO. The fifth section is represented as flow chart development in the proposed algorithm. The last section is explained about the results and discussion in this paper.

1.1 Motivation and Contribution

A tangent vector is a specific case flow solution. It is used as a predictor for subsequent solutions in this technique. By incorporating the CPF parameters with very minor alterations to their mathematical formulation, the Newton–Raphson approach is employed to correct obtained solutions. This paper proposes a creative solution to the issue of the loading point. This is supported by particle swarm optimization, a relatively new evolutionary approach. PSO was created by Kennedy and Eberhart as an optimization method stimulated by the natural performance of a flock of birds or insects. When it comes to chasing food, it works the same way. The key contribution of this study is to choose the best STATCOM position when weak buses are discovered. To acquire greater voltage magnitude and enhance loadability, a novel algorithm technique must be used to adjustthe STATCOM and place it in a certain location.

1.2 Continuation Power Flow Analysis (CPF)

The CPF tool can identify OPF solutions which jump with a convinced base load and continue until the VSL of the power network is achieved. The CPF technique is developed by rewriting the power flow equations (PFEs) to incorporate the charge factor, using the predictor corrector methodology. The loading parameter is the current load on the power system. Figure 1 demonstrates the fundamental knowledge of CPF.

2 Maximum Loading Point Formulation as an Optimization Problem

All system values, such as bus voltage, generator watt and wattles power, and transmission line power flow, can be handled within their limits at the system's maximum loading value. From a beginning operational point, a loading condition represents an increase in system load. The optimization function's goal is to maximise this factor. The fitness value is determined by the goal function.

$$\text{Thefitnessfunction} = \text{maximum}(\lambda) \tag{1}$$

Subject to the

The equality constraints of the wattful power balance equation

$$P_{Ti} = P_{Gi} - P_{Li} \tag{2}$$

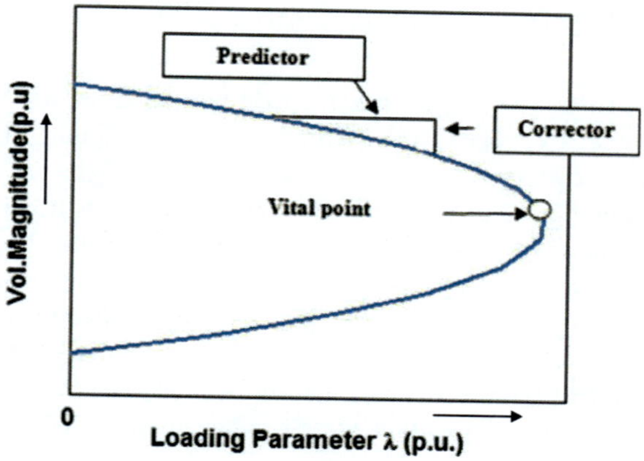

Fig. 1 CPF curve

$$P_{Ti} - P_{Gi} + P_{Li} = 0 \quad (3)$$

Here

$$P_{Ti} = \sum V_i V_j Y_{i,j} \cos(\theta_{i,j} + \delta_i - \delta_j) \quad (4)$$

Hence

$$\sum V_i V_j Y_{i,y} \cos\left(\theta_{i,j} + \delta_i - \delta_j\right) - P_{Gi} + P_{Li} = 0 \quad (5)$$

The equality constraints of reactive power equation

$$Q_{Ti} = Q_{Gi} - Q_{Li} \quad (6)$$

$$Q_{Ti} - Q_{Gi} + Q_{Li} = 0 \quad (7)$$

Here

$$Q_{Ti} = \sum V_i V_j Y_{i,j} \sin(\theta_{i,j} + \delta_i - \delta_j) \quad (8)$$

Hence

$$\sum V_i V_j Y_{i,j} \sin(\theta_{i,j} + \delta_i - \delta_j) - Q_{Gi} + Q_{Li} = 0 \quad (9)$$

Equations 5 and 9 can be reformulated based on loading factor λ because the load is varying. The new equation is

$$F_{Pi} = \sum V_i V_j Y_{i,j} \cos(\theta_{i,j} + \delta_i - \delta_j) - P_{G,i} + P_{L,i} + \lambda K_{L,i} \cos\phi_i \quad (10)$$

and

$$F_{Q,i} = \sum V_i V_j Y_{i,j} \sin\left(\theta_{i,j} + \delta_i - \delta_j\right) - Q_{Gi} + Q_{Li} + \lambda K_{L,i} \sin\phi_i \quad (11)$$

This term also permits the charge of a number of buses to be increased, and the necessary power factor is ideally suited for practical use.

2.1 Particle Swarm Optimization Algorithm Over View

In 1995, Eberhart and Kennedy were led by the social behavior of birds flying and fishing. PSO may be used in a range of disciplines, including social psychology, manufacturing, and computer technology. The people of PSO are known as 'swarm'; every member of the swarm is known as a 'particle'. The location and velocity of each element in the n-dimensional research room for the problem are recognized, and it is referred to as a possible solution. In each cycle, the speeds of the particles vary stochastically, taking into account its prior best location as well as the best place around it, computed by means of a predefined fitness function. This naturally leads to an optimal or almost perfect answer for every particle movement. Each particle has memory, and its current equal fitness value is reminiscent of its former perfect location. The fitness value, sometimes called Pbest, is kept. Gbest is the best value for the whole population as topological neighbors when a particle captures. The velocity and location of the particles will be updated if the two best values have been determined.

2.2 *Flowchart*

I. First, choose the test data, then set the particle index, maximum iteration, and power flow variable maximum and lowest values.
II. The particles should be initialized with maximum and minimum values that are given.
III. Decide on the number of iterations.
IV. Connect to STATCOM and set the weak bus by increasing loadability settings.
V. Create an initial population using the PSO method.
VI. Update the standards of Pbest and Gbest.
VII. If Pbest exceeds Gbest, the Pbest key develops the Gbest result. Else, Gbest will not change.
VIII. Determine the best solution for PSO-based boundary conditions.
IX. Verify that the Gbest solution and stop condition meet the convergence conditions. If not, proceed to step 6 (Fig. 2).

3 Simulation Test Results and Discussions

To evaluate the maximum loadability for IEEE 118 bus, the proposed PSO was used. A total of 100 iterations with various random values were performed to ensure consistency. A huge number of runs are carried out separately to determine the optimum parameter combination. The 118-bus system consists of 54 units and 99 loads, as is illustrated in Fig. 3. Firstly, the correct position and setup of STATCOM are established, and secondly, the maximum loadability calculated. The system performance is studied, which showed that the weak buses of the system are 38, 43, and 44. The parameter values are considered in this test system as shown in Table 1.

Fig. 2 Proposed flowchart

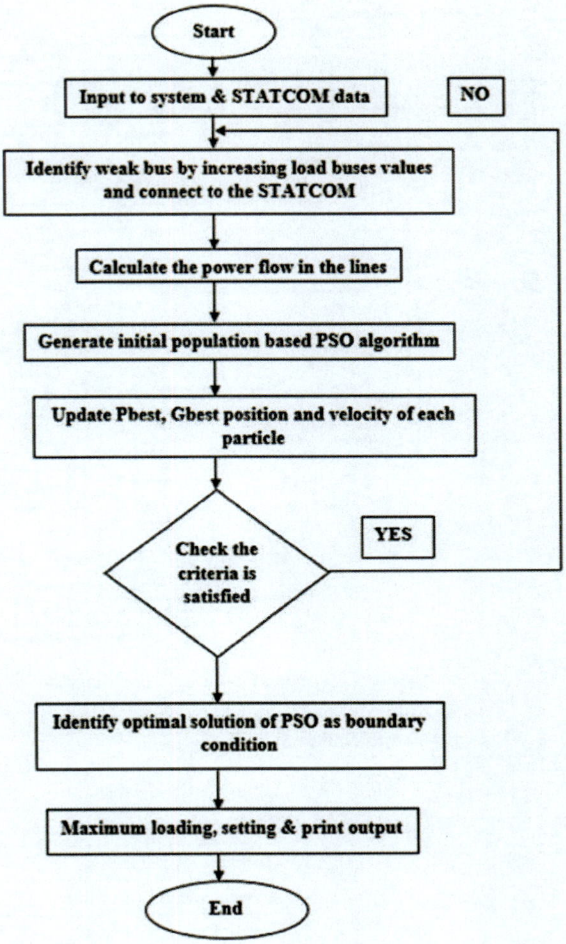

Figure 4 depicts the VPC of the 118-bus network without STATCOM. As demonstrated in Figs. 5, 6, and 7, the system's lowest value is 0.6896 p.u, 0.7178 p.u, and 0.6911 p.u at buses 38, 43, and 44, respectively. For bus 38, the associated real and reactive power is 1485.67 MW and 1016.36 MVAR. The watt and wattless power at bus number 43 is 1517.75 MW and 1226.56 MVAR, respectively. Similarly, the watt and wattless power for bus number 44 is 1460.23 MW and 1212.24 KVAR, respectively. This case's maximum loading parameter is 1.5667.

Figures 8, 9, 10, and 11 depict the system using STATCOM, as well as the voltage outline and load ability curve tuned by PSO. At bus 38, 43, and 44, the system value is 0.9712, 1.01, and 0.9801 p.u. For bus 38, the associated watt and wattles power is 1486.20 MW and 1017.26 MVAR. The actual and reactive power at bus number 43 is 1519.25 MW and 1230.25 MVAR, respectively. Similarly, the watt and wattless

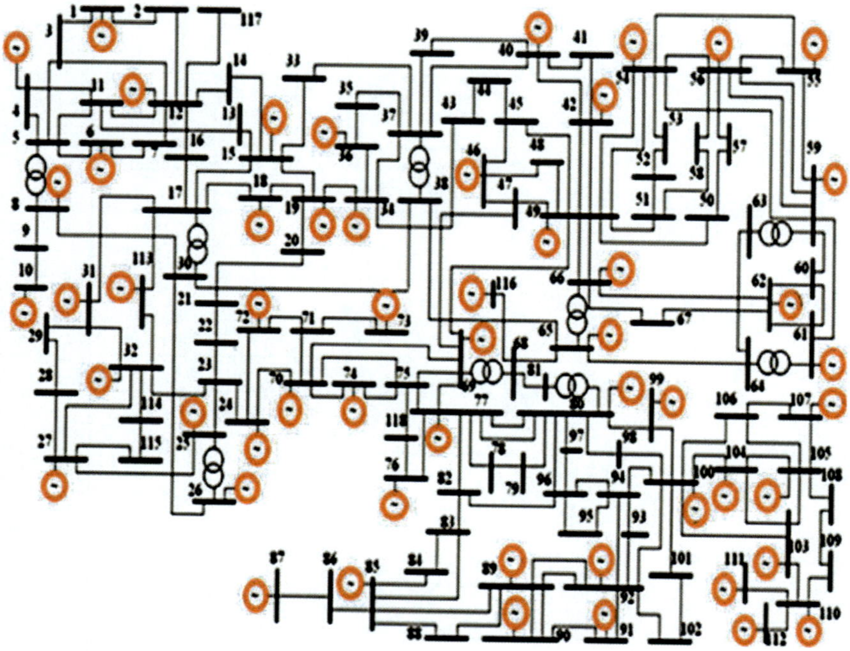

Fig. 3 118 node system is depicted in a single line diagram

Table 1 Optimum principles of PSO

Parameters	Optimal values
Total no. of variables	64
Size of population	50
Total no. of iterations	100
Inertial weight of initial	0.4
Inertial weight of final	0.9
Constant acceleration of each individual	2
Constant social acceleration	2
Rand1 = Rand2	0–1
Particle velocity	1
Λ	0–1.5686

Table 2 Various parameters value without STATCOM

Bus Nos	Voltage magnitude (MW)	Watt power (MW)	Wattless power (MVAR)
Bus no 38	0.6896	1485.67	1016.36
Bus no 43	0.7178	1517.75	1226.56
Bus no 44	0.6911	1460.23	1212.24

Table 3 Comparison of loading parameter and loads

Method	Maximum loading parameter	Total watt power losses (MW)	Total wattless power losses (MVAR)
Devoid of STATCOM	1.5667	4463.65	3455.16
STATCOM with PSO	1.5686	4467.20	3469.08

Fig. 4 Voltage outline devoid of STATCOM

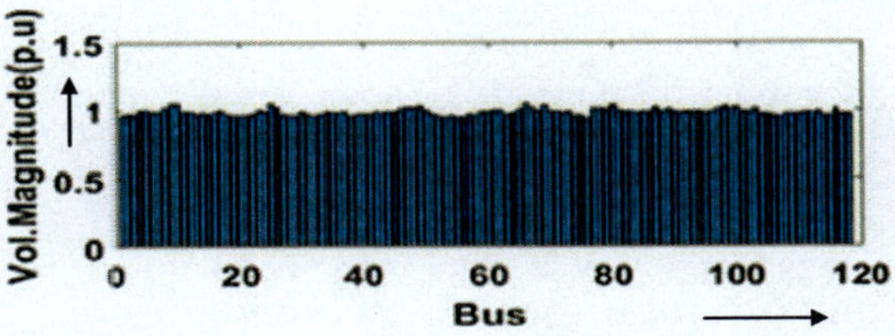

Fig. 5 Loadability of the network devoid of STATCOM (bus 38)

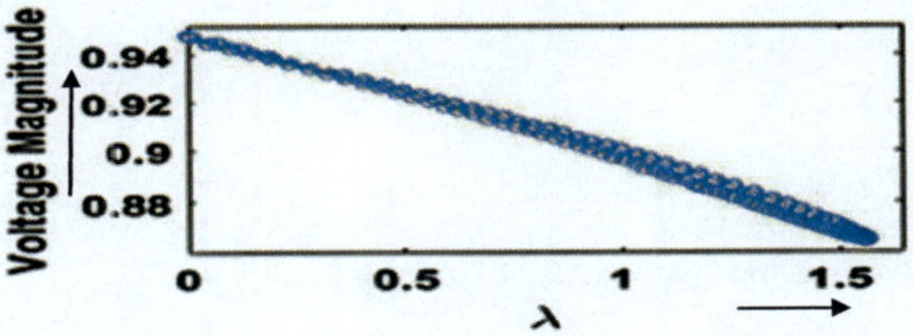

Fig. 6 Loadability of the network devoid of STATCOM (bus 43)

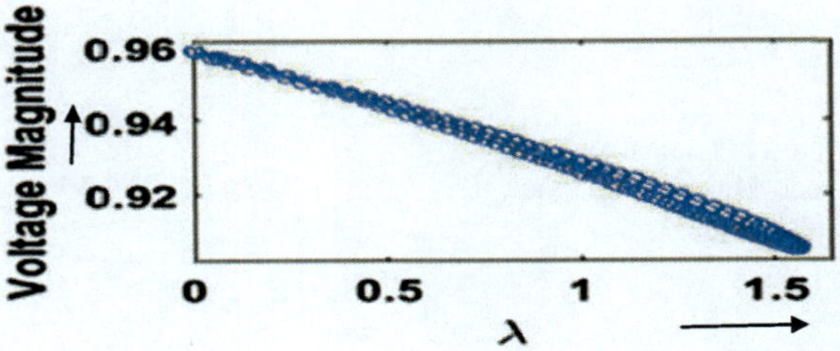

power at bus number 44 is 1461.75 MW and 1221.57 KVAR, respectively. This case's maximum loading parameter is 1.5686.

Fig. 7 Loadability of the network devoid of STATCOM (bus 44)

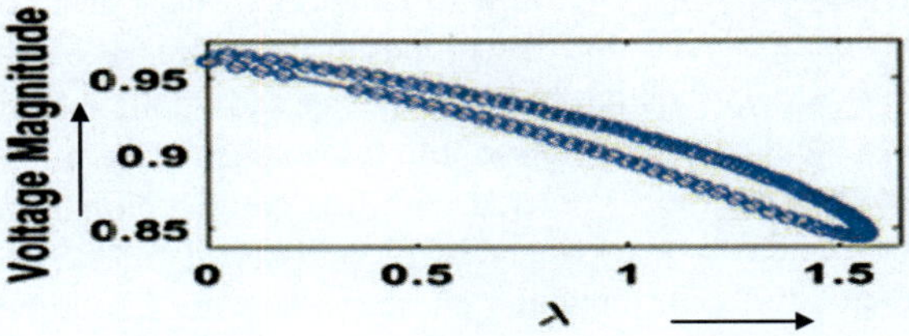

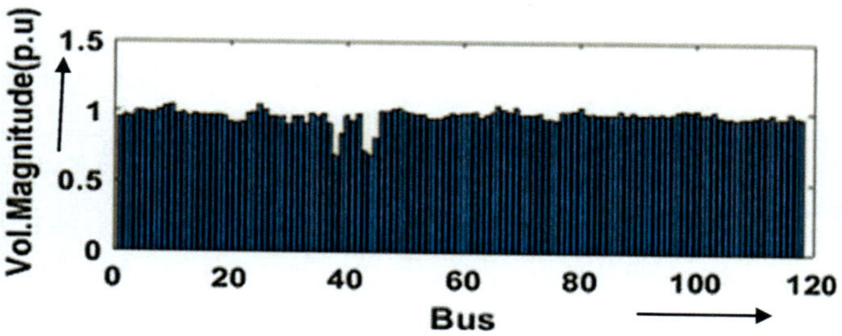

Fig. 8 Voltage outline with STATCOM adjusted by PSO

Fig. 9 Loadability of the node 38 including STATCOM adjusted by PSO

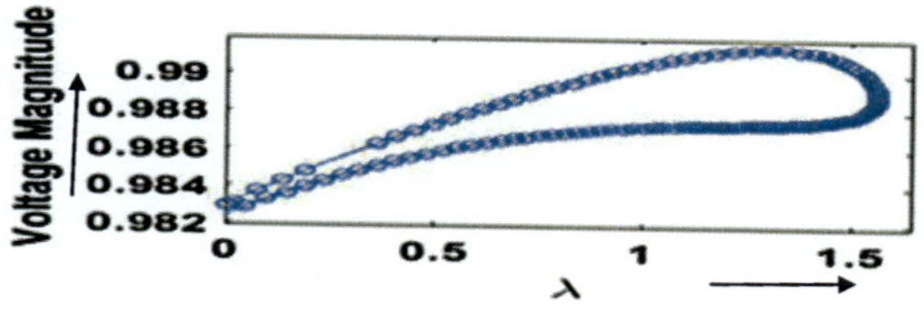

Fig. 10 Load ability of the node 43 including STATCOM adjusted by PSO

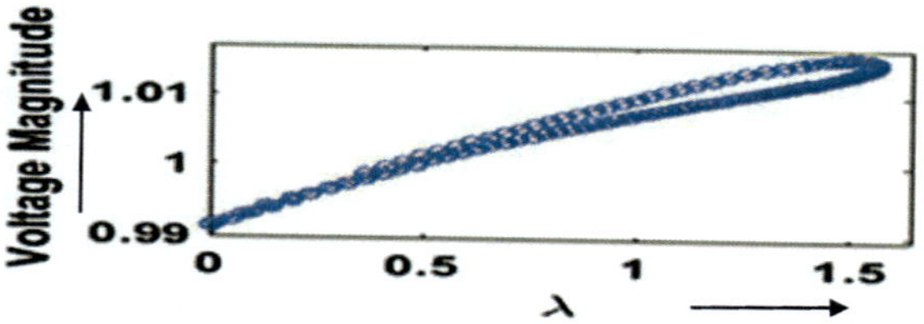

Fig. 11 Loadability of the node 44 with STATCOM adjusted by PSO

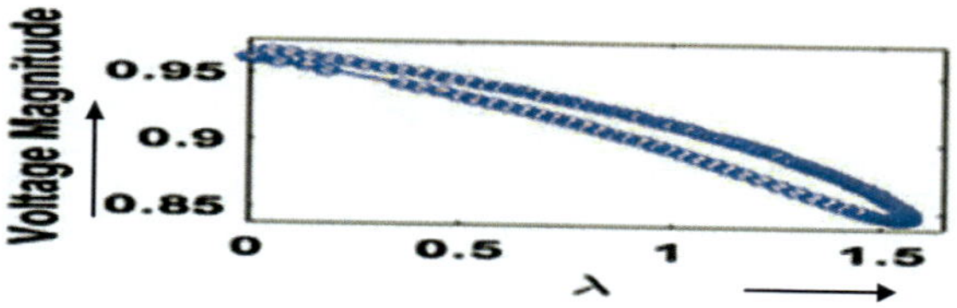

4 Conclusion

The continuing power flow simulations are used to improve the voltage of the network utilizing STATCOM. The proposed algorithm has been successful in determining the best site for the highest load capacity and stability. With this technique, it is possible to obtain an ideal solution through a series of iterations. As a result, the IEEE 118-bus test can be completed with less computational effort. The results suggest that the MLP can be enhanced by properly change the STATCOM by means of the PSO method. The forthcoming effort can be solved by hybrid algorithm.

References

1. Taher, M. A., Kamel, S., Kamel, S., Jurado, F., & Ebeed, M. (2018). An improved moth-flame optimization algorithm for solving optimal power flow problem. *International Transactions on Electrical Energy Systems, 29*, e2743.
2. Larik, R. M., Mustafa, M. W., Otuoze, A. O., Mohammed, O. O., & Sule, A. H. (2018). A new technique of load shedding to stabilize voltage magnitude and fast voltage stability index by using hybrid optimization. *ARPN Journal of Engineering and Applied Science, 13*(8), 2734–2745.
3. Rocha, L., Castro, R., & de Jesus, J. F. (2016). An improved particles warm optimization algorithm for optimal placement and sizing of STATCOM. *International Transactions on Electra Energy System, 26*, 825–840.
4. Ajjarapu, V., & Christy, C. (1992). The continuation power flow: A tool for steady state voltage stability analysis. *IEEE Transactions Power System, 7*, 416–423.
5. Irisarri, G. D., Wang, X., Tong, J., & Mokhtari, S. (1997). Maximum loadability of power systems using interior point nonlinear optimization method. *IEEE Transactions on Power Systems, 12*, 162–172.
6. EL-Dib, A. A., Youssef, H. K. M., EL-Metwally, M. M., & Osman, Z. (2006). Maximum loadability of power systems using hybrid particle swarm optimization. *Electric Power Systems Research, 76*, 485–492.
7. Shunmugalatha, A., & Mary Raja Slochanal, S. (2008). Optimum cost of generation for maximum loadability limit of power system using hybrid particle swarm optimization. *International Journal of Electrical Power & Energy Systems, 30*, 486–490.
8. Arya, L. D., Choube, S. C., Shrivastava, M., & Kothari, D. P. (2010). Loadability margin enhancement using-ordinated aggregation based particle swarm optimization (CAPSO). *International Journal of Electrical Power & Energy Systems, 32*, 975–984.
9. Gnanambal, K., & Babulal, C. K. (2012). Maximum loadability limit of power system using hybrid differential evolution with particle swarm optimization. *Electrical Power and Energy Systems, 43*, 150–155.
10. Saravanan, M., Slochanal, S. M. R., Venkatesh, P., & Abraham, J. P. S. (2007). Application of particle swarm optimization technique for optimal location of FACTS devices considering cost of installation and system loadability. *Electric Power Systems Research, 77*, 276–283.
11. Rath, M., Darwish, A., Pati, B., Pattanayak, B. K., & Panigrahi, C. R. (2020). *Swarm intelligence as a solution for technological problems associated with Internet of Things* (pp. 21–45). Swarm Intelligence for Resource Management in Internet of Things, Academic Press.
12. Wang, Y., Chen, H., & Zhou, R. (2000). A nonlinear controller design for SVC to improve power system voltage stability. *International Journal Electra Power Energy Systems, 22*(7), 463–470.
13. Rath, M., Pati, B., Pattanayak, B. K., Panigrahi, C. R., & Sarkar, J. L. (2017). Load balanced routing scheme for MANETs with power and delay optimisation. *International Journal of Communication Networks and Distributed Systems, 19*(4), 394–405.
14. Tibin, J., Sini, X., Chitra, S., Cherian, V. I., & Sreedharan, S. (2011, December). PSO based optimal placement and setting of FACTS devices for improving the performance of power distribution system. *Bonfring International Journal of Power Systems and Integrated Circuits, 1*, Spec. Issue, 2250–1088.
15. Behshad, M., Lashkarara, A., & Rahmani, A. H. (2009). Optimal location of UPFC device considering system loadability, total fuel cost, power losses and cost of installation. In: *2nd international conference on power electronics and intelligent transportation system*, 978-1-4244-4543-1.
16. Nagalakshmi, S., & Kamaraj, N. (2011). Secured loadability enhancement with TCSC in transmission system using computational intelligence techniques for pool and hybrid model Science Direc. *Applied Soft Computing, 11*, 4748–4756.

17. Chang, Y.-C., & Chang , R. F. (2013, April). Maximization of transmission system loadability with optimal FACTS installation strategy. *Journal of Electra Engineering and Technology, 8*(5), 991–1001.
18. Chang, Y.-C., Chang, R.-F., Hsiao, T.-Y., & Lu, C.-N. (2011, February). Transmission system loadability enhancement study by ordinal optimization method. IEEE Transactions Power Systems, 26(1), 0885–8950.
19. Acharjee, P., Mallick, S., Thakur, S. S., & Ghoshal, S. P. (2011). Detection of maximum loadability limits and weak buses using Chaotic PSO considering security constraints. *Chaos, Solitons & Fractals, 44*, 600–612.
20. Acharjee, P. (2012). Identification of maximum loadability limit and week buses using security constraint genetic algorithm. *Electrical Power and Systems, 36*, 40–50.

Virtual Reality in E-commerce: A Study

Aishwarya Nair, Ivy Chakraborty, Rajesh Kumar Verma, Chhabi Rani Panigrahi, and Bibudhendu Pati

Abstract Online shopping has made the hassles of stepping out of homes to purchase items a thing of the past. With various organizations running their businesses online and with the onset of the pandemic, relying on e-commerce sites has simplified our lives. Although these applications provide details of products through the use of images and text, consumers are still wary and skeptical about the quality or fit of a product before making the purchase. In this work, we have discussed the use of virtual reality (VR) in these applications, to create a shopping environment for the consumers from the comfort of their homes. We have also proposed an architecture framework to implement VR on e-commerce. We provide a use case of online shopping which involves a product such as apparels along with certain future research directions.

Keywords Virtual reality (VR) · Augmented reality (AR) · VR headset · Virtual embodiment (Avatar)

1 Introduction

In the era of technology and its vast use in a multitude of fields such as health care, education, and corporate sector, the average human is so highly dependent on it that they have gone way past the phase of manual or physical labor. It comes as no surprise that technology and its various advancements have made our lives so much simpler, and we rely on it every second of every day. The use of technology has expanded to various domains, and with each advancement, a new creation is introduced [1]. Data encryption, information systems, automation, artificial intelligence [1], quantum computing, etc., are just some of them. However, a rapidly flourishing and booming field is that of VR [2]. VR is a simulation generated using a computer

A. Nair · I. Chakraborty (✉) · R. Kumar Verma
Infosys Limited, Bangalore, India
e-mail: ivy1994c@gmail.com

C. R. Panigrahi · B. Pati
Department of Computer Science, Rama Devi Women's University, Bhubaneswar, India

B. Pati et al. (eds.), *Proceedings of the 6th International Conference on Advance Computing and Intelligent Engineering*, Lecture Notes in Networks and Systems 428,
https://doi.org/10.1007/978-981-19-2225-1_55

where a person can experience and interact with a simulated environment which is usually presented in a 3-dimensional artificial world. VR is applied in numerous fields [3] such as gaming, entertainment, education, military training, and even health care such as medical training. According to a report by Statista Research Department, the consumer VR market is predicted to reach $3.7 billion by the end of year 2021[4]. In the field of e-commerce, which accounts for over 3.4 billion users worldwide (according to a report by CSA, Worldometer), VR is yet to burgeon. A survey by UNCTAD has found that in the unprecedented rise of the COVID-19 pandemic, the online shopping purchases have surged by 6–10 points across most product categories [5]. This rise in the growth of online shopping proves its tendency to make a huge impact in the world of e-commerce and how the introduction of VR in this field can influence it for the better. A more rigorous study of VR can make a great advancement in how consumers can shop online on e-commerce sites through a more interactive experience.

There are numerous fields where VR can be implemented in different ways to ease human work. Our motivation to delve deeper into the world of VR stems from witnessing the sudden halt of the economy in the unprecedented rise of COVID-19. We have contributed our findings and research on the concept of VR and created a design to implement this technology in the field of e-commerce.

The rest of the paper is organized as follows. In Sect. 2, we present the related works in VR and fields similar to e-commerce. In Sect. 3, we present the brief history of virtual reality, its components along with features. Section 4 describes how the e-commerce shopping is done using VR. The proposed architecture framework is given in Sect. 5. Section 6 presents a business use case scenario and workflow of the design. Finally, Sect. 7 concludes the paper.

2 Related Work

The study of VR in various fields was a popular trend in the twenty-first century. Jung et al. [6] designed a model for the use of VR in rollercoasters which was then put to test at a Finnish amusement park. They introduced the social impact by measuring user satisfaction along with VR experience in the proposed model. Another study by Carlos Flavián et al. [7] explained the impact of VR, AR, and mixed reality by combining technological (embodiment), psychological (presence), and behavioral (interactivity) views to form taxonomy "EPI Cube." In their study of the proposed system, they have clearly distinguished the theory and the applications of three different realities—AR, VR, and MR. A text-mining approach was proposed by Loureiroa et al. [8] where the authors used a Bayesian model to analyze 150 articles and study VR in marketing. They have mainly focused on two fields, i.e., tourism and retailing, where five senses have been incorporated with VR experiences so that user is willing to experience the tourism and retailing again in the future. Xi et al. [9] reviewed 72 research papers that discuss the application of VR in shopping. In the proposed study, they mentioned several implications and aspects of VR in retail systems. Based on the conducted review by the authors, 16 future research agenda were proposed in the study.

Lu et al. [10] explained the use of AR in the field of e-commerce. The authors presented a prototype for AR e-commerce system whose result showed that how customer decision-making could be improved using AR. They concluded that the proposed AR system can also be improvised by uploading images and videos for avoiding portability of laptops and other gadgets for user convenience. Jain et al. [11] discussed different techniques like 3D and 2D display, body scanners, digital signage, and multitouch technology using VR, AR, and mixed reality to enhance shopping experience for users. They not only highlighted the limitations of e-commerce but also the integration of data mining to AR/VR techniques to provide recommended solutions in shopping experience. In the literature review by Swartout [12], we saw an overview of virtual humans, its behavior and future scope. Asadi et al. [13] designed an AR/VR platform for e-commerce. They discussed the UI design, system design, and the design of an AR/VR store in their work and elaborated the relevant components of a virtual store like salesperson, store design, sales desk, products, shelves, storeowner, and showroom as all in virtual mode.

3 Virtual Reality and Its Components

The history and origin of virtual reality dates back to 1838 when Charles Wheatstone performed a research to explain how the human brain perceives the 2-dimensional image captured by each eye as one single 3-dimensional object [14]. The two images were called stereoscopic images which were coined by the word stereoscope through which these images were viewed. The stereoscope produced two separate offset images, one for the left eye and one for the right. The brain then combines both the images to create a depth which causes the viewer to see the image in three dimensions. Obviously, the stereoscope created by Charles Wheatstone as shown in Fig. 1 was slightly complicated, which was later modified into a simpler device by Sir David Brewster in 1849, where he invented a lens-based stereoscope [14].

The evolution of the stereoscope over the years paved the way for the VR headsets currently in use. In the 1950s, cinematographer Morton Heilig invented the Sensorama (as shown in Fig. 2) with a stereoscopic 3D display including various other features [15, 16]. The 1960s saw the first head-mounted VR which was also developed by Morton Heilig which was called the Telesphere Mask (as shown in Fig. 3) [16, 17]. The stereoscope invented in the 1800s laid a foundation to the currently used VR headsets which works on the same principle.

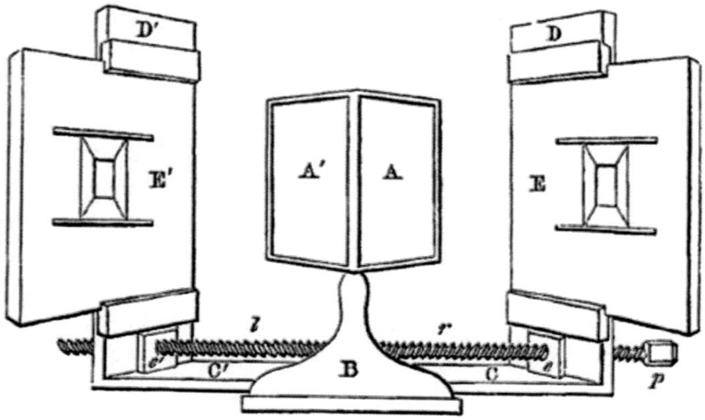

Fig. 1 Model of the stereoscope invented by Charles Wheatstone in 1838

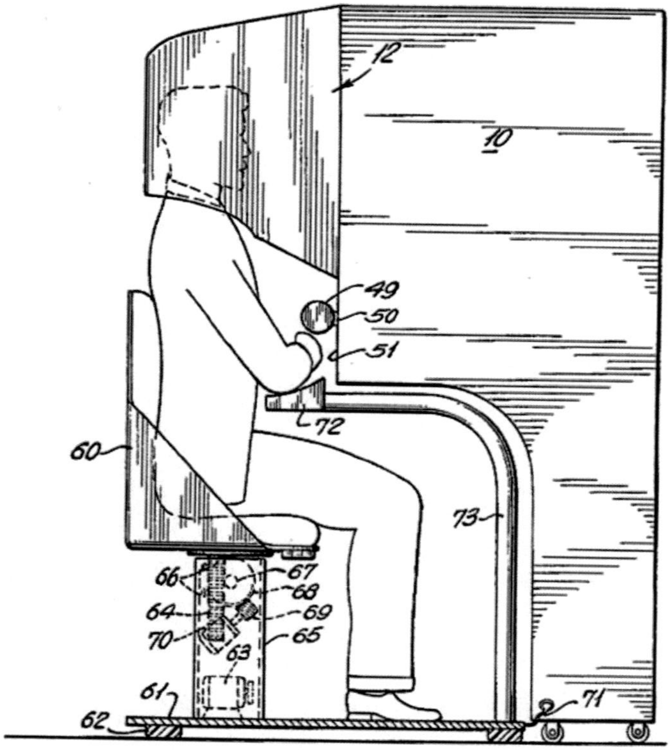

Fig. 2 Sensorama invented by Morton Heilig in the 1950s [14]

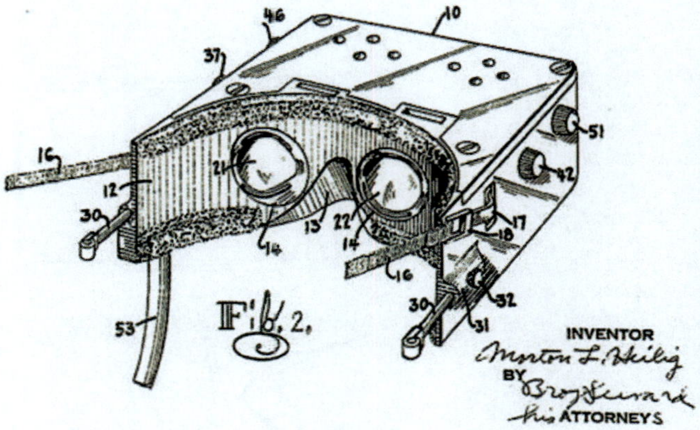

Fig. 3 Telesphere Mask invented by Morton Heilig in the 1960s [18]

3.1 Components and Features

In this section, we have presented the component used to implement VR along with their features.

3.1.1 VR Headset

E-commerce sites or mobile applications can use a VR headset to create a virtual experience for online shoppers. Understanding the working mechanism of the headset is one of the most pivotal segments in amalgamating VR and shopping apps.

A stereoscopic head-mounted display, stereo sound, and head motion-tracking sensors are components found in the headset. The head-tracking technology changes the field of vision depending on the movement [19]. It generally uses 360 degree videos. Users can use the mouse, the on-screen arrows or a Google Cardboard to look around. The Google Cardboard is a very basic VR headset device created by Google which can also be built by an ordinary person. The Cardboard Viewer can be used to experience immersion in a particular place which can be miles away from the location of the user. There are two parts in the Cardboard Viewer (as shown in Fig. 4):

1. A slot in the front where the smartphone is placed, and
2. The two glasses of plastic lenses.

The partition between the two glasses is a crucial element as it gives the user a feel of being "inside" the video, and this precipitates a 3D vision.

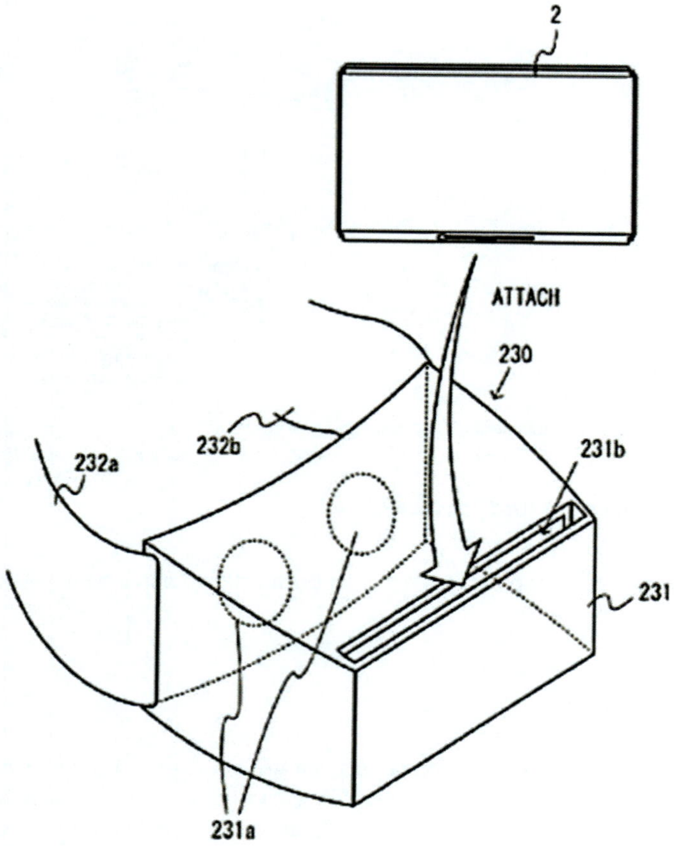

Fig. 4 Parts of a VR lens [20]

3.1.2 3D Vision

The sensation of depth from combining two slightly different pictures seen in each eye into one 3D image is called stereovision or stereopsis. When the eyes don't work together to make this 3D image, it is known as stereo blindness. Because our eyes are apart from each other by 3 inches, the view for each eye, differs in perspective. The brain merges the two views to create a sense of depth. The depth perception comes from our brain's ability to put together two 2D images to extrapolate depth. This is called stereoscopic vision [21]. Instead of using one picture, for 3D vision, two different pictures are used for each of the eyes that are slightly offset. The box hosts the image in a specific distance from the eyes and divides the images into two so that each eye focuses on the image [16]. This video uses stereoscopic display which stimulates our eyes to force our brain into believing that the image is three dimensional. The display device has a couple of sensors [22] that measures positions

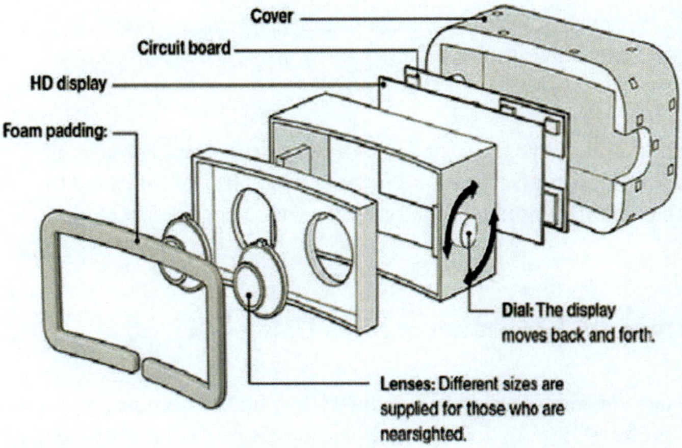

Fig. 5 Model of a modern VR headset [23]

and angle, to allow head tracking. The modern twenty-first century virtual reality headset comes with a variety of advancements and features added to it.

3.1.3 Features

Currently, VR headsets produced come with a few added features such as head tracking, motion tracking, and potentially, eye tracking.

Head Tracking

The modern VR headset as shown in Fig. 5 uses head tracking to increase immersion which is done by increasing the field of view. To display, a 360° view is highly expensive and, in most cases, unnecessary. Therefore, most headsets use a 100 or 110° view to display. The frame rate lies at a minimum of 60 frames per second. The latest headsets like Oculus have a frame rate of 90 fps, while PlayStation is capable of 120 fps. While wearing a VR headset, we want the display to change as and when we rotate our head. This is achieved through head tracking. 6DoF, which is an abbreviation for six degrees of freedom, is a system in which head tracking is performed using the three coordinate axes, X, Y, and Z. This measures head movements forward, backwards, sideways, and shoulder to shoulder. The components used in head-tracking system include magnetometer, accelerometer, and a gyroscope. To avoid a lag between head movements or display change, we need to keep a low latency which amounts to at least 50 ms.

Motion Tracking

Motion tracking in VR is still under development. A computer hardware sensor device which uses infrared sensors to track hand movements was developed by Leap

Motion which is used in Oculus dev kits. Similarly, Kinect 2 cameras can be used to track our body movements.

Eye Tracking

Like motion tracking, eye tracking is not a fully developed feature in VR headsets. But, the mechanism involves infrared sensors which track movement of your eyes in order to expand display widths required or to create depth of the field.

4 Shopping on E-commerce Sites Using VR

In the last decade, the e-commerce industry has been booming, and a majority of customers prefer shopping for products online from the comfort of their homes. While shopping through mobile applications has made our lives easier, it has several drawbacks. Users are sometimes skeptical while choosing a product only by judging its quality or fit from an image. One way to overcome this is by introducing VR in these sites so that users can see a 3D image of the product in a virtual environment. Products can range from groceries to furniture and apparels. Products under groceries include pantry, health and household, beauty, and luxury items. The application should create an environment for the customer to see the product while wearing the headset. Shopping furniture products can be made easier using VR. The customer is required to upload an image of the particular area or space where the furniture needs to be placed in the application. This image will be displayed in the virtual environment where the user can see how the particular piece would look in the room. In this work, we focus on apparel shopping in a VR environment.

4.1 Apparels

In this section, we discuss the design for shopping clothing items online.

Virtual Embodiment

Embodiment can essentially be defined as the illusion or sense of having a body. This illusion is achieved using computer-generated imagery (CGI) [15] where when the user wearing the headset looks down, sees a virtual body instead of their physical one. They see the same virtual body while looking into a mirror [24] in the virtual world. Three components of embodiment that can be considered are as follows: self-location, body ownership and agency. Self-location is when the position of our body is localized at a certain position in space. Body ownership describes the feeling or sense of the individual's "own body." For example, if the user's hand in the virtual environment is replaced with a robotic arm, it is perceived as a part of their own body in the virtual world. Agency is the feeling of "one's own actions." It describes the feeling of controlling one's own body movements by walking, running, turning, etc.

The technique used for virtual embodiment is based on CGI. It is achieved by using a VR headset and motion capture suit. The layout of the environment is divided into two sections—left and right (as shown in Fig. 6).

The virtual environment is designed in such a way that the body created using virtual embodiment is placed in front of a mirror for the user to see [24].

Right Section: On the right of the layout, there are four sections.

i. *Apparel List*: This section lists out the category of clothing, for example, trousers, t-shirts, etc. The user can select one of their preferred categories.

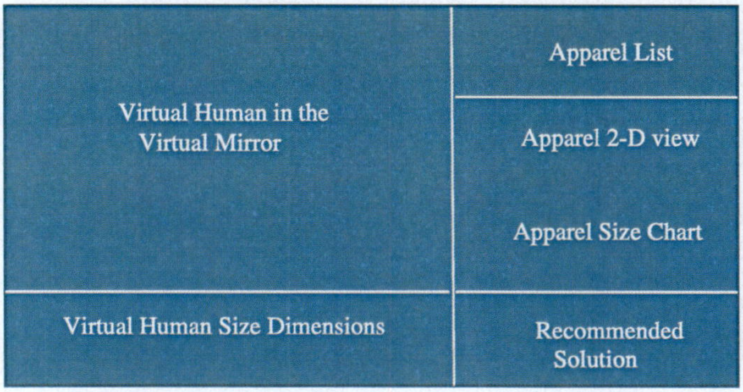

Fig. 6 Layout of VR environment view

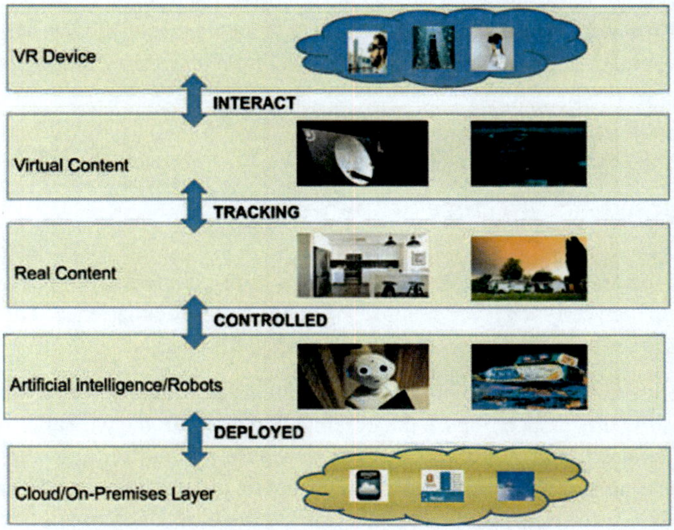

Fig. 7 Proposed system architecture

ii. *Apparel's 2D view*: A 2D image of all apparels under the above category is displayed to the user. The user can swipe through every outfit and select an apparel of his/her choice.
iii. *Apparel Size Chart*: In this section, the dimensions of the apparel as provided by the store/application are displayed in numbers to the user. This section can automatically recommend a size to the user from the input body dimensions or can also be selected manually.
iv. *Recommended items*: This section displays a collection of outfits/accessories that may complement the selected apparel. This section uses a recommendation engine like matrix factorization where the similarities are computed using content-based features [17]. While actual implementation of the system, we can use the Python language for the recommendation algorithm implementation. [25–28].

Left Section: The left side of the layout consists of the virtual human as seen in the virtual mirror, and below that we can see the dynamic body dimensions of the user.

i. *Virtual Human*: The virtual embodiment creates a virtual body using CGI. In the virtual environment, a virtual mirror is designed and placed in front of the user, and in this way, the user can see their virtual body as per the dimensions given by the user to the application.
ii. *Virtual Human Dimensions*: In this section, the user can see an animated body with the input dimensions. This is used to create a match between the user's body size and the size chart of the apparel.

The 2D image of the apparel and its size chart is coupled with the size chart of the virtual body and performs a 2D to 3D conversion of the apparel to perfectly fit the virtual body which is then available for the user to view through the VR headset. The algorithm for 2D to 3D conversion of images can be performed using multi-depth cues [29].

5 Proposed Architecture Framework

The proposed architecture framework as shown in Fig. 7 gives a brief layout of the relationship between each layer that is responsible for a VR experience. The different layers of the proposed framework are explained as follows.

Cloud/On-premises Layer: Several VR applications are employed on cloud or On-premises software. Depending on the organization's requirements, either or both of them can be used to create the VR experience. While cloud services range across the Internet and can be accessed from either public cloud or private cloud or both, On-premises software is installed locally. This is the base layer on which programs using artificial intelligence or robots are deployed.

Artificial Intelligence/Robots: The second layer of artificial intelligence or AI as we call it is an emerging field of computer science which uses machines and programs to replicate and perform the work and functionalities which is otherwise done by humans. AI requires the use of a large amount of data and algorithms to perform tasks automatically without the need for human intelligence. The data and algorithms used are deployed on the cloud/On-premises layer.

Real Content: The objects in the real world, i.e., real contents are monitored and controlled by AI and robots. This real-world content is manipulated to build the VR world.

Virtual Content: This layer involves the creation of the virtual content which is essentially what you see in your VR environment. Here, the real content is tracked and sometimes replicated to form a virtual content. This can be in any form like a video or an image.

AR/VR Devices: In the final layer, we have the physical devices used to see this virtual content. A VR headset is used to interact with the videos or images created to provide the user a 3D view of the virtual content.

6 Business Use Case and Workflow

In this section, a business use case, i.e., apparel shopping along with its workflow is presented. Apparel shopping is made quite easy where a user can just click on the 2D image of the product which pops out a 3D version of it. The user can change the perspective to view the product from different angles. For apparel shopping, the implementation of VR is required. The use case in Sect. 6.1 gives a detailed scenario of it.

6.1 Use Case: Apparel Shopping

Title: Online Apparel Shopping

Description: To purchase a t-shirt from an online store

Primary Actor: User/Customer

Pre-conditions: User is wearing the headset and can view the shopping environment having a mirror in front, a section of all apparels in the store, 2D image of the selected apparel for trial, size chart of the apparel, recommended solution consisting of complimentary apparels for the selected item. A view of the user's scaled down body dimensions are displayed below the mirror view. Selection is done using a controller/remote in hand.

Post-conditions: Environment having a list of all items added to the cart by the user. This is the checkout page.

Main Success Scenario:

Step 1: User clicks on a category from the "Apparel List" section.

Step 2: The 2D images of the apparels under the selected category are shown in the below section.

Step 3: User selects an apparel of their choice, and an enlarged image of the apparel is displayed.

Step 4: A size chart of the apparel is shown below the image displaying all the available sizes.

Step 5: The application automatically calculates the recommended size using the user's body dimensions which can be changed manually (or) user selects the size from the size chart.

Step 6: The apparel is then displayed on the virtual human body of the user in the mirror.

Step 7: User selects the apparel by clicking the "Add to Cart" button and taps on the cart icon.

Step 8: User is then taken to the checkout page containing images of selected apparel(s).

Status: Complete.

7 Conclusion

With increasing shopping applications, the need for trustworthy, good quality products that are displayed as images to match the product received upon delivery is huge. This paper gives a detailed design of merging online shopping applications with virtual reality to create an in-person experience for consumers. The proposed ideas can simplify the shopping process for consumers who can study product features and quality factors in detail. This can thereby also avoid unwanted cancellations and returns.

References

1. https://www.engpaper.com/technology-trends-2020.htm. Accessed on 1 Dec 2021
2. Lemaignan, S., Warnier, M., Akin Sisbot, E., Clodic, A., & Alami, R. (2017, June). Artificial cognition for social human-robot interaction: An implementation, ELSEVIER (Vol. 247, pp. 45–69). https://doi.org/10.1016/j.artint.2016.07.002
3. Cipresso, P., Giglioli, I. A. C., Raya, M., Alcañiz; R., & Giuseppe. (2018). The past, present, and future of virtual and augmented reality research: A network and cluster analysis of the literature. *Frontiers in Psychology, 9*, 2086, 1–20. https://doi.org/10.3389/fpsyg.2018.02086
4. https://en.wikipedia.org/wiki/Virtual_reality_applications. Accessed on 1 Dec 2021.
5. https://www.statista.com/topics/2532/virtual-reality-vr/#dossierKeyfigures. Accessed on 3 Dec 2021.
6. https://unctad.org/news/covid-19-has-changed-online-shopping-forever-survey-shows. Accessed on 3 Dec 2021.

7. Jung, T., Claudia tom Dieck, M., Rauschnabel, P., Ascenção, M., Tuominen, P., & Moilanen, T. (2018) Functional, hedonic or social? Exploring antecedents and consequences of virtual reality rollercoaster usage, Springer International Publishing AG (pp. 247–258).https://doi.org/10.1007/978-3-319-64027-3_17
8. Flavián, C., Ibáñez-Sánchez, S., & Orús, C. (2019). The impact of virtual, augmented and mixed reality technologies on the customer experience. *Journal of Business Research, 100*, 547–560. https://doi.org/10.1016/j.jbusres.2018.10.050
9. Loureiroa, S. M. C., Guerreiroa, J., Eloyb, S., Langaroa, D., & Panchapakesanc, P. (2018) Understanding the use of virtual reality in marketing: A text mining-based review. *Journal of Business Research* (pp. 1–17). https://doi.org/10.1016/j.jbusres.2018.10.055
10. Xi, N., & Hamari, J. (2021). Shopping in virtual reality: A literature review and future agenda. *Journal of Business Research, 134*, 37–58. https://doi.org/10.1016/j.jbusres.2021.04.075
11. Lu, Y., & Smith, S. (2008) Augmented reality E-commerce: How the technology benefits people's lives. *Human Computer Interaction* 215–238.
12. Jain, S., Hada, N., & Joshi, S. (2020). Shopping using virtual reality and augmented reality. *International Journal of Advanced Science and Technology, 29*(04), 7546–7552.
13. Swartout, W.R. (2006). *Virtual Humans, American Association for Artificial Intelligence* (pp. 1543–1545).
14. Asadi, A., & Hemadi, R. (2018). Design of an AR/VR platform for e-Commerce. In: *ICTD National Conference* (pp. 1–18).
15. https://www.vrs.org.uk/virtual-reality/history.html. Accessed on 1 Dec 2021.
16. Jones, S., & Dawkins, S. (2018). The sensorama revisited: Evaluating the application of multisensory input on the sense of presence in 360-degree immersive film in virtual reality, Springer International Publishing AG (pp. 183–197).
17. Ogle, K. N. (1952). On the limits of stereoscopic vision. *Journal of Experimental Psychology, 44*(4), 253–259. https://doi.org/10.1037/h0057643
18. https://itnext.io/what-are-the-top-recommendation-engine-algorithms-used-nowadays-646f588ce639. Accessed on 1 Dec 2021.
19. https://www.pebblestudios.co.uk/2017/08/17/when-was-virtual-reality-invented/. Accessed on 1 Dec 2021.
20. Loomis, J. M. (2016) Presence in virtual reality and everyday life: Immersion within a world of representation, immersion within a world of representation, Massachusetts Institute of Technology, Presence (Vol. 25, No. 2, pp. 169–174). Spring. https://doi.org/10.1162/PRES_a_00255
21. https://arstechnica.com/gaming/2018/01/nintendo-is-not-hopping-on-the-vr-bandwagon-any-time-soon/. Accessed on 1 Dec 2021.
22. Wade, N. (2016). [Vision, Illusion and Perception] Art and Illusionists Volume 1, Stereoscopic Vision., https://doi.org/10.1007/978-3-319-25229-2 (Chap. 16, pp. 335–354).https://doi.org/10.1007/978-3-319-25229-2_16
23. O'Riordan, A., Newe, T., Dooly, G., & Toal, D. (2018). Stereo vision sensing: Review of existing systems. In: *2018 Twelfth International Conference on Sensing Technology (ICST)* (pp. 178–184). https://doi.org/10.1109/ICSensT.2018.8603605
24. https://blogs.ucl.ac.uk/digital-education/category/our-views/matts-musings/page/2/. Accessed on 1 Dec 2021
25. Ro, Y. K., Brem, A., & Rauschnabel, P. A. (2018). *Augmented reality smart glasses: Definition* (pp. 169–181). Springer International Publishing AG.
26. https://towardsdatascience.com/recommendation-system-in-python-lightfm-61c85010ce17. Accessed on 1 Dec 2021.
27. Parida, S., Pati, B., Nayak, S. C., & Panigrahi, C. R. (2020). Offer based auction mechanism for virtual machine allocation in cloud environment. *Advanced Computing and Intelligent Engineering*, 339–351.
28. Das, H., Naik, B., Pati, B., & Panigrahi, C. R. (2014). A survey on virtual sensor networks framework. *International Journal of Grid Distribution Computing, 7*(5), 121–130.

29. Pati, B., Misra, S., & Saha, B. K. (2012, December). Advanced network technologies virtual lab: a human-computer interface for performing experiments on wireless sensor networks. In: *2012 4th International Conference on Intelligent Human Computer Interaction (IHCI)* (pp. 1–5). IEEE.
30. Ji, P., Wang, L., Li, D.-X., & Zhang, M. An automatic 2D to 3D conversion algorithm using multi-depth cues. In: *2012 International Conference on Audio, Language and Image Processing* (pp. 546–550). https://doi.org/10.1109/ICALIP.2012.6376677

Discovery of Periodic Rare Correlated Patterns from Static Database

Upadhya K. Jyothi, B Dinesh Rao, M. Geetha, and Harsh Kamlesh Vora

Abstract Finding the associations among the itemsets and discovering the unknown or unexpected behavior are the major tasks of rare pattern mining. The support measure has the main contribution during the discovery of low support patterns. As the association of low support patterns may generate a bundle of spurious patterns, other measures are used to find the correlation between the itemsets. A generalization of frequent pattern mining called periodic frequent pattern mining (PFPM) is emerged as a promising field, focusing on the occurrence behavior of frequent patterns. On the contrary, the shape of occurrence in the case of rare pattern mining is not much studied. In this paper, a single scan algorithm called *PRCPMiner is proposed to study the shape of occurrence of rare patterns. The proposed algorithm discovers periodic rare correlated patterns using different thresholds with respect to support, bond, and periodicity measures. The research shows the influence of these thresholds on the runtime performance for various datasets.*

Keywords Rare pattern mining · Periodic rare correlated patterns · Periodicity measure

---

U. K. Jyothi (✉) · M. Geetha · H. K. Vora
Department of Computer Science and Engineering, Manipal Institute of Technology, Manipal Academy of Higher Education, Manipal, Karnataka 576104, India
e-mail: jyothi.k@manipal.edu

M. Geetha
e-mail: geetha.maiya@manipal.edu

B. D. Rao
Manipal School of Information Sciences, Manipal Academy of Higher Education, Manipal, Karnataka 576104, India
e-mail: dinesh.rao@manipal.edu

B. Pati et al. (eds.), *Proceedings of the 6th International Conference on Advance Computing and Intelligent Engineering*, Lecture Notes in Networks and Systems 428,
https://doi.org/10.1007/978-981-19-2225-1_56

1 Introduction

Frequent pattern mining (FPM) is a vital field in data mining that insight into the significant frequent associations among the data items in the transactions. Support metric is the key measure used in FPM during the process of item selection and association. By considering the shape of occurrence, Tanbeer et al. [1] first highlighted the importance of behavioral measures. Since then the field of periodic-frequent pattern mining (PFPM) has emerged as a promising area. Started with the maximum periodicity measure [1–3], the importance is given to other measures like variance [4, 5], standard deviation [6], average periodicity [7, 8]. In recent years, unknown or unpredictable association discovery called rare pattern mining (RPM) has grabbed the attention of researchers. RPM discovers infrequent itemsets which appear less frequently in the database. The major focus of RPM is to generate patterns with a low frequency than the threshold value given by the user called *minSup*. The algorithms designed in this area can be divided based on traversing the search space level-wise or designing an efficient tree structure for the search exploration. Two different issues have to be handled by the RPM algorithms. First, the itemsets with a support value greater than *minSup* cannot be discarded as their supersets may become rare patterns. Therefore, an efficient strategy should be developed to minimize the search space. Second, because of low support values the possibility of spurious pattern generation increases. To filter the spurious patterns, other interesting measures are used along with the support values.

1.1 Motivation and Contribution

The existing RPM algorithms fail to focus on the occurrence behavior of the rare patterns. For example, "air conditioner" may be purchased only during summer by all the customers. This will aid in understanding customer purchasing patterns and in turn help in designing marketing strategies, managing off-shelf in market basket analysis. In genetic data analysis, instead of considering only gene occurrence considering time intervals in the DNA sequence help the scientists to study in detail the gene behavioral patterns. Viger et al. [6] associated periodicity measure in mining the rare correlated patterns in multiple sequences. The CORI algorithm [9] mines rare correlated patterns by considering support and bond thresholds, even though CORI algorithm is able to mine rare correlated patterns the occurrence behavior is not considered.
Example 1: The transaction database shown in Table 1 consists of 5 transactions with 6 unique items. Let *minSup* represents minimum support, *maxPer* represents maximum periodicity, and *minBond* represents minimum bond threshold value. Let *minBond* be set to 0.2 and both *minSup* and *maxPer* be set to 3. As shown in Table 1 item {D} appears in the transactions T_1 and T_5. Item {F} appears only at the end of the database. As these patterns satisfy support and bond thresholds, the patterns

Table 1 Transaction dataset—TD

TID	Items
*T*1	*A*, *B*, *D*
*T*2	*A*, *B*, *C*, *E*
*T*3	*A*, *B*, *C*, *E*
*T*4	*B*, *C*, *E*, *F*
*T*5	*B*, *C*, *D*, *F*

are mined in the CORI algorithm. But pattern {D} and {F} are non-periodic as their occurrence is not throughout the database. In this paper, the modification of the CORI algorithm called *PRCPMiner* (Periodic Rare Correlated Pattern Miner) is proposed to consider the occurrence sequence of the rare correlated patterns. The framework uses maximum periodicity along with support and bond measures to discard non-periodic rare patterns. As for our knowledge, this is the first framework that considers the periodicity measure along with the support threshold to mine rare patterns in the given static dataset. The major contributions of *PRCPMiner* algorithm are as follows:

- The transaction database is transformed into bit-vector representation. This transformed representation is used further for tree creation as well as computation of support, bond, and periodicity. From the literature work [9, 10], it is proved that bit-vector-based calculations are simple but effective.
- In order to filter large spurious patterns generated during periodic rare pattern generation, a bond measure is used. The literature work in [6, 8, 9, 11, 12] show that bond measure employed is simple and effective.
- Experimental studies show that *PRCPMiner* efficiently mines rare periodic correlated patterns from several datasets. The computational time of *PRCPMiner* is shown by considering various threshold values.

The remainder of the paper is arranged in the following way. Section 2 focuses on the literature survey. Section 3 defines the basic definitions of the *PRCPMiner* algorithm. Various modules of the *PRCPMiner* algorithm are presented in Sect. 4. Performance evaluation and analysis of results are presented in Sect. 5. Section 6 highlights the conclusion and shows the future direction.

2 Related Works

This section briefly discusses various frequent pattern mining algorithms that have considered the shape of occurrence along with the support measure. The section also highlights the different rare pattern mining algorithms designed to mine the rare patterns.

2.1 Periodic-Frequent Pattern Mining (PFPM)

Tanbeer et al. [1] constructed RP-Tree structure to search regular patterns by replacing the support measure used in FP-Tree with the concept of regularity. They also designed a single scan algorithm SDR-Growth [3] that discovers regularly occurring patterns in evolving body sensor data. Maximum periodicity is the measure used to discover the regular patterns in their approach. Rashid et al. [4, 5] introduced the concept of variance with the support threshold to find regular frequent patterns from the transactional database and senor data stream respectively. Kiran et al. [13–15] solved the "rare item problem" by setting multiple minimum support and maximum periodicity thresholds. They designed many algorithms where every framework is an improvisation of the previous work which is capable of finding frequent patterns with regular occurrence from transactional databases. PFPM algorithm is the contribution of Viger et al. [7] to mine periodic frequent patterns. Viger et al. [16] found a novel measure called *maxLa* that calculates the lability of itemsets to search frequent patterns that are stable and periodic. In the area of high-utility mining, Viger et al. [8] designed PHM algorithm, which contributed the novel measures minimum and average periodicity to enumerate periodic high-utility itemsets from the static database. MHUIRA and HUIIM are the contributions of Apmhawan et al. [17, 18] and the concept of regularity and irregularity are associated, respectively, during mining high-utility itemsets.

2.2 Rare Pattern Mining (RPM)

The level-wise exploration algorithms [19–21] follow Apriori method to mine the rare itemsets. Generation of vast number of candidate itemsets and multiple scans of the database to count the support are the drawbacks of these methods. As rare itemsets are dominant at the top part of the search space, Rarity [21] and ARANIM [20] follow the top-down approach to search the rare itemsets. On contrary, Szathmary et al. [19] designed a bottom-up algorithm to extract minimal rare itemsets. RP-Growth, modification of FP-Growth is a tree-based method designed by Tsang et al. [22] to mine rare patterns by removing noise components. A hyper-linked RPM algorithm is designed by Anindita et al. [23] to search rare patterns from the static dataset. Recently, Lu et al. [24] focused on the dual perspective by the negative representation of rare itemsets. They contributed NIIMiner a top-down depth-first method to find the entire set of rare patterns. As RPM results in a huge set of low support patterns, efficient techniques are required to filter the spurious patterns. The bond measure has attracted researchers and is used recently in many works to remove spurious low support patterns. Basker et al. [9] contributed algorithm named CORI, that is capable of finding rare correlated patterns in static data. They designed a single scan algorithm by transforming the input transaction database into bit-vector representation. Along with the support, a bond measure is used to find the correlation among the rare itemsets. Further, a tree structure is used to store support and bond measures of the itemsets. A depth-first search method is used for mining rare correlated patterns. But it does not consider the shape of occurrence of the itemsets. Viger et al. [6] first came up with the concept of periodicity in mining rare correlation patterns from multiple sequences. The first tree-based algorithm called *PRCPMiner* is proposed in this paper to mine PRCPs from the transactional database.

3 Definitions and Problem Statement

This section defines the basic terminologies and theorems related to *PRCPMiner* algorithm. Most of the terminologies are illustrated based on Example 1.
Given set of items $D = \{d_1, d_2, \ldots, d_n\}$, a transaction T = (tid,I) where *tid* represents unique transaction id and I is an itemset formed using D. Any itemset formed by combining k items of D are called as k-itemset. A transaction database TD consists of set of tuples $\{T_1, T_2, \ldots, T_m\}$. Any non-empty subset Q⊆I can also be written as Q⊆TD. The ordered transactions in which itemset I appears in TD are denoted as $T^I = \{T_g^i, \ldots, T_k^i\} \subseteq$ TD, where $g \leq k$ and $g, k \in$[1,m].

Definition 1: (Support of an itemset) The total count of transactions in which itemset I appears in TD is called Support(absolute) named as Sup(I,TD) = $| \{T \in \text{TD} \,|\, i_1 \in T \wedge i_2 \in T \wedge \ldots \wedge i_k \in T\} |$.

Definition 2: (Periodicity of an itemset) Given two consecutive transactions T_j^i and T_{j+1}^i in which itemset I appears, then *period* is calculated as $p^i = T_{j+1}^i - T_j^i$. Let $P^I = \{p_1^i, p_2^i \ldots, p_r^i\}$ represents the set of periods of itemset I. Then periodicity of itemset I denoted as Per(I,TD) = max($\{p_1^i, p_2^i \ldots, p_r^i\}$).

Definition 3: (Bond measure of an itemset) The total count of transactions in which at least one item of itemset I appears in TD denotes the disjunctive support called DisSup(I,TD)= $| \{T \in \text{TD} \,|\, i_1 \in T \vee i_2 \in T \vee \ldots \vee i_k \in T\} |$. The bond measure of itemset I is calculated as Bond(I,TD)= Sup(I,TD)/DisSup(I,TD).
For example, the item {D} is present in transaction T_1 and T_5. Therefore, Sup (D) = 2. As DisSup (D) is also 2, the Bond (D) results in value 1. The $P(D)$ = ($1-T_{\text{init}}$=1, $5-1=4$,$T_{\text{last}}-5=0$). Here T_{init} represents the initial transaction with value 0 and the last transaction in the database is represented by T_{last}. The maximum of these values is 4. Therefore, Per (D) results in 4.

Problem Statement: Consider a static database TD and three thresholds *minSup*, maxPer and *minBond*. The itemset I is said to be periodic rare correlated pattern (PRCP), if ((Sup(I,TD) $<$ *minSup*) $\wedge$ (Per(I,TD) $\leq$ *maxPer*) $\wedge$ (Bond(I,TD) $\geq$ *minBond*). The *MinSup* threshold value is used for selection of rare itemsets, whereas *maxPer* is a threshold value helps to identify periodicity of itemsets. As rare itemset extraction results in numerous spurious itemsets, *minBond* threshold value helps to discard itemsets co-occurred by chance.
The following properties can be defined using the above definitions.

Property 1: The support measure of rare patterns fulfills monotone property. For the given two itemsets $P \supseteq Q \wedge (\text{Sup}(Q,\text{TD}) < minSup)$, then $(\text{Sup}(P,\text{TD}) < minSup)$ since $\text{Sup}(P,\text{TD}) \leq \text{Sup}(Q,\text{TD})$ [9].

Property 2: The bond measure fulfills anti-monotone property. For the given two itemsets $P \subset Q$, then $\text{Bond}(P) \geq \text{Bond}(Q)$ [6].

Property 3: The periodicity measure fulfills anti-monotone property. For the given two itemsets $P \subset Q$, then $\text{Per}(P) \leq \text{Per}(Q)$.

4 The *PRCPMiner* Algorithm

This section demonstrates the working of the proposed framework *PRCPMiner*. It accepts three different thresholds named *minSup*, *minBond* and *maxPer* from the user. Utilizing these thresholds, *PRCPMiner* discovers PRCPs from the static database. Algorithm 1 shows the main steps of *PRCPMiner*.

Algorithm 1 PRCPMiner

Input: Transaction Database TD = $\{t_1, t_2, \ldots, t_m\}$, three threshold values *minSup*, *minBond* and *maxPer*

Output: Entire set of Periodic Rare Correlated Patterns - *PRCPList*

1: Scan the database *TD* and build the transformed bitset *BitSetTD*
2: *UniqueOneItemSortedSet* ← Find the support count, maximum periodicity. Using support count measure sort the items in ascending order using *BitSetTD*
3: Let root of the *PRCPTree* be 'NULL'. Add the resultant 1-itemsets in *UniqueOneItemSortedSet* to the root of *PRCPTree*
4: Recursively process each item in *PRCPTree* to discover PRCPs
5: Return *PRCPList*

Step 1—BitSet Transformation of the dataset: As the initial step, the database *TD* is scanned and transformed bitset representation named, *BitSetTD* is generated. Every item of *TD* shown in Table 1 is represented by the bitset form as shown in Table 2. Here the value '0' denotes absence and the value '1' indicates the presence of the item in the particular transaction represented by that column.

Step 2—Computation of support and maximum periodicity of 1-Itemsets: Using the *BitSetTD* information, the support and maximum periodicity information of all the 1-itemsets are obtained. The support count measure is used to sort the 1-itemsets in ascending order and is collected in *UniqueOneItemSortedSet*.

Step 3—Add the resultant 1-itemsets in the *UniqueOneItemSortedSet* **to the root of** *PRCPTree* **(Periodic Rare Correlated Pattern Tree)**: The root node of the *PRCPTree* is initialized with 'NULL' value. The resultant 1-itemsets in the *UniqueOneItemSortedSet* are added as the first layer children of the *PRCPTree* as shown in Fig. 1.

Table 2 Transformed representation of dataset shown in Table 1

Item	*T*1	*T*2	*T*3	*T*4	*T*5
A	1	1	1	0	0
B	1	1	1	1	1
C	0	1	1	1	1
D	1	0	0	0	1
E	0	1	1	1	0
F	0	0	0	1	1

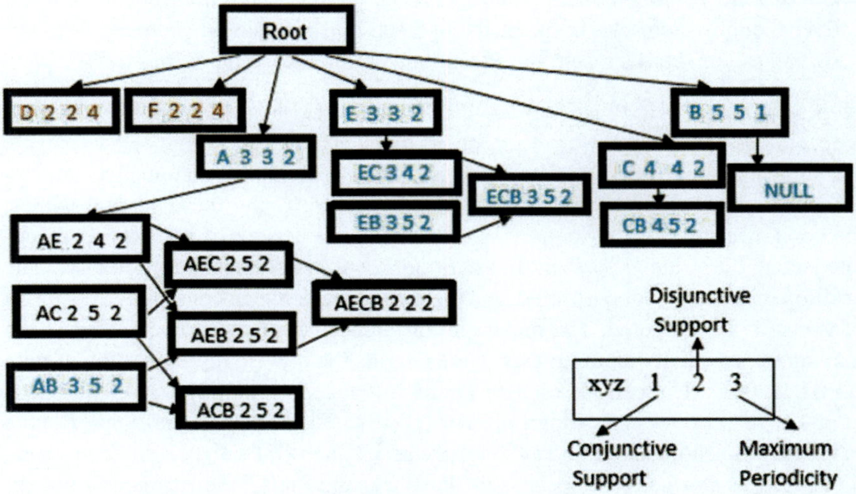

Fig. 1 *PRCPTree* built for the database shown in Table 1

Step 4—Recursively process each item in ***PRCPTree*** **to discover PRCPs**: This step is the major step in *PRCPMiner* to extract all supersets of periodic rare correlated 1-itemsets. Recursively, the sub-tree of every 1-itemset I of *PRCPTree* is constructed by performing a depth-first traversal method. Repeatedly, the candidates of size *n*-itemsets are constructed by the intersection of candidates of *n*-1-itemsets. The Sup(*n*,TD), Bond(*n*,TD), Per(*n*,TD) are computed for the generated itemsets and the computed values are compared with corresponding threshold values. The decisions are taken depending upon the result of comparison which is given below:

- **Condition 1: (Sup(*n*,TD) < *minSup*):**
  - **Condition 1 is TRUE**: If the candidate *n*-itemset is a rare itemset, then the periodicity and bond measures are compared.
  - **Condition 2: (Per(*n*,TD) ≤ *maxPer*) ∧ (Bond(*n*,TD) ≥ *minBond*)**

Condition 2 is TRUE: If Condition 2 is satisfied, then the candidate n-itemset satisfies both periodicity and bond measure. It results as a PRCP and stored in the *PRCPList*. Further, the depth-first search method is continued and sub-trees are built.
Condition 2 is FALSE: According to Property 2 and Property 3, both periodicity and bond measure fulfill anti-monotone constraint. Therefore, if either periodicity or bond measure is not satisfied, then supersets of n-itemset are not generating PRCPs. Hence, sub-trees are not constructed in this path.

- **Condition 1 is FALSE**: The n-itemset reaches this path, if it is a frequent itemset. Even though n-itemset is frequent, its supersets may result as rare itemsets. Without storing n-itemset as a result, the execution continues to build its sub-trees.

Step 5—Return *PRCPList*: All the PRCPs generated by following the recursive procedure as described in Step 4 are stored in the *PRCPList*.
The logical operations are performed on *BitSetTD* to obtain the conjunctive and disjunctive support as explained in the CORI algorithm. Using these support values, the bond measure of the 1-itemsets is calculated. The periods of itemsets are calculated using the value '1' present in the *BitSetTD* of the corresponding itemset. The position of value '1' is considered, and the difference between consecutive positions of value '1' is computed. The maximum difference computed is considered as the maximum periodicity of an itemset. For example, the bitwise representation of item {D} is '10001'. The periods of item D are 1–0 = 1, 5–1 = 4, 5–5 = 0, and maximum of these values is 4. Therefore, Per (D) is 4. The Sup($\{D\}$) and Sup($\{F\}$) are $<$ *minSup* threshold value. Even though items $\{D\}$ and $\{F\}$ satisfy rarity constraint, they result as non-periodic because the Per($\{D\}$) and Per($\{F\}$) are $>$*maxPer* threshold. The non-periodic itemsets are shown in *Red* color. As their supersets will not satisfy periodicity measures(Property 3), the sub-trees are not built. Figure 1 shows the *PRCPTree* built for the dataset shown in Table 1. The frequent itemsets are shown in *Blue* color. These itemsets are not discarded, and the sub-trees are built further to check for the possibility of generation of PRCPs. The final PRCPs generated for the dataset in Table 1 are shown in *black* color.

5 Performance Evaluation

The JAVA platform is used to implement the proposed *PRCPMiner* algorithm. The experiments are tested using the system with configuration Intel(R) Core(TM) i5-7400 CPU@3.00 GHz with 8 GB RAM running Windows10 Enterprise. The execution time taken by the *PRCPMiner* algorithm is tested using the two datasets downloaded from the "frequent itemset mining dataset repository". *Mushroom* is a dataset with total transactions of about 8 k, and *Connect* dataset is about 67 k transactions. The maximum transaction length of *Connect* is high. It will generate

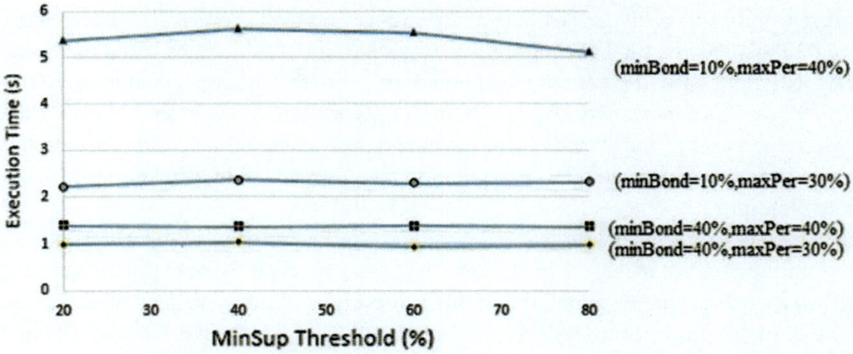

Fig. 2 Mushroom dataset

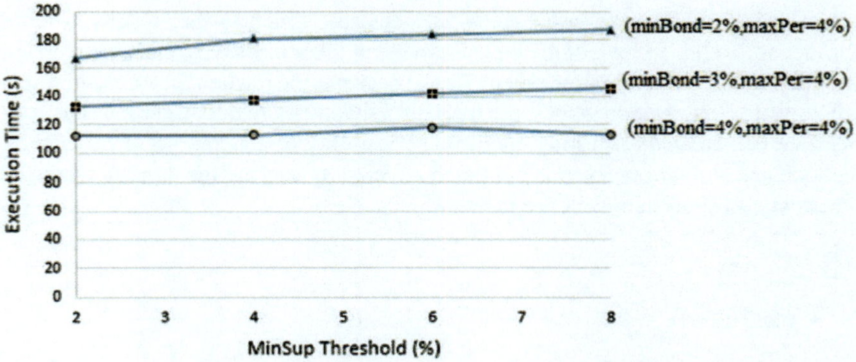

Fig. 3 Connect dataset

a huge set of rare itemsets because of which the algorithm is not able to complete the mining task. So the transaction length is restricted to a value of 15. The different experiments were carried out by keeping the *maxPer* threshold value constant 30 and 40% while testing the *Mushroom* dataset and 4% in case of *Connect* dataset. The *minBond* threshold is varied as 10 and 40% during the testing of the *Mushroom* dataset and 2–4% during the testing of *Connect* dataset. For all cases, the *minSup* threshold value is varied from 20 to 80% in case of *Mushroom* dataset and 2% to 8% in case of *Connect* dataset.

Figures 2 and 3 show the total execution time taken by the algorithm for *Mushroom* and *Connect* dataset, respectively. In these figures, the *X*-axis represents *minSup* threshold values and the runtime in seconds is represented by *Y*-axis. As this is the first algorithm that considers periodicity threshold in the mining of rare patterns, *PRCPMiner* is tested with variation in the threshold values considered as explained below.

Influence of ***maxPer***: It can be observed from Fig. 2, as the *maxPer* value is increased from 30 to 40%, the total execution time increases in both the cases when *minBond* is set as 10 and 40%. It is because the number of periodic 1-itemsets increases, which further results in more periodic rare itemsets generation. The number of sub-trees created increases as the periodic 1-itemsets and their supersets increase. Further, the number of periodicity and bond computation increase, which in turn affects the total execution time.
Influence of ***minBond***: Figures 2 and 3 show the influence of *minBond* threshold on the runtime performance. As Fig. 2 shows, when the *minBond* is increased by keeping the *maxPer* threshold constant, the number of correlated itemsets decreases. As the number of correlated itemset decreases, more sub-trees are pruned. This results in a decrease in the mining time.
Influence of ***minSup*** : When the *minSup* threshold value is increased by keeping the other two thresholds constant, the number of PRCPs increased in all the cases. This is because the number of rare 1-itemsets increases, which further results in more number of PRCPs. In spite of the increase in the number of PRCPs, there is not much variation in the execution time taken. This is because, even when the rare 1-itemsets are lesser for low *minSup* values, the frequent 1-itemsets are not discarded. The sub-trees are still created for testing the further possibility of generations of PRCPs. So the total execution time taken is not much affected by varying the *minSup* when the other two thresholds are kept constant.

6 Conclusion

A single scan algorithm called *PRCPMiner* is implemented in this paper to mine periodic rare correlated patterns. The algorithm introduced the periodicity concept during the mining of rare correlated patterns. Initially, the input transactions are transformed into bit-vector form. Then the *PRCPTree* is built which helps in pruning non-periodic and non-correlated itemsets. Influence of the three different thresholds *minSup*, *minBond*, and *maxPer* on the runtime performance of the *PRCPMiner* algorithm is studied. As *maxPer* threshold value is increased, the number of periodic rare itemsets increases. This further increases the computation time which causes an increase in the total execution time. Research shows that the number of correlated itemsets decreases when the *minBond* threshold value is increased. This results in the pruning of more number sub-trees which decreases the mining time. When the threshold *minSup* value is increased, the number of periodic rare correlated patterns increases. This is not affecting the execution time because the supersets of frequent items still need to be checked. The research carried out in this paper sets several future directions. The maximum periodicity threshold has the drawback of rejecting the itemsets even when a single period is not satisfying the threshold value set. Hence as a future work, other thresholds like variance and standard deviation can be considered for testing the occurrence behavior. The proposed algorithm can be extended further to handle the data stream.

References

1. Tanbeer, S. K., Ahmed, C. F., Jeong, B.-S., & Lee, Y.-K. (2008). Mining regular patterns in transactional databases. *IEICE Transactions on Information and Systems, 91*(11), 2568–2577.
2. Tanbeer, S. K., Ahmed, C. F., & Jeong, B.-S. (2010). Mining regular patterns in data streams. In *International conference on database systems for advanced applications* (pp. 399–413). Springer.
3. Tanbeer, S. K., Hassan, M. M., Almogren, A., Zuair, M., & Jeong, B.-S. (2017). Scalable regular pattern mining in evolving body sensor data. *Future Generation Computer Systems, 75*, 172–186.
4. Rashid, M. M., Karim, M. R., Jeong, B.-S., & Choi, H. J. (2012). Efficient mining regularly frequent patterns in transactional databases. In *International conference on database systems for advanced applications* (pp. 258–271). Springer.
5. Rashid, M. M., Gondal, I., & Kamruzzaman, J. (2013). Regularly frequent patterns mining from sensor data stream. In *International conference on neural information processing* (pp. 417–424). Springer.
6. Fournier-Viger, P., Yang, P., Li, Z., Chun-Wei Lin, J., & Kiran, R. U. (2020). Discovering rare correlated periodic patterns in multiple sequences. *Data & Knowledge Engineering, 126*, 101733.
7. Fournier-Viger, P., Lin, C.-W., Duong, Q.-H., Dam, T.-L., Ševčík, L., Uhrin, D., & Voznak, M. (2017). PFPM: Discovering periodic frequent patterns with novel periodicity measures. In *Proceedings of the 2nd Czech-China scientific conference 2016*. IntechOpen.
8. Fournier-Viger, P., Chun-Wei Lin, J., Duong, Q.-H., & Dam, T.-L. (2016). PHM: Mining periodic high-utility itemsets. In *Industrial conference on data mining* (pp. 64–79). Springer.
9. Bouasker, S., & Yahia, S. B. (2015). Key correlation mining by simultaneous monotone and anti-monotone constraints checking. In *Proceedings of the 30th annual ACM symposium on applied computing* (pp. 851–856).
10. Aryabarzan, N., Minaei-Bidgoli, B., & Teshnehlab, M. (2018). Negfin: An efficient algorithm for fast mining frequent itemsets. *Expert Systems with Applications, 105*, 129–143.
11. Bouasker, S., Hamrouni, T., & Yahia, S. B. (2012). New exact concise representation of rare correlated patterns: Application to intrusion detection. In *Pacific-Asia conference on knowledge discovery and data mining* (pp. 61–72). Springer.
12. Fournier-Viger, P., Chun-Wei Lin, J., Dinh, T., & Bac Le, H. (2016). Mining correlated high-utility itemsets using the bond measure. In *International conference on hybrid artificial intelligence systems* (pp. 53–65). Springer.
13. Uday Kiran, R., & Kitsuregawa, M. (2014). Novel techniques to reduce search space in periodic-frequent pattern mining. In *International conference on database systems for advanced applications* (pp. 377–391). Springer.
14. Uday Kiran, R., Kitsuregawa, M., & Krishna Reddy, P. (2016). Efficient discovery of periodic-frequent patterns in very large databases. *Journal of Systems and Software, 112*, 110–121.
15. Venkatesh, J. N., Uday Kiran, R., Krishna Reddy, P., & Kitsuregawa, M. (2018). Discovering periodic-correlated patterns in temporal databases. In *Transactions on large-scale data and knowledge-centered systems XXXVIII* (pp. 146–172). Springer.
16. Fournier-Viger, P., Yang, P., Chun-Wei Lin, J., & Kiran, R. U. (2019). Discovering stable periodic-frequent patterns in transactional data. In *International conference on industrial, engineering and other applications of applied intelligent systems* (pp. 230–244). Springer.
17. Amphawan, K., Lenca, P., Jitpattanakul, A., & Surarerks, A. (2016). Mining high utility itemsets with regular occurrence. *Journal of ICT Research & Applications, 10*(2).
18. Laoviboon, S., & Amphawan, K. (2017). Mining high-utility itemsets with irregular occurrence. In *2017 9th international conference on knowledge and smart technology (KST)* (pp. 89–94). IEEE.
19. Szathmary, L., Valtchev, P., & Napoli, A. (2010). Generating rare association rules using the minimal rare itemsets family.

20. Adda, M., Wu, L., White, S., & Feng, Y. (2012). Pattern detection with rare item-set mining. arXiv:1209.3089.
21. Troiano, L., Scibelli, G., & Birtolo, C. (2009). A fast algorithm for mining rare itemsets. In *2009 ninth international conference on intelligent systems design and applications* (pp. 1149–1155). IEEE.
22. Tsang, S., Koh, Y. S., & Dobbie, G. (2011). Rp-tree: Rare pattern tree mining. In *International conference on data warehousing and knowledge discovery* (pp. 277–288). Springer.
23. Borah, A., & Nath, B. (2017). *Mining rare patterns using hyper-linked data structure* (pp. 467–472).
24. Lu, Y., Richter, F., & Seidl, T. (2020). Efficient infrequent pattern mining using negative itemset tree. In *Complex pattern mining* (pp. 1–16). Springer.

Performance and Application of Digital Forensic Tools: A Comparative Study

Savali Deshmukh and Pramod Kumar Jha

Abstract Currently, computers and the Internet are used to conduct the majority of business transactions, communications and the automated control of industrial equipment, among other things. Working online makes the process more efficient and convenient. The risk of cyber-attacks has also increased significantly as a result of devices being exposed to the Internet on a daily basis. The Internet's speed, ease of use and invisibility, lack of geographical boundaries cyber financial crimes, stalking and bullying are becoming more commonplace, according to the FBI. A digital forensic investigation carried out with the assistance of software tools yields evidence against cybercriminals that can be presented in court. This review work aimed to evaluate and compare the performance and applications of ten online digital forensic tools. The conclusions, limitations of these tools and how after moral improvement, they can be used to assist digital forensics professionals in discovering digital evidence are presented.

Keywords Digital forensic tools · Cybercrime · Open-source software · Performance · Application

1 Introduction

Modern digital forensics deliver reliable computer analysis and digital evidence collecting for a wide range of applications in law and industry. As a result, prototype implementations are commonly included in research projects. For example, look at the DJI Phantom III Drone [1]. The findings on the patented encrypted file format were reported in their research work in which a reference manager to automate the

S. Deshmukh (✉)
School of Computer Science and Engineering, Vellore Institute of Technology—AP University, Amravati, India
e-mail: sadip.19bce7348@vitsap.ac.in

P. Kumar Jha
Centre for Advanced Systems, DRDO Hyderabad, Hyderabad, India
e-mail: pkj@cas.drdo.in

B. Pati et al. (eds.), *Proceedings of the 6th International Conference on Advance Computing and Intelligent Engineering*, Lecture Notes in Networks and Systems 428,
https://doi.org/10.1007/978-981-19-2225-1_57

process was also included. Although digital storage media (such as a USB memory stick or a hard disc drive) may be physically and visually examined, the data stored on these devices can only be analysed using specialised equipment and software capable of deciphering and displaying it understandably. While microscopic technologies may allow for manual data analysis on specific device types at a sector level, it is not practical to investigate media in this manner in most cases. And, when it comes to accurately interpreting and presenting digital evidence, forensic investigators rely on the digital software tools they use for the investigation [2].

In private, digital forensic techniques are frequently used to find a piece of evidence that can be utilised in court, reverse engineering of computer systems, data recovery, maintenance and troubleshooting [3]. Online digital forensics can be used by any user who has a clear idea about their needs. Specific tools are created for this exact reason, helping one choose which software would best solve one's requirement. Additionally, the creation of digital forensic competitions encourages tool development, with the Digital Forensics Research Workshop (DFRWS) challenges being a notable example. The DFRWS conferences have been challenging scholars and practitioners since 2005 to push the state of the art in developing forensic tools [4].

1.1 Introduction to Cyber Crimes

When dealing with a cybercrime scene, it is critical to pay close attention to digital evidence as the crime scene evidence is presented in an electronic form, which significantly distinguishes cybercrime from traditional crime. Further, it facilitates the criminal to store, hide, spread and delete information, making arresting cybercrime suspects more difficult [5].

According to [6], cybercrime covers the following:

1. Intellectual property theft
2. Damaging of service networks of a company
3. Financial fraud
4. The intrusion of hackers system
5. Distribution of execution virus and worms

Cybercrime can be split into three comprises or "3Ts" [7]:

i. Tools to commit a crime
ii. Targets of the crime (Victim)
iii. Material that is tangential to the crime

To detect and find evidence against a cybercrime, digital forensic can be used. Wang and his team devised a strategy of leveraging forensic toolkits to aid the collecting of robust digital evidence in order to keep the compelling clues from computer-based systems. As a result, vital tracks left at a cybercrime scene can be used to convict the perpetrators. To raise awareness of cybercrime, researches created a Web forensic framework based on four different sorts of patterns that provide them with proof of harmful Bot activity on Web services [8].

1.2 Introduction to Tools

Tools are not only made for a specific purpose but also for general use [9]. However, Lexico defines a tool as a device or implementation that is used to perform a specific function [10]. They have also been described as a self-contained tool and provide a particular amount of automation, i.e. user intervention is minimal, reduced and abstracted. For example, a tool should not need the user to determine sector numbers or translate virtual to physical addresses manually to access the disc. Individuals or research groups frequently create and use forensic tools in any computer language of their choice. Also, if a tool is automated, it can be employed in other programmes. Various forensic tools support us in obtaining the disc images and automating much of the analysis process as well, such as:

I. File fragments, hidden and deleted files and directories can be identified and recovered from any location
II. The file structure, headers and other aspects determine directly the kind of data each file contains
III. All the contents of the graphic files can be displayed
IV. Advanced searches can be conducted
V. Can exhibit the directory structure of the drive acquired graphically
VI. And producing the reports

The Autopsy tool, an upgraded version from the sleuth kit forensic tools with some add-ons, is used in the fields of law enforcement, military, corporate examination, recovery, data backup, training and in some commercial areas with restriction included or with the limited privileges over the problems. Wireshark is a packet sniffer and analyser and records on the local network all the network traffic and saves it for later study. The Metasploit framework is a forensic tool that may be used by both cybercriminals and ethical hackers to investigate network and server vulnerabilities. Nessus is a remote complete security scanning programme that checks for security flaws on a computer and informs you whether such vulnerabilities may potentially be exploited by malicious hackers to grant access to other networked computers. Nmap is a free network mapper that uses IP packets to search a network for live hosts, port scans, ping sweeps, OS detection and version detection. AccessData created FTK Imager, a data viewing and imaging application. Volatility is a memory forensics framework that is free (under the GPL licence) and can be used for incident response and malware investigation. Computer-aided investigate environment (CAINE) is a Linux distribution that provides a detailed forensic investigation and reporting environment, with a graphical user interface that is designed to let users examine, investigate and gather actionable findings. MAGNET RAM Record is an effective imaging technique because it allows investigators to extract and examine artefacts that usually only exist in local physical memory. Network Miner is a network forensics programme that utilises packet sniffing or a PCAP file to identify OS, sessions, hostname and open ports without putting any traffic on the network.

1.3 Motivation and Contributions

No research has been done on the diversity, availability or quality of the tools that have been published. As a result, this review work came up with the following study question: What factors influence the applicability and use of tools? To get an answer to this question, the research papers from a variety of digital forensics magazines and conferences and tools' performance and applications have been studied. Tools were tested for availability, usability, deployment, GUI, error prevention and handling and API integration. Along with this, the current challenges in the area of forensic tool development have been discussed in the paper.

2 Methodology

In this review work, the research and review publications on digital forensic tools' performance, applications and limitations were reviewed and analysed. The purpose of this investigation is to locate software mainly designed for research purposes, as well as to investigate any preceding work or other features of these tools. Figure 1 shows how the current work was written using the process of a detailed review. Cumulated research articles were reviewed explicitly, focusing on tools usability and limitations. Tools were tested on their performance features such as availability, deployment, GUI, error prevention and handling and API integration. Along with this, research articles were found where the application of any of these tools in digital forensic or any other field was proven. The result of the comparative study was separated and documented into two tables. With the help of these two tables, the various features of each tool, their application and their limitations are concluded.

3 Observations

After segregating the tool-specific research articles, the software tool's performance was tested depending on various key features. It is essential for a tool to be deployed so as to create a collaborative environment. This would make it easy for everyone to see all of the outcomes in near real time. Hence, there's no need to integrate the results anymore, and a single, unified report may be generated at any moment [11]. Autopsy provides the best collaborative environment. GUI is by far the most popular

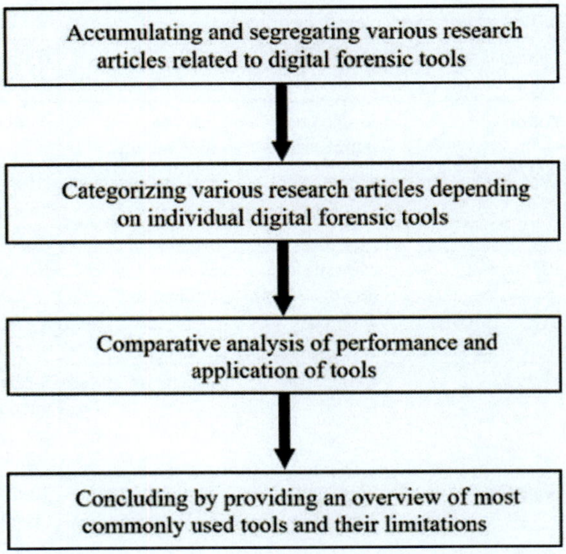

Fig. 1 Steps involved in the review

means used to interact with software today [12]. As it enables increased productivity, while also reducing cognitive stress. It is critical to effectively prevent and handle errors in requirements analysis and design to improve software productivity and reliability [13]. Since many software are developed to support a wide range of applications. It is critical for software tools to rely on the implementation of mature application programming interfaces (APIs) to facilitate the growth of software for artificial intelligence of things (AIoT) [14] (Table 1).

This research work also highlights the various applications in which these forensic tools were used. Table 2 gives the observations made by authors while they were using these software tools for their requirements. The various fields in which these tools are incorporated prove that the digital forensics community has a strong application orientation, which means we solve problems in practice rather than a theory. After reviewing the works, we have tabulated the application, conclusion and limitations of tools observed by the authors.

3.1 Limitations Observed

To be admissible in court, digital forensics must follow a specific collection, analysis and reporting process. Despite the expanding use of electronic forensics to help out in criminal cases and the necessity for practical tools, NIST's forensic tool testing programme remains the only one available. Also, digital forensic tools have no international standard.

Table 1 Performance analysis of tools

Name of software	Deployment	GUI	Error prevention and handling	APIs
Autopsy	Single and multi-user	Easy to understand	Not considered	Integrated
Wireshark	Single user	Can be improved	Uses macros based on kazlib's exception code	Lack of integration
Metasploit	Metasploit pro deploys single and multi-user	Easy to understand	Not considered	Integrated with Metasploit pro with REST API command
Nessus	Single user	Can be improved	It can find vulnerabilities but cannot prevent attacks	It supports API, but Nessus pro does not
Nmap	Single user	Zenmap is the GUI version of Nmap	To implement an exception handler Nmap, new try API method is used	Nmap API enables integration
FTK imager	Single user	Easy to understand	Not considered	Integrated
Volatility tool	Single and multi-user	Absence of GUI	Can detect bugs, viruses and malware but cannot prevent attacks	Apihooks are used to detect API hooks in-process and kernel memory
CAINE	Single user	Easy to understand	Built-in tools which can handle errors	Lack of integration
MAGNET RAM capture	Single user	Easy to understand	Not considered	Lack of integration
NetworkMiner	Single user and Xplico allows multi-user	Easy to understand	Not considered	Integrated

The lack of support, documentation, updates and the software's safety are all risks associated with using free tools/software. Tools used in the research were either poorly documented or not documented at all. The study highlights the poor user interface and developers' disinterest. Despite publication, it has been discovered that most tools were only used in cited works (2014) [25]. When limited test data is provided, or only specific tool versions or single picture format is inputted, the NIST tool testing requirements are "narrowly specified" [26]. There is also a lack of testing methods to analyse the tool reliability that exist for established technologies as well.

Because no single digital forensics tool can do everything due to the ongoing evolution of the field, researchers frequently build solutions to fill in the gaps left

Table 2 Application analysis of tools

Authors	Name of software	Application	Conclusion	Limitations observed
Negrão and Domingues [15]	Autopsy	Autopsy's SpeechToText modules identify and transcribe voice material	The detection and transcription can speed up the process of finding relevant information in audio files in forensic images	Saved amr files without headers were not detected by autopsy and had to be converted to WAV
Umar et al. [16]	Wireshark	Using an android-based live email service, they compared Wireshark and NetworkMiner forensic tools	NetworkMiner forensic tools have succeeded in getting more digital evidence than Wireshark	Wireshark cannot capture the receiving port
Tantawy et al. [17]	Metasploit	Metasploit modules are used to semi-automate attack injections	The authors demonstrated the need for an integrated approach to safety and security system design	Asynchronous communication does not slow down control algorithm performance or support Metasploit capabilities
Bairwa et al. [18]	Nessus	Identification of the underlying vulnerabilities	Nessus is a commonly used tool and shows the best scanning capabilities in comparison with the other tools for the selected vulnerabilities	It lacks functionalities and so cannot be integrated with another tool that acts differently and produces different results
Wiberg [19]	Nmap	In an active scan, Nmap scans the target using information files and SCADAScan options	The prototype application can take advantage of Nmap's extensive active scanning capabilities	When performing port recognition or service detection on SCADA devices, it reveals a flaw

(continued)

by existing tools. It is difficult to quantify how research has affected the real world since most of the study is academic-focused. This also leads to a trade-off between academic and field interests. Authors point out how digital forensic is applicable in various areas, and as a result, collaboration and transparency are required, possibly through programmes that distribute research-based tools to industrial participants who would not otherwise contact with academia [27].

Table 2 (continued)

Authors	Name of software	Application	Conclusion	Limitations observed
Dykstra and Sherman [20]	FTK imager	FTK imager was tested to find its effectiveness in remotely acquiring forensic evidence through cloud computing	Remote access to a hard disc and memory image was successful	Since a lot of trust and risk is involved, the authors do not recommend using EnCase and FTK for remote forensics in the cloud
Ghafarian and Wood [21]	Volatility tool	Memory analysis for obtaining operating system level data	Volatility is a set of commands that parse the memory tree structure and report memory activity in processes and their interrelationships	Even though both volatility and process monitor expose data, establishing a link between the suspect and Skype actions is challenging
James and Gladyshev [22]	CAINE	Both programmes, along with Deepthought, were run from a CD on five test systems with known cryptographic hash values to see if they changed data	CAINE and Deepthought showed no effect on the test discs data	Time is taken per device for enhanced preview processing
Faiz and Prabowo 2019 [23]	MAGNET RAM capture	Comparison between five software for the acquisition of the best random-access memory	Most artefacts captured, registry key and DLL by the magnet RAM capturer	It took the longest processing time in seconds
Song et al. 2019 [24]	Network Miner	Comparing machine learning and rule-based approaches like NetworkMiner for OS identification	Machine learning along with OS attribute values correctly identified the operating systems	When it comes to OS identification, IP and timestamp are frequently left out

A tool is primarily built based on the developer's individual demands and preferred language. The tool is not thoroughly tested or documented due to lack of robustness and maintainability. Insecure coding compromises security, dependability, flexibility and scalability. No technologies exist to extract IoT traces from mobile devices.

Collecting and analysing IoT memory is also difficult. Bluetooth, Zigbee and Z-Wave forensics tools must be added.

4 Conclusions and Future Work

Due to the fact that the vast majority of transactions and communication in today's world takes place online, digital security has become increasingly important. The demand for forensic-based approaches and tools has also skyrocketed as a result of this. Accurate computer analysis and digital evidence collection are required for various legal and commercial applications, and digital forensics technologies are critical in this regard. This review work provides a comparison of a number of free source digital forensic tools, which anyone can use depending on their requirements. In this comparison, the tools are evaluated on a variety of criteria, allowing users to choose the tool that best meets their needs and, as a result, provided superior forensic visualisation. Various performance features along with applications of these ten tools are tabulated. The work also discusses the limitations for forensic tools like reliability, usability, maintainability and the need of integrating IoT. These observations can be utilised to develop software that meets the needs of digital forensics professionals. The advancement of these forensic techniques will significantly aid in the discovery of digital evidence.

It is proposed to extend this comparison study of digital forensic tools on various other factors and across other forensic tools, which will give a better insight into these tools. It can further work to provide a set of guidelines for designing digital forensics tools.

Author Contributions Savali Deshmukh: Performed the analysis, wrote the paper. Pramod Kumar Jha: Conceived and suggested the analysis of tools.

References

1. Clark, D. R., Meffert, C., Baggili, I., & Breitinger, F. (2017). DROP (DRone open-source parser) your drone: Forensic analysis of the DJI phantom III. In *DFRWS 2017 USA—Proceedings 17th Annual DFRWS USA*, 3–14, https://doi.org/10.1016/j.diin.2017.06.013
2. Guo, Y., Slay, J., & Beckett, J. (2009). Validation and verification of computer forensic software tools-searching function. *Digital Investigation, 6*, 12–22. https://doi.org/10.1016/j.diin.2009.06.015
3. Hibshi, H., Vidas, T., & Cranor, L. (2011). Usability of forensics tools: A user study. In *Sixth international conference on IT security incident management and IT forensics*, IMF, 81–91, https://doi.org/10.1109/IMF.2011.19
4. Wu, T., Breitinger, F., & O'Shaughnessy, S. (2020). Digital forensic tools: Recent advances and enhancing the status quo. *Forensic Science International: Digital Investigation, 34*, 300999. https://doi.org/10.1016/j.fsidi.2020.300999

5. Wang, S. J. (2007). Measures of retaining digital evidence to prosecute computer-based cyber-crimes. *Computer Standards Interfaces, 29*(2), 216–223. https://doi.org/10.1016/j.csi.2006.03.008
6. Marcella, J. R. & Menendez, D. (2010). *Cyber Forensics*
7. Abdulhamid, M., Profile, S. E. E, & Waziri, V. O. (2013). Cyber crimes analysis based-on open-source digital forensics tools some of the authors of this publication are also working on these related projects: Nature inspired meta-heuristic algorithms for deep learning: Recent progress and novel perspective vie, July 2016
8. Rahman, R. U., & Tomar, D. S. (2020). A new web forensic framework for bot crime investigation. *Forensic Science International: Digital Investigation, 33*, 300943. https://doi.org/10.1016/j.fsidi.2020.300943
9. Sammons, J. (2015). Chapter 3—labs and tools. *Basics of digital forensics* (2nd ed., pp. 31–46). https://doi.org/10.1016/B978-0-12-801635-0/00003-6.
10. TOOL English definition and meaning _ Lexico.
11. Mohan, (2016). Digital forensic investigation using sleuth kit autopsy. In *National conference on information, communication cyber security at department of computer science and engineering Kalasalingam university*, 0–43, [Online]. https://www.researchgate.net/publication/302580146_Digital_forensic_investigation_using_sleuth_kit_autopsy.
12. Memon, M. (2003). Advances in GUI testing-4 challenges. *Advances in Computers, 58*, 159–161.
13. Liu, S. (2021). A three-step hybrid specification approach to error prevention. *Journal of Systems and Software, 178*, 110975. https://doi.org/10.1016/j.jss.2021.110975
14. Xu, Y., Wu, Y., Gao, H., Song, S., Yin, Y., & Xiao, X. (2021). Collaborative APIs recommendation for artificial intelligence of things with information fusion. *Future Generation Computer Systems, 125*, 471–479. https://doi.org/10.1016/j.future.2021.07.004
15. Negr, M. (2021). SpeechToText : An open-source software for automatic detection and transcription of voice recordings in digital forensics. *Forensic Science International: Digital Investigation 38*. https://doi.org/10.1016/j.fsidi.2021.301223
16. Umar, R., Riadi, I., & Muthohirin, B. F. (2019). Live forensics of tools on android devices for email forensics. *Telkomnika (Telecommunication Computing Electronics Control), 17*(4), 1803–1809. https://doi.org/10.12928/TELKOMNIKA.v17i4.11748
17. Tantawy, A., Abdelwahed, S., Erradi, A., & Shaban, K. (2020). Model-based risk assessment for cyber physical systems security. *Computers and Security, 96*. https://doi.org/10.1016/j.cose.2020.101864
18. Bairwa, S., Mewara, B., & Gajrani, J. (2014). Vulnerability scanners: A proactive approach to assess web application security. *International Journal of Computers and Application., 4*(1), 113–124. https://doi.org/10.5121/ijcsa.2014.4111
19. Wiberg, K. C., (2006). *Identifying supervisory control and data acquisition (SCADA) systems on a network via remote reconnaissance.* Security, *147*, [Online]. http://citeseerx.ist.psu.edu/viewdoc/download?doi=10.1.1.149.4823&rep=rep1&type=pdf.
20. Dykstra, J., & Sherman, A. T. (2012). Acquiring forensic evidence from infrastructure-as-a-service cloud computing: Exploring and evaluating tools, trust, and techniques. *Proceedings Digital Forensic Research Conference DFRWS 2012 USA, 9*, 90–98. https://doi.org/10.1016/j.diin.2012.05.001
21. Ghafarian, A., & Wood, C. (2019). Forensics data recovery of skype communication from physical memory, 857. Springer International Publishing
22. James, J. I., & Gladyshev, P. (2012). A survey of digital forensic investigator decision processes and measurement of decisions based on enhanced preview. *Digital Investigation, 10*(2), 148–157. https://doi.org/10.1016/j.diin.2013.04.005
23. Faiz, M. N., & Prabowo, W. A. (2018). Comparison of acquisition software for digital forensics purposes. *Kinetik Game Technology Information System, Computer Network, Computer Electronics and Control, 4*(1), 37–44. https://doi.org/10.22219/kinetik.v4i1.687
24. Song, J., Cho, C. H., & Won, Y. (2019). Analysis of operating system identification via fingerprinting and machine learning. *Computers & Electrical Engineering, 78*, 1–10. https://doi.org/10.1016/j.compeleceng.2019.06.012

25. "Limitations—5 *Dangers of free software.*"
26. Horsman, G. (2019). Tool testing and reliability issues in the field of digital forensics. *Digital Investigation, 28*, 163–175. https://doi.org/10.1016/j.diin.2019.01.009
27. Horsman, G. (2019). Raiders of the lost artefacts: Championing the need for digital forensics research. *Forensic Science International Reports, 1*, 100003. https://doi.org/10.1016/j.fsir.2019.100003

Design of Automatic Headlamp Control Using Light-Dependent Resistor Sensors and Night Vision Camera

S. Madhankumar, K. Abhinav Kumaar, S. Arunachalam, R. Bavan Kalyan, K. Hrithik, S. Rajesh, and Mohan Rao Thokala

Abstract Over the nighttime, a vehicle's headlights cause significant risk. Its high beam that comes from the opposite vehicle creates a temporary glare to drivers. The light intensity from the high beam is 1200 lumens which are comparatively higher than the light intensity from the low beam (700 lumens). For this circumstance, the headlight should be dimmed to avoid road accidents. To avoid such incidents, the automatic headlight dimmer circuit can be used. This circuit consists of light-dependent resistor (LDR) sensors and a night vision camera to sense the light from the approaching vehicle that comes in opposite direction. Then, the light signals are processed using an Arduino microcontroller and the output signals are given to the switching relays. Hence, it switches automatically from the high beam into the low beam and reduces the glare effect faced by the drivers coming from the approaching vehicle. It avoids human intervention in the dimming of headlights.

Keywords Headlight · Automation · Arduino microcontroller · Vehicle

1 Introduction

The requirement of a headlamp plays a major role in driving automobiles at night. The headlamp has two vision controls, one is a high beam which is used for long vision and the other one is a low beam which is used for short vision [1, 2]. The driver has the control to change the beams according to the driving condition [3]. The intensity of low beams is low compared to high beams [4, 5]. The need for

S. Madhankumar (✉) · K. Abhinav Kumaar · S. Arunachalam · R. Bavan Kalyan · K. Hrithik
Department of Mechatronics Engineering, Sri Krishna College of Engineering and Technology, Coimbatore 641008, India
e-mail: madhankumars@skcet.ac.in

S. Rajesh
Department of Mechanical Engineering, R. M. K. Engineering College, Tiruvallur 601206, India

M. Rao Thokala
Department of Electronics and Communication Engineering, Siddhartha Institute of Engineering and Technology, Hyderabad 501506, India

B. Pati et al. (eds.), *Proceedings of the 6th International Conference on Advance Computing and Intelligent Engineering*, Lecture Notes in Networks and Systems 428,
https://doi.org/10.1007/978-981-19-2225-1_58

changeover of low beams from high beams is to reduce the glare and dazzling effect faced by the driver in the approaching vehicle, especially in the two-way roads. Traditionally, the shifting of the high to low beam and from the low to the high beam is only done manually [6]. This paper clearly gives a solution of automatic switching of low beams immediately when the approaching vehicle is nearing. The entire work is very simple and changes according to environmental conditions.

And the average light intensity of a halogen-type headlight is between 700 and 2000 lumens when tested under the standard distance of 50 ft. Continuous exposure to these kinds of high lumens intensity lights leads to the Troxler effect. This effect causes temporary blindness.

1.1 Motivation and Objective

The automatic headlight control reduces the above-mentioned problems and gives pleasant and comfortable driving to drivers. In this paper, the light automatically switches from high to low intensity when an oncoming automobile is nearing the concerned vehicle. The circuit structure of the automatic headlamp control device is demonstrated in this article, along with a complete explanation of every component and its purpose. This circuit's real-time programs are run and tested. The goal of this project is to create an effective system that decreases human operators in headlights while using minimal resources to improve passenger safety.

2 Components

The components used for this system are LDR for light detection, night vision cameras for detecting the vehicle coming in opposite direction, an Arduino microcontroller for executing the particular function, relays for switching, a transistor for amplifying the signals.

2.1 SPDT Relays

A relay is an electrically operated switch that is used for switching bulbs in automatic headlight control. There are two types of relays used in this system for the automatic switchover. They are, normally closed (NC) and normally open (NO) contact relays. The NC contact relay is connected to the high beam which turns off the high beam bulb immediately when the approaching vehicle's light reaches the LDR. The NO contact relay is connected to the low beam which makes the low beam glow when the approaching vehicle's light reaches the LDR. The relays also help in transmitting the continuous flow of power to the headlights.

2.2 Light-Dependent Resistor

The LDR device and reacts depending on the intensity of light. It is also called a phototransistor or photodiode, and it is basically a type of capacitor that is controlled by light. The resistivity of a photoresistor reduces as the brightness of the beam increases. The resistance will be high in a dark surrounding, which will be recorded in a few mega ohms (MΩ), and low in a light environment, which will be measured in only several 100 Ω (roughly). If the incoming light on the LDR reaches a particular brightness, the photon captured by the semiconductors gives the required power to transfer the bound electron into the bandgap. The charged particles (as well as their entire companions) transmit energy, lowering impedance. The impedance spectrum and sensitivities of a phototransistor will vary greatly between machines. Furthermore, various light barriers may respond significantly differently to light within a specific spectral band. A photovoltaic detector may be classified into two types: intrinsically and extrinsically. An intrinsically semiconductor is composed of its own conduction electrons (ex, silicon). Because the only accessible protons in intrinsic systems are now in the conductive band, the light should have enough potential to stimulate the electrons over the whole bandgap energy. Defects, also known as doped silicon, are present in extrinsic systems.

2.3 Night Vision Camera

The night vision camera and device are made up of an infrared spectrum, ultraviolet lenses, and a display source. The terminals processor regulates the different wavelengths source, which emits illumination on the devices, and the cameras, which gather the picture of the item and transmit it to the terminal's processor for interpretation. The image recognition block diagram of a night vision camera is shown in Fig. 1.

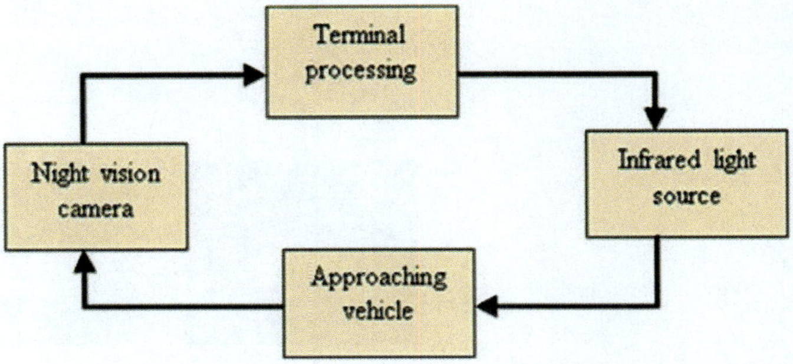

Fig. 1 Image processing block diagram of night vision camera

2.4 Voltage Regulator

The voltage regulator used in this circuit is LM7805 which is a three-pin voltage regulator. The three pins in this voltage regulator are the input pin, output pin, and ground pin. The need for a voltage regulator in this circuit is to maintain the threshold voltage throughout the circuit. It is a 5 V DC voltage regulator.

2.5 Arduino UNO Microcontroller

The Arduino Uno microcontroller is used in this project. It has fourteen electronic instruction ports (six of those are pulse width modulations outputting), six analogue ports, a 16 MHz porcelain resonant (CSTCE16M0V53-R0), a USB interface, a charging socket, a circuit sequential programmable interface, as well as a restart switch. To begin starting, attach the microprocessor to a laptop through a USB connection, or use an AC–DC converter or batteries. Figure 2 presents the circuit structure of the automatic headlamp control.

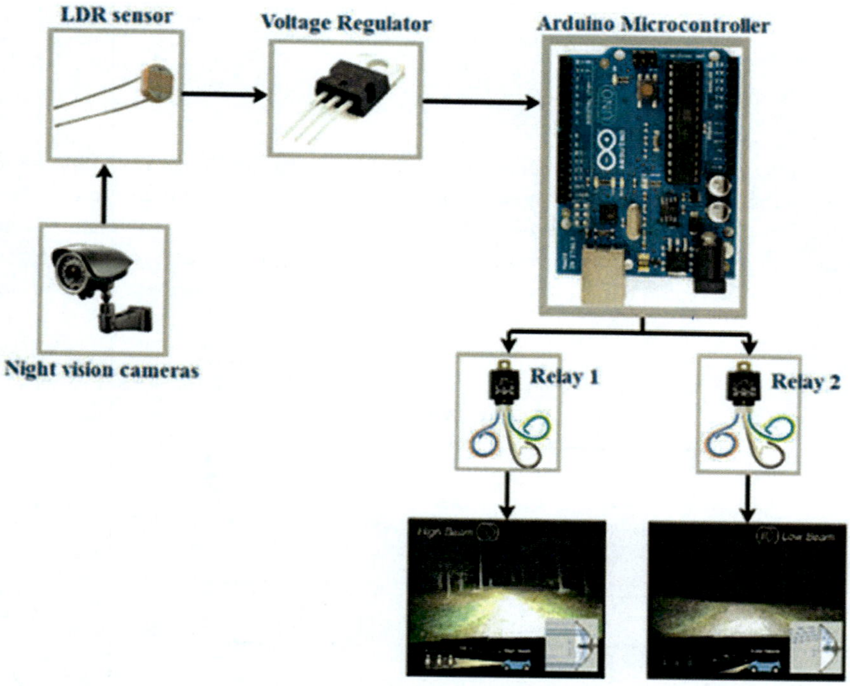

Fig. 2 Circuit structure of the automatic headlamp control

In this part, the elements utilized in autonomous headlamp control, such as SPDT relays, light-dependent resistors, night vision cameras, voltage regulators, and Arduino UNO microcontrollers, are explained in detail in terms of their unique tasks.

3 Design of Adaptive Headlamp Control

3.1 Working

The working of automatic headlight control is very simple as it looks. The LDR which is present in the circuit determines the intensity of light falling on it and varies the resistance depending on it. The low beam and high beam are switched accordingly depending on the intensity of light falling on the LDR. Then, a type of dynamic ultra-infrared imaging device is used to implement, which captures images of the pre-processed objects. Following image processing and morphology, the object characteristics are displayed. When the illuminance measured by the LDR exceeds a certain limit, the relay connected to the high beam closes, whereas the relay connected to the low beam remains open. The Arduino executes the uploaded code once the LDR is activated, and the sequential tasks are performed. The voltage controller is employed to provide a sufficient supply (i.e., 5 V) for the circuit to work. The relays receive voltage regulator signals and turn the lamps accordingly. Figure 3 presents the process flow of the device.

3.2 Principle

The basic principle behind this system is photoconductivity and infrared image detection using night vision cameras. This helps in getting the desired outcome with higher accuracy. The LDR works with the principle of photoconductivity. LDRs or photo resistors have a long response time. LDR varies the light depending on certain factors. The LDR with two resistors creates a potential divider network that provides the necessary current to the circuit. This balanced circuit activates the gate/base of the transistor. The circuit is activated when a voltage imbalance occurs in the circuitry as a result of a variation in the impedance of the LDR supplied by the beam of light. Figure 4 Circuit layout of automatic headlamp system control.

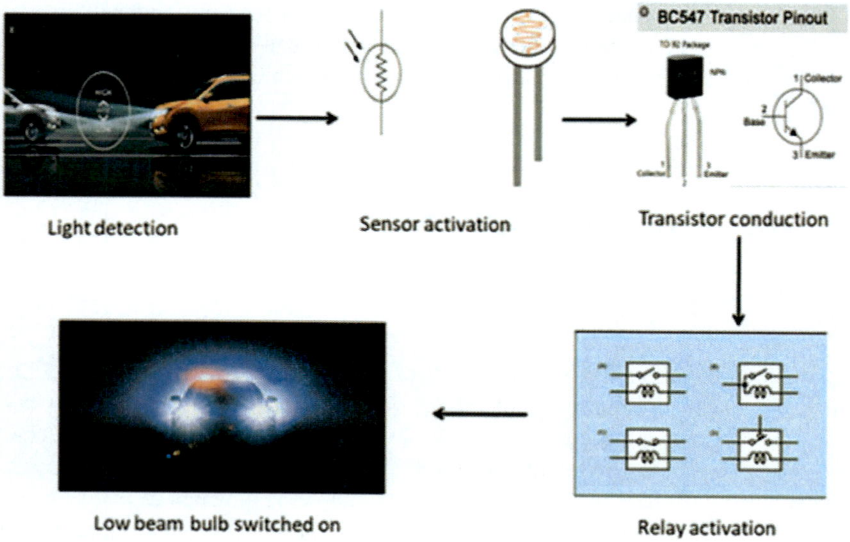

Fig. 3 Process flow diagram

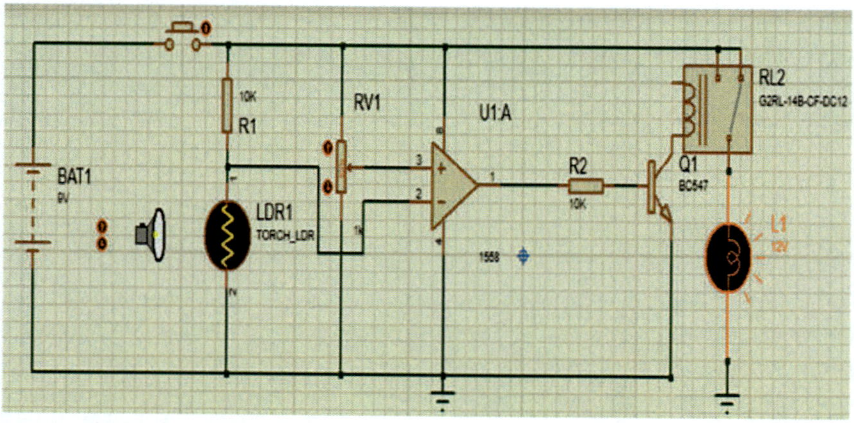

Fig. 4 Circuit layout of automatic headlamp system control

3.3 Simulation

Figure 5 shows the simulation of automatic headlight system (AHS) control and the required outcomes are achieved using Tinker cad software.

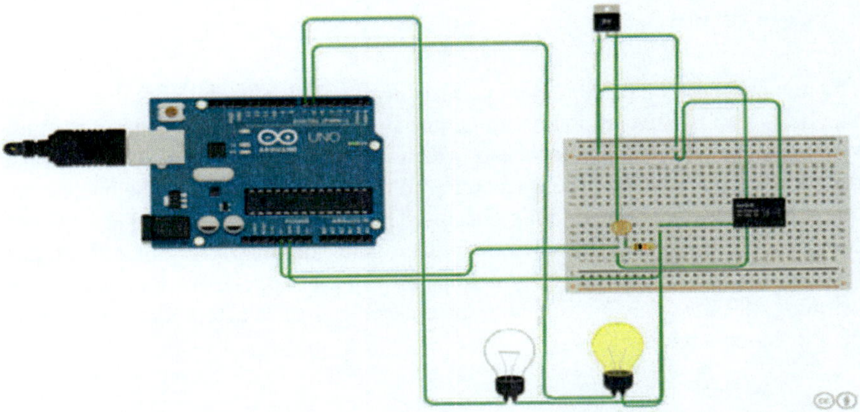

Fig. 5 Simulation of automatic headlamp system control using tinker cad software

3.4 Fabrication

Figure 6 shows the fabrication of the real-time circuit which has all the components connected. The circuit has been designed successfully and demonstrated to be a working model. There were a few criteria to be considered while placing the device in the real-time vehicle.

- It should be kept in a place where it is protected from rain, dust, etc.
- The circuit should be placed in the same line as that of the driver's eye contact, so it responds the similar way the driver would react to bright lights.
- The circuit should get a constant supply whenever the headlight is turned on.

Fig. 6 Fabricated model of automatic headlamp system control (video link)

4 Conclusion

The requirement of a headlamp plays a major role in driving automobiles at night. The headlamp has two vision controls, one is a high beam which is used for long vision and the other one is a low beam which is used for short vision. The driver has the control to change the beams according to the driving condition. The intensity of low beams is low compared to high beams. Thus, interfacing LDR and night vision cameras give the desired outcome to reduce glare and dazzling. Since the future involves the introduction of fully automated electric vehicles to minimize global warming, this idea may be used with the emerging electric automobiles to improve the project on a wide scale.

References

1. Chen, Y. S., & Chiu, S. C. (2018). New method of automatic control for vehicle headlights. *Optik (Stuttg), 157*, 718–723.
2. Karaman, S., Hacibekir, T., Kural, E., Öztürk, E. S., Demirci, M., & Güvenç, B. A. (2006). Design and hardware-in-the-loop simulation of an automatic headlight control system. *IFAC Proceedings Volumes (IFAC-PapersOnline), 4*, 746–751
3. Reagan, I. J., Brumbelow, M., & Frischmann, T. (2015). On-road experiment to assess drivers' detection of roadside targets as a function of headlight system, target placement, and target reflectance. *Accident Analysis & Prevention, 76*, 74–82.
4. Zhang, Z. F., et al. (2016). Research on lamp auto-control system based on infrared image processing. *Optik (Stuttg), 127*(3), 1144–1147.
5. De Santos-Berbel, C., & Castro, M. (2020). Effect of vehicle swiveling headlamps and highway geometric design on nighttime sight distance. *Mathematics and Computers in Simulation, 170*, 32–50.
6. Reagan, I. J., & Brumbelow, M. L. (2017). Drivers' detection of roadside targets when driving vehicles with three headlight systems during high beam activation. *Accident Analysis & Prevention, 99*, 44–50.

Design of LCL Filter in Front–End Inverters for Grid Interfaced Electric Generation

Bipin Singh, Arunima Verma, and V. K. Giri

Abstract LCL filter design is difficult; still know the figures and facts of it in power industry application. To the mitigation of distortion, reduce bulky inductor size and reduce grid harmonic current cause of high switching frequency in inverter by zero voltage switching. LCL filter is more attractive due to low inductor size over conventional inductor filter. Designed suitable LCL filter, enhanced power factor of grid due to low loss dissipation in damping resistor of filter with low EMI. In this paper, the design of filter with mathematical model to verify the design procedure is carried out. The performance of system is simulated under MATLAB/SIMULINK environment. Reduced total harmonic distortion is shown in the grid current by using various filters are observed in simulated results.

Keywords Voltage source inverter (VSI) · LCL filter · Total harmonic distortion (THD) · Zero voltage switching (ZVS)

1 Introduction

The generation of power from renewable energy, viz. solar, wind, etc., grid connected voltage source inverters (VSI) are more popular as the interface of power electronic [1].

However, high switched voltage source inverter using pulse width modulation (PWM) is generated grid currents distortion. Generally, *L* filter is used to connect inverter to grid. So, to minimize the cost of magnetic material and copper of *L*

B. Singh (✉) · A. Verma
Institute of Engineering & Technology Lucknow, Lucknow, Uttar Pradesh, India
e-mail: singhbipin08@gmail.com

A. Verma
e-mail: arunimadey@gmail.com

V. K. Giri
M. M. M. University of Technology Gorakhpur, Gorakhpur, Uttar Pradesh, India
e-mail: girivkmmm@gmail.com

B. Pati et al. (eds.), *Proceedings of the 6th International Conference on Advance Computing and Intelligent Engineering*, Lecture Notes in Networks and Systems 428,
https://doi.org/10.1007/978-981-19-2225-1_59

filter, LCL filter is becoming too attractive used. LCL filter value is chosen based on the size of filter, ripple current, and reactive power of capacitor [2, 3]. A passive damping resistor intended to ensure least power loss. Passive filter components design is required, and an exact calculation of ripple quantity difference is presented in different currents. Hence, to ensure the properly designed filter it must have: high power factor of grid, low voltage loss across filter, low loss in damping resistor, and minimum electromagnetic interference (EMI) [4].

The inverter circuit is connected grid with LCL filter is proposed in this paper, and with design of filter consideration is simulated. The control of system and grid synchronization are used by power balance theory. The power balance is implementing to generation of templets using the reference voltage of source/grid [5]. The different templates and operating condition, a mathematical equation are desired for the generation and investigated from SIMULINK results.

1.1 Motivation

In this section, a properly designed LCL filter plays the vital role in system stability. The design of LCL filter has to be carried out in a way, which takes care of better switching control with economical solution.

The larger capacitance value offers the economical way; whereas, the lower inductance value appears to be bulky and expensive too. Also, impedance of grid which has effect on the system stability and specially precaution requisite in the feature of LCL filter design.

The LCL filter resonance frequency varies with the grid impedance; hence, grid stiffness also changes. Control strategy and synchronization are required for operation of the solar photovoltaic system interfaced with grid.

The organization of the paper is follows as: Modeling of voltage source inverter with LCL filter are described with concisely in different time intervals and it average model in Sect. 2. Section 3 describes the filter design feature with LC and LCL. Section 4 describes simulation results and discussion followed by the conclusions in last Sect. 5.

2 Modeling of Voltage Source Inverter (VSI) with LCL Filter

The working operations of the proposed zero voltage switching three-phase four-wire inverter are presented in Fig. 1. The working operation could be analyzed on single phase as presented in Fig. 2 and explained as:

Duration-1 [$\mathbf{t_0 - t_1}$]: Prior to time t_0, switch S_1 is OFF and switch S_2 is still turned ON. Figure 2a shows the current direction through L_1, as assumed, and is right to left

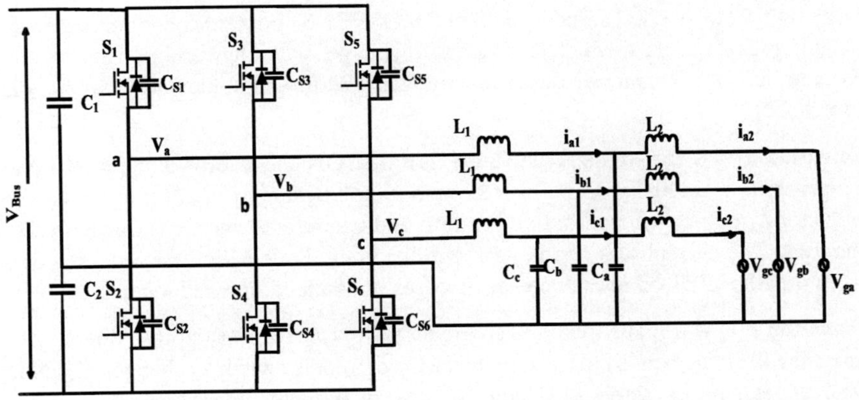

Fig. 1 Circuit diagram of three-phase four-wire grid connected inverter system

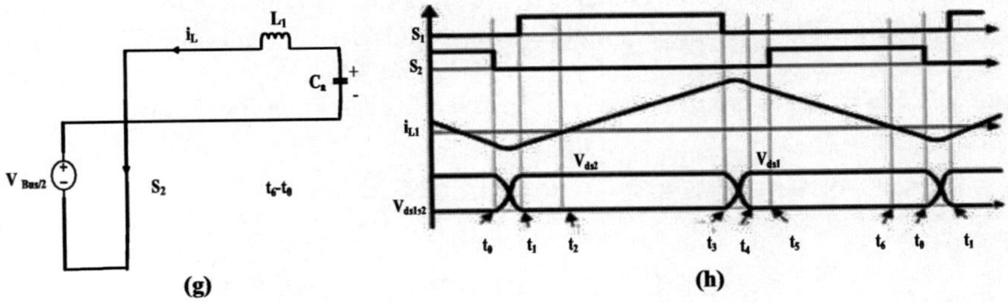

Fig. 2 Working intervals of a single-phase inverter

direction at t_0. Then MOSFET S_2 is turned OFF and parasitic capacitor C_{S2} voltage of low side S_2 starts rising due to the inductor current. As the voltage across switch S_1 decreases and C_{S2} charges, in this time, t_1 ends voltage across switch S_1 reaches zero.

Duration 2 [$t_1 - t_2$]: The MOSFET S_7 body diode will be conducting at t_1, and switch S_1 is ON with zero voltage switching. The flow direction of current from right to left decays linearly due to the $V_{bus}/2$ minus the voltage across L_1. At ends of this duration when the inductor current falls to zero.

Duration 3 [$t_2 - t_3$]: The switch S_1 is turning ON, and the direction of current increasing linearly through L_1 is changed from left to right. In this mode, power is delivery duration.

Duration 4 [$t_3 - t_4$]: MOSFET S_1 is turned OFF at time t_3. So the inductor current charges MOSFET parasitic capacitor C_{S1} while C_{S2} is discharging. When the voltage across parasitic capacitor C_{S2} zero drops, the body diode conducts since the direction of current L_1 does not change.

Duration 5 [$t_4 - t_5$]: The body diode of MOSFET S_2 continues conducting from the previous duration 4, which creates a zero voltage switching condition and switch S_2 is turned ON. This mode duration is typically short and at the ends switch S_2 is turned ON.

Duration 6 [$t_5 - t_6$]: At time t_5, switch S_2 is turned ON under zero voltage switching. The current is gradually decreasing through switch S_2 due to $V_{bus}/2$ plus across the inductor L_1 output voltage. In this duration, inductor-stored energy is transferred to the load. The current was flowing through the body diode MOSFET S_2 now flows from the MOSFET on resistance hence losses of conduction is reducing.

Duration 7 [$t_6 - t_0$]: The switch S_2 continues to flow the current and at t_6, direction of current will change once falls to zero. Once the current in switch S_2 changes direction from upper to lower as presented in Fig. 6, zero voltage switching is created for switch S_1.When switch S_2 current goes the negative threshold value, and repeats the cycle.

The average model neglects the series resistances of L_1 and L_2 in simplified model.

Three-phase synchronous reference frame average circuit model may be obtained by neglecting the large frequency components of AC phase currents and DC voltages which are presented in Fig. 3 [5]. By using KCL and KVL, differential equations of currents and voltages have obtained, respectively, as shown in Fig. 3, given as

$$I_1' = -\frac{Rd}{L1}I_1 + \frac{Rd}{L1}I_2 - \frac{1}{L1}V_{cf} + \frac{V\text{bus}}{L1}D - \frac{V\text{bus}}{2L1}\Gamma \tag{1}$$

$$I_2' = \frac{Rd}{L2}I_1 - \frac{Rd}{L2}I_2 + \frac{1}{L2}V_{\text{cf}} - \frac{1}{L2}V_g \tag{2}$$

$$V_{cf}' = \frac{1}{Cf}I_1 - \frac{1}{Cf}I_2 \tag{3}$$

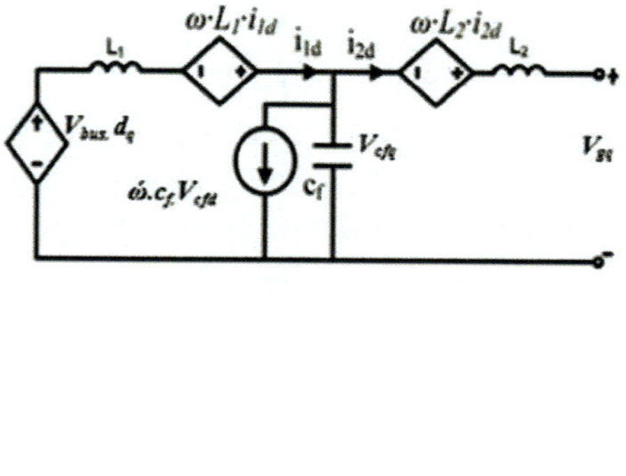

Fig. 3 Average circuit of three-phase synchronous reference frame

$$V'_{\mathrm{Bus}} = \frac{i\mathrm{bus} - i\mathrm{dc}}{\frac{C1}{2}} = \frac{2}{C1} i_{\mathrm{bus}} - \frac{2}{C1} D^T I_1 \tag{4}$$

Where $\mathrm{I}_1 = [i_{1a} i_{1b} i_{1c}]^{\mathrm{T}}$

$$\mathrm{I}_2 = [i_{2a} i_{2b} i_{2c}]^{\mathrm{T}}$$

$$V_{\mathrm{cf}} = [V_{\mathrm{ca}} V_{\mathrm{cb}} V_{\mathrm{cc}}]^{\mathrm{T}}$$

$$V_g = \left[V_{\mathrm{ga}} V_{\mathrm{gb}} V_{\mathrm{gc}}\right]^{\mathrm{T}}$$

$$D = [d_a d_b d_c]$$

$$\Gamma = [111]^T$$

Grid currents i_{2a}, i_{2b}, and i_{2c} are in controlled steady state sinusoidal and phase voltages of grid V_{ga}, V_{gb}, and V_{gc} can be given as:

$$\begin{bmatrix} V_{\mathrm{ga}} \\ V_{\mathrm{gb}} \\ V_{\mathrm{gc}} \end{bmatrix} = \begin{bmatrix} V_m \cos(\omega t) \\ V_m \cos\left(\omega t - \frac{2\pi}{3}\right) \\ V_m \cos\left(\omega t + \frac{2\pi}{3}\right) \end{bmatrix} \tag{5}$$

Where V_m amplitude of phase voltage
ω angular frequency of source power

To transform stationary coordinate into synchronous coordinate reference frame by Park's transformation is as follows [4]:

$$\mathrm{T} = \frac{2}{3} \begin{bmatrix} \cos(\omega t) & \cos\left(\omega t - \frac{2\pi}{3}\right) & \cos\left(\omega t + \frac{2\pi}{3}\right) \\ -\sin(\omega t) & -\sin\left(\omega t - \frac{2\pi}{3}\right) & -\sin\left(\omega t + \frac{2\pi}{3}\right) \\ \frac{1}{2} & \frac{1}{2} & \frac{1}{2} \end{bmatrix} \tag{6}$$

The equations of averaged model are given as follows: (Fig. 4)

$$\mathrm{I}'_{1\mathrm{dq}} = -\mathrm{WI}_{1dq} - \frac{\mathrm{Rd}}{L1} I_{1\mathrm{dq}} + \frac{\mathrm{Rd}}{L1} I_{2\mathrm{dq}} - \frac{1}{L1} V_{\mathrm{cfdq}} + \frac{V\mathrm{bus}}{L1} D_{\mathrm{dq}} \tag{7}$$

$$\mathrm{I}'_{2\mathrm{dq}} = -W I_{2\mathrm{dq}} + \frac{\mathrm{Rd}}{L2} I_{1\mathrm{dq}} - \frac{\mathrm{Rd}}{L2} I_{2\mathrm{dq}} - \frac{1}{L1} V_{\mathrm{cfdq}} - \frac{1}{L2} V_{\mathrm{gdq}} \tag{8}$$

$$V'_{\mathrm{cfdq}} = \omega V_{\mathrm{cfdq}} + \frac{1}{\mathrm{Cf}} (i_{1\mathrm{dq}} - i_{2\mathrm{dq}}) \tag{9}$$

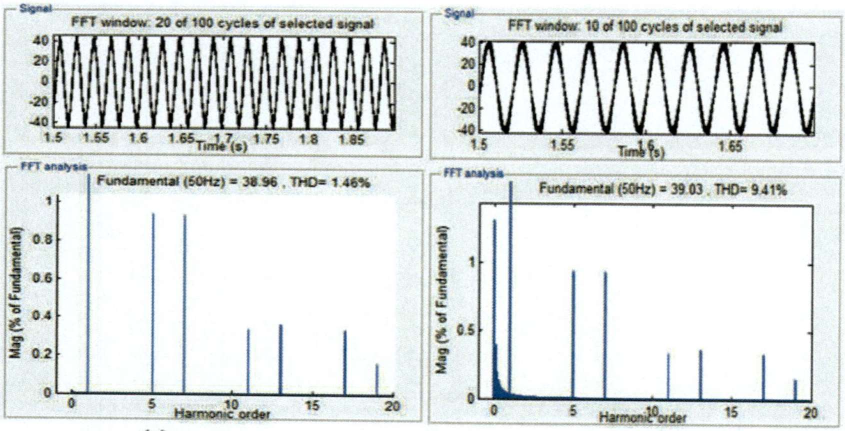

(a) (b)

Fig. 4 Grid current total harmonic distortion (THD), **a** LCL filter, **b** LC filter

$$V_{\text{bus}}^{'} = -\frac{2}{C1} D_{\text{dq}}^{T} I_{1\text{dq}} + \frac{2}{C1} i_{\text{bus}} \tag{10}$$

Where $\mathrm{W} = \begin{bmatrix} 0 & -\omega \\ \omega & 0 \end{bmatrix}$.

The working principle of three-phase four-wire inverter with LCL filter is analyzed and it averaged model to minimize the complexity of the system. The average model grid connected LCL filter is used to control the output by using transformation.

3 LCL Filter Design

The inverter output current harmonics is attenuated by electromagnetic (EM) interference filter because large switching frequency of inverter. The EM interference filter for utility interactive PV system consists usually L or LC filter and resulting higher inductance (L) value, expensive and bulky for low-rated systems [5]. The LCL filter is more effective than L filter, since, latter it has all frequency range only an attenuation of 20 dB/DEC.

The switching frequency of inverter has to be large for a correct attenuation of higher harmonics current. The higher switching frequency, resulting the higher switching losses. The variability of resonance frequency of grid inductance suffers the LC filters. Therefore, it is not appropriate for the grid. Hence, the actively damped the resonance problem of LCL filter correctly [6].

The step-by-step designed procedure of LCL filter with grid-interfaced converter is described [7]. The output of inverter side inductance at peak ripple current 15%

times the rated current, inductance L_1 is calculated as:

$$L_1 = \frac{V_g}{2\sqrt{6} f_s i_{\text{ripple peak}}} \tag{11}$$

$$C = \frac{0.05}{\omega_n Z_{\text{base}}} \tag{12}$$

$$Z_{\text{base}} = \frac{V_{\text{gLL}}^2}{P_n} \tag{13}$$

where V_g	rms grid phase voltage.
f_s	Inverter switching frequency
$i_{\text{ripple,peak}}$	Peak value 15% rated output current
L_2	Grid side inductance
L_1	Inverter side inductance
C	Capacitance of LCL filter
Pn	Inverter rated power
and ω_n	Operating frequency

The grid side LCL filter inductance L_2 is calculated by 20 dB attenuation of current ripple at the switching frequency:

$$L_2 = 0.8L_1 \tag{14}$$

4 Simulation Results of Filter

The proposed filter design method has been validated using MATLAB/SIMULINK in order to the performance of filter. Various parameters are used in MATLAB/Simulation can be referred as follows in given Table 1

DC voltage given to the VSI inverter, it feds to the grid through the different types of LC filter and LCL filter. The gating pulse of inverter is set to measuring of controller, viz. PI controller, abc–dq transformation and vice versa, PLL, and PWM. Total harmonic distortion is measured by FFT window in MATLAB/Simulation by using LC and LCL filter. The gating pulse of inverter is setting by the ON and OFF switching pattern. The inverter output is fed into grid and load. The grid current, load current, and injected current is observed by controller. THD analysis is done by FFT powergui tool in MATLAB/Simulation is shown in Fig. 4. Various grid currents compared with LC and LCL filter qualitatively in Table 2.

Table 1 System parameters

Parameter	Value/unit
DC voltage	700 V
DC link capacitance	4874.56 μF
Inverter side inductance (L_1)	1.4 mH
Output capacitance (C)	5 μF
Grid interface inductor (L_2)	1.4 mH
Grid base voltage (V_{LL})	380 V
Output frequency	50 Hz
Switching frequency	5 kHz
Sampling time	2×10^{-6} S
Proportional gain (Kp)	2.8
Integration gain (K_i)	0.03
Power factor	0.8
Loads	10 kVA

Table 2 Simulation results of filters

S. No.	Type of filters	THD (%)
1	LC	9.41
2	LCL	1.46

5 Conclusions

This paper presented the design of LCL filter to the grid connected voltage source inverter. The average model of VSI inverter for filter parameter design is derived. The VSI inverter is average modeled for filter parameter design is derived to calculate. With and without LCL filter grid connected to the inverter is simulated and results are compared with harmonic spectra in Table 2. Voltage source inverter is proposed to large quality sinusoidal output current, LCL filter is used to minimize the output current harmonics. The results of simulation show that the VSI inverter maintains a sinusoidal current into grid and improving harmonic reduction effectively. In the future, using this filter, output filter volume could be minimized with better performance of the system.

References

1. Kjaer, S. B., Andersen, G. K., Klumpner, C., & Blaabjerg, F. (2002). Control aspects of a LCL grid connected green power inverter. In *Proceedings IEEE Nordic workshop on power and Ind electronics*, pp. 1–7
2. Liserre, M., Blaabjerg, F., & Hansen, S. (2001). Design and control of an LCL filter based three-phase active rectifier. In *Proceedings IEEE 36th IAS conference*, pp. 299–307.

3. Zue, A. O., & Chandra, A. (2006). Simulation and stability analysis of a 100 kW grid connected LCL photovoltaic inverter for industry. In *IEEE power engineering society general meeting*, ISBN 1–4244–0493–2, Montreal, Que.
4. Jalili, K., & Bernet, S. (2009). Design of LCL filters of active-front-end two-level voltage-source converters. *Industrial Electronics, IEEE Transactions on, 56*(5), 1674–1689.
5. Wu, W., He, Y., & Blaabjerg, F. (2012). An LCL power filter for single-phase grid-tied inverter. *IEEE Transactions on Power Electronics, 27*(2), 782–789.
6. Sahoo, A. K., Shahani, A., Basu, K., & Mohan, N. (2014). LCL filter design for grid connected inverters by analytical estimation of PWM ripple voltage. *Applied Power Electronics Conference & Exposition, 3*, 1281–1286.
7. Geng, Y., Qi, Y., Zheng, P., Guo, F., & Gao, X. (2018). A virtual RLC active damping method for LCL-type grid-connected inverters. *Journal of Power Electronics, 18*(5), 1555–1566.

Improve Quality of Data Management and Maintenance in Data Warehouse Systems

Sakshi Hooda and Suman Mann

Abstract Data warehouses bring an organization's data together in one place for reporting and analysis. Top-level management receives key performance metrics, middle-level management receives analytical strength and bottom-level management receives consistent and accurate data based on analytical system information. Despite the numerous domain-specific challenges, data warehouses must be financially viable to be successful. We conducted a comprehensive literature review and read case studies of organizations that have successfully implemented data warehousing solutions in order to compile a list of best practices that can serve as guidelines for new practitioners and organizations new to this technology.

Keywords Data warehouse · Data partitioning · Indexing · View maintenance · Schema versioning · Incremental development · Refresh policy · Metadata management

1 Introduction

An operational or transactional database undergoes a lot of changes every day as transactions are processed. Any data a manager might try to access on a specific product, supplier, or customer has been overwritten by previous transactions. A data warehouse is a storage facility that houses all of a company's operational system data in one location. To summarize, data warehousing is a technology that collects data from a variety of sources and organizes it for later analysis and retrieval. Using data from a data warehouse, executives can make well-informed decisions [1]. For this purpose, the data warehouse serves as a multidimensional window into consolidated

S. Hooda (✉)
GGSIPU, New Delhi, India
e-mail: sakshi.hooda@gmail.com

S. Mann
Maharaja Surajmal Institute of Technology, New Delhi, India

B. Pati et al. (eds.), *Proceedings of the 6th International Conference on Advance Computing and Intelligent Engineering*, Lecture Notes in Networks and Systems 428,
https://doi.org/10.1007/978-981-19-2225-1_60

and generalized data. It keeps track of both the most recent and previous data during the analysis.

A company's operational database and data warehouse are usually kept separate. It is necessary to create systems that extract data from operational systems, convert it to a standard format and then store it in a data warehouse for end-users to use [2, 3]. Choosing "what to build" is one of the most difficult aspects of creating a data warehouse. The creation of a data warehouse is still considered a difficult task [4, 5]. We must focus on the details in order to achieve this. When we discuss requirements, we are talking about statements that tell the framework what it can and cannot do. Eliciting requirements, conducting extensive research and consulting with executives are all part of the prerequisites design process. Keep an eye out for the client's needs as well as the framework's limitations, when gathering prerequisite data. Customers can concentrate on their most pressing needs and objectives by first looking into the prerequisites. To be precise, a prerequisite must clearly and unambiguously record the client's needs and limitations. As well as double-checking that all of the framework requirements are correct. Executives are in charge of scheduling, coordinating and documenting requirements engineering activities [5, 6].

Requirements the difficulty users have in communicating their needs to developers adds to the complexity of engineering. There are many stakeholders in a data warehouse, each with their own set of requirements and expectations. As a result, the underlying communication problem gets even worse. Their requirements for long-term storage are constantly changing [7].

Data warehouses are becoming more popular as more businesses realize the benefits of decision and analytic-oriented databases. Commercial data warehouses, despite their rapid evolution, still face significant challenges due to poor data quality [8, 9]. Unresolved issues may force the development project to be delayed or even halted.

Researchers have been studying data quality issues and developing workable solutions for the past two decades [8, 10]. Few studies have attempted to identify the root causes of data quality issues that arise during the implementation of data warehouses. During the requirements engineering phase, data quality issues were thoroughly investigated, and solutions were developed for implementation. We believe that if a problem is discovered and addressed sooner rather than later, the outcome will be better.

Data is sent to the data warehouse from various operational and transactional systems. It is always possible that a minor flaw in the source system will result in significant changes in the data warehouse. Customer dissatisfaction at a datum distribution centre is frequently attributed to stale data [11]. The moment customers realize the information is of poor quality, their trust in the information distribution centre begins to erode. As a result, the undertaking group's partners, backers and members will all be working in the wrong direction. Customers will be hesitant to use data warehousing to make critical decisions if they do not trust it (Fig. 1)

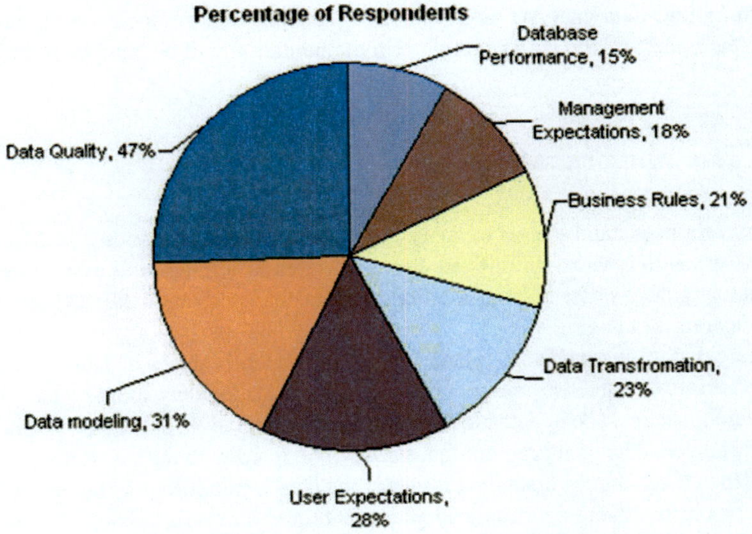

Fig. 1 Data quality is the topmost challenge in a data warehouse

2 Source Systems and Development Technology

The quality of the organization's existing data has a significant impact on the effectiveness of data warehouse systems. Data warehouses, which integrate data from all of the system's disparate internal systems, standardize and reconcile data before it is stored. Due to a lack of quality and uniformity, the task group at the distribution centre finds it extremely difficult to organize data. [12] (Fig. 2).

Standardized data, on the other hand, simplifies data processing, reduces problems and aids in the long-term development of a more fruitful framework. Equipment, programming, techniques and projects used to complete a task are just a few examples of information stockroom advancement. These are also factors to consider [13, 14].

To run the devices that aid in improvement, a lot of computing power is required. The data must be subjected to database performance optimization, multidimensional

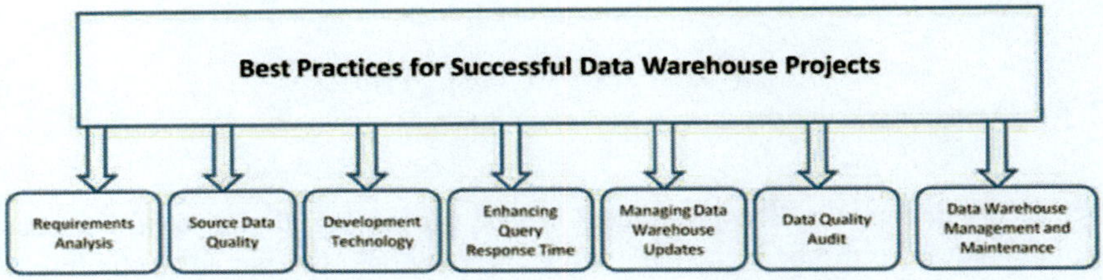

Fig. 2 Activities that must be performed for successful implementation of DW systems

modelling and data analysis as soon as it is loaded. If these tools are difficult to understand and use, the data warehouse implementation will be hampered. [15].

2.1 *View Maintenance*

Data warehouses can be used to analyse and report on large amounts of historical data. Users who create materialized views of warehouse data see it as a collection. A data warehouse's data and view definitions can change dynamically as user needs and constraints change.

Query definitions and actual relationships are frequently stored in data warehouses from a variety of database sources around the world. If the data source changes, the data warehouse must be updated to reflect the new information, particularly if the data is materialized. This is a problem that needs to be fixed with data warehouse views [16]. When data sources update their views, the data warehouse may be notified, but it may be impossible to determine what data is required for this to happen in the first place. To improve the efficiency of data warehouse queries, no access to the original data sources must be allowed, and some intermediate results of query processing must be stored as data warehouse relations in some materialized views. A well-known data warehouse challenge is improving query response time by materializing views. Remote data source updates must be propagated to the destined warehouse for a materialized view to be current. The authors of [17] proposed an incremental view maintenance approach based on a schema transformation pathway.

2.2 *Query Optimization*

Query optimization is done to efficiently handle and process complex queries. Querying is used to un-nest complex queries with nested sub-queries. Complex nested queries can be reduced to a single block query if certain syntactic requirements are met [18].

Using semi-join-like techniques, it is also possible to reduce the number of invocations and batch the execution of inner sub-questions.

2.3 *Denormalization*

Before the denormalization procedures can be carried out, the physical database design processes must be completed. When there are only two parties, a one-to-one relationship is useful; however, when there are many parties, it is completely useless. When it comes to data warehouse design, the star schema is sometimes a better option than others. In some cases, normalization can cause data retrieval to take longer [19,

20]. Because data warehouse systems rarely receive data updates, denormalization strategies can be used. In order to do so, the database designer must have a thorough understanding of the application's requirements as well as how frequently data will be updated.

The data warehouse development team must examine the following information before implementing a denormalization strategy:

- Cardinality, relevance and data volume of each table
- Data updates and retrieval patterns of the end-users
- Number of entities involved and numbers of rows accessed by each transaction
- Frequency of transactions.

In this section, we will read different strategies that are used to de-normalize the data warehouse data.

2.4 Collapsing Relations

A common denormalization strategy is to break up with a partner. As a result of the preceding example, two tables will be created: customer details (which includes the customer's ID) and customer preferences (including his ID, product preferences and purchase amount). When there are numerous queries that require joining two tables, it is preferable to combine the two connections into a single one [21, 22]. Because most records are dependent on essential/unfamiliar keys, this reduces the number of unfamiliar keys, the number of files, the extra space and the number of join tasks [23]. When a well-balanced relationship breaks down, there is no information duplication or change in the business view because there are fewer join activities. Even when there is a lot of redundancy in the data, relationships can be collapsed, but only if there are no other types of relationships.

2.5 Vertical Partitioning

Vertical partitioning divides each group of columns in a table into its own table (VP). Take a look at the PRODUCT table, which has columns for the product's ID, name, category, colour and price, to give you an idea. It not only keeps track of product details, but it also shows how much of each item is still available. If the queries posed on PRODUCT fall into two distinct categories in terms of association with the two attribute groups, the table PRODUCT should be split into two tables: PRODUCT and PRODUCT INVENTORY (ID, manufacturer, stock and price) [22, 24].

When long text fields have a lot of sections but only a few inquiries use them, vertical partitioning makes sense. There is no information duplication due to a customer-friendly vertical division (with the exception of the relating essential keys among the parcelled relations).

2.6 Horizontal Partitioning

In horizontal partitioning, the rows are divided based on key ranges. This is standard procedure for tables with a large number of rows. The SALES fact table in a data warehouse is an excellent candidate for horizontal partitioning because it contains millions of records spanning several years. For this reason, horizontal table partitioning is preferred, as it eliminates unnecessary data access to secondary storage (e.g. once or twice annually) [25]. Horizontal partitioning is useful when historical data is rarely accessed and most queries only look at current data. Horizontal partitioning allows for a natural separation of rows, such as different geographical regions. Vertical partitioning eliminates data duplication, and any data anomalies that may arise as a result.

3 Data Warehouse Management and Maintenance

In this section, we will read about different steps that when implemented results in successful deployment of a data warehouse system

3.1 Metadata Management (MM)

A data warehouse master model must be included in your warehousing design (MM). In a storage facility, various types of metadata must be managed. Images of source data sets, back-end and front-end equipment, stockroom blueprints and measurements, progressive systems, pre-defined inquiries and reports and total data are all examples of administrative metadata. Information store locations and content, information extraction, change and purging standards, information revival approaches and customer prowess are all examples of administrative metadata. Business metadata that serves as a navigation guide for the client characterizes end-client phrasing. Usage measurements, error reports and review trails, as well as data gathered during stockroom activity, are all examples of observational metadata. The cash at the distribution centre contains dynamic, filed and cleansed data. By storing and managing all stockroom data, the metadata warehouse assists with distribution centre set-up, operation and management.

3.2 Data Warehouse Management

To work together, efficient warehouse management software and metadata management are required. For the creation and revision of various types of documents, there are a variety of progress instruments available. Arrangement and examination instruments are used to investigate construction changes, invigorate rates and scope quantification. The storage capacity of a data warehouse, the time it takes to run an inquiry, the types of drill-downs and rollups used, as well as how often and how long each one takes to complete, should all be investigated.

The board instruments are used to quantify traffic by customers and employees, distribution centre workers and functional data sets, among others. Workflow management tools are used to track extraction, transformation and loading processes by running scripts that check for success or failure at each step and provide partial rollback, retry or recovery in the event of failure.

3.3 Data Warehouse Maintenance

Monitoring system changes, preventing resource loss, resolving system crashes and recovering lost data after a disaster are all part of data warehouse maintenance.

For a variety of reasons, such as intelligent design changes, information insurance rules, error detection frameworks and new applications, updating the data warehouse system are required. The only way to avoid this problem is to keep your network stable and trust the equipment and organizations you work with. This includes safeguarding against asset locking and workplace accidents on the product side.

If the system is to remain available and dependable, crashes must be repaired. To minimize the impact on users, data warehouse system auditors must act quickly to correct hardware or software flaws. They require information on how frequently the warehouse succeeds and fails. Auditors must also identify and evaluate disaster recovery procedures that aid in the recovery from crashes. To ensure rapid system recovery and availability, data duplication is required.

A final solution has been found to assist you in implementing and maintaining policies and procedures. Both architects and clients are interested in new ideas, methods or techniques for improving their current information stockroom framework. Creating rules and holding meetings to generate new ideas are just two examples of how this can be accomplished. Investigations have been conducted. To see if new strategies work, they are put to the test. There are safeguards in place to ensure that everything remains in check. The assignment was revised and updated in order to provide the most up-to-date information.

3.4 Documentation and Training

When it comes to information transfer, documentation and training are essential. Plans for user training must be included in an organization's strategy for documenting data and processes.

Before beginning work on the project, all of the project's specifications, designs and key terms should be thoroughly documented (glossaries). Documentation must clearly highlight data architecture, process architecture, system technical architecture, metadata and refreshment processes in order to be effective.

The results must also be documented and evaluated in order to determine if appropriate standards and recommended practices were implemented in the audited systems.

When it comes to teaching system users how to use it effectively, both documentation and on-the-job coaching are required. Professionals and non-professionals may be included in user groups; training must take this into account. The number and types of applications that can be built on the system grow and change as the system grows and expands. Creating discussion forums is an ongoing process to ensure that the expected criticism prompts new hypotheses about the current state of affairs and where it is all going. As a result, the customer is happier.

4 Miscellaneous Factors

- When employees are encouraged to help with data warehousing, they are more likely to do so. Data warehousing technology not only aids in the overthrow of political opponents, but also in the management of the resulting organizational changes.
- By better matching end-users' needs to the framework's requirements, assigning project tasks and undertakings to end-users will help ensure effective user participation [19]. This is critical if the system's requirements are unclear from the start. They will accept the data warehouse if they were involved in its development and if it meets their expectations when it is presented to them.
- The improvement of the task of the information distribution centre is influenced by technical and interpersonal skills. Complementary skills are required to identify and overcome all technical challenges in complex projects. You will need an experienced team on your side to succeed. [20].

5 Conclusion

Organizations that see information stockrooms as empowering agents should develop a comprehensive evaluation strategy in order to assess the impact of the information distribution centre. As discussed in the paper on best practices for data warehouse implementation, a data warehouse can perfectly align an organization's strategy and business objectives. According to the authors, users and business sponsors are critical to the system's success from the start. Modelling and design of data warehouses are also discussed. Despite a decade of attention, many issues remain that necessitate the use of new techniques, as applications and architectures rapidly evolve and become more complex.

When building a warehouse for your data, you should think about data quality, system quality and refresh policy. The paper also discusses table denormalization as a method for improving performance. While denormalization makes data retrieval faster, it also raises update costs, slows data modification and limits the design's flexibility. We will do more research if denormalization has an impact on capacity costs, memory utilization costs or interchange costs.

References

1. Gosain, A., & Mann, S. (2012). An object-oriented multidimensional model for data warehouse. In *Fourth international conference on machine vision (ICMV 2011): computer vision and image analysis; pattern recognition and basic technologies, 8350*, pp. 83500I.
2. Gosain, A., & Mann, S. (2014). Empirical validation of metrics for object oriented multidimensional model for data warehouse. *International Journal of System Assurance Engineering Management, 5*, 262–275. https://doi.org/10.1007/s13198-013-0155-8
3. Gosain, A., Khatri, S. K., & Mann, S. (2014). Multidimensional modeling for data warehouse using object oriented approach. In *Proceedings of 3rd international conference on reliability, infocom technologies and optimization*, pp. 1–6.
4. Winter, R., & Strauch, B. (2003). A method for demand-driven information requirements analysis in data warehousing projects. In *Proceedings of HICSS*. pp. 1359– 1365.
5. Mazón, J. N., Trujillo, J., Serrano, M., & Piattini, M. (2005). Designing data warehouses: From business requirement analysis to multidimensional modeling. In *Proceedings of int. work. on requirements engineering for business needs and IT alignment.*
6. Giorgini, P., Rizzi, S., & Garzetti, M. (2005). Goal-oriented requirement analysis for data warehouse design. In *Proceedings of DOLAP*, pp. 47–56.
7. SAP (2000). Multi-dimensional modeling with BW. Technical report, SAP America Inc. and SAP AG
8. Prashant, R., Suman, M., & Eashwaran, R. (2021). Efficient data cube materialization. *Advances in communication and computational technology, lecture notes in electrical engineering, 668*. Springer, Singapore.
9. Mann, S., & Siwach, M. (2020). Data model quality metrics of data warehouse: A survey. In *Proceedings of the international conference on innovative computing communications* (ICICC-2020).
10. Mann, S., & Phogat, A. K. Dynamic construction of lattice of cuboids in data warehouse. *Journal of Statistics and Management Systems*, 971–982.
11. Wixom, B. H., & Watson, H. J. (2001). An empirical investigation of the factors affecting data warehousing success. *MIS Quarterly, 25*(1), 17–41.
12. Jarke, M., Jeusfeld, M. A., Quix, C., & Vassiliadis, P. (1999). Architecture and quality in data warehouses: An extended repository approach. *Information Systems, 24*(3), 229–253.
13. Panigrahi, C. R., Tiwary, M., Pati, B., & Das, H. (2016). Big data and cyber foraging: future scope and challenges. In *Techniques and environments for big data analysis*. Studies in big data, *17*, pp. 75–100. Springer
14. Eder, J., Bebel, B., Koncilia, C., Morzy, T., & Wrembel, R. (2004). Creation and management of versions in multi version data warehouse. In *Proceedings of ACM SAC*, pp. 717–723.
15. Serrano, M., Calero, C., Trujillo, J., Luján-Mora, S., & Piattini, M. (2004).Empirical validation of metrics for conceptual models of data warehouses. In *Proceedings of CAiSE*, pp. 506–520.
16. Lechtenbörger, J., & Vossen, G. (2003). Multidimensional normal forms for data warehouse design. *Information Systems, 28*(5), 415–434.

17. Sarkar J.L., Panigrahi C.R., Pati B., and Das H. (2016) A novel approach for real-time data management in wireless sensor networks. In *Proceedings of 3rd international conference on advanced computing, networking and informatics*. Smart innovation, systems and technologies, *44*, pp. 599–607. Springer.
18. Curtis, M. B., & Joshi, K. (1998). Lessons learned from the implementation of a data warehouse. *Journal of Data Warehousing, 3*(2), 12–18.
19. Hartwick, J., & Barki, H. (1994). Explaining the role of user participation in information system use. *Management Science, 40*(4), 440–465.
20. Maish, A. M. (1979). A user's behavior toward his MIS. *MIS Quarterly, 3*(1), 39–52.
21. Goodhue, D. L., Wybo, M. D., & Kirsch, L. J. (1992). The impact of data integration on the costs and benefits of information systems. *MIS Quarterly, 16*(3), 293–311.
22. Rist, R. (1997). Challenges faced by the data ware-housing pioneers. *Journal of Data Warehousing, 2*(1), 63–72.
23. Hooda, Sakshi, & Mann, Suman. (2019). Distributed synthetic minority oversampling technique. *International Journal of Computational Intelligence Systems, 12*(2), 929–936.
24. Hooda, S., & Mann, S. (2020). Examining the effectiveness of machine learning algorithms as classifiers for predicting disease severity in data warehouse environments. *Argentina de Clínica Psicológica.*
25. Gosain, A., & Mann, S. (2010). Object oriented multidimensional model for a data warehouse with operators. *International Journal of Database Theory and Application, 3*(4).

Retraction Note to: A Survey on Applications of Machine Learning Algorithms in Health care

Deep Rahul Shah, Samit Nikesh Shah, and Seema Shah

Retraction Note to:
Chapter "A Survey on Applications of Machine Learning Algorithms in Health care" in: B. Pati et al. (eds.), *Proceedings of the 6th International Conference on Advance Computing and Intelligent Engineering*, Lecture Notes in Networks and Systems 428,
https://doi.org/10.1007/978-981-19-2225-1_9

The Editors have retracted this chapter at the request of the authors. After publication, the corresponding author contacted the Publisher stating that the article contained unusual and uninformative phrases (irresistible illnesses, bosom malignancy growth, credulous Bayes) that make it unsuitable for publication. The chapter was initially submitted without the phrases. The authors stated they used what they considered to be legitimate synonyms, but have since decided these could cause confusion for readers. All authors agree to this retraction.

The retracted version of this chapter can be found at
https://doi.org/10.1007/978-981-19-2225-1_9

B. Pati et al. (eds.), *Proceedings of the 6th International Conference on Advance Computing and Intelligent Engineering*, Lecture Notes in Networks and Systems 428,
https://doi.org/10.1007/978-981-19-2225-1_61

Author Index

B. Pati et al. (eds.), *Proceedings of the 6th International Conference on Advance Computing and Intelligent Engineering*, Lecture Notes in Networks and Systems 428,
https://doi.org/10.1007/978-981-19-2225-1

Author Index

Chinese Society of Aeronautics and Astronautics (Ed.): CASTC 2023, LNME, pp. 697–700, 2024.
https://doi.org/10.1007/978-981-99-8861-7